PENGUIN BOOKS

JANE GRIGSON'S VEGET

Jane Grigson was brought up in the north-east of England, where there is a strong tradition of good eating, but it was not until many years later, when she began to spend three months of each year in France, that she became really interested in food. *Charcuterie and French Pork Cookery* was the result, exploring the wonderful range of cooked meat products on sale in even the smallest market towns. This book has also been translated into French, a singular honour for an English cookery writer.

After taking an English degree at Cambridge in 1949, Jane Grigson worked in art galleries and publishers' offices, and then as a translator. In 1966 she shared the John Florio prize (with Father Kenelm Foster) for her translation of Beccaria's *Of Crime and Punishment*. It was in 1968 that Jane Grigson began her long association with the *Observer Magazine* for whom she wrote right up until her untimely death in 1990; *Good Things* and *Food With The Famous* are all based on these highly successful series. In 1973, *Fish Cookery* was published by the Wine and Food Society, followed by *The Mushroom Feast* (1975), a collection of recipes for cultivated, woodland, field and dried mushrooms. She received both the Glenfiddich Writer of the Year Award and the André Simon Memorial Fund Book Award for her *Vegetable Book* (1978) and for her *Fruit Book* (1982) and was voted Cookery Writer of the Year in 1977 for *English Food*. A compilation of her best recipes, *The Enjoyment of Food*, was published in 1992 with an introduction by her daughter, the cookery writer Sophie Grigson. Most of Jane Grigson's books are published in the Penguin Cookery Library.

Jane Grigson died in March 1990. In her obituary for the *Independent*, Alan Davidson wrote that 'Jane Grigson left to the English-speaking world a legacy of fine writing on food and cookery for which no exact parallel exists . . . She won to herself this wide audience because she was above all a friendly writer . . . the most companionable presence in the kitchen; often catching the imagination with a deftly chosen fragment of history or poetry, but never failing to explain the "why" as well as the "how" of cookery. Jane Grigson was married to the poet and critic, the late Geoffrey Grigson.

Jane Grigson's
VEGETABLE BOOK

With a new Introduction, Glossary,
and Table of Equivalent Weights and Measures
for the American edition

ILLUSTRATED BY
YVONNE SKARGON

PENGUIN BOOKS

PENGUIN BOOKS

Published by the Penguin Group
Penguin Books Ltd, 80 Strand, London WC2R 0RL, England
Penguin Putnam Inc., 375 Hudson Street, New York, New York 10014, USA
Penguin Books Australia Ltd, 250 Camberwell Road, Camberwell, Victoria 3124, Australia
Penguin Books Canada Ltd, 10 Alcorn Avenue, Toronto, Ontario, Canada M4V 3B2
Penguin Books India (P) Ltd, 11 Community Centre, Panchsheel Park, New Delhi – 110 017, India
Penguin Books (NZ) Ltd, Cnr Rosedale and Airborne Roads, Albany, Auckland, New Zealand
Penguin Books (South Africa) (Pty) Ltd, 24 Sturdee Avenue, Rosebank 2196, South Africa

Penguin Books Ltd, Registered Offices: 80 Strand, London WC2R 0RL, England

www.penguin.com

First published in Great Britain by Michael Joseph 1978
First published in the United States of America with a new Introduction
to the American edition, Glossary, and Tables of Equivalents by Atheneum 1979
Published in Penguin Books 1980
16

www.greenpenguin.co.uk

Mixed Sources
Product group from well-managed
forests and other controlled sources
www.fsc.org Cert no. SA-COC-1592
© 1996 Forest Stewardship Council
FSC

Penguin Books is committed to a sustainable future
for our business, our readers and our planet.
The book in your hands is made from paper
certified by the Forest Stewardship Council.

For Geoffrey

who introduced me to John Evelyn
and gardenage

'Every book is, in an intimate sense, a circular letter
to the friends of him who writes it. They alone take
his meaning; they find private messages, assur-
ances of love, and expressions of gratitude, dropped for them in every corner.' R. L. STEVENSON.

'Most people spoil garden things by over boiling
them. All things that are green should have a little
crispness, for if they are overboil'd they neither
have any sweetness or beauty.' HANNAH GLASSE.

CONTENTS

8   JANE GRIGSON'S VEGETABLE BOOK

ACKNOWLEDGEMENTS

Main debts of gratitude to other people, especially to colleagues, have been acknowledged through the text. I owe a wider, more general debt to Elizabeth David which can never be repaid. Without the specialized botanical knowledge of Geoffrey Grigson, his library and his own books – *The Englishman's Flora* and *A Dictionary of English Plant Names* in particular – this book would never have been written.

Observer readers have sent recipes and ideas; some of them have become friends and correspondents. I should also like to thank the following people: Joan Bailey of the London Library; Anna and Luca Benedetto for Italian recipes and information; Israel Ben Zeev of Agrexco, Tel Aviv; Betty Bolgar; Elisabeth Bond; Sue de Brantes; Deanna Brostoff; Anne and Helen Caruana Galizia, authors of *Recipes from Malta*; Ariane Castaing; Margaret Costa; Mimi Errington; Theodora Fitzgibbon; Yvonne Goldschmidt of the Israeli Embassy in London; Caroline and Sophie Grigson; Jean Grønberg; Mary and Richard Hatchwell who lent books and tasted dishes; Judith Jones of Alfred Knopf Inc, New York; Annette Macarthur Onslow; Gerald Mauri; Robin Milner-Gulland; Mary Norwak; A. R. Paske, for his knowledge of asparagus; Lisanne Radice for Polish information and recipes; Phyllis and Edward Shafer for help with Turkish and Chinese dishes; Mrs Agnes Short who tried to convert me to kale; Geneviève and Jean-Marie Stehli of Roc en Tuf; Baillie Tolkien who cooked cardoons; Mrs Viney of Palm's Delicatessen in Oxford Market, who supplied many of the ingredients in this book, and information on German recipes; Pauline Viola of the Danish Centre in London; Marjorie White who keeps the household going while books are written; Margarette Worsfold of *The Fruit Trades Journal*, for her enthusiasm and quick replies.

The disagreeable task of reducing an enormous manuscript to publishable size was carried out – with patience and good humour – by Gill Edden. I am grateful to her.

Roy MacGregor-Hastie has kindly given permission for me to quote part of his translation of *Seeds*, by the Romanian poet Lucian Blaga (p. 482), and was generous in answering questions.

INTRODUCTION

The first modern cookery book was written by a French court chef, La Varenne, who had started life as kitchen boy in the household of Henri IV's sister, Queen Henrietta Maria's aunt. *Le Cuisinier François* – the first of many editions appeared in 1651 – turned attention from the old style recipes with a gallimaufry of ingredients to a simpler style of recipe, of the kind we still make to-day, dishes like the one on p. 23 for artichokes and scrambled eggs. In spite of protests at French extravagance and praise of the simple plainness of English food, which is an older song than you might think, our cooks gradually took to the new style. That is to say the cooks of the aristocracy and manor houses and vicarages took to it, in the places where good vegetables were being grown in the kitchen garden.

Enterprising men like John Evelyn, who took the trouble to have their cauliflower seed from Aleppo and the best new celery from Italy, would obviously not be content to see the results of their labours added to a general patchwork of ingredients and stuffed under a pastry lid. Evelyn's *Acetaria* of 1699 is full of good vegetable dishes, so is Hannah Glasse's *Art of Cookery* that came out in 1747, and so – in spite of a sour comment by Coleridge – is *Domestic Cookery by a Lady* (Mrs Rundell), of 1808. This is what Coleridge said: ' . . . as to Vegetables why "the Lady" must have been all her life tethered in a Battersea Cabbage Garden, with a simple prospect of a Potato Field in the distance, and an occasional glimpse of a Turnip Waggon from over the Hedge – Covent Garden must have been a Terra Incognita, not even named in her Map'.

This is unfair to Mrs Rundell, but indicates expectation. Another interesting thing is that this category of food is at last defined as Vegetables. The word had been around for a long time, in the sense of animal, vegetable, mineral, but not until the 1760s did it come to be used in this new specific way, to describe 'herbs and roots grown for food'. The restricted meaning came to take first place in most people's vocabulary.

It shows that by the 1760s, for educated and prosperous English people, certain plants, their leaves, fruits and roots, had escaped from the black bag of antique medicine. Vegetables were – and still are – regarded as being good for you, but they had arrived as a pleasure on their own account. For the majority of the population, vegetables as a delight, to be eaten on their own, belong to this century, even to the period after the Second World War. One vivid comparison will show the old situation. In a book on life in the Lake District, William Rollinson remarks that apart from red cabbage and onions, garden vegetables were unkown there 'until the early 19th century' (presumably potatoes, cabbages, turnips and broad beans were grown as field crops). Yet those returning natives, the Wordsworths, were hard at work in 1800 sowing garden peas, French beans, spinach in the garden of Dove Cottage. 'Wm. stuck peas' is a frequent entry in Dorothy's Journal. They did not, it seems, try artichokes which were popular with 18th-century gardeners. I suppose Grasmere was too far north, though they might have flourished on that sunny slope behind the house.

Recently there has been something of a vegetable boom. It started in the fifties with Elizabeth David, who championed vegetables in their own right, not just as adjuncts to meat. As beef prices went up in the seventies, ecologists and vegetarians tried to turn the country to beans (I suspect their puritan fervour limited the number of converts, and undid some of Mrs David's good work). Unfortunately the trade has not profited intelligently from these moves towards vegetables. The magnificent choice at New Covent Garden is poorly reflected in shops outside London. Everything is left to foreign promoters, to the extent that peppers, aubergines, avocados can be bought everywhere, while seakale, our one native vegetable, and asparagus at a reasonable price, even fresh spinach are difficult to find. For commercial enthusiasm one must turn to seedsmen's catalogues. They have transformed a dull, middle-aged occupation into an exciting hobby.

So we are back with the private gardener, which could soon include everybody. Perhaps lively vegetables, the improvement of varieties, have always depended on him, since Adam put his food on the spade, with Eden in mind. Now we might extend the picture to include high-rise blocks, patched with vegetation on every balcony – Marmande and plum tomatoes in pots, herbs in window-boxes, courgettes and squashes trailed round the doors. Inside, there could be aubergine, pepper, chilli and basil plants on the window sill, jars of sprouting seeds, dishes of mustard and cress, with mushroom buckets and blanching chicory in the dark of broom and airing cupboards. In my most optimistic moments, I see every town ringed again with small gardens, nurseries, allotments, greenhouses, orchards, as it was in the past, an assertion of delight and human scale.

Jane Grigson.
Broad Town & Trôo, 1978.

ARTICHOKE

The artichoke is an edible thistle.
To John Evelyn it was the 'noble
thistle'; to a 19th-century writer
on food, E. S. Dallas, it was an
amusing moral lesson, 'It is good
for a man to eat thistles, and to remember that he is an ass'. John Evelyn
– as one might expect – is nearer the mark, for the artichoke was the
aristocrat of the Renaissance kitchen garden, as the asparagus was of
the Roman. It is sobering to realize that they are still the two finest
vegetables we can grow. Nothing we have developed since comes near
them for delicious flavour or for elegant form.

The artichoke above all is the vegetable expression of civilized living, of
the long view, of increasing delight by anticipation and crescendo. No
wonder it was once regarded as an aphrodisiac. It had no place in the
troll's world of instant gratification. It makes no appeal to the meat-and-
two-veg. mentality. One cannot attack an artichoke with knife and fork
and scoff it in three mouthfuls. It is first for admiration, then each leaf has
to be pulled away for eating and dipped in sauce. When the leaves have
gone, there is still the fibrous and tickley choke to be removed before the
grey-green disc – the *bonne bouche* – can be enjoyed.

Although the artichoke is easily grown – someone observed that it
flourishes, or can flourish as far north as the Orkneys – it has remained the
pleasure of people who mind about good food. At first it was the passion
of rulers, of the Medici in Florence, of François I and his court in
Touraine, of our Henry VIII. François's daughter-in-law, Catherine de
Medici, ate so many at a wedding feast in 1575 that she nearly burst (she
was given to over-eating: on another occasion she suffered from a surfeit
of melon). Henrietta Maria, queen of Charles I and daughter of Henri IV
and Marie de Medici, kept a garden devoted to artichokes at her manor in
Wimbledon. It was large – 44 perches – and its spiky vigour was enhanced
by the contrast of five dark bays of splendid growth. When a contemporary painter in France, Abraham Bossé, painted a series of *The Five
Senses* round about 1635, he chose an artichoke to embody Taste. He
depicted a dining room, with a young couple at table. The maid brings in a
fine melon, but the painting centres on a huge artichoke in a raised dish,
as the lady streches out her hand to remove the first leaf. It could have
been Henrietta Maria and Charles I enjoying an especially fine artichoke

from that Wimbledon garden, though it is really a scene from French court life under Henrietta Maria's brother, Louis XIII.

Although the name, from *al-kharsuf*, indicates an Arab origin, perhaps in gardens on the Barbary coast, the Italians must take the credit for developing the fine varieties of artichoke that captured the courts of Renaissance Europe. Italians would still take the prize for the largest repertoire of artichoke dishes; they make much of the young artichoke before stalk, fibre and leaf have hardened, so that it can be eaten whole. My first acquaintance with artichokes, in the mid-fifties before they were commonly on sale in our greengroceries, was in Venice where they have been growing for five centuries. We were visiting Torcello in the lagoon, walking along from the landing stage to the church by a canal, when a man swept into view punting a great boat load of artichokes. Presumably they were for the Rialto market. The blue sky swept by the strong wind, the orange and red bignonia trumpets pouring over every fence, the man's slow confident movements as he stood ankle-deep in the green and purple artichokes, made a first encounter of Medicean colour. We could understand how Rabelais – the first person in France to mention *artichaut* in print in 1534, though it seems to have been grown in Brittany around Tréguier, since about 1508 – put them as a final delight of the offerings to the Gastrolaters' god of good living. After the salads, which included asparagus, and the fish, came the dessert, ending with prunes, dates, walnuts, hazelnuts, parsnips – and artichokes.

The other great writer of the time, the poet Ronsard, loved artichokes too. Asparagus, artichokes, melons, he declared, are much better than the finest of meats: they should be served in great mounds. When he got fed up with studying, he sent out a friend to buy a picnic. 'Don't forget the wine. Remember that I hate meat in summer –

> Achète des abricots,
> Des pompons, des artichauts,
> Des fraises, et de la crème:
> C'est en Été ce que j'aime,
> Quand sur le bord d'un ruisseau
> Je la mange au bruit de l'eau,
> Étendu sur le rivage,
> Ou dans un antre sauvage.'

Apricots, melons, artichokes, strawberries and cream, the sound of water, a cool cave – could there be a better picnic for a hot day? This poem of 1554, 20 years after Rabelais was publishing his adventures of Pantagruel, shows that the artichoke was well settled into French life.

It took a little longer to establish the artichoke in England, though by 1629 when the great apothecary and herbalist John Parkinson wrote his *Paradisus in Sole* and dedicated it to Queen Henrietta Maria, he was able to say that even the youngest housewife knew how to cook artichokes and serve them with melted butter seasoned with vinegar and pepper.

A last curious thing about artichokes. Sheila Hutchins mentions it in *English Recipes*. She describes how popular they were until the end of the last century, then they seemed to vanish, so that now – she was writing in the sixties – they are appearing in the shops 'like some exotic foreign import'. She had heard 'English people on returning from holiday in Brittany where the vegetable is grown in quantity, protest after eating it in an hotel, that they had been fed on boiled water lilies'. I remember my mother, who had always been an adventurous eater, going out one day soon after I saw that boatload of artichokes at Torcello and returning with a couple for supper. We boiled them, then neither of us had the least idea what to do. We worked it out slowly. When it came to the choke, we worked it out the hard way. A quarter of a century later some green-grocers still feel the need to put a paper collar round artichokes explaining how to cook and eat them. This is why it seems important to go into some detail in the section following.

HOW TO TACKLE ARTICHOKES

There are a few general points that apply to nearly every method of preparing artichokes. Unless they are tightly closed, soak them upside down in a bowl of salted water for an hour, to dislodge earth or insects that might be concealed between the leaves. Provide yourself meanwhile with a piece of lemon to rub over any cut surfaces before they blacken. No lemon? Have a small bowl of wine vinegar to dip the cut parts into. When preparing artichokes in quantity, keep them in a big bowl or bucket of acidulated water (malt vinegar will do at that kind of dilution) until you have finished the last one; even if you have rubbed them with a lemon, they can still discolour slightly with a long wait.

Snipping the sharp points from the leaves with scissors or knife is an accepted thing to do. Why? There is no point in deforming such a beautiful object. I have never done this, and no one has pricked their finger and fallen asleep for 100 years at our table.

1) Basic method

When the artichokes are ready to come out of the salt water, put on a huge pan half full of acidulated salted water. Use 2 tablespoons vinegar to a litre (1¾ pt) of water. Cut the stalks from the artichokes, close to the leaves, and put them into the boiling water, cut end down. If the artichokes seem fresh and lively, put in the stalks as well; the inner part can often be scraped out and eaten. Once the water returns to the boil, allow 30 minutes. Remove a leaf from the base of the largest artichoke; if the nugget of flavour at the base is tender when you nibble it, the artichokes are done. If not, give them more time, testing every ten minutes. Drain the cooked artichokes in a colander, upside down.

If you want to eat them hot or warm, serve them with melted butter and lemon juice, or with hollandaise sauce. When cold, they taste most delicious when eaten with vinaigrette, mayonnaise or sauce tartare. Another good sauce is clotted cream, sharpened with lemon and seasoned with salt, pepper and a pinch of sugar.

Set the table with a large dish in the centre for discarded leaves. Provide everyone with a cloth napkin. Small bowls of water are also a good idea, for people to rinse their fingers. Unless you have space in the china cupboard, there is little point in buying special artichoke plates, though there is no denying that they do look pretty. If you make a runny sauce such as vinaigrette or melted butter, it is perfectly satisfactory to tilt the plates by sliding a knife underneath one end. The sauce gathers into a neat pool, so that mess is avoided. With firmer sauces such as mayonnaise or hollandaise you do not need to worry.

To eat an artichoke, pull away the leaves beginning at the bottom. Dip the tender base of each one into the sauce before chewing it from the leaf, which can then be discarded. The last inner rows of leaves will come away in one piece like a pointed egg cosy; nibble around the edge. This leaves you with a thick grey-green saucer of artichoke – the final reward. Alas, it is topped with a tight circular pad of whitish fibres. This is the 'choke'. If you are not careful, you will find out how apt the name is. The French call it the *foin*, or hay; I think our name is a better warning. The choke is quite easy to remove if the artichoke is properly cooked. You can do it by sliding a knife between the fibres and the artichoke bottom, but you risk losing quite a lot of the edible part. It is better to take wedge-shaped sections of the fibres between your thumb and a knife blade and pull them away with a flick of the wrist. You will be left with the artichoke bottom, the whole point of the enterprise. Eat it with a knife and fork, with plenty of sauce.

2) How to make artichoke cups

If you want to serve artichokes stuffed, you need to form them into cups.

Cut off the artichoke stalk and the two lower rows of leaves. Trim the exposed base part if it needs tidying up. Then slice across the top, about a third of the way down. Boil and drain the artichokes as above. When they are cool enough to handle, pull out the centre leaves to expose the choke, leaving several outer layers of leaves to form the cup. Using a pointed teaspoon, scrape away the choke. Pour a little well seasoned vinaigrette into the base, and leave the artichokes to cool down. The mayonnaise or salad can be piled into the centre – the flavours should be vivid not stuffy, which is why shellfish is so successful in this kind of artichoke recipe.

Artichokes cooked in this way have to be eaten with a knife and fork.

For serving stuffed artichokes hot, you have to cut the raw artichoke into shape. This is easier than it looks if your knives are sharp. Slice off the stalk and the two outer rows of leaves as described above. Then slice off the top third, or what you judge to be the right amount to make a nice shape; artichokes can vary in height or squatness – you have to go by eye rather than rule. Open out the inner middle leaves with your fingers, then tug and cut them away. You will now find that the uncooked bottom is hard enough to stand the vigorous scraping required to dislodge the choke – a strong teaspoon is the best utensil. Now you are ready either to boil the cups so that they can be served with separately cooked mixtures inside them, or to stuff and braise them with a breadcrumb mixture.

3) How to prepare artichoke bottoms

If you need cooked artichoke bottoms, this is simply a matter of boiling the whole artichoke, then dismantling it in the kitchen. The leaves should be scraped free of the edible part which can often be incorporated into the dish in some way.

If you need uncooked artichoke bottoms, cut away the stalk, then the rows of large leaves. The centre leaves can be pulled away and the choke discarded, as in method 2 above.

In Venice, where artichokes are cheap, you will see people sitting on the quays preparing artichoke bottoms for sale. Leaves and stalks go into the canal at flashing speed and the neatly turned bottoms drop into a bucket of acidulated water to await a sale.

4) How to prepare young artichokes

This is a paragraph for gardeners only (unless you happen to be house-keeping in southern Europe where such things can be bought). When the artichoke plants have to be pruned of their lateral buds so that the large ones can develop, don't throw them out. Sometimes the whole of their stalk can be eaten, they are so tender. Sometimes you may need to scrape away the tough outer layer.

5) *How to choose canned artichokes*

Don't. And don't be caught by cans labelled 'artichoke bottoms'. Look at the picture. They quite often turn out to be tiny whole artichokes, which are even more tasteless preserved in this way than the proper artichoke bottoms.

ARTICHOKE AND SHRIMP SALAD
FONDS D'ARTICHAUTS NINETTE

A fine way to start a special meal – the flavour of shellfish harmonizes beautifully with artichoke. Mayonnaise adds zest and richness.

> 7 cooked artichoke bottoms
> the cooked purée scraped from the leaves
> ¼ litre (8 fl oz) mayonnaise
> mustard
> fresh parsley, chervil, tarragon, chives
> 200 g (7 oz) shelled shrimps or prawns
> 6 prawns in their shells (optional)

Chop then mash one of the artichoke bottoms with the purée from the leaves. Mix with the mayonnaise and add a little mustard, then the chopped herbs. Fold in the shrimps or prawns. Pile this mixture up on the artichoke bottoms remaining and put a prawn in its shell on top of each. Serve chilled on a bed of lettuce.

If the artichoke bottoms are on the small side, and there is too much shrimp salad, pile the remainder in the centre of the serving dish and put the artichoke bottoms round it.

ELIZA ACTON'S SALAD OF YOUNG VEGETABLES

Put four or five cooked, trimmed and quartered artichoke bottoms into a salad bowl. Arrange over them cooked new potatoes and cooked young carrots, sliced thinly. Scatter with fresh chopped tarragon, chervil or any other herbs you might prefer, to make a thick layer. Just before you serve the salad, pour over a French dressing.

Young French beans, cut into lozenge shapes, asparagus tips or cauliflower florets may be added. If you make these additions, omit or reduce the quantity of the chopped herbs.

ARTICHOKE AND POTATO SALAD

Mix cubed artichoke bottoms with an equal weight of cubed new potatoes. Add the scraping from the artichoke leaves to mayonnaise, to dress the salad, and top with black olives, crumbled hard-boiled egg and chives.

ARTICHOKE PASTE

Mash the base and leaf pulp of cooked artichokes with some good olive oil, a little finely chopped garlic and lemon juice. Season with salt and pepper. Serve on slices of wholemeal bread and butter, or use for sandwiches, or to fill small pastry boats, or as part of a mixed hors d'oeuvre.

When we first went to live in France, I remember someone telling me that babies on the farms were often weaned on artichoke purée. She was buying high quality minced beef for her children at the time, and ended the remark with a loud sniff. I thought, 'lucky babies!'

BUTTERED ARTICHOKES

Cook and dismantle the artichokes. Scrape off the leaf pulp and spread it over the bottoms, then cut them into wedges or rough cubes. Just before the meal heat the artichokes through in butter, allowing them to brown hardly at all. Add a few tablespoons of juice from the roasting meat if you are serving the artichokes as a vegetable with meat. Then stir in some chopped parsley.

This kind of mixture can be put into shortcrust tartlet cases baked blind, and served round a joint of lamb or veal. It looks elegant, but is really an economy measure to make the most of a small quantity.

DEVILLED ARTICHOKES
ARTICHAUTS À LA DIABLE

The sharpness of the white wine vinegar sizzled with the butter is what gives the name to this recipe.

> *6 large artichokes, cooked*
> *6 eggs*
> *butter*
> *6 tablespoons white wine vinegar*
> *2 heaped tablespoons chopped green herbs: parsley, tarragon,*
> *chives*

Remove the artichoke leaves from the bottoms and scrape off the edible part. Discard the chokes. Use the scraped artichoke to make a wall round each artichoke bottom, to serve as a nest for the egg. Put the artichokes on a heatproof serving dish, cover them with foil and reheat in the oven or over boiling water.

Fry the eggs in butter, flipping over the whites to enclose the yolks neatly (or poach them, if you prefer). Place each one in a hot artichoke nest. Deglaze the pan with vinegar, sizzle for a moment or two, then whisk in a large knob of butter the size of an egg and add the herbs last of all. Pour over the eggs and serve at once.

FRICASSEE OF ARTICHOKE BOTTOMS

This is one of the best ways of serving artichokes. Do not be put off by thoughts of cholesterol – the modern equivalent of hell-fire – but adjust the other dishes in the meal to take account of the rich sauce.

> *9–12 cooked artichoke bottoms*
> *175 g (6 oz) butter, cut in pieces*
> *300 ml (½ pt) whipping or double cream*
> *lemon juice*
> *chopped fresh parsley and tarragon*

Prepare the artichoke bottoms and cut them into centimetre (½ in) pieces. Reheat them in a steamer, or wrapped in a piece of foil in a pan of boiling water, while you make the sauce.

Melt the butter in a frying pan. When it bubbles up, pour in the cream and stir steadily until you have a thick rich sauce. Add a squeeze of lemon juice and the herbs. Season if necessary. Put in the artichoke pieces and turn them over in the sauce, over a lower heat. Divide them between six little pots and serve with toast, baked bread or Dorset knobs.

Note If at any stage the sauce shows a tendency to separate (overheating causes this), just stir in 3 tablespoons of cold water. This will bring it back.

If you have only single cream, mix it with a teaspoon of cornflour before adding it to the butter.

ARTICHOKES AND EGGS ROC EN TUF
OEUFS BROUILLÉS AUX FONDS D'ARTICHAUTS

I have a friend in France whom I admire greatly. She is the complete 18th-century manor house wife. Her vegetable garden bulges with unusual things such as huge tufts of perpetual lettuce as well as the standard items. Her poultry yard quacks continually with ducks, geese, hens, guinea fowl, turkeys, pigeons. In the meadow beyond the garden a few sheep browse by the stream that supplies her trout. When we go to dinner we sit on chairs she has upholstered and embroidered. We drink cordials, apéritifs and wine that she and her husband have made. The things we eat have been grown, fed and cooked by her. One of the best meals we have eaten there started with this 17th-century dish that La Varenne describes in *Le Cuisinier François*. She was surprised when I commented on its antiquity. I think she had it from her mother-in-law, but she had added her own special seasonings. The recipe works well, too, with asparagus tips, young peas and many other vegetables of quality.

> 6 large artichoke bottoms, cooked
> 12 eggs, beaten
> 4 large sprigs parsley, chopped
> small bunch chives, chopped
> 100 g (3–4 oz) butter
> 1 tablespoonful Dijon mustard
> 2 tablespoons double cream
> salt, pepper

Cut the artichoke bottoms into centimetre (½ in) pieces, and reheat them if necessary. Mix the eggs with the herbs and scramble them lightly in the butter. Add mustard to taste gradually, then the cream and finally add seasoning to taste. Put the artichoke pieces over the base of a warmed serving dish and pour on the scrambled eggs. Set triangles of toast round the dish, or serve with baked bread, or French bread.

Note Be careful not to overcook the eggs in the first instance. They will continue to cook in their own heat, and in the heat of the serving dish.

ARTICHOKES BARIGOULE
ARTICHAUTS À LA BARIGOULE

This famous recipe from Provence has many variations. Sometimes small, unstuffed artichokes are cooked on a bed of vegetables, but for this country the second style suitable for larger artichokes is more practical. Why *barigoule*? Most French cookery writers glide over the problem but one of them, Maguelonne Toussaint-Samat, has this to say in *La Cuisine Rustique: Provence* 'Barigoulo or barigoule is a morel mushroom. It is possible that the method originated as a way of cooking morels. Perhaps trimmed young artichokes were thought to look like morels. Or did their shape remind people of a pot-belly, which in Spanish is *barigo*?'

Start by preparing six artichoke cups. If you are in a hurry cook them before you start on the stuffing, but it makes a better dish if you stuff them raw and leave them to simmer gently for a long time. The other ingredients are:

> 175 g (6 oz) mushrooms, chopped
> 100 g (3–4 oz) green streaky bacon, chopped
> butter
> chopped fresh parsley and chives
> 3 tablespoons breadcrumbs
> salt, pepper
> 6 long strips of pork fat cut very thinly (optional)
> oil
> 2 large onions, sliced
> 1 large carrot, sliced
> 1 small clove garlic, finely chopped
> pinch savory (pèbre d'ase)
> 200 ml (7 fl oz) dry white wine

Fry the mushrooms and bacon in butter until lightly browned. Mix in the herbs and breadcrumbs and season. Stuff the artichoke cups. Tie the fat strips over each one with button thread, if you are using them – they are not essential – and brown all over in oil. Put 2 tablespoons oil into a large pot that will take all the artichokes in one layer. Put the onion, carrot and garlic in a bed on top, with seasoning and savory. Pour over the wine, plus enough water to bring the liquid level slightly above the vegetables. Put in the artichokes. Cover and simmer until they are tender – about an hour and a half, if they started off raw. If they were already cooked, simmer a little more vigorously for about half an hour.

The liquor can be thickened slightly if you like with flour, cornflour or potato flour, but do not make it too gluey. I think it is better to reduce the cooking liquid slightly if it is on the watery side, rather than to make starchy additions.

OTHER ARTICHOKE STUFFINGS

Follow the *barigoule* method above for cooking the artichokes when stuffed, or simplify it by omitting the carrot and onion. Keep the liquid level in the pan at about 1–2 cm (½–¾ in), until the artichokes are tender. Then it can be reduced if necessary. Artichokes stuffed in this way can also be baked in the oven at 160–180°C/325–350°F/ Gas Mark 3–4, allowing one to two hours.

CARCIOFI ALLA RICOTTA   Mix 250 g (8 oz) ricotta cheese with an egg and 3 tablespoons grated pecorino or Parmesan cheese. Mix in 125 g (4 oz) chopped Italian salami. Pile into cooked artichoke cups. Bake in a generously buttered dish with a little water, at 190°C/375°F/ Gas Mark 5, until the stuffing is properly heated through and the top browned. Sprinkle the top with breadcrumbs and melted butter.

ARTICHAUTS FARCIS PROVENÇALE   For fresh, medium-sized artichokes, still at the tender stage. Cut off and peel the stalks. Chop the inner 'marrow' with some parsley, 2 cloves of garlic, 8 anchovy fillets or 12 large black olives or a mixture of both, salt, pepper and savory. Soften 6 tablespoons breadcrumbs with a little milk and add the chopped items. Fill into the artichokes, and complete the cooking *barigoule* style.

STUFFED COOKED ARTICHOKES
ARTICHAUTS FARCIS

For this style of stuffed artichoke you can use the bottoms alone with walls built up from the purée scraped from the leaves or – and I think this is easier – cooked artichoke cups with enough of their leaves left in place to form a wall. The quantity of stuffing you need will depend on the depth of the artichokes.

When the artichokes have been cooked in advance, they should be reheated in a steamer before the hot stuffing is added.

For a large party, fill the artichokes with a variety of stuffings: it makes a beautiful dish.

AUX FÈVES   My own favourite, a delicious mixture. Shell, cook and skin broad beans – for six artichokes you will need approximately 2 kg, or 4 lb. Sieve them into a clean pan, and reheat with a large lump of butter, 3–4 tablespoons cream and a squeeze or two of lemon juice. Finally mix in some chopped fresh savory. To eat artichokes cooked in this way, remove the outer leaves and dip them in the centre filling. Then eat the bottom with a knife and fork.

AUX ÉPINARDS   Cooked, chopped spinach, plus a few leaves of sorrel if possible, creamed with a rich béchamel sauce, or with butter, cream and nutmeg. Top with Parmesan. Brown under the grill.

AUX CHAMPIGNONS   Cook 250 g (8 oz) chopped mushrooms, a tiny clove of chopped garlic and 2 chopped shallots in a little butter. They should not brown, but the juices should be well evaporated. Add ¼ litre (8 fl oz) rich béchamel sauce and cream. Scatter with grated Gruyère or Parmesan and brown under the grill.

À LA ROYALE   Mix 2–3 sets of prepared, diced brains with 2 beaten eggs, 100 g (3–4 oz) grated Gruyère and chopped parsley and chives. Put into the artichokes, scatter with breadcrumbs and sprinkle on melted butter. Bake for 15 minutes in a hot oven, until the filling is just set.

ARTICHOKE GRATIN
GRATIN D'ARTICHAUTS

If you use this delicious way of stretching artichokes, five or even four large ones will do for six people, when the dish is being served as a first course. Cook the artichokes in the usual way. Remove the leaves and scrape off the pulp. Cut the bottoms into centimetre (½ in) dice and fry them very lightly in a little butter. Put them into a buttered gratin dish, with the pulp dotted about. Hard-boil three eggs, shell and quarter them and arrange them with the artichokes. Pour over the top about 300 ml (½ pt) mornay sauce or cream sauce, then run a little cream over the top, and sprinkle with either breadcrumbs or grated Parmesan or Gruyère according to the sauce you chose. Brown in the oven or under the grill and serve bubbling hot.

VARIATION Substitute 200 g (7 oz) chopped mushrooms, fried until light brown in a little butter, for the hard-boiled eggs.

VARIATION Slice the cooked artichoke bottoms and fry them lightly in butter. Put with the leaf pulp on a not too thick bed of cooked spinach. Pour mornay sauce over.

TOURNEDOS HENRI IV

Both taste and texture of artichoke marry well with steak. And when béarnaise sauce is added, you have one of the celebrated dishes of French classic cookery. The name of the dish refers to the sauce, which was developed from hollandaise at the Pavillon Henri IV, at St Germain-en-Laye in the 19th century. The Pavillon is still an hotel, an attractive place, converted from what remained of Henri IV's château. By family and upbringing he was a béarnais. Although he could not have known the sauce, artichokes were a fairly new and appreciated delicacy at court banquets in his lifetime.

> *½ kg (1 lb) tiny new potatoes, scraped*
> *butter*
> *6 artichokes*
> *béarnaise sauce (p. 557)*
> *6 slices bread*
> *6 tournedos, nicely trimmed*
> *salt, pepper*

First put the potatoes on to cook with a large knob of butter. Cover the pan and keep the heat low. Shake the pan from time to time. Leave until the potatoes are golden brown and tender. On a really slow heat, this can take up to an hour. At the same time, put the artichokes on to boil. When they are done, remove the leaves (keep them for another meal), and keep them warm. Make the béarnaise sauce.

15 minutes before the meal, remove the crusts from the bread, and fry the slices in butter. Put them on to a serving dish and keep them warm. Last of all, pepper the steaks and grill them to your liking and according to their thickness. Then – and only then – salt them. Place them on the fried bread. Fill the artichoke bottoms with sauce and put them on top of the steaks, with a little chopped fresh tarragon if possible. The potatoes should be drained and put round the edge. Any sauce left – this will depend on the size of the bottoms and the consistency of the sauce – can be served separately.

TOURNEDOS CHORON

Another classic recipe for steak with artichokes. This time the Choron variation of hollandaise is used (Choron was a chef from Caen in Normandy, who had the bright idea of flavouring the sauce with tomato).

> ½ *kg (1 lb) tiny new potatoes, scraped*
> *butter*
> *6 cooked artichokes*
> *6 tablespoons cooked peas, or asparagus tips, cut up*
> *sauce Choron (p. 557)*
> *6 slices bread, fried in butter*
> *6 tournedos steaks*
> *5 tablespoons good beef stock or jelly*
> *6 tablespoons dry white wine*
> *salt, pepper*

For a detailed explanation of the method, turn back to Tournedos Henry IV, on page 27.

Cook the potatoes slowly in butter until golden brown. Remove the artichoke leaves, discard the chokes and fill the bottoms with peas or asparagus; keep them warm. Make the sauce, fry the bread, and then fry the steaks in a clean pan. Salt them when cooked and place them on the bread, with the artichoke bottoms round the edge of the dish. Deglaze the steak pan with the stock or jelly and white wine, bubbling it to a good rich essence. Pour over the steaks. The sauce can be served separately or spooned over the steaks (the first way is better, I think).

Lamb noisettes can be prepared in the same way but omit the fried bread.

TINY ARTICHOKES FOR THE FIRST COURSE

Tiny artichokes are a familiar part of Italian *antipasti*. They appear with the rolled anchovies and capers, the olives, and other piquancies that are intended to rouse your appetite for the main part of the meal.

Prepare – see p. 19 – about two dozen, enough to fit closely into the bottom of a large heavy pan. Pour over them ¾ litre (1¼ pt) dry white wine with 125 ml (4 fl oz) lemon juice and 2 tablespoons white wine vinegar. Tuck in two bay leaves, a thinly sliced lemon, four cloves and a dozen peppercorns. Bring to the boil, cover and simmer until tender – allow at least 15 minutes.

Remove the artichokes and drain them upside down in a colander. Pack them closely into a jar, add three new bay leaves, eight fresh peppercorns (and a dried red chilli) and cover them with olive oil. Close the jar tightly. Put in a cool place for four days before using any. They keep well.

If you want to prepare small artichokes for immediate eating, cook them in equal quantities of water, white wine and olive oil. When they are tender, remove the lid and boil hard until the liquor is reduced to oil.

Really small tender artichokes can be eaten raw with a simple olive oil vinaigrette, as part of a dish of *crudités*, in the spring. This is popular in Provence, where the miniscule violet artichokes are known as the *primié grioù*, the first growth.

OTHER SUGGESTIONS

Artichauts aux beurre blanc. Serve hot artichokes with *beurre blanc* (page 356).
Artichoke and veal moussaka, see page 59.

ASPARAGUS

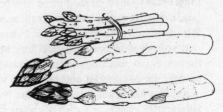

Many years ago now we made the journey from Venice to the Venetian Alps to spend a day in Bassano. Our particular reason was the paintings of the da Ponte family, better known as the Bassani, in the cathedral of their native town. The cathedral was closed. So was the museum. The wind blew from the hills. We consoled ourselves by sheltering in the covered wooden bridge, the famous bridge over the tumbling, ice-grey Brenta – and with the asparagus for which Bassano is also famous. It was thicker than our thumbs, going from white to pale yellow at the tips. Fat and edible to the last, it came with eggs, with olive oil and wine vinegar to make the dressing. The main course, no anti-climax, was a risotto of plump rice from the Po valley, well buttered, with quails on top.

I did not discover until I was preparing this book, how old the aparagus growing business is in north-eastern Italy. Those plump, pale asparagus stalks had ancestors right back to the end of the 16th century. In the hinterland of Venice, farmers discovered that asparagus was so profitable, although it had to be transported many miles to the main market at the Rialto, that 'in fields where they once grew corn and flax, they now raise asparagus'. Their profit had trebled. Giacomo Castelvetro, who made these observations in 1614, was astonished at the scarcity of asparagus in England and wondered why it cost so much in London. Three hundred and sixty-odd years later, I wonder why nothing has changed. Asparagus comes in from Hungary, Cyprus, the States, and what's the good? Asparagus needs to be eaten the day it is picked. Even asparagus by first class post has lost its finer flavour. Why gardeners all over the country do not automatically plan an asparagus bed, I shall never understand. The crowns available nowadays start producing in their third year; three-year roots are sold by some firms. Why grow row after row of cabbage, when cabbage is always in the shops in reasonably good condition, at a low price?

In France the smell of asparagus cooking is the universal savour of the village in May and June. In Germany, by Lake Constance, asparagus locally grown in Tettnang is served up in 20 different ways, in every kind of restaurant from homely to grand. One Düsseldorf restaurant runs to 209 asparagus dishes. Lucky people. Asparagus with melted butter and a

few fried breadcrumbs for contrast, asparagus with hollandaise sauce –
three ways – must be the limit of our ambitions. For the rare treat, there is
no need of a greater repertoire. Other recipes are included, though, for
the gardeners and for the lucky ones who are taking an early holiday
around Blois and in the Sologne, the greatest of France's asparagus
areas, or down in the south near Cavaillon on the Durance river, the great
centre of fruit and vegetables, and Lauris further upstream towards
Lourmarin.

HOW TO BUY AND PREPARE ASPARAGUS

With asparagus almost more than any other vegetable, you have to learn
your area and its resources to be able to buy it well. You are unlikely to
have much choice. The bundles on sale at any one time will most likely
have come from the same grower; picked on the same day, they will have
been graded to the same standard. The question is whether to buy it or
not.

First you must try to judge how long the journey has been. The
labelling will give the distance. But how long ago was it cut? If you shop
regularly, you will know whether it is new in. If it is local asparagus, the
greengrocer should be able to tell you exactly.

The asparagus season in England runs from as early in May as the
weather allows, until the end of June. Three hundred miles further south,
in the Loire region, people say that they stop cutting asparagus when they
start picking currants. This happens around 'le Saint-Jean', St John's Day
– our midsummer day – on June 24th. In those parts everyone with a
garden grows asparagus; in the markets there are big boxes of it, as well as
the kilo bundles. A 1 kg (2 lb) bundle of asparagus is enough for four as a
first course on its own, six if you are helping it out with other things.

If you examine the stalks, you can easily decide where the skin begins to
become inedibly thick. Remove it from that point downwards towards the
cut end, with a vegetable peeler. Rinse the asparagus as little as possible.
Cut off any damaged part and try to get the lengths of the asparagus much
the same. Any extra bits that you have to take off that are unblemished
can be cooked in with the bundle and kept with the liquor for making soup
later on. Tie the asparagus into one bundle or more.

If you have an asparagus boiler, cooking is simple. Remove the inner
lining and stand the asparagus in it, tips upright. Put about 6–7 cm (2½–3
in) of salted water into the outer pan, and bring it to a hard boil. Lower in
the inner container, keeping the heat high until the water boils again. Put
on the lid and leave until the asparagus is tender when the lower part of

the stalk is pierced with a pointed knife. The tips will have steamed to tenderness while the tougher stalks were boiled. Time will vary from 15 minutes to 45; it depends on the variety and quality of the asparagus. The thin asparagus that is sometimes sold as sprue, and that really looks like 'grass', is quickly done. ('Grass' is what asparagus is called in the trade, from the old name sparrowgrass.)

When you judge it to be ready, haul out the inner container by its handle with a wooden spoon and leave the asparagus to drain for a few minutes.

What if you do not possess an asparagus boiler? For most people, it is an unnecessary encumbrance in the kitchen. If you have some kind of tall pan, such as the French *faitout*, use that, standing the asparagus in a wire blanching basket to keep it upright. If you have nothing beyond the usual large kitchen pans, cut rather more from the stalks than you would normally do. Tie the asparagus in bundles and stand it in the boiling water, adding the trimmings. Put on your oven gloves and fix over the top a domed lid of foil, so that the asparagus tips that stand up above the rim of the pan are not harmed as they cook in the steam. Use the trimmings for soup.

Some growers recommend cooking the asparagus on its side in a roasting pan. Even if you stand the thick stalk end over the heat, so that the tips are right over to the side in the cooler water, this is not really satisfactory. You either have a good deal of waste in the stalks, or the tips disintegrate from overcooking.

Serve asparagus warm or cold, rather than hot from the pan, as it is eaten with the fingers. Another important point is to make sure it is well drained. Wet asparagus can be embarrassing to eat, as one's sleeves soak up the water that should have stayed in the kitchen. In the old days it was put on to a folded white napkin or a huge piece of toast, but this is unnecessary – especially if you have a pierced plate, or one of those meat dishes with a removable strainer tray in the middle, to serve it on.

The simplest sauces for warm asparagus are melted butter and vinaigrette (light olive oil and lemon juice). Hollandaise and maltese sauce, the cream sauces given on pages 553 and 555, all underline the delight of the occasion. This kind of treatment is enough for me, but some people add a few crumbs fried in butter and sprinkled over the tips, to provide a contrast. Italians put grated Parmesan over the tips and slip them under the grill until the cheese melts.

For cold asparagus, mayonnaise and vinaigrette are the best sauces. Cornish cream with a little lemon juice, a hint of sugar, some salt and pepper also makes quite a good dressing, though I think it is more successful with artichokes than with asparagus.

If you can only manage a few stalks for each person, arrange the ration on each individual plate, with an eye for gradation and length. Then chop some parsley finely, and put a 1 cm ($\frac{1}{2}$ in) band across each group of stalks 5 cm (2 in) down from the tip. This does not affect the flavour one way or another, but the tiny attention takes away any impression of meagreness. I saw this done at a restaurant in Vouvray. The few stalks looked special, moreover they were of exquisite quality, having come that morning from the sandy varenne behind the Loire levée at la Frillière, three kilometres away.

There are few nicer lunches than asparagus, the new potatoes that shared their pan, and eggs mollet, with melted butter and home-made bread. Add a glass of white Loire wine. It is always a good idea to cook some new potatoes in with the asparagus; they benefit from the flavour and help to keep the asparagus upright in the pan. If you do not want to eat them with the asparagus they can be served later in the meal, or kept for a potato salad on another occasion.

ICED SAUCE FOR COLD ASPARAGUS

This and the sauce following come from Nancy Shaw's *Food for the Greedy*, published by Cobden-Sanderson in 1936, in pink and white striped boards. Well-off cooks in those days were so in love with their new toy, the refrigerator, that they iced everything they could. It becomes excessive and, as far as cold asparagus is concerned, I prefer the sauce to be at normal room temperature rather than chilled.

> 3 large egg yolks
> 3 tablespoons melted butter
> pepper, salt
> 150 ml ($\frac{1}{4}$ pt) single cream
> 300 ml ($\frac{1}{2}$ pt) double cream
> 1 tablespoon wine vinegar

Whisk the egg yolks, butter, seasoning and single cream in a basin over a pan of simmering water. When it is thick – do not let it boil – remove the basin from the pan and allow it to cool. Whip the double cream until stiff, then fold it in. Flavour with the vinegar and extra seasoning. 'Ice it very well . . . no iced sauce must be frozen solid. It must only be half-frozen and easy to serve from the sauce boat.'

TARRAGON CREAM 'Season some thick cream with salt and pepper, and a dust of curry powder; and whip it until thick. Then flavour it by adding gradually a few drops of white tarragon vinegar and place it on ice for a short time before being served. To be sent up in a sauce boat for cold asparagus.'

HORSERADISH CREAM Flavour some lightly whipped double cream with finely grated horseradish (or a good proprietary brand of prepared horseradish), lemon juice, salt, sugar, all to taste. This is a popular sauce with asparagus in Germany.

CONSOMMÉ ARGENTEUIL

To the French, Argenteuil means asparagus. That is where it came from if you bought it in Paris in the 19th century, when the grand style of cookery reached its height. This recipe comes from the Loire, from the Hôtel de la Madeleine at St Benoît, one of the most delightful places you could hope to stay at. During the season a big shallow box of asparagus comes in every morning, just picked from the sandy fields by the river. Confident anticipation of the evening's meal adds an extra pleasure to the day's excursion.

If you cannot stomach the idea of using tapioca, remembering school puddings of the old days, reflect that the French call it *perles du Japon*. Such a beautiful name must indicate virtue, and as an addition to soup tapioca redeems itself.

> ½ kg (1 lb) asparagus, peeled, trimmed
> salt, pepper
> 3 tablespoons pearl tapioca (not flakes)
> 2 large egg yolks
> 6 tablespoons whipping or double cream

Cut the asparagus stalks into three sections each. Add the thickest pieces to 1½ litres (2¼ pt) boiling, salted water. Simmer steadily for 15–20 minutes, then add the centre pieces and after another ten minutes the tips. When everything is more or less tender – about another ten minutes, though this will depend on the asparagus – tip the whole thing into the blender, keeping back some of the liquid. Blend the asparagus until it is a purée, then pour it back into the pan through a sieve. Stir the whole thing up as you reheat it. When it boils, add the tapioca and simmer for 15 minutes, until the tapioca is transparent. Taste the soup and add more seasoning if necessary, more water if you think the flavour is too strong. Whisk together the egg yolks and cream and add to the soup off the heat. Stir for a few moments without allowing the soup to boil. Check the flavour again and serve.

If you are short of asparagus, put in a new potato or two, or reduce the initial quantity of water.

Should you be using the recipe in France, remember the blanched asparagus you buy there should be cooked in a *blanc*: mix a tablespoon each of flour and vinegar with the cooking water before bringing it to the boil. You will end up with a pale yellowish-fawn soup.

CREAM OF ASPARAGUS SOUP

The quality of this soup depends on the asparagus trimmings you use. If you take the precaution of cutting off the tougher stalks from the bundle, and cooking them in with the tips, you have the best basis for a quick soup. Another alternative is to keep only the asparagus liquid, with three or four whole stalks; this is the best thing to do if the asparagus is tender right to the end. After much experiment, I conclude that chicken or veal stock is quite unnecessary. Ariston men 'udor, as they have inscribed over the Pump Room at Bath: water is best. If the soup needs more body, use some new potatoes rather than more flour, or a couple of egg yolks with the cream. If the asparagus flavour is very thin, a glass of white wine will help, and a pinch of sugar.

> *trimmings from a bundle of asparagus*
> *100 g (3½ oz) chopped onion*
> *1 clove garlic, chopped*
> *60 g (2 oz) butter*
> *1 tablespoon flour*
> *½ litre (generous ¾ pt) asparagus cooking liquor*
> *salt, pepper, chopped parsley*
> *6 tablespoons whipping or double cream*

If the trimmings are uncooked, cut them into 2 cm (¾ in) lengths. Soften the onion and garlic in the butter in a covered pan. Stir in the flour, then the asparagus liquor, with the asparagus trimmings. Simmer until the maximum flavour has been extracted (this will take longer with uncooked than with cooked trimmings). Purée in a blender, then pass through the sieve to keep back any tough stringy bits. Taste and add more asparagus cooking liquor or water, with seasoning. Heat to just below boiling point, stir in the parsley and cream and serve with little cubes of bread lightly fried in butter.

ASPARAGUS SOLDIERS
ASPERGES À LA FONTENELLE

An attractive way of economizing on asparagus comes from Belgium. Fontenelle is in Hainault, to the south of Charleroi, and close to the French border.

Serve everybody with a boiled egg and a small bundle of cold or barely warm asparagus. Put on the table a large pat of butter and a loaf of brown bread, with salt and the pepper mill. Each person removes the top of his egg, seasons the nicely runny yolk with salt, pepper and a little knob of butter and dips the asparagus into it, nursery style. More bits of butter, more seasoning, may be added as the yolk goes down. Finish off the egg in the usual way with a spoon, eating it with bread and butter.

Note If the asparagus is cold, it will be easier to manage; if it is tepid, it will taste even better. Provide napkins of cloth, not paper.

ASPARAGUS AND BUTTERED EGGS

When asparagus is running short, and you have only four stalks or so per head, they will appear to best advantage in a tart – see recipe following – or with eggs.

Trim the asparagus before boiling it, so that you only cook the tender part. Meanwhile fry some bread, a slice for each person, or toast and butter it; or reheat some little pastry cases.

When the asparagus is cooked and drained, quickly scramble an appropriate number of eggs (three for every two people is about right). Do this with plenty of butter and keep them creamy. Put the egg on the bread or in the pastry cases and lay the asparagus decoratively across the top. Serve immediately.

If you want to serve asparagus in this way with drinks, make a number of tiny pastry *barquettes* in special oval tins. Fill them as above as late as possible before handing them round. If you must do them well in advance, cook the separate items and leave them to cool (put the egg saucepan in a bowl of iced water to prevent them over-scrambling). Assemble them when cold so as to reduce sogginess to a minimum.

ASPARAGUS TART

Bake a large pastry case blind until it is set but not coloured. Lay on it the boiled asparagus, trimmed into 2–3 cm lengths (about 1 in). Scatter them with 3 tablespoons grated Parmesan cheese. Mix together and season 200 ml (7 fl oz) single or whipping cream and three eggs. Pour over the asparagus. Bake at 180°C/375°F/Gas Mark 5 until just set – about half an hour. Serve warm for preference, for the best flavour.

ASPARAGUS 'PEAS'

This is an old way of treating thin green asparagus and the only bearable way, apart from soup, of presenting frozen asparagus.

Cook the asparagus in the usual way, then cut off and discard any tough stalks. With frozen asparagus you can use the lot, as they have already been well trimmed. Now cut your pieces into short lengths, so that they are virtually cubed into pieces the size of large peas. Weigh them – ½ kg (1 lb) will be plenty for four, can be stretched for six.

The next thing is to toast a slice of bread for each person. (If you can buy, or have thoughtfully brought back from France and frozen, a long brioche loaf, this gives a much finer result than even the best home-made bread.) Cut off the crusts and place the toast on a dish to keep warm.

For ½ kg (1 lb) of 'peas', melt 100–125 g (3–4 oz) butter in a frying pan. When it bubbles quickly stir in, with a wooden spoon, 150–200 ml (5–7 fl oz) whipping or double cream. Keep stirring over a moderate *not a fierce* heat, until the two ingredients have blended to a rich thick cream. Rapidly tip in the peas to heat through – no need to boil them, just make them good and hot – and divide them between the slices of toast. If you overheat the cream sauce it will turn oily; usually a couple of tablespoons of cold water will bring it back, but if it has gone too far for that, thicken it with a teaspoon of cornflour mixed with 2 tablespoons of water.

ASPARAGUS IN LOMBARDY AND EMILIA

Italians who live in Milan often make a spring-time journey on Sundays to a small town near Varese, close to the Swiss border. It is called Cantello and proclaims itself 'the capital of asparagus'. Here one eats the thin, green kind, more like our English asparagus. Local *amateurs* claim it has far more flavour than the fat white stalks of Bassano over the mountains towards Venice. Around Cantello the asparagus is boiled and arranged on a dish, heads together. Finely grated Parmesan is sprinkled over the heads, then some butter melted to noisette stage; a few minutes under the grill to colour the cheese lightly, and the asparagus is ready. You may think this would be too strong a finish for asparagus, but try it. If the Parmesan isn't overdone, it brings out the delicate flavour well.

Another way of arranging asparagus is to put it like the spokes of a wheel on to a large plate, heads to the centre. Pour the Parmesan and butter over in the same way. This practical arrangement means that a group of people can gather round the dish and pick up the stalks without fingers becoming greasy.

Italian cooks in Emilia make a version of our English asparagus 'peas' in cream sauce, *asparagi alla panna*. They bring rich unpasteurized milk slowly to the boil, then allow it to cool; the cream forms a crinkled skin on top and can be lifted off quite easily. Asparagus tips are mixed with this cream and seasoning, and reheated. Delicious. Here it is difficult to buy good unpasteurized milk, better to buy clotted cream ready made. When you reheat the asparagus with the cream, keep it under boiling point.

ASPARAGUS AND CHICKEN GRATIN

This English recipe was given me by a French friend who will not hear a word against our cooking. Her husband spent two happy years at Bury St Edmunds Grammar School 50 years ago and they return each year to visit friends he made there. As in the fricassee following, the exact quantity of the main ingredients is not important.

> about ½ kg (1 lb) thin green asparagus
> half a large roasted chicken
> 1 heaped tablespoon butter
> 1 heaped tablespoon flour
> 300 ml (½ pt) asparagus cooking liquor
> 150 ml (¼ pt) whipping or double cream
> 2 tablespoons breadcrumbs
> 2 tablespoons finely grated Cheddar
> 2 tablespoons melted butter

Cook the asparagus and drain it well, keeping the liquor which will be required for the sauce. Cut off any tough inedible parts and divide the rest into 5 cm (2 in) pieces. Dice the chicken neatly, discarding any gristle and bone. Layer them into a gratin dish. Make a sauce with the butter, flou., asparagus liquor and cream in the usual way. Have it nice and thick, remembering that the asparagus and chicken will exude a little more juice as they heat through. Pour this sauce evenly over the whole thing. Scatter the top with breadcrumbs and cheese, then with the butter, and bake in the oven at a moderate to high temperature until it bubbles and is nicely browned on top.

ASPARAGUS FRICASSÉE

Asparagus goes well with mild meats such as chicken or turkey, veal or lamb. The proportions of the main ingredients can be varied to taste and convenience – the more asparagus you can manage the better and finer the flavour.

> about ½ kg (1 lb) cooked, diced meat – chicken, turkey, veal or
>   lamb
> about ¼ kg (½ lb) tender cooked asparagus
> about ¼ kg (½ lb) cooked young peas
> salt, pepper, ¼ teaspoon sugar
> 1 heaped tablespoon butter
> 1 heaped tablespoon flour
> ¼ litre (8 fl oz) appropriate meat stock
> 125 ml (4 fl oz) dry white wine
> 125 ml (4 fl oz) whipping cream, or double and single mixed
> lemon juice, chopped tarragon or parsley
> large puff pastry vol-au-vent case

Sprinkle the meat with salt and pepper and set aside; it should be clear of fat, skin and gristle. Cut the asparagus into short lengths, mix with the peas and season with salt, pepper and the sugar. Make a sauce with the butter, flour, stock, wine and cream in the usual way, allowing it to reduce to a good thickness – it should not bind the main ingredients creamily. Taste the sauce and correct the seasoning. Finally add the meat and reheat it thoroughly before putting in the vegetables. Bring back to boiling point, add lemon juice to taste and herbs. Pile into the reheated vol-au-vent case and serve immediately.

This fricassee tastes best if the vegetables are cooked the same day. Meat from the day before is all right, a good way of serving up left-overs in style.

VARIATION Use cooked mild ham, about ¼ kg (½ lb) and rather more asparagus. Scatter the dish finally with fried breadcrumbs mixed with chopped parsley.

OTHER SUGGESTIONS

Asparagus and scrambled eggs Roc en Tuf, see page 23.

ASPARAGUS CHICORY

In recent years our most enterprising seedsmen have introduced many new vegetables into
this country. Some are only new
to our gardens; we have enjoyed
them on holidays abroad and then bought them from good greengrocers –
aubergines, peppers and courgettes are obvious examples. Some may
never have been on general sale, but we know about them in a vague way
– new kinds of squash, a different variety of endive. Occasionally, and this
is a brave leap, something entirely unknown appears in a catalogue –
asparagus chicory, for instance. It was first announced by Thompson and
Morgan of Ipswich in their 1976 catalogue and I hope they continue it.

In Italy asparagus chicory is a late winter and early spring vegetable. I
suppose this is why it remains unknown in our country, while Italian
vegetables of summertime and autumn – the holiday seasons – have made
such magnificent conquests. Asparagus chicory does not have that advantage, but it is good. Romans enjoy it so much that they have given it a
popular name, a pet name if you like, *puntarella*, an affectionate diminutive of *punta*, meaning tip or shoot, like an asparagus tip.

Cicoria asparagio has been taken straight into English by Thompson
and Morgan. This is sensible, but cumbersome. Sgaravatti, the seedsmen
in Rome, prefer another name, *cicoria di Catalonia*, or just *Catalonia*, so
that there is no chance of confusion with chicory or asparagus. If we
cannot manage *puntarella*, we might take to Catalonia, by analogy with
Calabrese, the green sprouting broccoli from Calabria. Whether this
variety of chicory came to Italy from Catalonia, whether it was first
developed there, I have not been able to find out. Botanical dictionaries
are silent about it. English books on Italian food do not even mention it.

HOW TO PREPARE AND COOK ASPARAGUS CHICORY

In cultivation the leafy plants are tied up so that the tips – the *puntarelle* –
grow up from the centre of a bundle. Then they are broken off, or cut off,
for eating. The thicker, tougher stalks are prepared in a most unusual

way. You shave off bits with a sharp knife, working from the broken end to the tip, as if you were making a pointed stake. These fine shreds come off in curls, like wood shavings.

You can eat asparagus chicory, tips, leaves and shreds, in a salad. One of our family who lives in Rome, and who first told me about *puntarelle*, says that they make an excellent Caesar salad instead of cos lettuce. Her mother-in-law, who came from Lombardy to Rome, prefers them cooked. 'We Lombards,' she says, 'boil them lightly and dress them with olive oil, lemon juice, salt and pepper. Or else we finish them with garlic and butter.' The first way she mentions explains itself. For the second way, chop a clove of garlic finely and sweat it very slowly in a large lump of butter in a covered pan for about five minutes, so that it melts and flavours the butter; put in the lightly cooked asparagus chicory and turn it over and over – it must not colour or fry in the proper sense of the word.

ASPARAGUS LETTUCE, see CELTUCE

ASPARAGUS PEAS

The asparagus pea is one of the
prettiest vegetables we eat. The
Latin name, *Lotus tetragonolobus*, is even more attractive
than the English one, for Lotus
was a nymph who turned into a tree to avoid the lusty attentions of
Priapus, and tetragonolobus, meaning four-lobed, describes the frills that
stripe the edible pods, giving them a look of four-square frivolity.
Another beauty of the plant is the red flower. Yet another is its willing-
ness to grow in poor soil.

Perhaps you can feel the 'but' coming, the worm in the paradisal plant.
Here it is – if you allow the pods to grow a millimetre too long before
picking them, they taste dull and turn unpleasantly thready. A friend who
once planted them by mistake for sugar peas, could not bear to eat any
more after the first fibrous mouthful. The ideal length is about 2½ cm (1
in) and I suspect she had let them grow a little too large. Picking any
vegetable when it is tiny is a real effort for English gardeners. Some
puritan streak insists on getting value for money, value meaning quantity
or size, not excellent flavour. Or perhaps it is the influence of the dropsi-
cal and inedible rotundities of Village Show vegetables that haunts us.
Nonetheless I have the feeling that if asparagus peas did not flourish in
bad ground, with little attention, they would never be grown at all in the
kitchen garden.

Thomas Martyn makes an interesting comment on them in his edition –
published in 1807 – of Philip Miller's *Gardener's and Botanist's Dictionary*: 'It was formerly cultivated as an esculent plant, for the green pods,
which are said to be still eaten in some of our northern counties' – I have
never heard this mentioned before – 'but they are very coarse. This plant
is now chiefly cultivated in flower-gardens for ornament. It is a native of
Sicily, where Ray found it on the hills above Messina; and was cultivated
in 1596 by Gerard. Parkinson calls it Crimson-blossomed or square-
codded Pease. Ray calls it square codded vetch. None of these speak of
the pods being esculent.'

Martyn's name for asparagus pea was Winged Bird's-foot
Trefoil.

HOW TO COOK ASPARAGUS PEAS

Top and tail the peas where necessary, then steam for about five minutes, checking on their progress after three minutes. They need hardly any cooking when they are picked at the right size. Mix them with a good knob of butter and a little cream after draining them, then eat immediately. The quicker you get them from pan to table, the better they will taste. They should never be left to keep warm while you eat another course or wait for someone late to arrive.

AUBERGINE

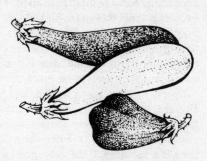

The aubergine so much belongs to the delights of Mediterranean food that it is surprising to learn of its original home in tropical Asia. Like ginger, or the cherry, it carries its history in its name. Aubergine, the French name we have adopted, goes back through Catalan *alberginera*, Arabic *al-bādinjān*, and Persian *badin-gan* to Sanskrit *vatin-ganah*. *Vatin-ganah* means an anti-wind vegetable – unlike Jerusalem artichokes or beans that unsociably provoke wind.

The other name of egg plant, once used here, now favoured more by Americans, refers to the shape of some varieties and goes back to the early days of aubergine in the north, when it was first grown for its decorative qualities. Egg plant, though, is a misleading name as the aubergine has several shapes and colours. Primitive forms picked and eaten in Thailand are often round and small. Other varieties in cultivation can be squat and pumpkin shaped, or long and oval. Colours go from white and whitish green, through dark green to yellow, purplish-red, purple and black. The most beautiful I have seen were on sale in Malvern and had been grown locally. They were long and swelling in form, ivory white in colour, with a soft sheen as if some 18th-century potter in Etruria had moulded them in Wedgwood creamware. Set with a few of the more common bloomy purple-black aubergines, in a yellow bowl, they looked too attractive to eat – but we did eat them in the end. Sadly they tasted no different. We had hoped, against reason, for a platonic hint of some heavenly aubergine flavour.

I am not complaining, as the earthy tenderness of the aubergine is unique among vegetable flavours. It gives depth to the mixed stews of Mediterranean cookery such as ratatouille and adds distinction to minced lamb in the Greek moussaka. There is even more to be said for exploiting the flavour of aubergine by itself or with only the assistance of a tomato or béchamel sauce. Aubergine fritters and grilled aubergines are two of the nicest things it is possible to eat.

HOW TO PREPARE AUBERGINES

Aubergines should look shiny and tight in their skins. Reject wrinkled ones firmly and any that are bruised. They are expensive enough for you to be firm about this and to pick out the ones you fancy without argument from the greengrocer. Size does not seem to affect their flavour and tenderness, as it does courgettes', so choice should be governed by the recipe you have in mind. For instance, if you want to make stuffed aubergines they need to be much of a size and shape.

Unless they are being stuffed, and you need the skin as a containing element, aubergines are best peeled. Remove the stem first, with the prickly calyx, then peel them downwards in long strips as thinly as possible. Slice, cube, or halve them according to the recipe, and place them in a colander. If the aubergines are halved, make deep slashes into them, being careful not to pierce through to the skin: this gives the salt a chance to penetrate. Sprinkle 1 tablespoon of salt over them for ¾–1 kg (1½–2 lb) – and leave them for an hour. If you weight them down with a bowl or dish they will more quickly exude their juice. Finally rinse them and dry them in a clean cloth. This salting process is not essential to the flavour but it does seem to stop the aubergines drinking up oil if you fry them. For dishes such as ratatouille or moussaka, it also means that the vegetable is thoroughly seasoned before you start cooking which is an advantage.

FRIED AUBERGINE AND AUBERGINE FRITTERS

The simplest way of cooking aubergine, after the slices have been salted and dried, is to dip them in flour and fry them in oil. Use olive oil for preference, as the two flavours marry well. Serve them as a vegetable with lamb, or better still on their own with the Greek garlic sauce known as *skordalia* (p. 561). If you want to make a contrast of flavour and colour (and to eke out expensive aubergines without appearing mean), cook slices of unpeeled, salted courgettes in the same way and serve them together.

Aubergine fritters are made by dipping the slices, salted or unsalted, into a light fritter batter (p. 568) and frying them either in a good depth of fat in a shallow pan or in deep fat. When they are golden brown, they are ready. Serve them simply with lemon quarters, or, again, with *skordalia*. You could put them with fish in batter – salt cod, for instance – but they are so delicious that most people prefer to enjoy them on their own, as a first course. There are few better things to eat, though I have lately come round to thinking that the best and simplest way with aubergines is to grill them.

GRILLED AUBERGINE AND AUBERGINE KEBABS

If you want to serve aubergine as a vegetable with roast meat, or on its own, peel the aubergine and cut it into wedges about 1 cm ($\frac{1}{2}$ in) thick. Alternate the pieces on skewers with young halved onions, or with pieces of older onion that have first been blanched for a few moments in boiling salted water. Brush them over with oil and pepper them well. Give them 20–25 minutes on a turning spit (if the kebabs include meat, they will need 25–30 minutes). When grilling over the fiercer heat of charcoal, 15–20 minutes should be enough, but test a piece of aubergine to make sure. Sprinkle them with salt when cooked.

In Provence, on the slopes of the Grand Lubéron in Vaucluse, they have a way of cooking aubergines whole by thrusting them into the hot charcoal or the embers at the edge of a fire. When the skins are blistered and charred, they split them open and eat them with a seasoning of salt, pepper, lemon juice and olive oil. An anchovy is added, too, sometimes. You can quite well copy this method on the grill of an outdoor barbecue, with the burning coals close beneath. In winter you might thread them on to an electric spit, or keep turning them under a grill. Choose medium-sized aubergines for this treatment.

MELITZANES SALATA
AUBERGINE SALAD

If you like the smoky, very slightly bitter taste of grilled or baked aubergines, this is another delicious recipe. When serving it to visitors, make it as part of a mixed hors d'oeuvre, in case they do not share your enthusiasm. In the eastern Mediterranean, Cyprus for instance, it is put on the table as part of the *meze*, the little dishes of cold and hot food served either with drinks, or as a meal on their own.

> 2 large aubergines, about $\frac{1}{2}$ kg (1 lb)
> 1 clove garlic
> 3 tablespoons olive oil
> lemon juice, salt, pepper
> chopped parsley

Bake the aubergines at 200°C/400°F/Gas Mark 6 until they are soft. This gives a more even result than grilling them and you do not have to watch them. Split them and scrape the pulp from the skin into a basin, in which you have crushed the garlic. Mix it well, adding the oil drop by drop until you have a thick, smooth paste; you may not need more than 2 tablespoons of oil, let taste and consistency be your guide. Season with lemon juice, salt and pepper. Stir in a tablespoon of chopped parsley. Serve chilled with slices of hot bread, or with tomato salad, a fish hors d'oeuvre and so on.

SWEET-SOUR AUBERGINE SPANISH STYLE
MERENJENA PRETA ALA SPAÑOLA

An unusual and savoury dish from *New Food for All Palates*, by Sally and Lucian Berg (Gollancz, 1967), served with a saffron rice salad. The slight bitterness of the lemon slices complete with their peel makes a good contrast to the soft richness of the aubergine.

> *2 medium aubergines (approx 400 g, 14 oz)*
> *olive oil*
> *7—8 shallots, or 250 g (½ lb) sliced onion*
> *about 2 lemons*
> *2 large sweet red peppers, seeded*
> *60 g (2 oz) demerara sugar*
> *5 tablespoons cider or wine vinegar*
> *1 heaped tablespoon seedless raisins*
> *1 rounded tablespoon butter*
> *½ teaspoon salt*
> *1 large clove garlic, crushed*
>
> SAFFRON RICE
> *250 g (8 oz) rice*
> *pinch saffron*
> *1 tablespoon olive oil*
> *2 large teaspoons chopped spring onion*
> *1½ tablespoons toasted pine kernels*
> *2 teaspoons chopped sweet red pepper*
> *lemon juice, salt, pepper*

Slice the aubergines lengthways just under 1 cm (½ in) thick, salt them and leave them to drain in the usual way, then rinse and dry them. Heat the oil in a large pan, just enough to cover the base, then put in a layer of the shallots or onion – if they do not quite make a layer, slice some more. Next put in a layer of aubergine slices side by side, without overlapping. On each slice put a thin slice of lemon and two or three strips of pepper. Repeat until the aubergine is used up. Pour over carefully, so as not to dislodge the lemon and pepper, a tablespoon of oil and 125 ml (4 fl oz) water. Cover and simmer for 15 minutes.

In another pan put a further 125 ml (4 fl oz) water and the remaining ingredients; simmer for ten minutes. Add this sauce to the aubergines and cook them for a further 30–45 minutes, taking care that the aubergines do not disintegrate. Remove the pan from the heat and, when the contents are tepid, lift out the aubergine slices, complete with their decoration, and arrange them on a large flat dish – a meat dish is ideal. In the pan there will be quite a lot of liquid left; boil it down until it has thickened a little. Correct the seasoning and pour it over the aubergines.

To make the saffron rice, put the rice and saffron into the pan with 300 ml (½ pt) water. Bring it to the boil, cover the pan and let it simmer for ten minutes. The rice will have absorbed most of the water and should be nearly cooked. If it seems almost tender, put the pan, still covered, at the side of the stove in a warm place. In a further 20 minutes, the rice should be done. If you are in a hurry, and the rice is not cooked after the ten minutes, add a little more water and continue to simmer it gently until the rice is tender. So much depends on the brand you use. Allow the cooked rice to cool, then mix in the remaining ingredients. Put the rice round the edge of the aubergine dish, or serve it separately if you like. Chill the whole thing thoroughly before putting it on the table.

A delicious and beautiful dish for a cold buffet, or for a cold lunch in the garden on a hot day.

PARMIGIANA DI MELANZANE

A delight of eating in Italy is the *rosticceria*, the cooked meat shop. There you may buy your food by weight, as much or as little as you like, and repair to the high stools and bar at the side to eat underdone beef and salads, or pasta, or a square section cut from one of those crusty baked dishes Italians cook so well. In the north it will be *lasagne al forno*, around Naples this appetizing *parmigiana*. It's the smell of such food, cooked in vast shallow tins, that pulls one towards the *rosticcerie*, the rich heady fragrance of tomatoes and olive oil and Parmesan cheese. Italian members of the family point out that only in Naples in small, scruffy eating places, does *parmigiana* taste exactly as it should, 'I expect they never clean the pans. Perhaps that explains the mature flavour.' Nonetheless I continue to make *parmigiana*, even if it lacks the high authenticity of Naples and the right, first quality Mozzarella cheese.

olive oil
½ litre (¾–1 pt) tomato sauce (p. 311)
250 g (8 oz) Italian Mozzarella or other soft cheese
75–100 g (2–3 oz) finely grated Parmesan

Flour and fry the aubergines in oil without colouring them. Remove when tender and drain on kitchen paper. Into a wide shallow ovenproof dish, or tin, put a layer of aubergines, then tomato sauce, then thin slivers or slices of the cheese with plenty of pepper and a scatter of Parmesan. Repeat until the aubergines are finished, and the sauce. Scatter the remaining Parmesan over the top and sprinkle lightly with olive oil. Bake for about an hour at 160–180°C/325–350°F/Gas Mark 3–4 until the cheese has turned to a golden crust on top and the sides are bubbling.

You can cook this kind of dish at a lower temperature, for longer, if this is more convenient. It depends on what else you are using the oven for. You can also cook it more quickly, though it is better if given a good chance to blend and mature its flavours in a lingering heat.

Parmigiana freezes well. Prepare it up to the final baking, then stow it away. To cook it, start with a high temperature for about 20 minutes or half an hour, then reduce it and leave for a further hour.

Note As this dish is inclined to be over-rich in oil, I fry ⅓ to ½ of the aubergine slices only, and blanch the rest for 2 minutes in boiling water.

AUBERGINE GRATIN FROM PROVENCE
TIAN D'AUBERGINES

Provençal cooks make a similar dish to the neapolitan *parmigiana* above, though there are a few differences. Firstly the aubergines are sliced across into rounds, peel and all, before they are salted. Secondly they are blanched just for a minute in water acidulated with 1 tablespoon of lemon juice to a litre (1¾ pt) and not fried. Thirdly, the cheese is confined to a top dusting of grated Gruyère mixed with chopped fresh basil.

This lighter version is delicious cold, where *parmigiana* would be rather heavy.

A second version substitutes béchamel sauce for tomato. Make it with olive oil instead of butter (see the onion gratin on p. 351), and enrich it finally with an egg before assembling the gratin. The slices of aubergine should again be cut into rounds, complete with peel, and salted. Then they should be lightly fried as for *parmigiana*. Scatter a little grated Gruyère over the top if you like. It is best eaten hot on its own, or with lamb; I think that aubergines go particularly well with lamb, better than with any other meat.

AUBERGINE KUKU
KUKUYE BĀDEMJĀN

Among the deaths, golden weddings, murders, and the bargain sales, our local newspaper in France manages to insert a daily recipe, just below the weather forecast and the day's Great Thought. It is usually good, often a little unexpected, and sometimes excellent – like this aubergine dish. It's a variation of the usual style of Persian kuku, see p. 471, and comes from Mr Miam Baghi, chef to the Peacock Throne of Persia. He makes it often for the Empress Farah. It has a certain flavour of the Thousand and One Nights. Serve it with drinks, cut into small diamonds, or as a first course with yoghurt for sauce, or take it on a picnic.

> 4 aubergines, about ¾ kg (1¾ lb)
> 2 large onions, chopped very coarsely
> olive oil
> 4 large eggs
> 2 heaped tablespoons flour
> salt, freshly ground black pepper
> lemon juice

Slice the aubergines in half lengthways, sprinkle with salt, and leave for an hour before wiping them dry. Cook them with the onions in oil – you may need two pans for this – until the vegetables are golden brown and soft. Scoop out the aubergine flesh and mash it with the onions (discard the skins). A certain amount of oil is bound to be included, but drain off any conspicuous surplus. Mix the vegetables with the remaining ingredients, using plenty of black pepper and a little lemon juice. Smooth out into a square or oblong pan, so that the mixture is between 2 and 3 cm (about 1 in) thick, and has room to rise slightly. Bake for about 30 minutes at 200°C/400°F/Gas Mark 6, or until the top is nicely browned and the edges crisp. Cut into diamonds to serve.

be reheated, and made into appetizing dishes such as the Basque eggs on p. 383.

Ratatouille is a provençal word, first recorded in the 18th century, though it is probably a good deal older than that. It is a cross between *tatouiller* and *ratouiller* which, as the dictionary says, are expressive forms of *touiller*, an old verb from the Latin *tudiculare*, meaning to stir and crush.

> ½ kg (1 lb) aubergines
> ½ kg (1 lb) courgettes
> ½ kg (1 lb)tomatoes, skinned
> 2–3 sweet peppers
> 2–3 large onions, sliced
> 4 tablespoons olive oil
> 2 large cloves garlic, chopped
> salt, pepper, sugar
> vinegar
> ¼ teaspoon coriander seeds, crushed
> chopped fresh basil or parsley

Slice the unpeeled aubergines and courgettes. Put them in a colander, sprinkle with a teaspoon of salt and leave for an hour to drain. Pat them dry with kitchen paper or a cloth. Chop the tomatoes roughly. Remove the stalks and seeds from the peppers and cut them into strips.

Cook the onion slowly, without browning it, in the olive oil with the garlic. As it softens, add the aubergines and peppers. Cover and simmer for 20 minutes. Put in the tomatoes and courgettes. Season with salt, pepper, and a little sugar and vinegar if the tomatoes lack flavour as they often do in the north. Cook steadily without covering the pan until all wateriness has disappeared – about 50 minutes. Ten minutes before the end, add the coriander. The vegetables should retain a certain identity, so do not crush them to a purée, although they should be stirred vigorously from time to time. Serve hot or cold, sprinkled with basil or parsley. This quantity is enough for eight or ten.

Now 125 g (4 oz) of black olives are sometimes added to *ratatouille* about five minutes before the end of the cooking time.

BOUMIANO

A provençal dish of the *ratatouille* family, with an added piquancy of anchovies. The name is the Vaucluse dialect word for *bohémienne*, meaning a gipsy dish, or rather a dish in the gipsy style. Unlike ratatouille, *boumiano* depends on two vegetables only, aubergines and tomatoes.

> ½ kg (1 lb) ripe tomatoes, skinned
> ½ kg (1 lb) aubergines, unpeeled
> salt
> 2–3 tablespoons olive oil
> 1 fat clove garlic
> 8 anchovy fillets
> 1 heaped teaspoon flour
> 2 tablespoons milk
> pepper, sugar
> breadcrumbs
> extra olive oil

Cut the tomatoes across, sprinkle the cut sides with ½ teaspoon salt and turn them upside down on a pierced dish or colander. Cut the aubergines into 1 cm (½ in) dice; put them into another colander and sprinkle them with ½ teaspoon salt. Leave the vegetables to drain for an hour, then pat them dry with a cloth or kitchen paper.

Heat 2 tablespoons oil in a frying pan with the clove of garlic left whole. Put in the vegetables. Stir them about, crushing them down as they cook, until they are soft. Add the extra spoonful of oil if necessary. Meanwhile in another pan, a small one, crush the anchovy fillets in a little of their oil over the heat. Stir in the flour, then the milk, to make a savoury roux.

When the vegetables are reduced to a lumpy stew, without wateriness but nicely moist, stir in the anchovy mixture. Taste and correct the seasoning, adding pepper and a hint of sugar to bring out the flavour of the tomatoes.

Turn into a gratin dish, sprinkle with a layer of breadcrumbs and scatter a little oil on top. Bake at 180–190°C/350–375°F/Gas Mark 4–5 until golden on top; or place under the grill.

before being stewed in a tomato sauce.

> 1 kg (2 lb) tomatoes, peeled, chopped
> 1 onion, chopped
> olive oil
> 125 ml (4 fl oz) white wine vinegar
> 2 tablespoons sugar
> salt, pepper
> 1–1½ kg (2–3 lb) aubergines, diced, salted
> 1 head celery, sliced
> 16 green olives, stoned, coarsely chopped
> 1 heaped tablespoon capers
> 4 chopped fillets anchovy or 1 tablespoon pine kernels
> chopped parsley

Make a tomato sauce by stewing the tomatoes and onion in a very little oil. When it becomes pulpy, add vinegar, sugar and seasoning. Leave to simmer to a dark reduced sauce. Meanwhile deep-fry first the aubergines, then the celery – alternatively you can blanch the celery until it is almost tender, which I think is better though it is not the usual thing to do.

Add these vegetables to the tomato sauce and leave to stew for about half an hour. If the whole thing gets too dry, add a little water. Stir from time to time so that the vegetables don't stick to the pan. Add the olives, capers and anchovies or pine kernels and stew for another 15 minutes. Check the seasonings and general flavour, making any adjustments. Keep until next day and serve as a first course, scattered with parsley. In fact it keeps well in the refrigerator, in a covered jar, so can be made well in advance.

STUFFED AUBERGINES
MELANZANE RIPIENE

One of the simplest and best versions of stuffed aubergines, from Elizabeth David's *Italian Food*.

3 large aubergines
6 anchovy fillets
9 black olives, stoned
small bunch parsley
2 cloves garlic
1 tablespoon capers
pepper, oregano or marjoram
100 g (3–4 oz) breadcrumbs
milk or water
150 ml (¼ pt) olive oil

Halve the aubergines. Remove half their flesh, leaving six firm aubergine shells to arrange in a baking dish. Chop the flesh you have cut away, with the anchovies, olives, parsley, garlic and capers. Season with pepper and oregano or marjoram, just a pinch. Mix the breadcrumbs with just enough milk or water to make them into a crumbly paste rather than a wet poultice. Add the chopped ingredients. You should not need to add salt, as the anchovies will probably give enough, but sprinkle the aubergine shells inside with a very little salt before you put in the stuffing.

Pour the oil over the aubergines. Bake at 160°C/325°F/Gas Mark 3 for about an hour, until the aubergines are tender. Serve hot.

STUFFED AUBERGINES TURKISH STYLE
IMAM BAYILDI

'The Imam fainted' – presumably with pleasure at this delicious way of cooking what is, in Turkey, a common vegetable, as common there as cabbage is in Ireland. Another holy man, not a prayer-leader this time, but a scholar and divine fool, Nasreddin Hodja, had a different reaction to aubergines. One hot day he got off his donkey and went to rest in the shade of a walnut tree, by a field of aubergines. 'How strange,' he thought sleepily, 'that God put large aubergines on the end of tiny stalks and plants, and tiny walnuts on large trees.' As he dozed, a walnut fell on his bald head. 'Now I understand,' said the Hodja, rubbing the sore place, 'All is for the best . . .'

6 aubergines
salt

plenty of chopped parsley

PLUS
125 ml (4 fl oz) each olive oil and water
pepper, salt
1 teaspoon sugar
juice of a lemon
bayleaf
1 clove garlic, whole

Prepare the aubergines by slicing off a cap at the stalk end and hollowing them out from the top. Salt the aubergines inside and drain them upside down in a colander. Chop the pulp and cook with the onions and garlic in the oil, until they are soft, but not brown. Add the tomatoes, allspice and seasoning. As the tomatoes turn to pulp, raise the heat and boil hard to evaporate wateriness. Add the parsley. Fill the stuffing into the aubergines, after drying them out with kitchen paper. Replace the caps and stand them close together upright in a saucepan. Pour in the remaining ingredients. Cover and simmer for an hour until the aubergines are tender and the liquid much reduced. Serve chilled.

I admit that this is not the usual way of preparing Imam Bayildi, but I find the higher proportion of stuffing to aubergine is more satisfactory. Moreover the stuffing cannot wander out of the aubergines into the cooking liquid. However, if you would prefer to try the normal way, this is what you do:

Make five slashes into each aubergine, longways from stalk to stern, being careful not to cut right through. This enables the aubergines to be spread out, like an opening book. Salt them and turn cut side down into a colander to drain for an hour.

Cook the filling, using three-quarters of the quantity of the ingredients listed. Towards the end, mix in a tablespoon of currants or raisins.

Rinse any salt grains from the aubergines, and dry them as best you can, putting kitchen paper down into the slashes. Spoon the filling into the slashes, then range them in a flameproof or baking dish that holds them closely but easily. They should not keel over. Pour the oil, water etc. round the aubergines, and complete the cooking either on top of the stove, or in the oven at 150°C/300°F/Gas Mark 2. Allow an hour, then test the aubergines and give them longer if necessary. They should be really tender. Cool and chill before serving.

MOUSSAKA

Moussaka is made from layers of aubergines and meat stew; *papoutsakia* (the following recipe) is made from the same ingredients, but the stew is put into halved aubergines so that they end up looking like little slippers. In both cases, a rich béchamel sauce is spread over the top.

In both cases, too, the results are appetizing and piquant. Or they should be. If a few things are not understood, either dish can be a disaster in the hands of a northern cook. This is because we are sadly inclined to wateriness and slop in our food. Although there is some excuse for this, it can quite well be avoided, once once realizes that words such as 'tomatoes' or 'cheese' mean quite different things in, say, Greece, and in our country.

By comparison with the rich, firm Mediterranean tomato, ours is a sad, mushy affair. There are occasions when canned tomatoes should be used instead; there are other occasions when fresh ones plus some brown sugar and a dash of vinegar would have a better effect. Our own hard cheeses, though excellent for English cookery, are too moist in texture and not piquant enough in flavour for more southern dishes. If the local delicatessen does not sell *feta* or *kefalotiri*, Parmesan is the correct substitute.

When cooked meat is used, choose *under*cooked beef or lamb. If you are starting with raw meat, never buy butcher's mince; ask for a lean piece of lamb, beef or veal, and then chop or mince it yourself at home. This avoids excessive oiliness. For delicacy of flavour use veal rather than lamb or beef – the expense can be reduced by substituting cooked brains for a quarter of the weight.

The béchamel for the cheese sauce must be made with butter, flour and milk – never use margarine or milk powder, whatever the manufacturers may say. The egg enrichment is essential; it thickens the sauce, without making it gluey, and it adds the beautiful golden colour which is part of the charm of the dish.

A last thought – when building up a *moussaka* in layers, begin and end with a vegetable layer. And whenever you are making a composite dish of this kind, see that each element is correctly seasoned before you combine them. It makes all the difference whether you are constructing a

MEAT SAUCE
400 g (good ¾ lb) lamb or beef, rare, from a cooked joint, chopped and minced or ½ kg (1 lb) raw lamb or beef, chopped or minced or 400 g raw veal, minced, plus 125 g (4 oz) cooked brains, chopped
125 g (4 oz) chopped onion
2 good tablespoons butter
½ kg (1 lb) tomatoes, skinned, chopped, or a medium can 400 g (14 oz) tomatoes
½ teaspoon dried rigani or oregano
black pepper, cayenne
1 glass red wine
brown sugar
wine vinegar
salt
60–90 g (2–3 oz) feta *or* kefalotiri *cheese, grated*
2 tablespoons grated Parmesan
2 tablespoons breadcrumbs

CHEESE SAUCE
generous ½ litre (1 pt) thick béchamel sauce
2–3 large eggs
60–90 g (2—3 oz) feta *or* kefalotiri, *grated*
extra grated cheese

Fry about half the aubergine slices in oil until just browned and just tender. Drain well. Blanch the rest for 2 minutes in boiling water.

To make the meat sauce, cook the lamb, beef or veal with the onion in butter until browned. Add the brains if used, then tomatoes, rigani or oregano, and peppers. Simmer for 15 minutes to make a rich moist stew; if it is watery, raise the heat. If it seems dry, add a little water. When the sauce is cooked, add the red wine and boil hard for a few moments, then correct the seasoning with sugar, vinegar and salt. If the basic ingredients were full of flavour, you will not need the sugar and vinegar. Remove the

pan from the heat. Pour off any surplus oil, or blot it away with kitchen
paper, and add cheeses and breadcrumbs.

To make the cheese sauce, heat the béchamel but do not allow it to
come near boiling point. Whisk in the eggs, then the cheese. Mix
two tablespoons of this sauce into the meat mixture to give it smoothness.

To assemble the *moussaka*, choose a wide but fairly deep dish and
grease it with a butter paper or a very little oil. Starting and finishing with
aubergine slices, put the aubergine and meat stew into it in layers. Spread
the sauce over the top, and sprinkle it with the extra cheese. Bake at
150°C/300°F/Gas Mark 2 until bubbling; if you start with warm ingredients, this will take 30–40 minutes, if you are starting with a dish you
have assembled in advance, allow a further ten or 15 minutes. The
moussaka may need a few minutes under the grill, to complete the
browning to an even spread of golden-brown blisters. Watch this carefully, as the appetizing brown colour can rapidly change to an unpleasant
black.

Moussaka can also be made with sliced potatoes, p. 415 or courgettes
instead of aubergines. There is also an elegant version using 18 artichoke
bottoms, which are boiled and sliced instead of being fried; use veal or
veal and brains, rather than lamb or beef.

MELITZANES PAPOUTSAKIA
AUBERGINE SLIPPERS

Use the same ingredients as for *moussaka*, but buy six aubergines, evenly
sized, each weighing about 250 g (8 oz). Slice the aubergines in half,
cutting down from the stalk end. Sprinkle the cut sides with a little salt,
and leave them turned down to drain in a colander for an hour. Wipe
them free of moisture. Fry them gently in olive oil until they are nicely
browned – particularly on the cut side – and soft. Scoop out the pulp with
a pointed spoon, being careful not to pierce the skins, and chop it roughly;
put this pulp with the meat. Season the empty aubergine skins, the
'slippers', and put them in a lightly oiled gratin dish. Choose one that
they fit into closely, for mutual support; if they wobble about in space,
the filling will bubble over the edges and spoil the appearance of the
dish.

Make the meat stew as for *moussaka*. Include the aubergine pulp and
fill the aubergine skins, leaving a little space at the top for the cheese
sauce and the final sprinkling of cheese. Bake in the same way as *moussaka*.

AVOCADO

The avocado, or avocado pear, was originally a native of tropical America. The name comes, via Spanish, from the Aztec *ahuacatl*, and has nothing to do with lawyers, or with egg drinks from South Africa – or with pears for that matter. Nowadays

many varieties grow in other regions of the world. In this country we import them from Israel in the winter and South Africa in the summer.

A Ghanaian friend told us that in Accra there is a tree in every garden, like apple trees in some parts of England; everyone gets bored with the fruit during the season, as we get bored with endless Bramley windfalls in the autumn. In Kenya it is much the same. Barrow-loads are put by the road, at the end of gardens, with notices saying Help Yourself. The stray dogs make up their dietary deficiencies, becoming sleek and glossy on the tumbled fruit that no one bothers to pick up. Certainly avocados are good for you, being so nutritionally endowed that many Israeli babies are weaned on them. The price would have to come down before any baby of my acquaintance got the chance, but I pass on the idea.

Here in this country we have known about avocados since the 17th century, when they were given delightful nicknames such as midshipman's butter, or subaltern's butter, as if in those days the children of our harsh services kept healthy when they were in the West Indies on windfall avocados, like the stray dogs of Kenya today. We ate them in England from time to time before the Second World War if we were well off and lived near a grand greengrocer. In other words they had not progressed very far in three centuries. Yet it took the Israelis ten years only, perhaps less, to get them into most of our supermarkets. By the seventies, they had become as much a cliché start to a meal as grapefruit.

If you visit an avocado-growing kibbutz, you will see that the groves of trees are kept at a height around ten metres by pruning. As you stand admiring the dense, dark leaved growth, a head will suddenly pop out eight or nine metres up, like a Jack in the Green. Then another to the right, lower down. Then again, a minute later, both heads are at normal level as the pickers bring their mobile lifts to earth, to roll out the harvest. Then they are graded and taken rapidly to refrigerated ships, where the even coolness prevents them ripening further, to arrive at our markets hard and firm, but not too far away from the soft, brown-splashed ripeness that makes them so delicious to eat.

As you will see from the following section I feel that there are few recipes to improve on *avocado nature*, with an olive oil vinaigrette. There are a few magnificent partnerships, avocados with pineapple, or with grapefruit, with spiced or salt beef, with wholemeal bread and butter, but I cannot bear to see a few cubes of avocado wasting their substance in a vast fruit salad. No doubt if avocados covered the orchard grass, I should be happy creaming them into meat stews and soups, as they do in Latin America, to avoid waste. Here, they demand more considerate treatment, as star items for special meals.

HOW TO CHOOSE AND PREPARE AVOCADOS

Judging by the hard unripe avocados one occasionally gets in restaurants, and in restaurants that pride themselves on their food, some people still do not know that they should be kept until the whole surface is slightly yielding, with an extra softness around the stalk. If you hold an avocado in your hand and gently feel round the skin with a slight pressure, you can soon tell if it is right.

This does not mean that you should never buy a hard avocado. It is often convenient to do so, if you can only shop once or twice a week. In a warm kitchen it will soon ripen; I allow two or three days, and if it gets to the proper state before it is needed, I put it into the vegetable drawer of the refrigerator. Green-skinned avocados often become patched with brown. This is a sign that they are ready. The warty, purple-brown-skinned avocados can only be tested by feeling them. Do not let the lumpy skin put you off; these dark, pebbled fruit of the Hass variety have the best flavour of all.

If you see really bruised looking avocados, you may be able to get them at a reduced price that makes them a bargain. Eaten the same day, they are really good and buttery. Mash them to a cream with seasoning and vinaigrette and serve the paste on buttered wholemeal bread or in sandwiches. Or use them to make avocado butter to serve on steaks.

If you are ever in an avocado-growing country in the season, keep an eye open for tiny, freak avocados. They look like small courgettes, about 5 cm (2 in) long or less. They have no stones, but thin quills instead that are easily pulled out. Peeled and sliced, these delectable miniature avocados make excellent salads.

The one problem with avocados, after settling the matter of ripeness, is that they discolour when cut. It does not matter from the flavour point of view, but the appearance loses its freshness. Always have ready a wedge of lemon to rub over the new surfaces. No lemon? Then use orange or citrus juice of another kind, or some vinaigrette, according to the recipe you are following.

To halve an avocado, cut down from the stalk, then go round the stone. The two parts can then be twisted free. For slices, remove the peel which comes away neatly with the aid of a sharp knife, and rest each half, cut-side down, on a board, so that it can be cut lengthways into pearlike slices.

To cut avocados into circles and – inevitably – rings, peel the whole thing then slice it across, working carefully when you get to the stone so that the rings are as even as possible. Should you need chunks of avocado, it is best to slice the fruit across in this way first, rather than lengthways, then you can divide circles and rings rather as if you were dealing with a pineapple.

Although it is waxier and less fibrous, an avocado has something in common with the base of a globe artichoke. Both, for instance, are deliciously partnered by shellfish and mayonnaise or mornay sauce. One French chef has invented recipes using avocado with meat, in something of an artichoke style. Apart from an avocado butter with steak, they are not altogether successful. But it is a hint worth knowing, especially if avocados ever become cheap enough for experimenting.

AVOCADO VINAIGRETTE

Prepare a good vinaigrette with a robust olive oil and wine vinegar. Season it with salt, pepper, a hint of sugar and plenty of chopped parsley and chives.

Halve the avocados, as above. Rub the cut surfaces with lemon, and serve with some vinaigrette poured into the cavities, extra vinaigrette in a small jug and thin slices of wholemeal or rye bread and butter.

mayonnaise, and pile them in the centre of the meat. You can extend the avocado with diced, cooked potato, and chicory, but avocado should be the dominant item.

GRAPEFRUIT AND AVOCADO SALAD

A good marriage and one of the most beautiful of vegetable dishes. For six or eight people, allow three large pink grapefruits, and four large avocados.

Peel the grapefruit first, then remove the tough skin from each segment and discard it. Leave the neatest slices whole, cut the damaged ones into the best looking pieces possible. Make an olive oil vinaigrette and pour a little into a large plate. Peel and slice the halved avocados longways. Turn the slices in the vinaigrette and arrange them on a wheel shape on a large round flat serving dish. Pile the grapefruit in the centre, drained of its juice (drink it yourself – cook's perks), using the neatest segments to make a rose effect. Pour on a little vinaigrette, scatter with a little parsley and serve cold but not chilled. The salad should be moistened, but there should be no liquid swilling around.

Note If you have to substitute the usual yellow grapefruit, sprinkle it with sugar before making up the salad. It is not as sweet as the pink kind.

AVOCADO AND PINEAPPLE SALAD

These fruit make an exceptionally good mixture, ideal for the first course of a celebration. Choose a fine ripe pineapple. Cut off the top plume of leaves and set it aside. Hollow out the centre carefully with a knife and pointed spoon, then cut it into fairly even pieces. Leave them in sieve or colander to drain.

Next prepare a mayonnaise with 2 large egg yolks and a generous ¼ litre (½ pt) oil. Use a light flavoured olive oil, or a mixture of half olive oil and half tasteless salad oil; the flavour should be mild.

Peel and cube two avocados, sprinkle them with the juice of a lemon, two teaspoons ground ginger, a good pinch of cayenne and half a teaspoon of paprika. Mix with the drained pineapple and enough mayonnaise to bind the mixture. Pile the salad into the cavity of the pineapple shell and top with the plume of leaves. Chill in a closed plastic bag, so that no smell escapes to dominate the refrigerator. Serve later the same day, but not too much later or the salad will become wet. If you wish to prepare the three different ingredients many hours ahead, keep them separately in the refrigerator until the last moment, covering each bowl and storing the pineapple shell in a bag.

HAROLD WILSHAW'S AVOCADO SALAD

> *1 ripe avocado*
> *lemon juice, olive oil, salt, pepper*
> *250 g (8 oz) shelled broad beans*
> *soured cream, chopped parsley*

Peel and dice the avocado, sprinkling it immediately with lemon, oil and seasoning. Cook the beans, then skin them – this is essential. Arrange beans and avocado on a plate, preferably a bright pink plate to show off the different greens. Pour over a little soured cream and sprinkle with parsley.

There is no need to restrict this salad to June. It works well with frozen broad beans.

AVOCADO MOUSSE

Although this avocado mousse makes a good start to a meal, it is better still as the centre piece for a cold main course. Turn it out on to a large dish – I use a big white meat plate with a deep pink rim – and surround it with salad greens, hard-boiled egg, tomatoes and either shellfish and firm white fish, or chicken cut into neat pieces.

> *300 ml (½ pt) chicken stock*
> *1 packet gelatine*
> *2 large or 3 smallish avocados, peeled, mashed*
> *1 level dessertspoon chopped mild onion, ·or spring onion*
> *salt, pepper, lemon juice*
> *1–2 teaspoons Worcester sauce*
> *1–2 teaspoons chilli or Tabasco sauce*
> *mayonnaise made with 2 egg yolks, 150 ml (¼ pt) olive oil, wine*
> *   vinegar*

> *150 ml (¼ pt) double cream, whipped*
> *2 egg whites, stiffly whipped*
> *150 ml (¼ pt) vinaigrette*
> *4–5 tablespoons chopped parsley*
> *4–5 spring onions, chopped*
> *16–20 black olives, stoned, halved*

Heat a little of the stock and dissolve the gelatine in it. When clear, add the remaining stock, avocado and onion, beating the ingredients to a smooth mixture. Flavour with seasonings and sauces, adding the Worcester sauce and chilli or Tabasco gradually to taste. Remember that the flavours will be muted by the mayonnaise and cream.

Put this mixture into the fridge, until it begins to set, and has the consistency of thick egg white. Fold in the mayonnaise, then the cream and finally the egg whites. Check the seasoning again. Pour the mousse into an oiled decorative mould – preferably one with a hole in the centre – and chill until set. If you wish to leave it overnight, press cling film over the surface as soon as it is firm.

Turn the mousse out on to a dish. Mix the vinaigrette with parsley and spring onion. Pour some over the mousse, the rest over any surrounding food. Decorate the mousse with black olives, left in halves, or cut into smaller pieces. The centre of the mousse can be filled with the olives if you like, or hard-boiled egg.

AVOCADO SOUP
SOPA DE AGUACATE

A delicate soup from Mexico, the home of the avocado. The only problem is that you cannot keep it hanging around, or you will lose the fine flavour; the avocado cream and stock should be heated together at the last minute. The complementary advantage is that you can produce a first-class soup in minutes, for unexpected visitors.

> *2 ripe avocados, peeled, stoned*
> *250 ml (8 fl oz) whipping cream, or half single and half double*
> * cream*
> *1 litre (1¾ pt) chicken stock*
> *1 glass dry or medium sherry*
> *parsley, salt, pepper, lemon juice*

Reduce the avocados to a purée with the cream. Get it as smooth as possible by using a blender. Heat the stock and pour it on to the cream. Return to the pan and bring to just below boiling point. Add sherry, chopped parsley and seasoning if required. A very little lemon juice can be used to bring out the flavour.

This soup can be served chilled. Use a good home-made stock of light flavour, and blend it with the avocado and cream without heating it. Instead of sherry, try a seasoning of ground allspice and chilli sauce, with a little sugar. Sprinkle with chopped chives just before serving.

AVOCADO SAUCE FOR FISH AND SHELL-FISH

Avocado makes a good sauce for salmon and salmon trout in particular, though it also goes well with crab and lobster, or tomato and egg salad.

> 2 ripe avocados
> juice of a small lemon
> 1 large garlic clove, crushed
> 300 ml (½ pt) soured cream
> salt, pepper
> 2 spring onions, chopped (see recipe)

Peel, stone and mash the avocados with the lemon juice and garlic. Mix in the cream gradually. Taste and season with salt and pepper. Finally mix in the onion, if the sauce is to be served with cold fish. If the sauce is for hot fish, omit the onion and warm the basin over a pan of simmering water until the sauce is reasonably hot, but nowhere near boiling point.

GUACAMOLE

A spiced creamy sauce from Mexico, the avocado's homeland. It can be used as a dip, with toast, biscuits, pieces of raw vegetable, potato crisps. It can be served as a sauce with fish or chicken. Mexicans eat it with tortillas rolled round a well seasoned meat filling. The proportions of the various ingredients can be altered to suit your taste; the quantities below were given me by Elizabeth Lambert Ortiz, the expert in Mexican cookery. She often uses canned jalapeño or serrano chillis, rather than a fresh one: their flavour is milder and more aromatic from the canning liquid. Mix together:

> *2 large avocados, peeled, mashed*
> *1 rounded tablespoon finely chopped onion*
> *2 medium tomatoes peeled, seeded, drained and chopped*
> *1 fresh green chilli, seeded and chopped (or chilli powder to taste)*
> *1 clove garlic, finely chopped*
> *1 tablespoon fresh chopped green coriander, or 2 tablespoons chopped parsley*
> *lemon juice, salt, pepper, sugar to taste*

Put into a bowl, cover with plastic film if you wish to keep it overnight, and serve well chilled, with drinks or as a first course.

AVOCADO AND SMOKED SALMON

This is a good combination. There are two ways of preparing it. The first is to slice a couple of peeled and stoned avocados, brushing them over with lemon juice or a lemon-flavoured vinaigrette. Then arrange them on a large dish, with 125 g (4 oz) of smoked salmon cut into strips in the centre. Serve with brown bread and butter.

The second way is to turn the avocados into a paste by mashing them with six tablespoons of vinaigrette, two tablespoons of mayonnaise and two tablespoons of cream, lightly whipped. Season and flavour with lemon juice if necessary. Cut the smoked salmon into squares and fold them into the avocado mixture. The advantage of this method is that you can use smoked salmon trimmings, which works out cheaper; mix the smoked salmon in just before serving. The paste can be made in advance, covered with plastic film and kept in the refrigerator. Again serve with thin brown bread and butter.

Instead of smoked salmon, you can fold in pieces of shellfish.

BAKED AVOCADO WITH SHRIMPS, PRAWNS, OR CRAB

Here is the most successful way of serving avocados hot. The flavour is not lost in the brief cooking and blends deliciously with the shellfish and cheese sauce.

> *3 large avocados*
> *lemon juice*
> *300 ml ($\frac{1}{2}$ pt) thick béchamel sauce*
> *2 heaped tablespoons grated Cheddar*

1 heaped tablespoon grated Parmesan
3 tablespoons double cream
175 g (6 oz) peeled shrimps or prawns or crabmeat
salt, pepper
breadcrumbs, melted butter

Halve the avocados and remove the stones. Enlarge the cavities, but leave a good firm shell behind. Cube the avocado you have cut away. Sprinkle it with lemon juice, and brush more lemon juice over the avocado halves, to prevent discoloration.

Heat two-thirds of the sauce, which should be very thick indeed as it is a binding sauce. Keep it well below boiling point. Leave the pan on the stove while you stir in the cheeses, gradually, to taste. The flavour should be lively, but not too strong. Mix in the cream and shellfish, with seasoning, and the avocado cubes. If the mixture is very solid, add the remaining sauce. You need to strike a balance between firmness and sloppiness; in the final baking the sauce should not run about all over the place, but keep the shellfish and avocado cubes nicely positioned.

Put the avocado halves into a baking dish. Divide the stuffing between the cavities, mounding it up. Scatter on the breadcrumbs and pour a little butter over them. Bake for 15 minutes at 200°C/400°F/Gas Mark 6 and complete the browning under the grill if necessary. Do not keep the avocados in the oven any longer than this, as they do not improve with prolonged heating.

BAKED AVOCADO AND CHICKEN OR TURKEY

Follow the recipe above, substituting 250 g (8 oz) diced cooked chicken for the shellfish. If you wish to use a larger quantity of chicken, scoop out all the avocados, and put the mixture into a gratin dish. You will also need to increase the quantity of sauce, cream and cheese.

Try this as an after-Christmas dish for left-over turkey.

AVOCADO PANCAKES
CRÊPES SURPRISES AUX AVOCATS

Light crisp pancakes, not too much rum and sugar, combine to show off the virtues of avocado in a tactful delicacy.

3 ripe avocados, or 2 huge ones
100 g (3–4 oz) sugar
4 tablespoons rum

BATTER
100 g (3–4 oz) flour
30 g (1 oz) sugar
1 tablespoon oil
1 tablespoon cream
2 eggs
pinch salt
milk

Peel, stone and slice the avocados. Put them into a bowl, mix the sugar and rum and pour it over them. Turn the pieces gently so that all surfaces are moistened. Leave for an hour.

Make a batter from the ingredients listed, keeping it on the supple and liquid side. Cook 12 pancakes in the usual way and roll each of them round a spoonful of the drained avocado slices. Put the pancakes side by side on a heatproof dish. Pour over the rum juices left in the bowl, set under a very hot grill for three minutes and serve.

AVOCADO FOOL

I used to think that the word fool as applied to mixtures of fruit and cream – or custard – derived from the French *fouler*, to crush. It would have been appropriate. But it seems I was wrong, and the word means what it says more or less, being a synonym for something of small consequence. Two similar names, given to other light and quickly made puddings, are trifle and whim-wham. Affectionate names, denoting pleasure and fun and an easy simplicity. Lime juice is by far the best flavouring for avocados used in this way, but lemon juice can be used instead.

3 avocados
2 limes or 1 large lemon
icing sugar
150 ml (5 fl oz) whipping or double cream

Peel the avocados and cut them up into the goblet of a blender, or sieve through a *mouli-légumes*. Slice the limes or lemon in half, and cut one thin slice from the middle; divide each lime slice into three, or the one lemon slice into six wedges and set aside. Squeeze the juice from the rest of the limes or lemon, and add it to the avocados, together with a tablespoon of icing sugar. Blend the avocados to produce a smooth cream, or mix the sieved avocado well to an even mash. Whip the cream until stiff, fold in the purée and add more sugar to taste.

Divide between six glasses, and chill for two hours. Decorate with the reserved lime or lemon wedges, sticking them into the top of each fool like a tiny fan. Serve with sponge or shortbread fingers.

BATAVIAN ENDIVE, or SCAROLE

Batavian endive, or batavia, is a broad-leaved variety of endive, much grown for winter salads around the Mediterranean and in France. At first glance, it looks like an extra sturdy lettuce. At second glance, you will see that the leaves are toothed and curling in a decidedly un-lettuce-like way. Some batavia have a bluish-green tone, a closed-up shape like a loose cabbage. Others will be far shaggier, often opening right out to show a yellowish-white blanched centre inside the deeper green outside leaves. The flavour is good, more positive than lettuce, juicy and less bitter than the curly, mop-headed endives (q.v.) of summer and autumn. The texture is even crisper than a Webb's Wonder lettuce, which makes it particularly good for the kind of salad that is dressed with hot bacon pieces and vinegar (see p. 284). You will see from the brief suggestions following that it goes best with strong-flavoured or crisp ingredients when you want to make a mixed salad.

Some English writers are sniffy about cooking batavia, as if that were what the lesser breeds across the Channel get up to. One feels the phrase 'mucked-up fancy foreign stuff' lurking at the back of their minds. Nevertheless let us bravely continue to contemplate the cooking of batavia. I reflect that as the Italians and French grow a lot more of it than we do, they are likely to know more about eating it than we do. The supreme recipe for batavia, an altogether exceptional recipe, I do not include in this section. This may seem strange, but it can be made with chicory, so I put it in that section to make sure that no one missed it. Batavia is not easily bought outside the centre of London.

HOW TO CHOOSE AND PREPARE BATAVIA

Do not be put off if the outside leaves look damaged or tough. They can be cut away, to leave plenty of vigorous inner part. If you do not wish to use all the batavia at once, break or cut off what you require, and put the rest away in a polythene bag in the refrigerator without washing it (it keeps fresher this way). Then prepare the leaves you need as if they were lettuce.

NORMANDY SALAD

Cut the batavia in shreds and mix in a salad bowl with two Reinette apples, peeled, cored and diced (or use Cox's Orange Pippins instead). Make a dressing of a quarter of a teaspoon of Dijon mustard, four tablespoons double cream, two tablespoons lemon juice, salt and pepper. Mix into the salad and scatter with chives.

BATAVIA SALAD WITH ROQUEFORT

Mash 60 g (2 oz) each of butter and Roquefort together, add half to an olive oil vinaigrette, whisking the whole thing together hard, and put this into a bowl with the batavia and a small handful of walnuts. Toast eight small rounds of bread and spread them with the remaining Roquefort mixture straight from the grill. Mix into the salad while they are still warm.

CREAMED BATAVIA

Blanch shredded batavia until just tender. Have ready about 300 ml (½ pt) béchamel sauce, flavoured with nutmeg and into which you have beaten an egg yolk and 3–4 tablespoons cream. Drain the batavia well and mix the sauce in slowly, stopping when the shreds are bound together smoothly. Do not swamp them. Reheat over a low to moderate heat, stirring all the time, without boiling. Serve with toast or croûtons, or with beef, veal or chicken.

OTHER SUGGESTIONS

Batavia in Waldorf salad, as an addition, when celery is on the short side, p. 189.
Batavia in Caesar salad, as a substitute for cos lettuce, see p. 320.
Batavia in Winegrower's salad, as a substitute for lamb's lettuce, or with it, see p. 284.
Stuffed batavian endive, see p. 213.

BEANS FROM THE NEW WORLD

French, scarlet runners,
flageolets and haricots,
fresh, semi-dried and dried

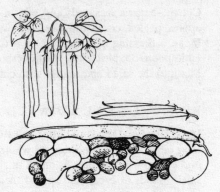

Beans, one of our greatest bles-
sings from the New World, are a
splendidly versatile and varied
vegetable. They begin as tiny
green strings, which can be cooked without topping, tailing or stringing,
to provide one of the best delicacies of early summer. Slightly larger
beans, that snap – sometimes they are called snap beans – as you break
them, make an ideal companion for roast and grilled meat, especially for
lamb and venison. Towards the end of summer the haulms, i.e. the pods,
dry and wither, leaving the fat kidney beans inside for serving in their
semi-dry state. Last of all, one of the final chores of summer is the shelling
and drying of beans to tide us over to the next season.

This progression is not supplied by exactly the same bean. Different
varieties of the genus *Phaseolus* – the name means a small swift sailing
boat, on account of the pod – have been developed that are at their best at
one particular stage in the cycle. In Europe the great bean gardeners have
been the French. They have perfected the tiny *deuil fin précoce*, with its
purple splashed pod that turns green when cooked, pale waxy looking
beans, the yellow *haricots beurre* and many other *mange-tout* varieties – in
France *mange-tout* signifies young pod beans rather than sugar peas. As
the name implies, everything can be eaten. A favourite midsummer bean,
a dual purpose bean as it can be eaten when young in the pod or
semi-dried, is the flageolet with its pale green and white shaped kidney
beans. The *Comtesse de Chambord* gives white haricot beans for drying; if
you are a gardener you can grow them for yourself, and make up your
own winter store.

The beans from America soon showed their superiority as a year round
food. Broad beans, the beans of antiquity, are a treat in the spring, so long
as you skin them. Once they get old and dry, they sit heavily in the
stomach, and are not interesting to eat. The new beans were not too
different in general shape and type from those we were used to, so there

was not the prejudice against them that there was against, say, tomatoes. They could share the broad bean niche in our diet. Soon they began to oust the broad bean as John Parkinson noted in 1629; he was looking back, 'Kidney beans boiled in water and stewed with butter, were esteemed more savoury meat to many men's palates than the common Garden Bean' – i.e. the broad bean – 'and were a dish more oftentimes at rich men's tables than at the poor.' He named several varieties, Roman beans and French beans, which give an idea of their route to this country from America, and sperage-beans which shows how much their flavour was enjoyed. It still is, for as Elizabeth David says, in *French Provincial Cooking*, she would as well eat a dish of tiny green beans in the early summer as go to the money and trouble that asparagus entails.

Among these beans, the one I do not much care for is the scarlet runner. Unless picked very young the pods are large, rough and stringy. Their flavour is different from the run of French beans, and they have to be sliced to disguise their tough pod, which means that much of the flavour disappears into the cooking water. The scarlet runner was introduced by John Tradescant, the great collector and gardener to Charles I and Henrietta Maria. It was first grown for its scarlet flowers, 'ladies did not . . . disdain to put the flowers in their nosegays and garlands'. Philip Miller, keeper of the Chelsea Physic Garden, and author of the famous *Gardener's Dictionary* first published in 1731, seems to have been the first to bring the scarlet runner 'into much repute as an esculent Kidney-bean; and I well remember' – this is Thomas Martyn, who edited the 1807 edition – 'his speaking much in praise of it more than fifty years since, and that he preferred it at his own table'.

Perhaps Philip Miller was able to ensure that his runner beans were picked young. The giant monstrosities that I see on sale, at least in Wiltshire, make me feel that early gardeners had the right idea when they kept the scarlet runner to decorate a trellis with its brilliant flowers, or to please their wives in a bouquet.

HOW TO COOK FRENCH BEANS

Tip them into a large pan half-full of lightly salted water at a rolling boil and boil them hard and briefly – eight minutes should be enough – with the lid off the pan (this method helps to keep the bright green colour). Older beans will need longer, but they should be allowed to keep a hint of their original crispness; this does not mean they should be under-done.

Another way which I prefer when the beans are tiny and barely thicker than a bootlace, is to steam them over boiling water. This, too, is the best way with beans that you have put down in the freezer (or with bought frozen beans).

There are several ways of finishing them:

1. Stir in a good lump of parsley (maître d'hôtel) butter before serving them.
2. Reheat the drained beans briefly in butter and add a few tablespoons of double cream with a squeeze of lemon juice. Use no more than enough to coat the beans nicely. Add a little parsley.
3. Stew a chopped spring onion or very young white onion or shallot and a small crushed clove of garlic in butter. Add the drained beans and stir them for a few minutes. Again, add parsley.
4. Drain the beans and mix them with just enough olive oil to make them glisten. Serve with lemon quarters, or with lemon squeezed over the beans. They can be served cold as a salad, if you like, but put on the oil while they are hot.

BEATRICE SALAD

Steam 1 kg (2 lb) of young French beans. Mix them with an olive oil vinaigrette and leave them to cool, turning them over occasionally. Arrange them in a bowl and on top put a really large sliced tomato – skin it first – and a couple of sliced hard-boiled eggs, with some cress. Pour a little vinaigrette over the tomato and eggs.

This makes a good first-course for a summer lunch – especially if the beans are from the garden.

GREEK STEWED BEANS

The standard Greek way of cooking green beans is in a stew with onion, tomato and sometimes a hint of cumin.

> 1 kg (2lb) French beans, topped and tailed
> 2 medium onions, sliced thinly or chopped
> 1 clove garlic, crushed with salt
> 175 ml (6 fl oz) olive oil
> ½ kg (1 lb) skinned chopped tomatoes
> 3 tablespoons chopped parsley
> salt, pepper, 1 teaspoon sugar
> ½ teaspoon ground cumin (optional)

Cut or snap the beans in two. Soften the onion and garlic in a heavy saucepan in the oil, until they begin to soften and turn yellow. Put the beans on top, then the tomatoes mixed with the parsley. Sprinkle with salt, pepper and the sugar. Cover and cook over a moderate heat until tender – about half an hour. Keep an eye on the liquid level, the tomatoes and oil should provide enough for the beans to cook in, leaving only a little thick sauce. Stir in the cumin slowly to taste at the end. Serve hot with lamb and rice, or cold as a salad.

FRENCH BEAN AND ALMOND SOUP

Split almonds fried to an appetizing golden-brown in a little butter make a good garnish for French beans as a vegetable. With a fresh bean soup they are even more successful.

> *375 g (12 oz) French beans*
> *100 g (3–4 oz) chopped onion*
> *1 large clove garlic, finely chopped*
> *butter*
> *1 heaped tablespoon flour*
> *1 litre (1¾ pt) water or light chicken stock*
> *salt, pepper*
> *½ teaspoon chopped fresh summer savory*
> *lemon juice*
> *60 g (2 oz) almonds, blanched and split*

Top and tail the beans, stringing them if necessary. Cook the onion and garlic gently in two tablespoons of butter, without browning them at all. When they are soft, stir in the flour, then just over half the water or stock. Simmer for a few moments, then put in the beans, seasoning and the savory. When the beans are just tender – do not overcook them, or you will lose the best of their flavour – purée the soup in a blender. Use the rest of the water or stock, plus extra water, to dilute the soup to the consistency you like. Reheat, adding more seasoning to taste if necessary and a squeeze of lemon juice to bring out the flavour. Keep the soup just below boiling point.

Fry the almonds in a tablespoon of butter, stirring them about until they are nicely browned. Mix into the soup with a little extra knob of butter and serve immediately.

SOUPE AU PISTOU

The great event of our summer in France is the garlic and basil fair at

Tours. It is always held on St Anne's day, July 26th. Why, I do not know, except that it is a convenient saint's day at a time when garlic, shallots and onions have been lifted and dried and when basil is at its best. It occupies a small square to the north of Les Halles, the main market place, and spills out along the street connecting them. Tresses and bunches of garlic are bought for the year; I always choose the violet garlic of Bourgeuil for its long keeping and juicy quality. Another speciality is shallots, the grey kind, the copper tawny variety known as *cuisse de poulet* and the pinkish ones for storing. Crowds gather round the mountain ponies and monkeys and dwarfs selling herb flavoured Alpine sweets (nasty – I am sure they must be good for you). The bars are open and overflowing; towards midday there are queues at the Breton stalls that sell *charcuterie*, huge rounds of bread and yellow cake.

African leather sellers, in striped robes, try to distract one from the brilliant basil, pot after pot, carpeting the pavements with tufted green. The leaves drip from the continual watering that keeps them lively in the hot sun. Most of the basil is bushy, magnificent, not nourished by a lover's head though some of the pots are large enough. It is good for drying. For present eating, for the real kingly flavour, I buy the broad leaved basil from a grower who signs his place with a board giving his recipe for *soupe au pistou*, provençal basil soup. Perhaps in Provence the flavour will be stronger from the stronger sun; but we do not complain for this basil from just outside Tours has enough aromatic presence to add an extra delight to the fine tomatoes that go into our salads and this soup. To me, the other essential ingredients for this mixed vegetable soup are beans: green beans and haricot beans, that is why I give the recipe here.

> 125 g (4 oz) green beans, cut in 1 cm (½ in) lengths
> 1 or 2 courgettes, diced
> 100 g (3–4 oz) potatoes, scraped, diced
> 125 g (4 oz) white part of leeks, diced, or spring onions
> 125 g (4 oz) carrots, scraped, diced
> 125 g (4 oz) young turnips, peeled, diced
> 125 g (4 oz) celeriac, peeled, diced, or celery
> 125 g (4 oz) flageolet or haricot beans, soaked if necessary and
>   partly cooked, see note below
> 2 tomatoes, peeled, chopped coarsely
> three sprigs basil, chopped
> salt, pepper
> 100 g (3–4 oz) shell pasta

PISTOU
good bunch basil leaves
3 large cloves garlic
1 tomato, peeled, chopped
75 g (2½ oz) grated Parmesan or Gruyère
olive oil
salt, pepper

Put all the vegetables in the first list into a large pan, with the basil, seasoning and enough boiling water to cover the vegetables generously. Simmer for half an hour. Add the pasta and continue cooking until it is tender but not too soft.

Meanwhile make the *pistou*, in a mortar, or blender. Crush the basil and garlic together, add the tomato and cheese and enough olive oil, gradually, to make a mayonnaise consistency. Season.

The *pistou* can either be mixed into the soup in its tureen, or it can be served in a separate bowl for people to help themselves. Serve a bowl of grated Parmesan or Gruyère as an extra seasoning.

Note Monsieur Couratin's recipe specifies the half-dried haricot beans that are around at that time of the year in France. In England one can either use dried flageolet or haricot beans, but they will need soaking and half an hour's cooking in unsalted water. If you have some cooked haricot beans left over from an earlier meal, they should be added with the pasta, so that they do not disintegrate.

STIR-FRIED FOUR SEASON BEANS
CH'AO SSU CHI TOU

In other words French beans with water chestnuts. It is a pity that water chestnuts have to be bought in cans, since their flavour is much diminished. Still their crispness remains. If you cannot find them locally, substitute blanched almonds.

½ kg (1 lb) French beans
6 teaspoons oil
1 scant teaspoon sugar
10 water chestnuts, sliced, or 60 g (2 oz) blanched almonds
4 tablespoons chicken stock
1 teaspoon cornflour mixed with 3 extra teaspoons chicken stock

Cut the beans into 5 cm (2 in) pieces. Heat a heavy frying pan. Add the oil, then the beans. Do not overheat, as the beans should not brown. Keep them moving over the heat for three minutes. Add the sugar, water chestnuts or almonds and the stock, stirring all the time. Cover and leave to simmer for two minutes. Taste and add a little salt. Mix up the cornflour liquid and add it to the pan. Stir until the beans are lightly glazed. Serve at once.

In winter this is a good way of cooking frozen beans, the long whole kind not sliced runner beans. Be prepared to adjust the times according to the tenderness of the beans. They should be cooked but still slightly crisp. Serve with rice.

Note 'Four season' refers to the long growing period of French beans like *quatre-saisons* roses and strawberries in France.

STIR-FRIED GREEN BEANS

For another way of cooking young green beans, turn to the stir-fried recipe in the section on mange-tout peas, p. 327. It includes soya sauce and sherry, which gives the beans a stronger flavouring.

HAMBURG BEANS AND PEARS

The Germans are great ones for fruit with vegetables, fruit with meat. I have often heard French friends joke about the way we eat 'jam' with pork, duck, hare and so on, but we have barely scratched the surface by comparison with the Germans' skill at this kind of partnership. (I should also add that these French friends often sit down to their favourite local dish of pork with prunes and a cream sauce flavoured with redcurrant jelly, *porc aux pruneaux de Tours*.) As to the bacon in this dish, try to find a source of the real German *geräuchter bauchspeck*. This will probably be a delicatessen, rather than butcher or grocer. It is worth it for the beautiful genuine flavour of cured pork.

>   375 g (12 oz) smoked belly bacon
>   2 tablespoons butter
>   ½ kg (1 lb) French beans
>   3 ripe pears, peeled, diced
>   ½ litre (1 pt) beef or veal stock
>   salt, pepper, chopped parsley

Cut the bacon into strips, leaving on the rind, and cook it gently in the butter until transparent. It should not brown. Put in the beans, pears and stock, bring quickly to the boil, cover and cook until the beans are tender. Strain off such liquor as remains for using up another day. Correct the seasoning, adding plenty of pepper. Serve sprinkled with parsley, with boiled or baked potatoes and butter.

FRENCH BEAN MOUSSAKA
MUSSAKES SELEN FASSUL

A Bulgarian version of the *moussaka* on page 57. Instead of aubergines use ¾ kg (1½ lb) small French beans, 250 g (½ lb) sliced carrots and three large green peppers, sliced and seeded. These vegetables need only be boiled together until barely cooked; they should not be fried. Layer them into the pot with the minced meat in the usual way. Instead of the béchamel and egg sauce, beat together two tablespoons of flour, four eggs and 250 ml (8 fl oz) yoghurt. Season with salt and pepper and pour over the top of the dish when it is almost done. Allow a further 20 minutes for the sauce to set and colour slightly. You can, if you prefer, stick to the usual sauce.

GREEN AND WHITE BEANS IN THE TOURAINE AND ANJOU MANNER

As neighbouring provinces, Touraine and Anjou have much in common in their styles of cookery. I like their way of mixing fresh green beans either with pale green, half-dried flageolets or white haricot beans. This summery looking mixture can be carried through into winter, using frozen green beans and dried haricots or flageolets.

GARNITURE TOURANGELLE For lamb especially, but also delicious with venison or beef steaks. Cook equal quantities of flageolets or white haricot beans (soak them first if necessary) and green French beans in separate pans. Drain. Cut green beans into 2–3 cm (1 in) lengths, or diagonally into diamonds. Mix together and bind them lightly with velouté sauce. Or bind them with a béchamel sauce improved with cream and seasoned with nutmeg. Sprinkle with parsley.

ANJOU SALAD (SALADE ANGEVINE) Take equal quantities of cubed new potatoes, flageolets or white haricot beans, and green beans cut in 2–3 cm (1 in) lengths and mix with a walnut oil vinaigrette.

TOURAINE SALAD (SALADE TOURANGELLE) Use the same vegetables. Mix them while still hot and pour on some vinaigrette. Leave them to cool, then drain them. Mix in mayonnaise, which will slip easily round the dressed vegetables.

HARICOT BEANS, FRESH AND DRIED

The word haricot is confusing. It was first used by French and English cooks in the late middle ages for a dish of cut-up meat, from *harigoter* meaning to cut up. It still has this meaning in both languages, a haricot of mutton. When the new beans came to France in the 16th century they were called *fèves*, like broad beans; then it seems that an attempt was made to distinguish between them by adding the Aztec *ayacotl* which by 1642 had slipped into the familiar word *haricot, fève de haricot*. Eleven years later it was also recorded in English as 'aricot'.

To the French, haricots are still the green beans from America. To us they have become the small dried white beans, as opposed to the larger butter beans, on sale in every grocery. The old simple choice between haricot and butter beans has recently been extended by the health food craze. We can now make acquaintance with red kidney beans, aduki Japanese beans, navy, Lima, brown, black or black-eyed beans, Boston or Borlotti beans. If you are catering for the family in France, you will come across huge Belle de Soissons and Arpajon beans, or even *mojhettes* or *braisins* from the Marais if you are in Poitou or the Vendée. At the end of the summer pale green and white flageolets beans are sure to catch your eye in the market. One French seed catalogue I have lists 47 varieties of green string beans, plus ten varieties of beans for drying. The best English catalogue only lists 26 in all, including four for drying. A friend told me that in Latin America there are nearly 40 varieties of dried beans alone on sale in the shops. You can understand that in this country we have barely started on the possibilities.

Naturally the varieties vary slightly in flavour and texture – it is fun trying as many as you can to see which you like best – but methods of cooking remain the same, with minor adjustments of time and seasoning.

If the beans, for instance the flageolets, come more or less straight from the pod, remember that they have already dried out a little before being picked (the French call them *demi-sec*). They do not need soaking, but they will take much longer to cook than string beans – allow 45 minutes at least. It is not always easy to know how dry these beans are, so bite one cautiously. If it breaks without much persuasion, it does not need soaking.

When beans are properly dried out for storage, they pour from the bag with a dry rustling and are hard to bite. They will need soaking. The longer they have been in store, the more soaking they will need. This is why it is essential to check the beans in your larder and make a point of using them up before the autumn every year. Really old beans will remain tough through hours of soaking and stewing.

The usual soaking period is overnight – or at least this is the time given in many recipes. Often they will be soft enough in six hours. If you pour boiling water over them, they can be ready in three hours. Quickest of all is to cover them generously with water in a pan, and bring them slowly to a vigorous boil; cover them and turn off the heat – an hour and a half or two hours later they will be ready for cooking.

As a rough and ready guide, reckon that your beans will double in weight after soaking – a little more or a little less. Therefore when they reach this weight, they can be cooked.

There is no magic about soaking. It is merely a device to shorten cooking time which saves firing and forethought. The dried vegetables we buy these days are of such quality and so well prepared that they need less soaking and attention than they did in the past. No need, either, for the pinches of bicarbonate of soda and so on, unless you are specifically instructed to add them.

To cook the beans, pour off the soaking water (unless they are the half-dried kind, of course). Pour on enough fresh water to cover the beans by 2–3 cm (1 in) at least. Tuck in aromatics and flavourings appropriate to the recipe – herbs, pot-herbs, cured pork, perhaps a ham bone – but no salt. Salt has a hardening effect.

Cover the beans and simmer on top of the stove, in the oven, or even in the plate-warming oven of a solid fuel cooker until they are tender. Allow an hour at least, if they are cooking fairly fast, but check after 30 minutes. In the plate-warming oven they may need 24 hours. Beans are good tempered. I suppose that is one of the reasons why they have been so popular with so many different peoples. There is a general feeling in favour of two or three hours of slow cooking, when mixed stews are concerned. Beans for a salad, which need to keep their shape, can be simmered more briskly for the shortest possible time. When you come to the end, add salt. Give the dish five or ten minutes longer for the salt to make an impression, then serve or finish the cooking as indicated in the recipe.

Chick pea, lentil and dried pea recipes can be adapted for haricot beans – and vice versa.

CREAMED BEANS OF THE VENDÉE
MOJHETTES À LA CRÈME

The *mojhettes* of the Vendée are large white haricot beans, in size some-where between our haricot and butter beans, grown in the moist heat of the Marais fields at Arçais and Damvix. The Marais is one of France's most hidden places. On the high road which runs from Nantes to Niort, a noticing traveller might glance seaward and wonder what the sunken, tree-filled basin below him concealed. Once it was the gulf of a lazy sea, which came and went so mildly that monks and then Dutch engineers of the time of Henri IV were able to reclaim the land. In winter much lies under water, patterned by squares of trees, causeyed roads and village mounds. In summer it returns to green life, the island-fields enclosed in canals so that cows, men, machinery can only reach them by punt. Going along these canals, disturbing pallid carpets of pondweed, one can scarcely believe that this silent land has been devoted for centuries, and with energy, to market gardening. Charentais melons, pumpkins, Swiss chard, salad greens, angelica, artichokes, rose-skinned garlic, cabbages – and these fine beans all come from here.

> ½ kg (1 lb) newly shelled beans, or ¼ kg (½ lb) soaked dried beans
> 60 g (2 oz) butter
> 60 ml (2 fl oz) oil
> 1 medium carrot, quartered
> 1 medium onion, halved
> 10 cm (4 in) stalk celery
> 1–2 large cloves garlic
> 2 sprigs thyme, salt, pepper
> 75 g (2½ oz) melted butter
> 150 ml (¼ pt) double or whipping cream
> lemon juice

Stew the beans in butter and oil for 10–15 minutes in a covered pan. Add the vegetables, garlic, thyme, and plenty of boiling water to cover the beans easily. Stew until they are tender, topping up the water if necessary. Drain the beans and remove the aromatics and flavouring vegetables. Salt them. Heat the melted butter in a frying pan, stir in the cream vigorously and when it boils, season and add lemon to taste. Mix in the beans and turn into a heated dish. Serve with an endive salad, dressed with a walnut oil vinaigrette; mix in some small cubes of bread fried in walnut oil with a crushed clove of garlic – or rub the bread with garlic before you cut it up and fry it.

FASOULIA

The best bean salad I know is headed simply *Fasoulia*, the Greek for beans. It's a recipe of Elizabeth David's, from her *Mediterranean Food*, and I have never come across another one quite like it. The usual style of bean salad, simple and pleasant, is cooked beans dressed with olive oil and lemon juice or wine vinegar, with various herb flavourings. Good, but not as good as this.

The unusual detail is the simmering of the soaked beans in olive oil. Try to do better than olive oil BP for cooking vegetables; it has had all the guts knocked out of it, which is reasonable for a medicine but rotten for cooking. Greek Minerva is an obvious choice with its green rustic flavour, or the golden-green Tuscan olive oil when you can get it.

> *250 g (8 oz) small dried beans, soaked*
> *125 ml (4 fl oz) olive oil*
> *2 cloves garlic, crushed*
> *1 bay leaf*
> *1 sprig of thyme*
> *1 dessertspoon tomato concentrate*
> *lemon juice*
> *sliced raw onion*
> *salt, pepper*

Drain the beans. When the oil is hot, put in the beans and cook gently for ten minutes, with the garlic, herbs and tomato paste. Pour in enough boiling water to cover the beans by 2–3 cm (1 in). Stew for up to three hours, until the beans are tender and the water has slowly reduced to a thick brown sauce. Squeeze in lemon juice to taste – add onion rings, salt and pepper to taste. Turn into a dish and leave to cool.

HARICOT BEAN SALADS

The straightforward salad is cooked haricot beans, with an olive oil and vinegar or lemon dressing. Chopped onion in one or other of its forms is scattered over the top, with parsley to give zest.

Tuscans dress up their bean salads with some slices of raw sweet onion and pieces of canned tunnyfish. In Provence an anchovy sauce of some kind (see p. 189), or strips of salted anchovy, are popular. The Turks make a splendid and beautiful salad called *piyazi* – the beans are covered with chopped peppers and tomato, slices of hard-boiled egg and onions, black olives and parsley make a final decoration. Lemon juice then olive oil are poured over as dressing.

HARICOT BEAN PURÉE

Haricot beans can be used instead of chick peas to make a delicious purée with tahina (sesame seed paste). The recipes are identical.

Soak and cook the beans. Drain and purée in a blender with olive oil, salt, lemon juice and garlic. Use a little of the bean cooking liquor, if the purée sticks in the blender. Gradually add half the weight of the cooked beans in tahina paste; taste from time to time, you may prefer to use less – or more.

Put into a bowl, scatter with cayenne, and serve with pitta bread, spring onions, and the Eastern salad on p. 381.

BAKED BEANS

It can be something of a surprise to find that there are many versions of baked beans outside America, and that the Bostonians may well have learned their famous dish from a French cassoulet rather than a Red Indian bean mash.

CASSOULET DE CASTELNAUDARY A simple version, that is worth attempting if you pot your own goose (*confit d'oie*) or bring back a tin from a holiday in south-west France. The other ingredients are to hand in this country.

Soak and cook a kilo (2 lb) of haricot beans with rolls of pork skin, or a piece of salt pork or bacon, until they are tender but not splitting. Drain off and keep the liquid.

In another pan reduce three large sliced onions and three crushed cloves of garlic to a golden softness in some goose fat or lard or bacon fat. Sprinkle on a tablespoon of flour, moisten with a litre (1¾ pt) beef stock and add a bouquet garni with four chopped tomatoes. Boil steadily for 20–30 minutes.

In a frying pan, cook a wing and thigh joint of potted goose in its own fat until lightly browned, with a small garlic sausage.

These are the three elements of the cassoulet. Take an earthen or stone-ware pot and layer in the beans with their pork skin or sliced cured pork, and the potted goose and garlic sausage, sliced. Pour over the onion and tomato sauce. Bake in the oven for an hour. 15 minutes before the end of cooking time sprinkle the top with breadcrumbs. Then give the whole thing another half hour, cutting in the crust three times and adding a further thin sprinkling of crumbs each time. You should end up with a fine golden crust on the top.

The bean liquid that you set aside can be used to moisten the cassoulet if it becomes too thick.

The major variation on the basic theme is the addition of lamb. It can either be a boned and rolled piece of leg, stuck with garlic and braised until it is three-quarters cooked, or some boned and cubed shoulder that has been browned and stewed until just over half-cooked.

Pig's trotters, pork sausage of the Toulouse type (the closest is our Cumberland sausage), cured pork and sausages of various kinds can all be added. The trotters should be cooked with the beans; the sausages with the goose if they are preserved – should they be fresh they can be fried briefly and added to the cassoulet as it is layered into the pot.

BAKED BEANS AND DUCK A recipe from Evan Jones's *American Food*, that is not unlike cassoulet with duck instead of goose. It is much sweeter than the French would like and in this it comes close to Boston baked beans. The recipe came from the author's mother; her family moved south from Canada to the Maritime provinces, then to Minnesota.

$\frac{1}{2}$ kg (1 lb) small dried beans
$1\frac{1}{2}$ litres (2$\frac{1}{2}$ pt) beer
$\frac{1}{2}$ litre ($\frac{3}{4}$ pt) beef stock
2 bay leaves
1 medium onion, chopped
3 lumps of ginger in syrup, chopped
1 lemon, sliced and chopped
1 heaped tablespoon maple sugar
125 ml (4 fl oz) blackstrap molasses
2 tablespoons Worcester sauce
1 tablespoon Dijon mustard
$\frac{1}{2}$ teaspoon ground cumin
1 teaspoon dried savory
125 g (4 oz) diced salt pork or green streaky bacon
1 duck, 2–2$\frac{1}{2}$ kg (4–5 lb)

Soak the beans in a litre (1$\frac{3}{4}$ pt) beer overnight. Mix the remaining beer with beef stock and combine in a large baking dish with the beans and their soaking liquor. Add everything else except the duck. Cover and bring to the boil. Bake for three hours at an oven temperature of 180°C/350°F/Gas Mark 4, adding boiling water every so often to keep the beans covered by 2–3 cm (1 in).

Cut the duck into six pieces and remove the lumps of fat carefully from the carcase. Chop the fat roughly and heat it in a large frying pan, so that it melts. Brown the duck pieces in it for about ten minutes, so that they are completely sealed. Pour off the duck fat and put the joints into the baked bean dish, pushing them down. Pour 125 ml (4 fl oz) water into the frying pan to deglaze it, boil for a minute, then add to the beans and duck. Continue to bake for another three hours, topping up as before with water to keep the beans just covered. They should have absorbed all the fat from the duck and most of the liquid as well.

BOSTON BAKED BEANS You can use haricot beans for this, although Dutch brown beans would be better if you grow them, or navy beans.

Soak and simmer 1 kg (2 lb) beans until they are half tender, with an onion. Drain them and take out the onion.

Into the bean pot put three or four onions, each stuck with two cloves. Tip in the beans. Mix eight tablespoons of molasses, 250 g (8 oz) soft dark brown sugar, three teaspoons mustard powder, with a generous ½ litre (1 pt) water. Pour this over the beans, add plenty of pepper, and push a nice piece of scored salt pork down into the middle – use a piece of green streaky bacon if you cannot get salt pork. It should weigh something between 300 and 375 g (10–12 oz). Cover and bake in the oven at 120°C/250°F/Gas Mark ½ for four hours. Take off the cover from time to time and check the liquid level; it need not cover the beans, but should just be visible – top it up with boiling water if necessary.

When the four hours is up, remove the cover and leave the pot for a further hour to brown on top. You can add a sprinkling of brown sugar if you like, to help this along.

BEANS WITH CORN AND PUMPKIN
POROTOS GRANADOS

Elizabeth Lambert Ortiz, the great expert on Latin American cookery, remarks that although this dish comes from Chile, it has 'decided Indian overtones, since tomatoes, beans, corn and marrow all originated in Mexico and gradually spread over the whole South-American continent'. The beans she specifies in the original recipe are cranberry beans or navy beans. The nearest we can buy is the small dried haricot bean on sale in every grocer's. If you can get to an Italian shop, ask for saluggia beans, which another cookery book suggests. They are pinkish, with a tan fleck. The combination of corn and pumpkin is also popular in the Basque country and Aquitaine.

> 250 g (8 oz) small dried beans
> 2 tablespoons mild paprika
> 4 tablespoons corn oil
> 2 medium onions, finely chopped
> 2 green chillies, seeded, chopped
> kernels from 2 large ears of corn, or 250 g (8 oz) frozen sweetcorn
>   kernels
> ½ kg (1 lb) pumpkin, cut in cubes
> ½ kg (1 lb) tomatoes, skinned, chopped
> ½ teaspoon dried oregano
> salt, pepper

Soak and boil the beans in unsalted water until tender. Drain them, but keep their liquid. Meanwhile heat the paprika with the oil and fry the onions slowly until they are soft. Add the remaining ingredients, and simmer steadily for five minutes. If this mixture seems at all dry, add a little of the reserved bean liquid, then add the beans. Cover and simmer for a further 15 minutes. The pumpkin will disintegrate and thicken the sauce as it does in the Argentinian *carbonada* on p. 424. The hot chillies add liveliness to the dish – this can be emphasized with cayenne pepper if the chillies have been badly stored, with a consequent loss of flavour.

BEAN SPROUTS see SOYA BEANS

BEET GREENS, or SPINACH BEET

Beet greens or spinach beet is closely related to beetroot (q.v.) and sugar beet, but only the leaves and stalks are cooked.

This is the *beta* of the Romans, or very close to it. According to Pliny the leaves can be added to salads with lentils and beans, they can be eaten raw, or dressed in the same way as cabbage. The best thing is 'to stimulate their insipidity with the bitterness of mustard'. Apicius must have concurred; in the cookery book that bears his name, and certainly includes dishes he ate, boiled beet greens are to be served with mustard, oil and vinegar. He also gives a recipe for barley soup (see p. 114) that includes beet greens with other leafy vegetables. I am afraid that Pliny's word insipidity sums up beet greens for me, insipidity verging on dull nastiness.

Modern gardening books say 'cook like spinach'. The trouble is that beet greens fall far behind spinach. If you think that by cooking them like spinach, they are going to taste like spinach, you are in for a sad disappointment. The French are clever at amalgamating such greenery with pork, into the homely peasant *farcis*, the green herb pâtés, of Poitou and the Vendée, or into the *caillettes* of the south. Recipes for these and other dishes, methods of preparation and so on, are to be found under Swiss chard, to which beet greens are also related.

BEETROOT

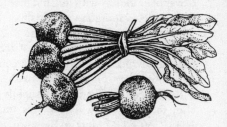

We do not seem to have had much success with beetroot in this country. Perhaps this is partly the beetroot's fault. It is not an inspiring vegetable, unless you have a medieval passion for highly coloured food. With all that purple juice bleeding out at the tiniest opportunity, a cook may reasonably feel that beetroot has taken over the kitchen and is far too bossy a vegetable. I have never heard anyone claim it as their favourite.

One would expect, I suppose, the best recipes to come from the north of Europe, as the beetroot is said to have been developed by German gardeners in the Middle Ages. Enthusiasm for it seems to have spread most successfully from Germany to the north and east, where one finds good herring and beetroot salads and the supreme dish of all as far as beetroot is concerned, borshch in its various forms. It can be a delicate pink and creamy chilled soup, speckled with chives, suitable for the most elegant of dinners. Or it can be a great one pot meal for a huge family, with duck, beef, sausages and several other vegetables floating in a rich dark red soup, attended by bowls of soured cream and little cheese patties. I do not understand why the English and Italians, and even more the French who understand such things so well, have not developed some striking beetroot soup of their own.

HOW TO CHOOSE AND PREPARE BEETROOT

When you buy beetroot, make sure none of the whiskery roots are broken off and that there is at least 5 cm (2 in) of unbroken stalk at the top. All one's efforts with beetroot are aimed at preventing the purple juice bleeding away. It is also sensible from the cooking point of view to choose even-sized beetroot.

At home, rinse them delicately under the tap to dislodge any earth. Try to avoid scrubbing, or you risk damaging the skin, tiny roots or stalk. If the beetroot are very whiskery, or uneven in size, you will find it more convenient to boil them. Use plenty of water, with a tablespoon of salt to 1 kg (2 lb) beetroot. Cover and simmer steadily for one to two hours. Smoother beetroot can be baked – this is the better method from a flavour point of view. Place them in a large ovenproof dish with plenty of room.

Cover over loosely with foil, but mould it closely to the dish so that there is as little evaporation as possible. Bake for three or four hours at 140–150°C/275–300°F/Gas Mark 1–2. With either method you should not pierce the beetroot to see if they are cooked, as you do baked potatoes; wrinkle the skin and if it comes easily away from the root, it is done. Remove from the water or the oven and leave to cool.

If you need the oven on at a slightly higher temperature for something else, the beetroot will take less time. But avoid temperatures over 180°C/350°F/Gas Mark 4. Beetroot need time and a low steady warmth to cook through evenly.

You can buy ready prepared and frozen beetroot of reasonable quality, though I find that the very small ones used are often too sweet and lack the flavour of tennis-ball sized beetroot. You should also be careful when buying ready prepared beetroot that it has not been over-vinegared.

BEETROOT CONSOMMÉ

> 1½ litres (2½ pt) clarified beef, chicken or duck stock, see p. 550
> 2 large boiled beetroot, skinned
> salt, pepper, lemon juice
> 150 ml (¼ pt) soured cream
> chopped parsley and chives

Put the stock into a large pan. Chop the beetroot and set aside three heaped tablespoons. Add the rest to the stock and bring to just below boiling point. Keep it at this heat for about half an hour – it should give an occasional burp, or show a few small bubbles, but that is all. The point being to make an infusion of the stock and beetroot. Just before serving add the reserved beetroot to freshen the colour and leave for a further few minutes. Strain into a hot tureen, check the seasoning and add a little lemon juice. Mix the cream with the herbs and serve in a separate bowl, along with a plate of warm *vatrushki*.

RUSSIAN CURD CHEESE PATTIES (VATRUSHKI)

Recipes for *vatrushki*, one of the best Russian delicacies, vary a good deal – sometimes a brioche dough is used, sometimes plain shortcrust, sometimes a soured cream dough. Flaky pastry is a good choice, as it keeps the *vatrushki* light, and is quick to make. Served with clear vegetable soups, like this beetroot consommé or the clear root vegetable soups on p. 525, they make a sustaining first course for a dinner party, or the main course for supper. The characteristic shape of *vatrushki* depends on pinching up the edges of the pastry discs to form cases for the curd filling.

flaky or puff pastry made with 300 g (10 oz) flour

FILLING
*250 g (8 oz) curd cheese (*not cream cheese*)*
125 g (4 oz) melted butter
1 large or 2 small eggs
salt, pepper

GLAZE
1 beaten egg

Roll out the chilled pastry and cut into 8 cm (3 in) rounds. Raise the edges by pinching them up to form cases. This will give an evenly fluted appearance around the outer edge.

Mix the filling ingredients together. Put some into each case, smoothing off the top. Brush over the top and sides with beaten egg. Bake for 20 minutes at 220°C/425°F/Gas Mark 7, or until puffed and nicely browned.

Note If you cannot buy curd cheese, use sieved cottage cheese instead. Or Gervais squares or *petits Suisses* mixed with sieved cottage cheese half-and-half, to make a slightly richer mixture than cottage cheese alone.

CHILLED BEETROOT SOUP

Follow the recipe for beetroot consommé above, but strain the consommé into individual bowls and set in the refrigerator to chill. Mix the cream with the herbs, and swirl a spoonful of this mixture into each bowl of soup.

RUSSIAN BORSHCH

There are any number of recipes for borshch. This is a modern, simplified recipe from *Russian Cooking*, by F. Siegel, published in Moscow by Mir in 1974. This time the soup is thickened with flour, which makes it more substantial than the light summer recipes following. Fresh beef stock adds weight to the flavours, too. Do not feel bound by the quantities given in these recipes, or by the lesser ingredients. Use what you have in the quantities you like, as you would for more familiar soups.

> 375 g (¾ lb) shin beef, cut into small pieces
> bouquet garni
> ½ kg (1 lb) beetroot
> 2 tablespoons red wine vinegar
> 3 tablespoons tomato concentrate, or 2–3 fresh tomatoes, chopped
> 2 tablespoons melted butter
> 250 g (8 oz) root vegetables, including a carrot, an onion and Hamburg parsley if possible
> ½ kg (1 lb) cabbage, shredded
> 3 good tablespoons butter
> 1 tablespoon flour
> salt, pepper, sugar
> beetroot stock (see p. 95)
> soured cream
>
> OPTIONAL
> raw or cooked potato, diced
> chipolata sausages, grilled
> boiled ham, diced

Put the beef into a large pan with 2½–3 litres (4½–5½ pt) water and the bouquet garni. Bring to the boil, skim and simmer slowly, covered, for an hour and a half.

Next wash and shred the beetroot into a pan (no need to remove the peel unless it is coarse). Use a mandolin slicer if possible, as it produces good matchstick pieces; the coarse blade of a grater is the next best thing. Add to it the vinegar and tomato concentrate (keep fresh tomatoes until later on) and the melted butter, plus ¼ litre (½ pt) of liquid from the stewing beef. Cover and simmer for an hour.

Prepare and slice the other root vegetables, cutting them into shreds where appropriate. Fry them in two tablespoons of the butter, until they are slightly golden, then tip them into the beetroot pan. Simmer for 15 minutes, then season with salt, pepper and a tablespoon of sugar.

Meanwhile strain off the meat stock, put it back into the rinsed pan and bring it to the boil. Put in the cabbage and the raw potato if used. After ten minutes, stir in the beetroot and vegetable mixture. Add the bay leaf, fresh tomato and appropriate seasoning. Mash the flour with the last of the butter and use it to thicken the soup (*beurre manié*, page 552). Finally add any of the cooked optional ingredients and the beetroot stock to improve the colour. Reheat without the soup boiling. Serve with soured cream.

SIMPLIFIED VARIATIONS

1) Use beef stock, if you do not want the trouble of cooking the meat. Stock cubes should be well diluted, below the strength recommended on the packaging. Otherwise the chemical additions, monosodium glutamate for instance, will spoil the freshness of the flavour.
2) Use cooked beetroot, when raw is not available. Shred it and heat it through slowly in some of the beef stock. Add vinegar with discretion, as beetroot boiled commercially has often been seasoned with vinegar already.
3) Instead of the final thickening of *beurre manié*, stir the flour into the fried root vegetables, then moisten with beef stock. Stir the last of the butter into the soup with the beetroot stock.
4) For a spendid supper dish or lunch for a large winter party, substitute 1 kg (2 lb) or more brisket of beef for the shin and add a duck, three-quarters roasted, with its giblets, an hour before the end of the cooking time.

If you own a freezer and can count on prepared beetroot and beef stock, borshch can be made quite quickly.

BEETROOT STOCK

When beetroot is cooked for a long time in soup, its vivid jewel colour turns to a brownish tone, the purple becoming a little rusty at the edges. From a flavour point of view this does not matter much, but the glory has gone. Russian cooks counteract this by preparing a simple beetroot stock to be added to the soup just before serving and I think it is worth the trouble. Uncooked beetroot is the ideal but cooked can be used instead, though the flavour tends to be less lively and the colour less rich. Such an addition is unnecessary with the consommé on p. 92, because it is itself a grand version of beetroot stock.

> *1 large beetroot, washed*
> *3 tablespoons wine vinegar*

Grate the beetroot into the pan, so that none of the juices are wasted. Pour enough boiling water (or stock if you like) over the beetroot to cover it. Add the vinegar. Bring slowly to the boil, then take it off the heat and leave to infuse for 20 minutes. Strain into the soup just before serving. The soup should be very hot indeed, at boiling point, so that no further reheating is needed once the stock has been added. It is boiling that ruins the colour.

SUMMER BORSHCH

> ½ kg (1 lb) beetroot, uncooked
> 1 large carrot
> 1½ litres (2½ pt) water
> 2 tomatoes, peeled, chopped
> 90 g (3 oz) spring onions, chopped
> 1 stalk celery, chopped
> 4 large new potatoes, scraped, diced
> 150–175 g (5–6 oz) sliced courgettes
> bouquet garni
> salt, pepper, 4 cloves
> beetroot stock (p. 95)
> soured cream

Cut the stalks and leaves from the beetroot. Discard any that look floppy or brown and cut the rest into pieces. Blanch them for two minutes in boiling salted water, then drain them and set aside.

Wash the beetroot and grate it into matchstick strips into a large pan. Grate the carrot and add it to the beetroot with the water. Cover, bring to the boil and simmer for 20 minutes. Then add the chopped stalks and leaves, the remaining vegetables, the bouquet and seasonings. Cover again and simmer until the vegetables are cooked. Refresh the colour with beetroot stock and remove the soup from the heat before it boils again. Either mix in a few tablespoons of soured cream, or serve it separately.

Note Dill weed is often used to flavour borshch.

BEETROOT SALAD WITH ANCHOVY DRESSING
BETTERAVES À LA PROVENCALE

Beetroot profits from the rich piquancy of anchovies. The dressing is a useful one to know for other vegetable salads.

½ kg (1 lb) boiled, peeled beetroot
250 g (½ lb) boiled firm or waxy potatoes
2 hard-boiled eggs, peeled
chopped parsley

DRESSING
2 medium onions, chopped
4 tablespoons olive oil
1 tin anchovies in oil
1 teaspoon wine vinegar
¼ teaspoon Dijon mustard
pepper

Slice the beetroot and put it into a shallow bowl. Peel and slice the potatoes into half-circles and arrange them in a ring round the edge, slipping the straight edge down between the beetroot and the edge of the bowl. Mash the eggs to crumbs with a fork, mix them with a heaped tablespoon of parsley and set them aside.

For the dressing, cook the onions in a tablespoon of oil in a small covered pan, so that they become soft without browning. Cool and pound with the anchovies, their oil and the remaining ingredients (use a blender if possible). Adjust the seasonings. Spread this dressing evenly over the beetroot. Scatter the egg on top with extra parsley if necessary. Serve chilled.

BEETROOT AND ORANGE SALAD

Slice ½ kg (1 lb) beetroot into a bowl. Dress it with an olive oil and lemon juice vinaigrette, well seasoned. Leave for a couple of hours or longer. Meanwhile with a knife peel two large oranges, taking off the inner white skins as well as the pith and peel. Slice them as thinly as possible, and remove pips. Drain the beetroot, then put it round the edge of a shallow serving dish. Arrange the orange in the middle, and pour a little dressing over it. Sprinkle the whole thing with chopped chives or spring onions. Serve lightly chilled.

A little chopped garlic can be added to the dressing.

BEETROOT AND POTATO SALAD

Use equal quantities of sliced beetroot and diced potato, dressed separately then arranged on a dish as in the recipe above. Scatter with one or two chopped hard-boiled eggs, mixed with parsley and chives.

BEETROOT AND APPLE SALAD

Use ½ kg (1 lb) beetroot, sliced, 2–3 Cox's Orange Pippins, cored, cubed and separately dressed with olive oil vinaigrette. Arrange as above. Scatter with thinly cut onion rings.

ITALIAN BEETROOT SALAD

Dress the beetroot with an olive oil and wine vinegar mixture, scatter with chopped fresh mint and add thin onion rings.

SCANDINAVIAN BEETROOT AND HERRING SALAD
SILDESALAT

Sildesalat is a favourite around the Baltic Sea and in Norway. The quantities of each ingredient can be varied, according to your taste and what is available.

> 3 salt herring fillets, or matjes herring, or bloaters or kippers,
>   filleted
> ½ kg (1 lb) boiled beetroot, diced
> vinaigrette
> 250–375 g (8–12 oz) cooked potato, diced
> 2 Cox's Orange Pippins, unpeeled, cored, diced
> 1 medium onion, cut into thin rings
> 2 hard-boiled eggs, quartered
> chopped parsley

Cut the fillets so that you have six nice strips, and if they are on the strong side soak them. Bloater and kipper, English alternatives, need no soaking or cooking. Put the beetroot in a bowl, and pour over a little vinaigrette. Mix the potato and apple in another bowl and add vinaigrette. Leave for an hour or two.

About one hour before the meal, arrange the beetroot on a dish after draining it. Surround it with the drained potato and apple. Or vice versa. Place the herring strips on top. Scatter with onion, arrange the eggs and finish with a good layer of parsley. Chill.

You might prefer to use a Scandinavian cream dressing. Arrange the beetroot, potato and apple on the dish, without soaking them in vinaigrette. Mix together a teaspoon each of German mustard and sugar, with a dessertspoon of wine vinegar, salt and pepper. Whip 175 ml (6 fl oz) double cream, and fold in the vinegar mixture thoroughly. Pour over the vegetables. Top with herrings, onion, egg and parsley. Chill.

GRATIN OF BEETROOT
BETTERAVE AU GRATIN

A dish of mottled purple-red and gold creaminess, with a good flavour. Serve it with toast or bread to mop up the juices, or in a smaller quantity with a firm fish. *Sole au betterave* rings with brave effrontery, but it works well: beetroot has a subtle flavour when not soused in malt vinegar. Other fish to try are brill, cod, haddock, whiting.

> *6 beetroot, boiled, skinned*
> *butter*
> *4 tablespoons grated Parmesan, or 2 of Parmesan and 3 of Cheddar*
> *salt, pepper*
> *cream*
> *breadcrumbs*

Cube the beetroot (or chop it very coarsely for serving with fish). Butter a gratin dish, and sprinkle a third of the cheese over it. Put in half the beetroot with seasoning, then another third of the cheese, and repeat. Pour in enough cream to come to the top of the beetroot (obviously it pays to pack the beetroot into the dish closely), being careful not to dislodge the cheese on top too much. Scatter over a few breadcrumbs and dot with little bits of butter. Bake for about 15 minutes in a moderate to fairly hot oven, until the juices bubble at the side, and the top is splashed with golden-brown.

MRS CONRAD'S BEETROOT AND POTATOES

Not long after the 50th anniversary of Joseph Conrad's death, in 1974, I came across *Home Cookery* by his wife, Jessie. It was her second cookery book, published long after Conrad's death, but happily she included his preface to her first one. He enjoyed his wife's cooking and in their early, hard up, lodging house days he must have appreciated her frugal skill. He concluded that of all the books, only those on cookery are morally above suspicion. Their one aim, he wrote, is 'to increase the happiness of mankind, to add to the cheerfulness of nations'. As negative proof, he adduced the theory that the ferocity of the Red Indians was caused by their wives' lack of culinary skill. 'The Seven Nations around the Great Lake and the Horse tribes of the plains were but one vast prey to raging dyspepsia.'

Mrs Conrad doesn't say so, but I wonder if she had this recipe from her husband, or from someone she met when they were visiting Poland, his native country? It seems to belong more to that part of Europe, with its sweet-sour sauce, than to English cookery, and it is improved by adding dill weed.

> 60 g (2 oz) butter
> 2 small onions, sliced thinly
> 4 or 5 medium-sized boiled beetroot
> 1 tablespoon flour
> 300 ml (½ pt) milk, or milk and single cream
> pepper, salt
> 3–4 tablespoons sugar
> 1–2 tablespoons white wine vinegar
> 2–3 tablespoons double cream
> mashed potato

Melt the butter in a frying pan. Add the onion, cover and cook gently until soft and yellow. Peel and slice the beetroot and place it on top of the onion. Sprinkle over the flour, and stir in the milk and cream. Simmer for 10–15 minutes to cook the sauce properly, turning the beetroot about with a wooden spoon. Add salt and pepper, then the sugar and vinegar gradually, a little bit of each at a time, until the sauce tastes piquant and sweetish, without being over sugary. Stir in the cream. Have ready a hot serving dish, bordered with the hot mashed potatoes, and pour the beetroot mixture into the middle.

If you want to add dill weed, do this when you stir the cream into the sweet-sour sauce.

POLISH BRAISED BEETROOT WITH STUFFED EGGS

The *polonaise* garnish of fried breadcrumbs, chopped hard-boiled eggs, parsley and *noisette* butter makes something special of familiar vegetables – chicory, salsify, cauliflower, cabbage. These stuffed eggs are a variation on this theme, making a larger, more substantial dish. The crisp surface of the fried eggs contrasts deliciously with the soft egg and cream mixture inside, the whole thing setting off the vegetable. The eggs can be prepared well in advance, with only the final frying left until the last minute.

STUFFED EGGS
6 eggs
3 tablespoons double cream
chopped chives and/or parsley
butter, salt, pepper
white breadcrumbs

BEETROOT
1 kg (2 lb) boiled beetroot
2 tablespoons lard
1 onion, chopped
1 tablespoon flour
150 ml (¼ pt) single cream
lemon juice
fresh dill weed and parsley
piece of horseradish root

Hard-boil the eggs and run them under the tap until cool. With a very sharp, thin-bladed knife cut the eggs in half lengthways, being as careful as possible not to damage the shells; the edge is bound to be a little uneven, but this does not matter too much. Scoop the whites and yolks of the eggs into a bowl and set the shells carefully aside. Mash the whites and yolks with a fork, until they are crumbly. Mix in the cream and herbs and a tablespoon of melted butter. Season to taste. Put this mixture back into the shells, smoothing it off. Press the smooth sides of the filled shells on to the breadcrumbs, so that they are fairly thickly coated. Set aside until the beetroot is nearly ready, then fry the eggs bread-crumbed side down in butter, until crisp and golden.

Skin the beetroot and shred it into matchstick strips on a mandolin slicer, or coarse grater. Or chop it coarsely. Melt the lard and soften the onion in it without browning. Stir in the flour, cook a moment or two, then moisten with the cream and add the beetroot with any juice. Heat through, sharpen the flavour with lemon juice to taste and stir in chopped dill weed and parsley, with salt and pepper. Pile it on to a hot dish, surround with the eggs and grate horseradish over the top in fine light shreds – not a lot, but enough to add its earthy piquancy. Serve immediately.

Note If you are not able to get horseradish root, flavour the sauce with a good proprietary brand of prepared horseradish. Keep the flavour light.

BEETROOT IN THE DUTCH STYLE

This recipe, from Sheila Hutchins' *What's Cooking* (1954), is a winner. Serve it with duck, ham, roast pork, sausages, or on its own with triangles of fried bread.

> *2 large cooked beetroot, peeled*
> *1 large onion, chopped finely*
> *60 g (2 oz) butter*
> *4 cooking apples, peeled, cored, chopped*
> *salt, pepper, grated nutmeg*
> *chopped chives*

Chop the beetroot coarsely. Cook the onion until soft in the butter in a covered pan. Add the beetroot, then the apple and simmer for 20–30 minutes until they have blended into a thick purée. Season well, adding enough nutmeg to give a subtle, not quite identifiable flavour. Turn into a dish or into individual ramekins, depending on how you are serving it, and sprinkle with chives.

OTHER SUGGESTIONS

Lamb's lettuce and beetroot salad, see page 284.

BROAD BEANS

In Europe we should feel particularly comfortable with broad beans. They have been bobbing about in our cooking pots certainly since the Bronze Age, perhaps even earlier. They are 'the bean' for century after century, the star of the 'bean feast', and most of our bean phrases until scarlet runners, haricot, kidney and butter beans, arriving from the New World, made a distinguishing adjective necessary. Small, early beans were coarser eating than our improved modern varieties (Dioscorides described them, in the 1st century A.D., as 'windy, flatulent, hard of digestion, causing troublesome dreams').

More lyrically, the white and purple flowers, spreading over the fields in spring, provided a sweet smelling world for lovers. 'My love is as sweet as a bean field in blossom', wrote John Clare, who met his Mary where the 'bean fields were misted wi' dew'. Once they were a sweetness of spring round many villages. Nowadays they are not so common.

Another thing we have lost, not such a sad loss, is dried broad beans. To buy them, you must go to the Middle Eastern delicatessen, where you will find the favourite small dried brown bean of the Egyptians, *ful medames*. They are soaked, boiled until tender and dressed with an olive oil vinaigrette and hard-boiled eggs. Include plenty of garlic, and supply lemon wedges, too.

We relied on dried broad beans before Columbus went to America, for protein and stodge in the early spring. By the regulations of the Sherburn leper hospital, founded in 1181, and a most humane institution, the inmates were given two rasers of broad beans 'for baking' at Quadregisima, the first Sunday in Lent. This works out at a gallon per person, half a bucket full. Presumably it is no accident that Lent with its fasting coincides with the lean time of the year, when stores are nearly exhausted. It seems that the lepers of County Durham did better than many other people of the time, and were never driven to eating the grass or dying of starvation by the community bread ovens as the poor did in France, even if they got thoroughly sick of baked beans.

Luckily for us to-day, broad beans freeze well. They may lose the finest freshness but they taste closer to the real thing than, say, frozen peas. If you skin them after cooking, they are a delightful luxury.

HOW TO CHOOSE AND PREPARE BROAD BEANS

Few things taste better than a dish of new young vegetables, lovingly cooked. No need for meat or apologies. Broad beans are one of the first vegetables of spring, appearing in the greengrocer's in mid-May or early June. By this time they are already becoming large. Only gardeners have the best, as they are free to pick broad beans when the pods are 5–7 cm (2–3 in) long. These young beans should be topped and tailed and the string removed, as if they were French beans. They can then be boiled and eaten whole, with melted butter. Or they can be eaten raw (p. 106).

Medium- and large-sized beans have to be shelled. Wear rubber gloves to do this, or your skin will blacken with the juice from the pods. If the pods are in fair shape, the best bits can be cooked separately for a purée or soup. Serve the purée with grilled bacon and new potatoes, or with grilled lamb.

Deciding the quantities of broad beans to buy can be tricky, as one cannot be sure how well filled they are. The safest thing is to allow a generous 2½ kg, or 5 lb, for six people. Should any be left over, they can be eaten as a salad.

The shelled beans should be boiled in the minimum of salted water, with a sprig or two of savory or parsley. Savory is the favourite herb in France, but be cautious if you have never tried it before; the flavour is slightly bitter, and it can be overdone. My mother always included a couple of pods with the beans – she did the same with peas – and certainly this helps the taste of the cooking liquor if you mean to use it for sauce or soup. Once the beans are cooked, tip them into a sieve with a bowl underneath to catch the juices. Run the beans under the cold tap, then slip off the white skin from the largest ones. This is a bore, but it makes all the difference between eating beans-to-do-you-good and eating beans for pleasure. The small brilliant green kernel of flavour intrigues people who have never eaten them skinned before.

Once the beans grow large, the pods pocked and dented with black, it becomes essential to skin all the beans if they are to be enjoyed at all. Sometimes these older beans are mealy – then the only way out is to sieve them into a thick mass and serve them with plenty of butter and crisp bacon bits, like the bean pod purée I mentioned above. Small pieces of bread fried in butter are a great help, too, with this kind of food, or chopped, fried celery. They add crispness and zest, to lighten the solid nourishment. An attractive way to make a special dish is to put a good spoonful of the purée into a tartlet case and top it with a poached egg, or an egg mollet, and melted butter with parsley and chives.

BROAD BEAN SOUP

125 g (4 oz) chopped onion
1 clove garlic, chopped
60 g (2 oz) butter
½ kg (1 lb) shelled broad beans
2–3 bean pods, black part removed
chopped fresh savory, or sage, or parsley
salt, pepper, sugar
6 tablespoons double cream
lemon juice, chopped green onion stalk

Soften the onion and garlic in butter without letting them colour. Add the beans, a litre (1¾ pt) water, the pods and a few sprigs of whichever herb you choose. When the beans are cooked, remove a tablespoon of them and peel off the white skins. Set aside. Sieve or blend the soup. Reheat with the skinned beans, adding more liquid if necessary to dilute the consistency. Season to taste, with a pinch of sugar. Add a little more of the chopped herb, the cream and a few drops of lemon juice to bring out the flavour, then the onion stalk. Serve with little cubes of bread fried in butter.

BROAD BEAN POD PURÉE AND SOUP

If your broad beans are young and vigorous, but beyond the stage of eating raw, you can turn the pods into a purée for serving with lamb, pork or bacon, or for diluting into soup. This is better than throwing them away, or adding them all to the compost heap. The flavour is earthy and strong, so treat the purée more as a sauce than a vegetable; a tablespoonful or two with meat is agreeable, more than that is overwhelming.

First top and tail the pods, cutting out any black patches. Cook them in boiling salted water with savory if you like the flavour. When they are tender, strain off and keep the liquid.

TO MAKE THE PURÉE Put the pods through the coarse plate of a *mouli-*

légumes, using a little of the cooking liquor if necessary but not too much. Reheat and dry off any wateriness. Add a huge knob of butter or some cream, or both, extra seasoning if required and chopped parsley or savory to taste.

TO MAKE THE SOUP Blend and sieve the pods, using the cooking water to moisten the purée. Or put them through the coarse, then the medium plate, of a *mouli-légumes*. Reheat and dilute with water or ham liquor, to taste. Finish with cream beaten up with an egg yolk. Serve with small cubes of fried bread. Scatter with chopped chives or the green of spring onions.

Note Pea pods can be treated in the same way.

CREAMED BROAD BEANS

If the beans are young and green and tender, not very many to the pod, this is a great treat.

Shell 2–3 kg (4½–5½ lb) broad beans – this may give you ½ kg (1 lb).

Chop half a dozen spring onions or two tablespoons onion, and stew in 100 g (3–4 oz) butter for five minutes. Add the beans. Jam on the lid and leave to cook a little harder for a further five minutes. By this time beans and onions are likely to be cooked, but give them a minute or two longer if necessary.

When they are tender, stir in 4–6 tablespoons whipping or double cream, and a tablespoon of butter. Raise the heat, leave the lid off and cook the juices to a rich creamy sauce. If you overdo this, and it begins to turn oily, put in a tablespoon or two of cold water; this will bring back the consistency. Season with a little lemon juice. Add chopped parsley.

As you will not have a large quantity, it is worth dividing it between six small ramekins, or six heated shortcrust pastry cases. This makes the dish look as special as it tastes. Provide good white bread to eat with it.

BROAD BEANS ON ISCHIA

In May one year we were taken to Ischia to sample the wines. At one vineyard crouched into a scoop of the mountains, they had put up trestle tables covered with cloths. Spread out on the whiteness were huge majolica dishes of blue and white, of swirling green and yellow, holding salami and *prosciutto crudo* in Parma style, cut paper thin. Wide baskets were filled with yellowish country bread, and tiny broad beans, just picked, for eating raw. There were slices of Provolone cheese as well, and a great deal of white wine from the vineyard.

It is difficult to manage so fine a setting, but the food is easy for a summer party – so long as you may command broad beans of perfect freshness, no longer than 8 cm (3 in) long, and home-made bread. Don't forget sea salt crystals to go with the beans. You eat them pods and all. The cured flavours of Parma ham and salami contrast with the fresh earthy flavour of the beans. Buttery cheese, bread and wine soften pleasantly their direct strength of flavour.

BOILED BEANS AND HAM OR GAMMON, WITH PARSLEY SAUCE

When buying your piece of cured pork, ask how long it should be soaked. The Danes claim that their gammon needs no soaking. Nor should salt pork from the brine tub. Cures nowadays are far milder than they used to be, because refrigeration and careful hygiene mean that less salt is used. Most curing nowadays is a cosmetic, rather than a necessity.

> *piece of ham, gammon, bacon or salt pork*
> *3 kg (6 lb) broad beans, or 1½ kg (3 lb) if really small*
> *1 kg (2 lb) new potatoes*
> *butter*
> *parsley sauce, made with some ham liquor (p. 555)*
> *breadcrumbs*

Calculate the cooking time of the pork. Up to 2 kg (4½ lb) allow an hour to the kilo, plus 30 minutes. From 2½–5 kg (5–10 lb), allow 40 minutes per kilo, plus 20 minutes.

Soak the meat if necessary, then drain it and put into a large pot with enough cold water to cover it by 2 cm (¾ in). Bring slowly to the boil and count the cooking time from boiling point. Keep the liquor at a simmering boil and take off any scum that rises after about ten minutes. Half-way through, check the water – if it is unpleasantly salty, pour half away and replace with fresh boiling water.

An hour before the end of cooking time, remove ¼ litre (8 fl oz) of the liquor to use for the parsley sauce and put in the broad beans, whole if they were tiny, shelled if they were the larger ones, together with the potatoes tied loosely in a muslin cloth. If there is no room for the potatoes, they must be cooked separately.

Half an hour before the end of cooking time, remove the piece of pork from the pot. Check the beans and potatoes, removing the whole thing from the stove if they are ready. Strip the skin from the pork, press breadcrumbs into the surface and pour melted butter over them, being careful not to dislodge too many. Place in the oven for 30 minutes to develop a crust at 180–190°C/350–375°F/Gas Mark 4–5, or at a higher temperature for 20 minutes if this is more convenient.

This half hour gives you the chance to finish off the parsley sauce and the potatoes and to peel the white skin off the largest beans.

Put the pork on a hot meat plate. Surround it with the beans and potatoes with a little parsley sauce poured over them. Serve the rest of the sauce in a sauceboat.

Note Peas can be used instead of beans. Serve with a cream sauce.

BROAD TOWN BEANS

Here is a light version of beans and bacon I find appropriate to June.

> 1½ *kg (3 lb) broad beans*
> *butter, flour, milk, cream*
> 125 *g (4 oz) Polish* sopocka, *Danish* hamburgerryg, *or Italian salami, cut paper thin*
> *chopped spring onion or chives, and parsley*

Shell and cook the broad beans. Drain them, reserving their liquid. Now make a velouté sauce with butter, flour, an equal quantity of broad bean liquor and milk, and two or three tablespoons of cream – you need a generous ¼ litre (½ pt). As the sauce simmers, remove the skins from the beans – at least from the largest ones, if you cannot manage them all. Put the beans into the finished sauce to reheat.

Cut the cured pork or salami slices in half, and arrange them round the edge of a hot serving dish of the ovenproof kind. Place the beans and their sauce (keep back a little sauce, if the mixture is on the liquid side) in the middle. Put into the oven for five minutes. Serve with wholemeal bread and salty butter, if possible of the farmhouse type.

BROAD BEANS IN THE TOURAINE STYLE
FÈVES À LA TOURANGELLE

A particularly good French variation on the beans and bacon theme. The

sauce goes a strange, grey-green colour, unusual in cookery, but beautiful.

> 2½ kg (5 lb) broad beans
> bouquet of parsley and tarragon
> 20 small new onions (very large spring onions do well)
> 100 g (3–4 oz) butter
> 150 g (5 oz) lean bacon, chopped
> 2 egg yolks
> sprig of chervil
> 3 tablespoons cream
> chopped parsley and savory

Shell and cook the broad beans with the bouquet of parsley and tarragon
and a few of the better pods. Brown the onions lightly in the butter, then
add the bacon. When it begins to colour, lower the heat, cover the pan and
leave to finish cooking.

When the beans are ready, skin the largest ones unless the outer skins
are tender. Discard pods and bouquet. Beat up 150 ml (¼ pt) of the
strained bean cooking liquor with the egg yolks.

Add the beans to the cooked onions, stirring them well round together.
Check the seasoning. Stir the egg mixture into the pan with the chervil,
and keep the heat very low, so that the sauce thickens to double cream
consistency. Stir in the cream. Check the seasoning again and fish out the
chervil. Turn into a warm serving dish and scatter with chopped parsley
and savory. Serve immediately with wholemeal bread, unless you have
good light bread in the French style.

BROAD BEAN SALAD WITH MUSHROOMS

To be on the safe side, allow 1½ kg (3 lb) beans in the pod for 250 g or 8 oz
of shelled beans.

> 250 g (8 oz) cooked broad beans
> 250 g (8 oz) firm small mushrooms
> 5 tablespoons olive oil
> 1 tablespoon wine vinegar
> 1 teaspoon French mustard
> salt, pepper, sugar
> 4 rashers bacon, crisply cooked
> chopped chives and parsley

Skin the cooked broad beans if they are large. Put them into a bowl. Slice the mushrooms thinly and add them to the beans with the vinaigrette ingredients. Stir up gently but thoroughly and put into a serving dish. Chill briefly. Serve sprinkled with the bacon and herbs.

BROAD BEANS AND RICE

Sometimes this Middle Eastern dish is made with dried broad beans and turmeric to emphasize the yellow colour. Sometimes crushed coriander seed will be added and cooked with the garlic. Sometimes the beans and rice will be cooked and served separately. It can be eaten cold with yoghurt as sauce. It can be well buttered and put on the table with grilled or roast lamb and chicken.

> *2 large onions, sliced*
> *2–3 cloves garlic, finely chopped*
> *6 tablespoons clarified butter*
> *1 kg (2 lb) broad beans, shelled*
> *250 g (8 oz) rice*
> *salt, pepper*
> *butter (if being eaten hot)*

Soften the onion and garlic in the butter, then stir in the broad beans for a minute or two. Mix in the rice, and stir another minute. Pour in ½ litre (good ¾ pt) water, cover the pan with a double muslin cloth and jam on the lid. Simmer slowly for about 20 minutes. By this time the broad beans and rice will be tender and the liquid absorbed (though it is wise to check after 10–15 minutes' cooking, to make sure that it has not disappeared too quickly; if it has, add a little more). Leave over an even lower heat for another ten minutes for the rice to become a little dryer.

Turn out and serve with a good lump of butter mixed in. Or else allow it to cool and put a bowl of yoghurt on to the table as well, and a green salad.

OTHER SUGGESTIONS
Artichokes stuffed with broad beans, p. 26.

BROCCOLI or SPROUTING BROCCOLI

Broccoli are among the best variations on the cabbage theme. Inasmuch as they are flowering shoots, they come closer to cauliflower but taste much nicer. They are as near to a fine vegetable as the brassicas ever get, with the exception of Chinese leaf. This means you can serve good fresh broccoli as a course on their own, with hollandaise or maltese sauce, and everyone will be delighted. As you may have gathered from the name, they were developed in Italy, broccoli being the plural diminutive of *brocco*, meaning a sprout or shoot.

The broccoli we are most used to in this country is the purple-flowering kind. Italians enjoy the white as well and in recent years have added an excellent green-flowering variety that comes in larger clumps than the purple or white, with less leaf, and in shape closer to a small cauliflower. It was first developed in Calabria, according to one authority, as a more resilient kind for canning and freezing. We have been eating it for years as frozen green broccoli, but the fresh that is becoming more familiar in shops and supermarkets under the name of calabrese is far superior. Being fresh it tastes better and, being unblanched, it keeps its beautiful bluish-green colour and tight juicy look.

The man who did most to popularize purple-sprouting broccoli in this country was Stephen Switzer. He designed gardens for the aristocracy, and was famous for his books on the subject. In the 1720s he was running a seed business in London. He got so fed up with writing down instructions for each customer who bought his seeds for 'foreign kitchen vegetables', that he had a small pamphlet printed telling his clients how to grow and cook Italian broccoli, Spanish cardoons, celeriac, Florentine fennel and other items. Evelyn had mentioned the delicate shoots of cabbage in *Acetaria*, in 1699, and Switzer quotes his opinion that the best broccoli seeds come from 'the sea–coast about Naples and Other Italian places'. There was a snag. The Italians often mixed the seeds of these delicacies with turnip seed; they did not regard it as a sin 'to cheat heretics'.

At that time the seeds of the different coloured flowering broccoli – yellow, green and purple – were mixed together indiscriminately, but the purple was the best. Gather it while young, says Switzer, 'not above six or eight inches long, before they come to a flower, and about the bigness of a man's little finger at most'. Next to asparagus, broccoli gives us the best boiled salad we have, 'at least much better than any other kind of sprouts that grow, and is in season at such a time, when nothing else equal to it can be got'.

HOW TO PREPARE AND COOK BROCCOLI

Purple-sprouting broccoli are often sold hugger-mugger in boxes, without the kind of treatment that it received from the shopkeepers of ancient Rome who bunched the stalks attractively, as they did asparagus. If they have a limp and suffering look, buy something else instead. All young sprouts and shoots need to be fresh, whether broccoli or asparagus, bean sprouts or Brussels sprouts.

When you get the broccoli home, pick them over carefully, cutting away tough ends and damaged or coarse leaves. If you want a particularly fine dish of broccoli, strip off all the leaves, just to leave the stalks and the closed flower-heads. Rinse, then peel any hard skin from the stalks. Tie into bundles and stand upright in a metal basket. Heat a large saucepan with 3–5 cm (1–2 in) of salted water. When it comes to a rolling boil, lower in the basket. Sprinkle the tops with salt and pepper. Put on the lid or a piece of foil if the tops come above the rim of the pan. Leave until the stalks are tender and the flowering part has been steamed to perfection. Haul out the basket and leave it to drain.

Calabrese is easier to deal with as it is given grander treatment, particularly in supermarkets. Three or four of the large heads will be trimmed, and covered with plastic film. The price will be higher than for familiar purple-sprouting broccoli, but so will the quality. There will be no waste and you will need do no more than rinse it. The heads are so large, that there is no point in tying them in bundles. They can stand up by themselves in the basket. Cook them in the same way as broccoli.

Unless you live in Italy, or grow them yourself in this country, you are not likely to come across white-sprouting broccoli. It looks like skinny and premature cauliflower, with a green stalk, but tastes more like the purple-sprouting kind. Prepare and cook it in exactly the same way. There is one particularly good Roman recipe for it – modern Roman – as you will see later in the section.

A number of cauliflower recipes can be adapted to the various broccoli: soups, cauliflower *polonaise* and gratins are the most obvious examples.

APICIUS' RECIPES FROM
DE RE COQUINARIA

With such brief instructions as one finds in Apicius' book, modern commentators are bound to differ in their translation and interpretation. The English edition by Barbara Flower and Elisabeth Rosenbaum, translates *cymae* and *coliculi* as cabbage, which is misleading, not to say downright wrong as Pliny makes it quite clear that such things were similar to our broccoli. The French edition by Jacques André takes account of this and on the whole I prefer his versions of the recipes as he knows more about present day Mediterranean cooking which still contains vestiges of the classical Roman style. Recipe no. 6 is a good example of this survival. South Italian and Sicilian recipes for vegetables and fish sometimes make use of the cereal/pine kernel/raisin combination, often with chopped anchovy fillets which takes care of the *garum* flavouring (see recipe 7). M. André suggests semolina in the recipe as a modern version of *alica*; the English edition gives 'spelt-grits'; I use fried breadcrumbs, following the Sicilian style. Do not be horrified by the many flavouring ingredients mentioned – think of the number of spices in a good curry powder or the Algerian *ras el hanout*; use them in delicate quantity. *Passum* and *caroenum* were reductions of wine, to make a concentrated flavouring. *Passum* was sweet, so a fortified wine can be used instead, and *caroenum* was red or white wine boiled down by between a third and a half.

In these recipes, the broccoli shoots were stripped of their leaves before being cooked; sometimes the leaves were chopped and added to the sauce, as in recipes 1 and 2.

1) Cook the broccoli and serve with cumin, salt, 'old wine' and olive oil. If you like, add pepper and lovage, mint, rue, coriander. Or add leaves of broccoli, liquamen(i.e. *garum*), wine and olive oil.
2) Split the cooked broccoli. Chop the tender parts of the leaves, cook them with coriander, onion, cumin, pepper, *passum* or *caroenum*, and a little olive oil. Pour over the broccoli to serve.
2) Put the boiled broccoli in a shallow dish, and dress with *garum*, olive oil, wine, cumin. Sprinkle with pepper, chopped leek, cumin and chopped green coriander leaves. Serve cold or hot.
4) Have ready some boiled leeks. Add them to the cooked broccoli, with *garum*, olive oil, wine and cumin; then heat them through together.
5) Drain the broccoli and return them to a dish that you can put over the heat. Season with *garum* olive oil, wine, cumin. Add some stoned olives (black are best) and bring slowly to the boil.

6) Drain the broccoli, arrange them on a hot serving dish. Scatter over the top fried breadcrumbs, mixed with pine kernels and raisins. This is very good, rather like the *polonaise* treatment that goes so well with cauliflower (browned crumbs, hard-boiled egg and parsley) from the point of view of texture, but with the mild sweetness of pine kernels and raisins. Add the nuts to the crumbs when they are nearly brown enough; plump and heat the raisins with a little boiling water, then drain them and add to the mixture.

7) PATINA OF BROCCOLI Boil and drain the broccoli, then place it in a pan with the following mixture: pepper, lovage, coriander, savory, onion pounded with wine, *garum*, vinegar and olive oil. Heat through then bind the mixture with beaten eggs, cooking them for barely a minute or two. If you like, serve the broccoli on a bed of fish or chicken fillets.

Use anchovy essence or a little crushed anchovy in place of *garum* or, if you have such a thing, one of the south-east Asian fish sauces of similar type.

A patina was a deep dish, of varying shape, in which the ingredients were cooked and it gave its name to the recipe (casserole of pork, terrine of game). A patina could be anything from a purée of vegetables, as in this recipe, to a complex pâté baked in a pastry crust, or the filling for a sweet turnover.

The English edition of Apicius adopts the suggestion made by a German scholar that this recipe should be thickened with cornflour and sprinkled with thyme and pepper. The eggs are better and I cherish this recipe as one that has survived the centuries. A similar mixture, using artichokes instead of broccoli is a favourite of a neighbour of ours in France, see p. 23.

8) BARLEY SOUP Soak some chick-peas overnight (unless you decide to use canned ones). Put them into a pan with green lentils, dried split green peas and pearl barley. Cover generously with water and simmer until tender. Add some olive oil and a good chopping of fresh leeks, coriander leaves, dill weed, fennel leaves, Swiss chard or other beet leaves, mallow and tender broccoli. In another pan cook some broccoli. Meanwhile pound together fennel seed, oregano, asafoetida, lovage, then add a little *garum* (see no. 7). Pour this into the finished soup and last of all add the separately cooked broccoli cut into small pieces.

Barley soup is a similar mixture to some of the soups of southern France to-day. You will have to leave out the mallow (a different variety from our wild mallows, still eaten as a vegetable in parts of the Eastern Mediterranean) and probably the asafoetida, which is the English interpretation of the Latin *silphium* or *laser*, a herb much used in ancient cookery that remains unidentified. It was getting scarce when Pliny was writing about it, and cooks then had to make do with an inferior Persian kind that is pretty certainly asafoetida. It can be bought in the form of essence – very strong, so use it by the drop – from chemists. The English editors observe that it is particularly good with fish, if used with great care, and is still used in Middle Eastern cookery.

RED HOT BROCCOLI
BROCCOLI AL PEPERONCINO

A lively Roman way of cooking white flowering broccoli. It works well, too, with cauliflower; in fact it is the best way I know with cauliflower, which can be a disappointing vegetable. Although Romans serve *broccoli al peperoncino* with the meat course, as a *contorno*, it is quite delicious enough to eat on its own.

Two points – you must use an oil of good flavour. Olive oil is the right thing, but if you are a devotee of walnut oil you will find that its nut flavour blends well with the chilli and pepper. The other thing is the chilli – Romans use two dried ones, complete with seeds, to 1 kg (2 lb) of vegetable. I find this just right, hot without too much of a blaze. If your family are not used to this kind of thing, start by using one chilli only. I also make an addition of my own, of half a sweet red pepper. It adds an extra pleasantness to the flavour and gaiety to the appearance.

> *1 kg (2 lb) white sprouting broccoli, or 1 large cauliflower*
> *2 dried red chillis*
> *½ sweet red pepper, chopped coarsely*
> *olive oil*

Trim the broccoli or cauliflower of dead leaves and stringy or tough stalk, then boil in salted water until just cooked, no more. Drain and run under the cold tap. Separate the cauliflower into florets and cut the stalk into thick strips.

Meanwhile chop up the chillis and put with their seeds and the red pepper into a large frying pan. Add enough olive oil to cover the base in a thin layer. Heat gently for about ten minutes to extract some of the pepper flavours into the oil. Raise the heat to moderate and put in the broccoli or cauliflower. Stir about to reheat; the vegetable should stew rather then fry, it must not colour. Transfer the vegetable to a hot dish, with the pepper and chilli bits, leaving behind any surplus oil. Serve immediately.

BROCCOLI SALAD

Prepare, cook and place the broccoli on a draining plate. Arrange on a serving dish while still warm, and pour over some olive oil vinaigrette. Leave to cool before serving. I think that this kind of cooked vegetable salad is best served cold rather than chilled. Greeks prefer it to be tepid, with the dressing on p. 559.

BROCCOLI ASPARAGUS STYLE

Prepare and cook the broccoli, allowing 1½ kg (3 lb) for six people if it is going to need trimming. With calabrese, 1 kg (2 lb) should be enough for a small first course. Arrange it on a hot serving dish, with the stalks lying neatly. Serve with it a jug of melted butter seasoned with pepper and lemon juice, or a bowl of hollandaise or maltese sauce (p. 556).

A late winter and spring dish that I am particularly fond of, is a mixture of hot cooked vegetables arranged on a huge dish, with calabrese as a main item, accompanied by plenty of hollandaise sauce. It can include mange-tout peas, potatoes, carrots, cauliflower, leaf spinach and so on. Unfortunately it is a nuisance to cook, unless you are resourceful with steaming and packages of foil all sharing a large pan, as each item should taste fresh and good on its own.

BROCCOLI WITH EGGS

This has been a favourite recipe with the English, since the early 18th century when Italian broccoli first came into the country. It is, as you will see, particularly good for calabrese, or fine large heads of purple-sprouting broccoli. The small wispier kind is less effective.

> about ½ kg (1 lb) trimmed broccoli
> 6 eggs, beaten
> salt, pepper
> 125 g (4 oz) butter
> slice cut across a cottage or round loaf

Cook the broccoli, then sort out one or two fine large pieces, splitting the rest into about eight sprigs. Keep them warm. Meanwhile trim the huge slice of bread to fit into the base of an attractive round serving dish. Toast it on both sides, put it into the dish and keep it warm, too.

Finally season and scramble the eggs in the butter (this deliciously large amount is Mrs Rundell's idea; other writers suggest a large piece, which usually means more of an egg-sized lump). Be sure to keep them on the liquid side, as they will cook further in their own heat. Pile on to the toast, making a depression in the centre to take the large head(s) of broccoli. Put the sprigs in a wreath round the edge.

BROCCOLI AND CHICKEN GRATIN

A similar dish in style to the asparagus and chicken gratin on p. 38, but with a different flavour altogether. Quantities of chicken and broccoli – or calabrese, if you like – can be varied, so long as the chicken is not driven into nullity by too much of the strong tasting vegetable. Aim to have an even, gappy layer of green with the pale chicken showing through, before you cover the whole thing with the sauce.

> *Half a roasted chicken*
> *½ kg (1 lb) purple sprouting broccoli or calabrese, cooked*
> *½ litre (¾ pt) thick béchamel sauce*
> *½ litre (¾ pt) lightly seasoned chicken stock*
> *1 dried heaped teaspoon dried tarragon*
> *4–6 tablespoons dry white vermouth or white wine*
> *6 tablespoons whipping or double cream*
> *salt, pepper, grated nutmeg*
> *2 heaped tablespoons grated Parmesan*
> *50 g (scant 2 oz) breadcrumbs*
> *1 heaped tablespoon butter*

Remove the meat from the chicken carcase, which can be used to make the required stock. Cut it into convenient strips and pieces. Arrange them in the lightly buttered base of a large shallow ovenproof dish. Boil down the sauce, stock and tarragon to half quantity, then add 4 tablespoons vermouth (or wine – but vermouth is better), cream and nutmeg, salt and pepper to taste. Take the sauce off the heat; the consistency should be on the thick side.

Add cheese gradually to bring up the flavours rather than to give an emphatic cheese taste. Pour over the chicken and broccoli, scatter the breadcrumbs on top. Melt the butter and sprinkle it over the crumbs. Bake in a hot oven until the gratin bubbles at the edges and the chicken has a chance to heat through properly.

BRUSSELS
SPROUTS

Brussels sprouts are an elegant
miniature cabbage. Their tight
rosettes encrust the thick stalk
of the plant and are shaded by a
high top plume of large leaves. They are something of a mystery vegetable. It seems they were being grown around Brussels in the Middle
Ages; market regulations of 1213 mention them. They were ordered for
two wedding feasts of the Burgundian court at Lille in the 15th century (in
those days the dukedom of Burgundy stretched north to the coast of
Flanders).

Then silence. They do not seem to have caught on in Burgundy as a
whole, which one might have expected. Nor did they appear in French
and English gardens until the end of the 18th century. Cookery books
around the turn of the century, whether by chefs or housewives, do not
mention them. As far as I have been able to find out, Eliza Acton was the
first in England to give a recipe in her *Modern Cookery* of 1845. In fact
she gives several suggestions in one recipe, including the Belgian style of
pouring a buttery sauce over them, or tossing them in butter and a
spoonful or two of veal gravy; she says that this is the Belgian mode as
served in France, which makes one conclude that she had eaten them
when she had spent a year there as a young girl round about 1820. Alexis
Soyer, too, gives a recipe in his *Modern Housewife* of 1849. This bears out
the opinion in the *Treasury of Botany* that they had only become at all
popular in the 1840s.

In America, Thomas Jefferson planted them for the first time in his
garden at Monticello, in 1812. He had been ransacking the gardens and
farms of Europe for new vegetables, since his time in Paris as Minister to
France, from 1784–1789. Had he encountered Brussels sprouts when he
was there, he would surely have included them among the many plants he
introduced to the United States on his return.

The great success of Brussels sprouts in this country has been in
modern times. We serve them now with beef, game, poultry, and especially with the Christmas turkey, when they are often embellished with
chestnuts. We could eat them more on their own, baked in a sauce
mornay, or with plenty of maître d'hôtel butter (p. 551) and sippets of
toast or fried bread. I am not so devoted to Brussels sprouts raw in salads,

though I make an exception for them mixed with grapefruit – just. Otherwise their cabbagey nature and texture is too pronounced and reminds me vividly of the dreaded coleslaw. If you want to make a good cabbage salad, you must use Chinese leaf (q.v.).

Sometimes in January and February you will see sprout tops on sale – this is the plumey head of the stalk that sheltered the clustered sprouts. For cooking directions, turn to the section on spring and winter greens.

HOW TO PREPARE BRUSSELS SPROUTS

When you come to deal with the sprouts, cut away the mess before you put them into a bowl of well salted water. This encourages miniature forms of life to come out of the sprouts. Some people cut a cross in the base, as one does with a cauliflower or cabbage, but I think this is a waste of time. They cook perfectly well without it, and the cutting can lead to a loss of shape and flavour.

Drain the sprouts and cook them in a tightly covered pan with 2–3 cm (1in) of boiling salted water. After eight or ten minutes they will be ready, unless they are mammoth sized. The base part should be nutty textured, not exactly crisp, but slightly resistant in an agreeable way. They should now be drained and finished in the various ways described in the recipes following.

If you are concerned about loss of greenness, start them in enough boiling water to cover them properly, with a very little salt. Boil them hard without a lid on the pan. The idea is that the water evaporates gradually, providing plenty of steam to cook the sprouts lying on the top; you should keep an eye on the pan as the sprouts should not be burned. Taste them and add seasoning, with plenty of pepper (essential to all of the cabbage family). They can be finished according to most of the following recipes.

PROFESSOR VAN MONS' BRUSSELS SPROUTS

In 1818 the Royal Horticultural Society was addressed by Jean Baptiste Van Mons, Professor of Chemistry and Rural Economy at Louvain. He was talking about Brussels sprouts, a fairly new vegetable to his audience. Here is one of the recipes he gave: cook and drain the Brussels sprouts. Put them into a hot serving dish and keep them warm. Cook 60–125 g (2–4 oz) butter, according to the quantity of sprouts, in a frying pan until it turns a delicious nut brown colour (not dark brown or black). Quickly stir in two to four tablespoons of wine vinegar, so that the mixture sizzles together for a few seconds. Have ready a good chopping of green herbs – parsley, chives, tarragon, fennel, thyme. Mix them into the *beurre noisette* off the heat, and pour over the sprouts.

BRUSSELS SPROUTS IN THE OLD STYLE
CHOUX DE BRUXELLES À L'ANCIENNE

Turn the cooked, drained sprouts in a good slice of butter. For half a kilo (1 lb), sprinkle them with a teaspoon of flour, then pour on 150 ml (¼ pt) cream and mix well. Cook a minute or two longer, stirring. Serve with snippets of toast.

CREAMED BRUSSELS SPROUTS
CHOUX DE BRUXELLES À LA CRÈME

A good recipe for when the sprouts are on the large side.

Cook 1 kg (2 lb) sprouts for only five minutes. Drain and run them under the cold tap. Drain again and chop them finely.

Make a cream sauce with 30 g (1 oz) butter, a level tablespoon of flour and 150 ml (¼ pt) whipping or single cream. If you have no cream, you can use top of the milk. Add this to the chopped sprouts with a little ground nutmeg. Simmer in a heavy pan for 10–15 minutes, or put into the oven in a shallow baking dish, or in individual dishes.

Melt 30 g (1 oz) butter in a small pan when the sprouts have finished cooking. Stir in a couple of tablespoons of double cream when the butter is bubbling. Keep stirring and when the mixture thickens pour it over the top of the sprouts.

Scatter over the whole thing some very small cubes of bread that have been fried in butter.

BUTTERED SPROUTS
CHOUX DE BRUXELLES SAUTÉS

A popular French style of finishing sprouts. Cook and drain them well. Melt a large knob of butter in a frying pan that will take the sprouts comfortably in a single layer (this is not a suitable recipe for large quantities). Put in the sprouts and cook briskly for four or five minutes. Shake the pan from time to time to turn the sprouts. Season with plenty of pepper.

Have ready a couple of tablespoons of maître d'hôtel butter (page

551). Put the sprouts into a hot serving dish and dot the butter over them, so that it melts in their heat.

BRUSSELS SPROUTS WITH CHESTNUTS
CHOUX DE BRUXELLES AUX MARRONS

Deal with the chestnuts first. If you are using dried chestnuts, weigh out a quarter of the quantity of sprouts you intend to cook. Put them in a bowl and pour on boiling water. Leave overnight. Next day simmer the drained chestnuts in enough stock to cover them, until they are tender. Then drain them.

If you are using chestnuts in their shells, you will need half the weight of the sprouts. Nick them and boil them in plenty of water for three or four minutes. Turn off the heat, remove a few chestnuts at a time for peeling, leaving the rest to keep hot in the water. They are likely to be tender enough after boiling, but if you feel they are on the hard side, complete their cooking in stock as above.

Finally cook the sprouts in the usual way. Drain them and return them to the rinsed out pan with a large knob of butter and the chestnuts. Gently mix them together over a moderate heat until both are hot again.

BRUSSELS SPROUTS WITH ALMONDS
CHOUX DE BRUXELLES AUX AMANDES

Cook the Brussels sprouts, turn them in hot butter, put on to a hot serving dish and scatter with blanched split almonds fried golden-brown in butter.

BRUSSELS SPROUTS WITH BUTTERED CRUMBS

Cook the Brussels sprouts, drain and put on to a hot serving dish. Scatter generously with breadcrumbs fried golden-brown in butter. Particularly good with game.

GRATIN OF BRUSSELS SPROUTS
CHOUX DE BRUXELLES AU GRATIN

The unorthodox method of the velouté sauce, the unconventional finishing of the gratin, bring out the good qualities of what can be a difficult vegetable. This is not, I think, a dish for reheating. Every vegetable loses something when it is eaten a few hours after it was first cooked; sprouts coarsen in flavour. If you need to catch up on time, make the sauce in advance.

> ½ litre (¾ pt) thick béchamel sauce
> ½ litre (¾ pt) lightly seasoned chicken stock
> 1 heaped teaspoon dried tarragon
> ¾ kg (1½ lb) trimmed Brussels sprouts
> 4–6 tablespoons whipping or double cream
> 2 heaped tablespoons grated Parmesan
> 2 heaped tablespoons grated Gruyère or Cheddar
> salt, pepper, cayenne
> 50 g (scant 2 oz) breadcrumbs
> 1 heaped tablespoon butter

Put sauce, stock and tarragon on to boil in a wide pan, until they reduce by half. Stir from time to time. You do not need to tip the sauce into the measuring jug to check the reduction; it has to be on the thick side, the consistency of double cream that pours without being glue-like. Meanwhile cook the sprouts until just done. The stalk parts should have a certain bite to them. Fit them into a buttered gratin dish in a close, even layer. Stir the cream into the reduced sauce, take it from the heat and add the cheeses gradually to taste. The amount you need will depend on the original strength of the chicken stock. Add seasoning. Pour it over the sprouts. Place under a moderate grill, or in the oven if you have it on. Fry the breadcrumbs in the butter until crisp. When the gratin is bubbling at the edges, scatter on the crumbs and serve immediately.

Chestnuts can be added to the sprouts to make a more sustaining dish.

BRABANT PURÉE
PURÉE À LA BRABANÇOISE

Brabant is the province which centres on Brussels, so it is natural that it should give its name to a recipe for sprouts. In Belgium they are often served with pork, either as a vegetable or in this purée form.

> ½ kg (1 lb) trimmed Brussels sprouts
> 2 tablespoons lard or pork fat
> 1 large onion, chopped
> ½ kg (1 lb) peeled, diced potatoes
> salt, pepper, grated nutmeg
> 60 g (2 oz) butter
> milk

Cook the sprouts for five minutes, then drain them. Meanwhile melt the lard in a heavy pan and soften the onion in it, without allowing it to colour. Mix in the sprouts and potatoes, then pour on ½ litre (¾ pt) boiling water and season with a level teaspoon of salt, a few grinds of black pepper and a pinch of nutmeg. Simmer, covered, for 30 minutes, until the vegetables are really tender. Drain off the liquid. Put the vegetables through the *mouli-légumes* to make a purée and return it to the pan. Reheat, mixing in the butter and a little milk to make a good consistency that is soft, but not at all sloppy.

You can also use this recipe as a basis for soup. Purée the vegetables with their cooking liquor. Reheat in the pan, adding more water, and about 250 ml (8 fl oz) milk to get the dilution you prefer. Add extra seasoning. Stir in the butter just before serving. Do not be tempted to use all milk to dilute the soup, or it will seem heavy and cloying. Milk should always be added to vegetable soups with great discretion; water is often much more satisfactory.

BRUSSELS SPROUT SOUP

The flavour of Brussels sprouts in soup can be too cabbage-like and strong, unless it is softened with potato, celeriac or with potato and a stalk of celery.

> *1 large onion, chopped*
> *60 g (2 oz) butter*
> *375 g (12 oz) trimmed Brussels sprouts*
> *250 g (8 oz) peeled diced potato or celeriac, or potato and a stalk*
>   *of celery, chopped*
> *6 tablespoons cream*
> *chopped parsley*
> *croûtons, or fingers of toasted Cheddar on bread*

Sweat the onion in the butter in a covered pan. When it begins to soften, stir in the quartered sprouts and the other vegetable(s). Cover again and cool for a further three or four minutes. Make sure nothing catches or browns. Pour in a generous litre of water (2 pt) and simmer until the vegetables are tender. Put through the medium plate of the *mouli-légumes*, then through the fine plate if you like a smooth soup. Or use the blender, though to my way of thinking this makes too emulsified a soup; if you are prepared to blend the soup gradually, in small quantities, you may be successful in getting a good consistency, but beware of total smoothness.

Reheat in the rinsed out pan to just below boiling point. Off the heat, stir in the cream and parsley. Serve with the croûtons or toasted cheese.

CABBAGE

Cabbages have been eaten in this country since the time of the Celts and Romans. These were probably of the loose headed, green leaved type. Then round about the 1570s we seem to have had the first Savoys – not from the Savoie but from Holland and as Evelyn pointed out, they were 'not so rank but agreeable to most palates'. The name of Savoy cabbage was introduced into the English language by Henry Lyte in his *Niewe Herball* of 1578. This was not an original work, but a translation of Dodoens' *Cruydeboeck*, and the name was a direct translation of the Dutch *Savoyekool* (*kool* as in coleslaw, etc.). Sir Anthony Ashley, of Wimborn St Giles in Dorset, is recorded as planting in England the first cabbages from Holland, which may have been Savoys or perhaps the white, smooth, hard-packed cabbage that we still call Dutch cabbage to-day.

Cabbage as a food has problems. It is easy to grow, a useful source of greenery for much of the year. Yet as a vegetable it has original sin, and needs improvement. It can smell foul in the pot, linger through the house with pertinacity, and ruin a meal with its wet flab. Cabbage also has a nasty history of being good for you. Read Pliny, if you do not believe me.

How the Celtic inhabitants of Europe regarded the virtues of the cabbage is not known. Wild cabbage is native to the sea coasts of northern France and Great Britain, and they ate that before the Romans arrived. Wild cabbage is very nasty indeed; the rest of the diet must have been stodgy and dull for the Celts to have tolerated it. In *The Englishman's Flora*, Geoffrey Grigson remarks that wild cabbage used to be sold in Dover market, and that it needed several washings and two boilings before it could be eaten. The garden varieties of cabbage brought in by Roman settlers can only have seemed an improvement, as being somewhat less bitter than the wild sea cabbage and larger-leaved. Whether they regarded it as medicine or food, one cannot know.

As Ireland settled into farming, cabbage asserted itself and became an even more important part of people's diet than elsewhere in Europe. So much so that it came to have a share in celebrations at Hallowe'en, the great festival of the year's end, in the form of colcannon (page 134).

Another favourite Hallowe'en dish was champ, a splendid buttery version of mashed potatoes mixed with greenery including cabbage. On the Isle of Man at Hallowe'en mummers would go round and bang on doors with cabbages and turnips stuck on the end of sticks. Then they sang and sang until silence was bought by the householder, with a scone or a potato.

HOW TO CHOOSE AND PREPARE CABBAGE

Cabbage, in a number of varieties, is with us the year round. Usually firm solidity is the thing to look out for, although young cabbages such as 'spring greens' and some early Primo cabbages are much looser. I am a recent convert to the tight round cabbages, and prefer to buy them still in their dark green outer leaves. Dutch white cabbage is always sold trimmed though it seems to taste all right so long as it is not overcooked. Nonetheless if there is any choice in the matter, go straight for the mild and crisp Savoy, with its dark wrinkled outer leaves that look as if they were fresh from some porcelain factory. It is the ideal cabbage for raw salads – always excepting the delicious Chinese leaf – on account of its mild flavour and crunchy texture.

Because of the compact form, cabbage should be sliced before it is rinsed. First cut away the outer leaves and the stalk; discard the withered stringy parts, and slice the rest of this hard part rather thinly. Then slice the inner heart across. Rinse the two lots of cabbage separately and briefly.

If you intend to stuff a cabbage whole, washing it properly is impossible. Rinse the outside as best you can in a bowl of salted water and hope that insects and grubs have found it as difficult as you have to get at the centre. A really firm, unblemished exterior usually denotes a wholesome interior. The problem about stuffing individual cabbage leaves is to get them away from the cabbage without tearing their tight roundness. You can blanch the whole thing, as you do for stuffed cabbage, and then remove the leaves – but this means that you have to use up the inside of the cabbage the same day. It is more convenient to cut away as many leaves as you require, plus a few extra to allow for damage, and blanch them for three minutes on their own; they should be pliable enough to roll without breaking, but not completely cooked. In cutting away the leaves, it helps if you first cut the stalk right out and ease the leaves from the base end.

There are a good number of exceptions to the brief cooking rule, but one has to emphasize it in a book for this country as most of us have such appalling memories of overcooked cabbage in childhood. In France and Germany, long cooked cabbage dishes are much approved. I suppose that *perdrix aux choux* could be described as the best of all cabbage recipes: in it the cabbage is given time to absorb the meaty juices of the partridge.

Many peasant dishes depend on the lengthy cooking of cabbage. Soup-stews such as *potée* and *garbure* require the slow blending of all the vegetables together for the right flavour; nonetheless you may prefer to keep back half the cabbage and put it into the pot at the end of cooking time, so that the soup may also benefit from its crispness. The Christmas cabbage from Schleswig, and a similar dish from the French Marais, are thick purées of cabbage that must first be thoroughly cooked, then squeezed of all juices: this essence or soul of cabbage is finally enriched with butter or cream and is very good.

SAUERKRAUT, CHOUCROUTE or salted cabbage, is described, with recipes, on pages 146 to 153.

MICHAEL SMITH'S APPLE AND CABBAGE SOUP

Apple and meat stock soups have been part of the English repertoire from the Middle Ages to Eliza Acton. They continue still in parts of France. The apple gives a refreshing and light acidity to the soup that is most agreeable. Here is a modern version, with cabbage, by Michael Smith, from his *Fine English Cookery* (Faber paperback, 1977):

> ½ large white cabbage
> 3 medium onions
> 4 large green apples
> 60 g (2 oz) butter
> 1 clove garlic, crushed
> 1¾ litres (3 pt) chicken or vegetable stock
> salt, pepper
> 1 teaspoon caster sugar
> 1 tablespoon chopped green ginger root

Shred the cabbage finely and slice the onions. Peel, core and cut up the apples evenly. Melt the butter in a heavy pan, add the cabbage, onion and apple and stir them up well. Cover the pan and set over a low heat to sweat for about ten minutes. Shake the pan, so that nothing catches or colours.

Add the garlic and stock. Simmer until the cabbage is tender then put through the blender or *mouli-légumes*. Season with salt, pepper and sugar. Reheat, and serve with a little ginger in each soup bowl.

Note If you blend the soup, you may need to add a little more stock when you reheat it. On the whole, I think the *mouli-légumes* is to be preferred.

CABBAGE SOUPS OF SOUTHERN FRANCE
POTÉE, GARBURE, SOUPE AUX CHOUX

Cabbage soup is a favourite item of peasant food in southern France, particularly in the central and south-western part that curves from Atlantic to Mediterranean above the Pyrenees. When I first began to think about cookery, this surprised me. I had always thought that cabbage soup was a Protestant form of torture, entirely appropriate to wet northern climates and plain living. Of course the Auvergne and part of Languedoc and the Limousin are wet too, even wetter than England, and they have their Protestant pockets, but at least their *potées, garbures* and *soupes aux choux* are enriched with collections of cured pork joints and various devices to enrich the basic mixture of cabbage, potato and water.

Indeed the two words, cabbage soup, may cover few or many ingredients. In 1364–5, boys training to be priests at Trets in Provence, boys receiving a privileged education by the standards of the time, were served cabbage soup on 125 days of the year. Put another way, they could look forward to – or anticipate – cabbage soup one day in every three. We only have the housekeeping documents for this school. We do not know the recipes that were used, nor do we have any of the boys' letters home. Was that soup just cabbage and water (there were no potatoes in Europe in the 14th century)? Or did it include a little meat or some meaty bones (meat was served on 217 days of the year)? Was it padded with bread so that the ladle stood upright in the huge pot? Was it full of subsidiary details and pleasant devices and aromatic herbs? Let us hope, for the sake of those long forgotten children, that it was something like the *potées* and *garbures* of the following recipes – though I have my doubts.

There is little difference between a *potée* and a *garbure* and a *soupe aux choux*. They are all vegetable soups, with the dominating vegetable cabbage, flavoured with more or less pickled pork according to the resources of the household. In Guyenne, *confit d'oie* or *de canard* is sometimes added; this is the goose or duck that has provided *foie gras* – it is jointed, salted, cooked and preserved in its own copious fat. In this country one may buy it in tins. You can make it yourself, but our geese and ducks do not have the same quantity of fat as the force-fed birds destined for the production of *foie gras*. Other districts will put in smoked boiling sausages, or fried fresh chipolata sausages.

. All these things are stewed in a large pot slowly for several hours, from two to four hours according to the housewife's convenience and the heat of her stove. They make an ideal farm lunch, as everything can be pushed into the pot at breakfast time, with water to cover generously, and be left to stew gently without supervision until half an hour before the midday meal, when special details may be added, such as *farcis* or the sausages and potatoes. A modern solid fuel stove or oil-fired cooker is ideal for this kind of dish; it can stew on top or in the oven.

When mealtime comes, the soup with a little of the vegetable will be drunk first, sometimes thickened with slices of bread and flavoured with a little grated cheese. After that the meat and the main of the vegetables will be eaten, together with the rolls of *farci* or the maize meal dumplings know as *miques*.

I am quoting various lists of ingredients, to give you an idea of the kind of thing you put in, but remember that you can vary them as you like. The Auvergne is said to be the *fons et origo* of such soups, so let us start with a couple of recipes from those parts.

POTÉE D'AUVERGNE

> 1½ kg (3 lb) bacon hock
> half a pig's head, salted or fresh
> ½ kg (1 lb) salt pork, preferably spare rib chops or loin chops
> 2 firm green cabbages, or 1 large one
> ½ kg (1 lb) carrots, sliced
> 4 small turnips, diced
> ½ root celeriac, diced, or 3 stalks celery, chopped
> bouquet garni
> salt, pepper, grated nutmeg
> 4 potatoes, preferably Desirée, sliced
> smoked sausages or one long smoked sausage

If the salt meat is likely to be very salty, soak it in tepid water for a couple of hours, or overnight if the cure was a long one. The average salted meat in shops in this country should need no soaking. Slice and blanch the cabbage for five minutes in boiling water, then drain. Put into a big pot, cover with boiling water generously, add the other vegetables, bouquet and seasonings. Half an hour before the soup is to be served, put in the potatoes and sausage.

SOUPE AUX CHOUX AUVERGNATE

> ½ kg (1 lb) salt pork
> salted pig's hock, or bacon hock
> ½ kg (1 lb) breast of mutton or lamb
> cabbage and vegetables as above
> 1 onion stuck with 2 cloves
> bouquet garni and 2 cloves garlic
> salt, pepper
> 4 potatoes (approx ½ kg, 1 lb), sliced
> smoked sausages

Follow the above recipe, but do not blanch the cabbage as you want its full flavour. Serve the meat with pickled gherkins. If you like, the vegetables when extracted from the soup may be quickly fried in a little lard, to give them some richness. They are then served with the meat in the usual way.

To this and the recipe above, you may add lentils – a couple of handfuls. They should be the slatey-green lentils of the Auvergne, not the orange ones that turn rapidly to a purée.

GARBURE BÉARNAISE

> 1 large cabbage, sliced
> 4 large potatoes, left whole
> ½ kg (1 lb) carrots, sliced
> 2 large onions, sliced
> 4 small turnips, diced
> a bundle of 3 leeks tied together
> 100–150 g (3–5 oz) haricot beans, soaked
> 1 chilli pepper, fresh or dried
> bouquet garni
> 1 kg (2 lb) piece salt pork or green bacon
> a drumstick and thigh of confit d'oie, or confit de canard, with its
>   surrounding fat

Boil everything together, apart from the confit, with plenty of water in the usual way. Add the confit half an hour before the meal.

GARBURE GASCONNE

> *2 onions, sliced*
> *2 turnips, sliced*
> *4 potatoes, cubed*
> *half a firm green cabbage*
> *goose fat (duck fat or pork dripping make a reasonable substitute)*
> *a piece of salt pork*
> *a joint of* confit d'oie, de canard

Fry the vegetables in some goose fat until they colour very lightly. Keep stirring them about, over a good heat. Add 3 litres of water (generous 5 pints), and put in the pork. Simmer in the usual way. Half an hour before the meal, add the *confit*.

When everything is cooked, slice the pork and *confit* thinly. Into a large ovenproof pot, layer the meat and vegetables with thick slices of bread. Pour the liquid over the top, sprinkle with grated Gruyère and brown lightly in a very hot oven or under the grill.

I am not one for these solid peasant soups, thickened with bread. Instead a few slices of toasted bread, sprinkled with grated cheese, floated on top of the soup, seem a better idea. The bread should be rye or coarse country bread; a light wholemeal works well, too.

POTÉE AU FARCI

Make a *potée* or *garbure* according to one of the recipes above, but before you start remove half a dozen or more of the largest cabbage leaves. Blanch these leaves in boiling, salted water for a few moments so that they become supple. Make a stuffing – this is the *farci* proper – by soaking the crumb of three thick slices of bread, about 150 g (5 oz), in milk to make a paste. Add about the same weight of minced pork or gammon with a rasher of bacon (or use left-over chicken or beef), a heaped tablespoon of finely chopped onion, 2 chopped cloves of garlic, plenty of chopped parsley and 3 large egg yolks, with seasoning. Roll up knobs of this stuffing in the cabbage leaves and tie them with thread. Or make one large package by overlapping the leaves. These bundles are known as green chickens in some parts of south-western France. Put them into the soup about 45 or 30 minutes before it is ready, instead of sausages. They are to disguise and make up for a shortage of meat.

To serve, slice one or two of the little rolls and add them to the soup. Eat the rest with the meat and vegetables.

BUTTERED CABBAGE

Slice and wash a large firm cabbage. Put 2 cm (¾ in) salted water on to boil. Place the sliced outer leaves and stalk in the pan, give them three minutes vigorous boiling, tightly covered, then add the sliced heart and boil for five minutes, or until you judge the cabbage is cooked but still a little crisp. Drain off the water thoroughly. Melt a really large knob of butter the size of an egg in the rinsed cooking pan, put in the cabbage, plenty of black pepper and a little nutmeg. Replace the lid and set over a low to moderate heat for a few minutes. Shake or stir the cabbage about, so that it is completely mixed with the butter and seasonings. Serve immediately. The best cooked cabbage in the world will flop and loose its charm, if it is left to hang about in the oven. If there is to be a delay, hold back the buttering process until just before the meal.

With this kind of attention, cabbage keeps that nutty fresh flavour that can be so good. Wateriness is the great sin against cabbage. Butter and a good grinding of black pepper at the last moment, just before serving, are what save it.

CREAMED CABBAGE FROM THE AUVERGNE AND SCHLESWIG

I suppose one should expect vegetable recipes to be similar, at any rate in northern Europe, but I was surprised to come across an almost identical recipe from the centre of France and the very north of Germany. Two damp districts would naturally have plenty of cabbages, but the way of cooking them is unusual. The Schleswig finish is the more interesting one. It goes back to the days when sugar and spices were great luxuries for ordinary people, to be kept for special occasions when one could justify extravagance, such as Christmas. This cabbage with sugar and cinnamon is eaten with sausages or pork chops, and goes with them well as pork benefits from spicy sweetness. In the Auvergne, too, cabbage and pork are frequent partners, but sweetness is not as popular as it is with more northern people, though sometimes chestnuts do supply it.

1 large firm cabbage

AUVERGNE FINISH
about 150 ml (¼ pt) thick béchamel sauce
up to 100 ml (3 fl oz) double cream
salt, pepper

SCHLESWIG FINISH
125 ml (4 fl oz) double cream
salt, pepper
1 tablespoon sugar
1 teaspoon ground cinnamon

Cook the cabbage in boiling salted water in the usual way. Leave it to drain and cool in a colander. As soon as it can be handled, squeeze it and turn it with your hands until all moisture is driven out and you have a small heap of cabbage left. Put it through the fine blade of the mincer, pouring off any extra moisture that comes out.

For the Auvergne style, mix in the very thick sauce to bind it – don't swamp it. Then add cream on the same principle. Season to taste and reheat.

For the Schleswig style, mix in the cream. Reheat and season the cabbage, and serve with the sugar and cinnamon mixed together in a small bowl. People can then help themselves according to their own tastes.

DR KITCHINER'S BUBBLE AND SQUEAK,
OR FRIED BEEF AND CABBAGE

Dr William Kitchiner published *Apicius Redivivus, or The Cook's Oracle*, in 1817. It is the raciest, most opinionated, least practical cookery book ever written. 'Every individual who is not perfectly imbecile,' declared the doctor, 'and void of understanding is an Epicure in his own way – the Epicures in boiling of Potatoes are innumerable.' He did not despise cheap dishes, family food, but had no time for puritanism. He tells the tale of Descartes and a lively Marquise who was surprised to see a philosopher enjoying some delicious food – presumably she considered plain living would have been more appropriate to so high a mind. 'Do you think,' said Descartes, 'that God made good things only for fools?'

Dr Kitchiner was not strong on vegetable cookery. He belonged to the plenty-of-water school. But he knew how to choose vegetables: 'I should as soon think of roasting an Animal alive, as of boiling a Vegetable after it is dead'. Here is his masterpiece from *The Cook's Oracle*, a recipe for bubble and squeak, fried beef and cabbage. It starts with a quotation from Peter Pindar (which is followed by a tune):

> 'When midst the frying Pan, in accents savage,
> The Beef so surly, quarrels with the Cabbage.'

'For this, as for a Hash, select those parts of the joint that have been least done; – it is generally made with slices of cold boiled salted Beef, sprink-led with a little Pepper, and just lightly browned with a bit of Butter in a fryingpan: *if it is fried too much it will be hard*.

'Boil a Cabbage, squeeze it quite dry, and chop it small; take the Beef out of the frying-pan, and lay the Cabbage in it; sprinkle a little pepper and salt over it; keep the pan moving over the fire for a few minutes; lay the Cabbage in the middle of a dish, and the Meat round it.'

This was served up with Wow Wow sauce, a velouté made with beef stock, sharpened with a tablespoon each of vinegar, mushroom ketchup and port wine, and a teaspoon of made mustard. Finally you add plenty of chopped parsley and two or three pickled gherkins. There is also a list of extra and alternative sharp flavourings, that would certainly have con-vinced any foreigner reading the book that he was right about the un-subtlety of our cookery. The name, Dr Kitchiner's invention I think, and typical of his humour, comes rather from the noise made by those who first encounter the piquant sauce than from the sound of the ingredients cooking.

COLCANNON, THE NORTHERN BUBBLE AND SQUEAK

In Ireland, Colcannon was a favourite dish for fast-day celebrations, when meat could not be eaten. At Hallowe'en, the eve of All Saints' day, a wedding ring would be pushed into the crusty mass; whoever found it would be married within the twelvemonth. It is best to make it from freshly cooked vegetables, best of all from cabbage and potatoes straight from the garden.

> *1 bowlful cooked potatoes*
> *1 bowlful lightly cooked cabbage*
> *1 large onion*
> *dripping, lard or butter*
> *salt, pepper*

Push the potatoes through a *mouli-légumes*, sieve or ricer. Chop the cabbage. Mix the two together thoroughly. Cook the onion in the fat in a frying pan – a non-stick one, if possible. When it is soft and lightly browned, press in the potato and cabbage to form an even layer. Sprinkle with salt and pepper. When it is nicely coloured underneath and crusted, cut into pieces – with a wooden spatula if you're using a non-stick pan – and turn them over to form a fresh layer. Repeat until you have a green and white marbled cake, specked with crisp brown bits. (At this stage the wedding ring was pushed into the colcannon.)

Turn it on to a heated dish to serve, with lightly fried slices of beef if you like. We sometimes have it with sausages. Or have it on its own with a bit of butter.

Note Brussels sprouts and other forms of cabbage can also be used.

STUFFED CABBAGE LEAVES LIMOGES STYLE
CHOU FARCI À LA LIMOUSINE

The simplest of the stuffed cabbage leaf recipes, from one of the great chestnut districts of France, is the best. The country round Limoges, on the rainy edge of the Auvergne, the Central Massif, has many Spanish chestnut trees. In the moist air of spring the smell of flowers can be oppressive.

This combination of crispness, softness and piquancy is perfectly balanced. It makes a good supper dish, or first course, and the small cabbage rolls can be used to decorate and enhance a roast chicken or turkey. At Christmas time they can be prepared the previous day and kept in the refrigerator until an hour before the meal.

> *36 cabbage leaves*
> *750 g (1½ lb) chestnuts, peeled, or 350 g (¾ lb) dried chestnuts, soaked overnight, cooked*
> *12 rashers smoked streaky bacon*
> *125 g (good 4 oz) butter*

Blanch the cabbage leaves in boiling salted water for two minutes, then drain and pat them dry. Spread out the blanched leaves and divide the chestnuts between them (it does not matter if they have broken up a little). Cut the bacon into 36 pieces and lay them on top. Roll up and place closely together in a large pan in layers. Pour in enough water to come within 1 cm (½ in) of the top. Dot with a third of the butter, put a plate on top and cover. Bring to the boil and simmer for 25 minutes. Remove the lid so that the liquid evaporates to a smallish amount of juice – a further ten minutes. Place the rolls on a dish or round the bird. Whisk the remaining butter into the pan juices and then pour it over the cabbage rolls.

Note The bacon provides enough salt.

CABBAGE LEAVES WITH CARAWAY CHEESE STUFFING

Here is a slightly different way of cooking stuffed cabbage leaves. You can if you like follow the method of the previous recipe, but the flavour will not be quite so good. The creamy filling does not need long cooking and because it is richer than a rice stuffing, you should find 24 little rolls quite enough for six people.

> 24 large cabbage leaves
> 175 g (6 oz) chopped onion
> 1 heaped tablespoon butter
> ¾ kg (1½ lb) curd cheese
> ½ teaspoon thyme
> up to 2 heaped tablespoons caraway seeds
> 1 heaped tablespoon paprika
> salt, pepper
> 2 eggs
> oil

Blanch the leaves in boiling salted water for seven minutes, until they are just tender and pliable. Drain and dry them on a clean cloth or kitchen paper. Cook the onion gently in the butter until soft. Mix the cheese with the thyme and add the caraway seeds gradually, to your taste. Mix in the warm, soft onion and the paprika. Season with salt and pepper. Beat in the eggs. Fill and roll up the cabbage leaves. Fry them until lightly brown in a little oil – allow 15 minutes for this until they are thoroughly heated through. Serve with a tomato sauce, coarsely sieved.

Note Curd cheese must be used, or cottage cheese with a little cream. Full fat cream cheese is no good, as it melts in heat.

STUFFED CABBAGE
CHOU FARCI

Every French housewife has her own version of this recipe and variations on her version as well. It's a dish that comes in handy for using up the last meat from the *pot-au-feu* or Sunday's *rosbif*, and any other little bits left in the larder. Better still, it can be made with sausage-meat from the pork butcher, eked out with cooked rice or breadcrumbs, or with chopped spinach and chard in provençal style. Chicken livers can be added for elegance, so can sliced mushrooms cooked in butter. Having made and eaten a good number of *choux farcis*, I conclude that the simplest ingredients are best so long as they are of good quality. In this country, for instance, Cumberland sausage provides a delicious filling for a stuffed cabbage, otherwise I am inclined to stick to rice fillings in the Greek style and leave out meat altogether.

> *1 large firm cabbage*
> *½–¾ kg (1–1½ lb) stuffing, bound with 2 small eggs if it is very loose*
> *½ litre (¾ pt) stock, or half wine/half water*
> *1 onion, sliced*
> *1 carrot, diced small*
> *2 cloves garlic (optional)*
> *bouquet garni*
> *salt, pepper*
> *square of thin pork back fat, or 3 rashers very fat bacon*

Trim the cabbage. Put on a large pan of salted water to boil and when it rolls vigorously add the cabbage, cover the pot and leave for about ten minutes. The cabbage should be softened enough for you to open it out. Drain it, run it under the cold tap and drain again.

Have the stuffing ready. Gently open out the cabbage from the centre – cut a little away if you like – and put a tablespoon of stuffing into each leaf, working outwards. Using your hands, mould the cabbage back into shape as best you can. Tie it together with string. If you intend to cook the cabbage in a respectable, tight-fitting pot that can be put on the table, tying it up with string is quite adequate. If you are going to transfer the cabbage from its pot to a serving dish, then put two long wide straps of double foil into the pot, at right-angles to each other, so that you can raise the cabbage at the end without too great a risk of disaster. Alternatively you could use a chip or vegetable-blanching basket.

Into the cooking pot put the liquid, onion, carrot, garlic if used, bouquet and seasoning, plus any extras you may decide to add (tomatoes, chopped leek, bits of bacon or ham). Put the cabbage on top, and place the pork fat or bacon over it to keep it basted. Fasten the pot closely using foil as well as the lid to make a good seal. Put into a slow oven, 150°C/300°F/Gas Mark 2, for three hours, more or less. This is a patient dish, that will suit itself to your convenience.

Before serving the cabbage, discard the fat or fatty bacon and cut away the string. Taste the cooking liquor and adjust the seasoning. If you like a little sharpness, add capers or olives and heat for a further five minutes. If you want to make a meaty addition, put in sliced garlic sausage, frankfurters or chorizo, and give them a further ten minutes. A less orthodox but successful addition is a couple of bacon or gammon chops, cut into strips.

Serve with plenty of good bread and a bottle of *vin ordinaire*.

PROVENÇAL STUFFED CABBAGE
SOU FASSUM (CHOU FARCI)

Here is a different version, giving another method of stuffing the cabbage and cooking it. String will not do this time; you need a piece of doubled muslin (an old muslin nappy is ideal).

> 1½ kg (3 lb) firm round cabbage
> 2 litres (3½–4 pt) beef stock

> STUFFING
> 175–200 g (6–7 oz) salt pork belly or green streaky bacon, chopped
> 3 tablespoons olive oil
> 2 medium onions, chopped
> ¾ kg (1½ lb) high quality sausages
> 2 tomatoes, skinned, chopped
> 100 g (3–4 oz) rice, or breadcrumbs
> 60 g (2 oz) raisins
> 125 g (4 oz) peas (optional)
> 1 large clove garlic, finely chopped
> salt, pepper

Cut out the hard stalk of the cabbage and remove any damaged leaves. Blanch the cabbage for ten minutes in plenty of boiling salted water.

Spread out the muslin. Drain the cabbage and run it under the cold tap until you can handle it. Cut away the outside leaves. Lay them on the muslin overlapping each other in a couple of layers, to make a big circle. Squeeze the cabbage heart as free of moisture as you can, chop it and put it into a basin.

Now attend to the stuffing. Cook the pork or bacon in the oil until it looks transparent. Add the onion and stew for ten minutes – cover the pan if you like – without browning anything. Add to the cabbage. Skin and break up the sausages and mix them into the cabbage, then mix in the remaining stuffing ingredients. Use your hands rather than a spoon. Form the stuffing into a large ball and place it in the centre of the cabbage circle. Mould up the cabbage leaves to enclose the stuffing and conceal it. Tie the muslin firmly on top.

Bring the beef stock to the boil in a large pan. Put in the cabbage. Bring back to the boil and adjust the heat to give a steady simmer. Put the lid on the pan and leave for two and a half to three hours.

To serve, lift the muslin package to a colander, untie it and carefully remove the cabbage to a deep serving dish. Spoon a little cooking liquor round it.

In Provence, the cooking liquor provides the soup. It is poured over slices of bread in individual soup plates, and the cabbage is cut into wedges like a melon and eaten with it. You could separate the two courses – I think this is easier and pleasanter – or keep the liquor for soup next day. Any cabbage left over is eaten cold in slices with a vinaigrette: rather a hefty dish. I find stuffed cabbage is better eaten hot.

STUFFED CABBAGE IN THE TRÔO STYLE

I went into our neighbour's house in France one Saturday, and remarked on the wonderful smell. Madame Glon was making a *chou farci* as her grand-daughters were coming to supper. She lifted the lid of the red enamelled pot to show me. I expected to see a round cabbage, swollen out with the filling, in the style of the previous recipes. But all I could see was a flat layer of cabbage, bubbling and heaving gently. She soon convinced me that her quick method was just as good as the classic one; certainly it is now one of our favourite dishes. I have tried it with various spicy and aromatic additions, tomatoes, bits of bacon, herbs and so on, but reject them all in favour of the Trôo simplicity of cabbage, good sausagemeat and butter. It has a pure directness that is just right, and cannot be improved. One can use either a Savoy or Dutch cabbage; or a Chinese cabbage for greater delicacy. Adjust the blanching and cooking times accordingly.

> 1½–2 kg (3–4½ lb) cabbage
> ¾ kg (1½ lb) good sausages or sausagemeat
> salt, pepper, butter

Cut the cabbage across into slices, and blanch it for five minutes in hard-boiling, salted water. Drain it and run it under the tap to prevent further cooking at this stage. Butter an ovenproof or flameproof pot (according to whether you intend to cook the dish on top of the stove or in the oven). Put in a third of the cabbage. Remove the skin from the sausages if necessary, and put half over the cabbage. Then repeat, and cover with a final layer of cabbage. Season each layer as you go. Dot over with butter and cover tightly. Bake at 150°C/300°F/Gas Mark 2, for two and a half hours, or simmer on top of stove very gently. With a Chinese cabbage, you would need two to two and a half hours.

CABBAGE IN THE DUTCH STYLE

The important thing about this recipe is to buy good sausages (indeed it is important for all cabbage recipes in which they appear). I use a Cumberland sausage that is made in Cheltenham. Happy are those who can use a Cumberland sausage made in Cumberland, in Carlisle, or in Keswick. It is Justin de Blank's recipe and naturally he prefers his own sausages. He points out that the recipe works with ordinary cabbage, Dutch white cabbage and even spring greens. Not being an enthusiast for spring greens, I go for Dutch or Savoy cabbage, as crisp and firm as possible.

> ¾ kg (1½ lb) sausages
> 1 large cabbage
> 2 heaped tablespoons flour
> ½ litre (¾ pt) milk
> salt, pepper
> 2 heaped tablespoons each grated Parmesan or Cheddar, and
> Gruyère

Grease a roasting tin with lard, put in the sausages after pricking them and bake at 180°C/350°F/Gas Mark 4 until brown all over. Turn them occasionally. They should lose a fair amount of fat. Drain them well.

Meanwhile slice and blanch the cabbage until it is half cooked, but still crisp. For this dish, unlike the French dishes, the cabbage should retain a certain bite. Drain it well, too.

Mix the sausages and cabbage together in a gratin dish and return them to the oven while you make a thick sauce, with three tablespoons of fat from the sausages, the flour and the milk. Allow it to cook thoroughly, about 10–15 minutes, stirring most of the time. The sauce really has to be thick, as the cabbage will still be exuding some moisture however well you drained it.

Season the sauce and add most of the cheese. Pour it over the cabbage and sausage, mixing everything gently together. Scatter the top with the remaining cheese and return the dish to the oven. Turn up the heat to

220°C/425°F/Gas Mark 7 to brown it. 'Just time for another quick drink, then on to the table bubbling hot.'

CABBAGE IN THE STYLE OF POITOU
LES CHOUX À LA POITIVINE

A good recipe, especially if you are cooking a goose, a farmer's version of the classic French *perdrix aux choux*. The cabbage is cooked long and slowly with the meat, so that the two flavours melt together. The proportions of cabbage to goose or pork can be varied.

> 200 g (7 oz) fat smoked streaky bacon in a piece
> 1 goose, jointed, or a piece of boned roasting pork from fillet or loin
> 1 very large firm cabbage, sliced
> salt, pepper, nutmeg
> 1 clove garlic, chopped
> 1 large onion, sliced

Chop the bacon into little bits then put it into a large frying pan and heat until the fat runs. Raise the heat and brown the goose or pork. Remove them to a plate and pack in the cabbage a little at a time, adding more as it begins to collapse in the hot juices; it should catch the heat slightly, but not come anywhere near burning. Put half the cabbage into an enamelled iron or stoneware casserole, then lay the meat on top with seasoning, garlic and onion, and the last of the cabbage on top of that. Cover closely with foil and a lid. Either leave the pot on top of the stove to simmer or put it into the oven as for cabbage in the Trôo style. Test the meat after an hour and a half. A solid joint of pork will almost certainly need two hours. There is no point in hurrying this dish, it should gently cook together. Taste the cabbage and correct the seasoning.

If you used pork, remove the joint from the pot, cut it up and arrange it in neat slices on top of the cabbage. Serve with plenty of bread.

PARTRIDGE WITH CABBAGE
PERDRIX AUX CHOUX

When made in the grand classic style, *perdrix aux choux* is the least plebeian way of eating cabbage. This is not difficult if you are well supplied with partridges. Pick out an elderly bird and follow the recipe as given below, using this bird alone. At the end of cooking time, it is discarded and used up in croquettes (Escoffier's suggestion) or in making a secondary stock. Two or three freshly roasted young partridges are then placed on top of the cabbage, which is permeated with the rich flavour of the elderly bird.

For town housewives who are lucky to have a brace of uncertainly tender partridges, this grand treatment is not possible. The thing is to keep an eye on the pot and to remove the partridges when they are cooked, leaving the cabbage to simmer on in the juices. They can be cut up and reheated in the sauce at the end. The recipe can be adapted to other game birds. Or you could use guineafowl, with good game rather than beef stock.

> 1 large Savoy cabbage, trimmed
> 1 brace partridges
> 2 small onions, each stuck with a clove
> 175–200 g (6–7 oz) fat salt pork, or fat green streaky bacon
> 2–3 tablespoons lard or bacon fat
> 2 wide pieces of pork skin
> 175–200 g (6–7 oz) carrots, peeled
> salt, pepper
> ½ litre (generous ¾ pt) beef stock
> 100 ml 3½ fl oz) dry white wine (optional)
> 250 g (½ lb) chipolata sausages
> 1 heaped teaspoon arrowroot or cornflour
> 1 tablespoon butter

Quarter the cabbage, without quite cutting through. Blanch it in plenty of boiling salted water for 15 minutes, then run it under the cold tap and drain it well. Stuff the partridges with the onions, cut the pork or bacon into strips and brown them all in the fat. Put a strip of pork skin into an ovenproof pot that will just hold the cabbage and partridges. Put the cabbage on top and gently open it out to make a nest for the birds and fried pork or bacon. Arrange the carrots in with them. Press the cabbage up and over the partridges, if you can – i.e. if the cabbage is large enough and the pot close-fitting. Place the second piece of pork skin on top, add seasoning, stock, and wine if used. Cover tightly, jamming on the lid with foil. Place in the oven at 150°C/300°F/Gas Mark 2 or slightly under and leave for two or three hours. Remove the partridges when they are cooked – check after one hour. 40 minutes before the end of the cooking time, brown the chipolatas quickly in the fat left from browning the birds and add them to the pot.

To serve, remove the cabbage to a serving dish, draining it back into the cooking pot as best you can. Put the partridges, whole or cut up, in the centre and surrounded by the sausages. Cut the carrots into slices and add them, too. If you like the pork skin, as I do, cut the two pieces into strips and arrange them with the other things; if not, discard them.

Last of all attend to the sauce. Skim the fat from the cooking juices, taste them and see how salty they are. If they are on the mild side, boil them down to concentrate the flavour; as the dish is moist in itself, you do not need a lot of sauce. You can now serve the juices as they are, or you can thicken them with arrowroot or cornflour and finish them with the butter. If the juices were just right, or on the salty side, pour 300 ml ($\frac{1}{2}$ pt) into a small pan, and thicken them with arrowroot or cornflour and stir in the butter last of all. Pour some of the sauce over the birds; serve the rest in a sauceboat.

SAUTÉED PHEASANT WITH CABBAGE

This very delicious recipe comes from the American cookery writer, James Beard. I use it as adapted by Julia Drysdale in her *Game Cookery Book* (Collins, 1975). Instead of pheasant, you could use tender partridge or guineafowl; wood pigeons would also do, except that they would need a much longer cooking time, after being browned, with some game stock added to the pan. Reduce it before adding the cream, when the pigeons are tender.

> *1 pheasant, jointed*
> *4 tablespoons butter or bacon fat*
> *1 medium cabbage*
> *salt, pepper*
> *1 teaspoon slightly crushed juniper berries*
> *250 ml (8 fl oz) double cream*
> *$\frac{1}{4}$ teaspoon paprika*

Brown the pheasant pieces in the fat. Cover the pan, lower the heat and leave to stew for 20 minutes. Slice the cabbage thinly and blanch for ten minutes; drain well and add to the pheasant with seasoning and juniper berries. Cover again and cook for a further ten minutes. Pour in the cream and simmer for a further five minutes, with the lid off the pan if there is a lot of juice. Sprinkle with paprika just before serving. Boiled potatoes go well with this dish.

PELMENI WITH CABBAGE FILLING

Pelmeni are small pastries from Siberia and northern Russia. The dough circles are folded over and the two points pinched together, rather in the shape of Italian *tortellini*, sometimes called 'navels of Venus' from their plump curled shape. The thing about pelmeni is that they must be frozen before they are cooked. In *Russian Cooking* (translated by F. Siegel and published by Mir), it says that 'when the frosts set in, pelmeni are made in quantities running into thousands. They are laid on floured boards, carried outside to freeze, then packed in large sacks and stored away until needed.' For anyone with a deep freeze, they are a useful item – the icy cold is said to give them 'a peculiar sapid flavour'.

350 g (12 oz) plain or strong flour
1 large egg
100 ml (¼ pt) water or milk
1 rounded teaspoon salt

FILLING
350 g (¾ lb) boned pork spare rib
60 g (2 oz) onion
250 g (8 oz) chopped cabbage
60 ml (2 fl oz) water
1 heaped teaspoon salt
freshly ground black pepper

Mix the flour with the egg, water or milk and salt to make a firm dough. Knead lightly and put into the refrigerator for half an hour. Then roll thinly and cut into 5 cm (2 in) rounds.

To make the fillings, mince the pork and onion and mix with the remaining ingredients, adjusting the seasonings to taste. Remember that fillings for pancakes and pastries need to be strongly seasoned to get through the blandness of the dough. Always bear in mind the proportion of filling to dough and act accordingly.

Put a small knob of filling on one side of each circle of dough. Fold the free side over and pinch the edges together to form a half-moon shape. Gently curve the pelmeni until the two points can be drawn together. Freeze on trays in the usual way, then store in plastic bags. This quantity is enough for four to six people, depending on appetite and the rest of the meal.

BOILED PELMENI

The pasta principle. Add the pelmeni gradually to 4½ litres (8 pt) boiling salted water. Remove them with a perforated spoon as they rise to the top, and keep them warm until all are cooked. Serve with plenty of butter, or butter and wine vinegar, or soured cream, with grated Parmesan or mustard to give an edge.

FRIED PELMENI

Fry them while still frozen, in butter, until nicely browned on both sides. Serve in individual dishes with soured cream over them.

BAKED PELMENI

For pelmeni which have been either boiled or fried. Divide them among individual dishes, pour some soured cream over them, sprinkle them with grated Parmesan and a little melted butter and brown them in the oven.

For other fillings, see below and p. 152–5, 168.

CABBAGE KULEBIAKA

After the tiny, light Russian mouthfuls of pelmeni, here is the large and splendid pasty, the kulebiaka. A popular filling is buckwheat kasha and salmon, but cabbage and egg make a cheaper and less heavy pie with plenty of character.

> *1 large onion, chopped*
> *150 g (5 oz) butter*
> *1–1½ kg (2–3 lb) Savoy or other crisp cabbage*
> *4–5 hard-boiled eggs, chopped*
> *about 2 teaspoons dried dill weed*
> *2 tablespoons chopped parsley*
> *salt, pepper, sugar*
> *puff pastry made with ½ kg (1 lb) flour*
> *beaten egg to glaze*

Cook the onion without browning it in half the butter. Cut the coarse stalks and leaves from the cabbage, slice it, put it into a colander and pour boiling water over it. Press out the moisture as soon as it can be handled, then add it to the onion to cook down further in its own juice. When it is just done, but still a little crisp, drain it well. Add the egg, herbs and seasonings to taste.

Roll out the pastry into two almost equal oblongs. Put the smaller one on to a baking sheet lined with foil (to help you transfer it to a serving dish later). Pile the filling on to it, leaving a 2–3 cm (1 in) rim, which should be brushed with the beaten egg. Put the second oblong on top. Roll up and press the edges firmly together, cutting away any surplus pastry. Make a central hole, and decorate the pie with pastry leaves. Brush over with egg. Bake at 220°C/425°F/Gas Mark 7 until the pastry is well risen and nicely browned – about 25 minutes – then give it a further quarter of an hour at 190°C/375°F/Gas Mark 5. Before serving pour the remaining butter, melted, into the pastry through the central hole. Serve with a jug of soured cream or melted butter.

SAUERKRAUT

To know that cabbages were once the only vegetable people had in some parts of Europe, does not mean much. To experience it is another matter. We spent a fortnight in northern Donegal in the summer of 1950. We left with the impression that all the Irish ever ate was cabbage, potatoes, poor quality lamb, and Victoria sponges. No wonder they took to whiskey. The village street smelt intensely of cabbage every day at noon. This in a village of fishermen. Why did we not eat lobsters and crabs and other fish? Surely those fishermen must have caught a whiting or two?

Germans and Austrians made something better of their lot by turning their endless cabbages into *sauerkraut*. Cabbages were stashed down with salt and left to ferment into sour cabbage, which is what the name means. The Romans did similar things to their leafy and stalked vegetables, to keep them over the winter. *Sauerkraut* has the advantage of being anti-scorbutic, which mattered when fresh greenery and citrus fruits were not available; now Germans and Austrians make it because they like it, not because they need it. My daughter stayed in a guesthouse near Hallstat in Austria, where the prehistoric salt mines are, and the proprietors made their own *sauerkraut*. It was so delicious, so far above the kind we get in tins and jars, that she went round to all the tables after meals and polished off the remains that less enthusiastic English visitors had left in the dishes.

Sauerkraut is given style and flourish by adding extra ingredients when it is being prepared for the table. One popular and unexpected item is pineapple. More usual additions are different joints of cured pork and sausages, turning the *sauerkraut* into a meal, a germanic version of the *potées* and *garbures* of south–western France. In the top corner of France that borders Germany, *sauerkraut* or *choucroute* is popular, too, and vast dishes of the cabbage with pork and sausages and potatoes are known as *choucroute garnie*. This kind of thing is hefty, savoury, powerfully filling, very much a well-off farmer's dish, though it is also popular with prosperous town's people as well. The poor cannot afford such lavish additions, but take their *sauerkraut* as it is, or in soup, or with some small meaty adjunct to give an extra flavour. It is not everybody's taste. Goethe who loved fine food and knew a great deal about it, loathed *sauerkraut*. I am not much given to it myself, but it has its passionate admirers.

SAUERKRAUT can be bought in bottles and cans, and occasionally from the barrel in German delicatessens. It needs to be rinsed well, and then soaked to remove the harshness of the salting and fermenting. Taste it after 15 minutes in water; it may be all right. If not, give it another ten minutes or so. Drain it well in a colander before using it.

If you grow cabbages, you can make your own *sauerkraut*, starting off with 3 kg (6 lb) cabbage and a 4½ litre (1 gallon) stoneware jar.

Method

Cut each cabbage into four pieces, remove the damaged outer leaves and the hard core of stalk. Weigh out 2½ kg (5 lb) of this prepared cabbage. Next measure 60 g (2 oz) of sea salt or pure rock salt – do not use free-running table salt ever for pickling anything, because of its chemical additions.

Shred the cabbage finely, on a mandolin slicer if you have one or with a sharp knife. Pack the cabbage and salt into the jar in layers. Pat the layers down with a jam jar. Finish with a sprinkling of salt. Rinse and sterilize in the oven (same process as for jam jars) a plate that will fit into the mouth of the jar and keep the cabbage pressed down below the brine that forms from the juices and the salt. Skim the brine daily and put a clean, sterilized plate on top. Stir the cabbage about from time to time when you have skimmed it, and always keep the jar covered over with a cloth to prevent dust getting into the brine; this should not be an air-tight seal, just a covering. Store the jar in a room with a steady temperature above 15°C (60°F) and below 21°C (70°F); 18°C (65°F) is the ideal. As the cabbage ferments, the brine will bubble. In two to three weeks this will stop, which means that fermentation is completed. Now the *sauerkraut* is ready for use, though it will taste better if you leave it for another month. Put a clean plate on top of the cabbage, cover the jar tightly this time, and move it to a cooler place where the temperature will not exceed 3°C (38°F). You can see why *sauerkraut*-making is a winter affair.

When you need some of the cabbage, remove it from the jar with a handled vessel or ladle, so that your hands never come in contact with the brine. Let the liquid drip back into the jar. Make sure that the cabbage remaining is weighted down below the surface of the brine.

The flavour of *sauerkraut* can be improved by adding caraway seeds (one tablespoon for the quantity of cabbage given above) and juniper berries (one dessertspoon). Polish housewives sometimes put in a cooking apple – one small one would be enough, the usual proportion being 12 normal sized apples to 40 kg (100 lb) cabbage.

Chinese leaf can be pickled in the same way. Because its juices are sweeter, it makes a much finer-tasting *sauerkraut*.

A warning. No, two warnings. The fermenting cabbage can smell. Marike Hanbury-Tenison tells a story of visiting friends one autumn in Zürich. As they made their way up the stairs to the flat, she could not help noticing the awful stink. 'Oh, that's the *sauerkraut* in the cellar. It'll be over in a month' – i.e. when the fermenting had stopped and everyone had fastened down their barrels.

The instructions above are partly based on a most useful book, *Putting Food By*. It was written by Ruth Herzberg, Beatrice Vaughan and Janet Greene, and first came out in the States in 1973. Dent are the publishers in this country. It deals with freezing, canning, preserves, drying, root-cellaring, salt and smoke curing, rendering lard, making sausages and soap and cottage cheese. It is an invaluable book for anyone who grows their own produce and has the space for storing things. Let me take my second warning from this book, 'If the brine gets slimy from too much warmth it's best not to tinker with it: do the simplest thing and decant the batch on the compost pile – and wait until cooler weather to start over again.'

Serve beer rather than wine with *sauerkraut* dishes.

HOW TO PREPARE AND COOK SAUERKRAUT

These instructions apply both to canned and fresh *sauerkraut*. Rinse it well, squeezing out the briny juices. Taste a bit, and if it is too strong, put it into a bowl of tepid water. After 20 minutes, taste it again. It should be all right, but change the water and continue to soak it if it still seems overpowering. Drain and squeeze it again before cooking it.

Instructions about cooking *sauerkraut* can seem puzzling. Some books say 15 minutes, some say two hours. It will be tender after 15 minutes, but if you are making a Polish *bigos* or an Alsatian *choucroute garnie*, you want the *sauerkraut* to become as mild as possible and all the different flavours to blend together into a general succulence. This kind of recipe needs two hours' cooking.

You can always eat *sauerkraut* raw, too, after rinsing and soaking it.

A standard Polish way of cooking *sauerkraut*, is to put a kilo of it (2 lb) into a pan, with a generous ½ litre (1 pt) of boiling water. When the whole thing returns to the boil, cover the pan, lower the heat and leave to simmer for 15–30 minutes. Halfway through the cooking time, you can add a couple of tablespoons of tomato concentrate, or a small packet of soaked dried ceps (*funghi porcini* from Italy are the usual ones at the delicatessen), with their soaking liquor. In a separate pan, soften two chopped onions in a good tablespoon of lard, stir in a tablespoon of flour and moisten with some of the liquid from the *sauerkraut* pan. When you have a thick bubbling sauce, put it into the *sauerkraut* and complete the cooking.

If you have not flavoured the dish with tomato or mushrooms, stir in some caraway seed to taste, and a little brown sugar. Turn into a hot serving dish and top with bacon chops, fried sausages, frankfurters, or a piece of roast pork.

The German way is to stew the *sauerkraut* slowly for two hours with just enough boiling water to cover and a level teaspoon of caraway seeds for 1 kg (2 lb) cabbage. Pepper the dish well. Drain off the surplus liquor, leaving the *sauerkraut* moist. Serve with frankfurters or cured pork and with boiled or mashed potatoes.

Sometimes the *sauerkraut* is briefly fried in lard in the pan, before the boiling water is added. Peeled and sliced cooking apples are sometimes added to the *sauerkraut*, as well as the caraway seeds.

SAUERKRAUT SOUP
SOUPE À LA CHOUCROUTE

Sauerkraut soup can be very simple indeed: just ½ kg (1 lb) *sauerkraut*, washed and pressed, then stewed for a couple of hours with 2½ litres (4–5 pt) lightly salted ham or beef stock. At the end add a good knob of butter. Serve with bread croûtons, toasted or fried. This is for *sauerkraut* enthusiasts.

To my way of thinking, it needs one or two more ingredients to tone the *sauerkraut* down a little. Start by browning a large chopped onion in a little bacon fat. Stir in ½ kg (1 lb) washed and pressed *sauerkraut*, and 250 g (8 oz) of chopped lean pork. When everything is coated with the fat, pour in 2 litres (3½ pt) lightly salted ham or beef stock. Simmer for one to two hours. Thicken with two generous tablespoons of *beurre manié* (page 552). Check the seasoning (a little sugar is sometimes a good idea).

Serve with a chopped slice of boiled smoked ham or gammon and croûtons, or put in pieces of game, instead of the ham, allowing them to heat through in the soup for about 15 minutes. Sliced boiled potatoes can be added as well. It is easy to work out variations of this kind of soup, by looking at the recipes for *bigos* and *choucroute garnie*: one uses less *sauerkraut*, less meat and more liquid.

SAUERKRAUT SALAD
SAUERKRAUTSALAT

The simplest way of making this is to rinse and press ½ kg (1 lb) of *sauerkraut*, soaking it as usual if it is too salty and sour for comfort. Put it into a bowl and rapidly grate in one or two Cox's Orange Pippins – do not remove the skin, but avoid the core. Stir together, then add olive oil and pepper to taste. It must be olive oil, rather than one of the tasteless oils. Walnut oil may be used, though this is a French rather than a German idea. Add a hint of sugar if it seems necessary.

SAUERKRAUT SALAD WITH PINEAPPLE
SAUERKRAUT MIT ANANAS

This pretty and delicious salad goes well with ham, game, duck, pork. You can use tinned pineapple, well drained of syrup, but fresh pineapple is much better.

Buy a large ripe pineapple. Slice off the leafy plume and hollow out the fruit with a metal spoon and knife. Cut the pieces up as evenly as you can and keep the juice. Mix with one and a half times the weight of *sauerkraut*, measured after you have rinsed, soaked and pressed it as free of moisture as possible. Add a little olive oil. Put it into the empty pineapple case, piling it up, and perch the rosette of leaves on top.

In some versions of the recipe, the *sauerkraut* is first simmered in canned, unsweetened pineapple juice, until it is tender. Then it is drained and mixed with the pineapple pieces and oil.

If you are not sure which version to choose, mix a little *sauerkraut* and pineapple and taste it. If the whole thing seems too chewy, then cook the *sauerkraut*.

GOOSE AND SAUERKRAUT

A good mixture much favoured in Alsace, Germany and Eastern Europe. The tart chewiness of the cabbage goes well with the rich goose and its crisp skin. Try it, too, with duck.

A goose on St Martin's day, November 11th, was once common in most of northern Europe, including England. I do not think there is much of a saintly connection. St Martin of Tours, our great local saint in France, is figured in many churches as a painted statue on his horse, cutting his cloak in half for a chilly beggar. It is most likely the coincidence that this great feast day of the past occurs when the geese grow fat enough to be eaten. These are the large stubble geese, not the small green goose that was a favourite summer dish with young green peas.

In Alsace the bird is stuffed with sausagemeat mixed with chopped fried onion, and served with *sauerkraut* that has been stewed with a little goose fat and water, with six or more slices of lean smoked streaky bacon added halfway through the cooking.

In Germany the goose is stuffed with apples, quartered eating apples, and soaked stoned prunes. The *sauerkraut* will be cooked without the bacon.

In more Eastern countries, the *sauerkraut* is turned into a stuffing. Take 1½ kg (3 lb) *sauerkraut*. Rinse, soak and press it dry. Brown a large chopped onion lightly in some of the rendered goose fat, add the *sauerkraut* and stir it round well for five minutes. Mix this into five peeled and grated Cox's Orange Pippins, and one large peeled, grated potato. Season with three heaped teaspoons caraway seeds and plenty of black pepper. Add salt if the *sauerkraut* was mild. Stuff the goose and roast slowly for two and a half hours for a 4–5 kg (10–12 lb) goose, at 160°C/325°F/Gas Mark 3. Put a tumbler of water into the pan and place the goose on a rack.

HUNTER'S BIGOS

A favourite Polish way of cooking *sauerkraut* is with a variety of meats, so that everything blends into one delicious and peppery richness. Cooking time and quantities are variable, so long as the *sauerkraut* is cooked for a good two hours at least. A frugal *bigos* can be made with left-over cooked meat. One Polish cookery book lists pork, beef, veal, fatty bacon, smoked sausages and frankfurters, tongue, ham, poultry and game – above all game for its wonderful flavour. The meat is always cooked partly before being added to the *sauerkraut*, so using left-overs need not be as nasty as it sounds. Do not be tempted to use stewing cuts for the following recipe, which will provide food for a dozen people.

> *3 kg (6 lb)* sauerkraut, *rinsed and soaked if necessary*
> scant ½ litre (¾ pt) light beef or veal stock
> 4 cooking apples, peeled, sliced
> 4 large onions
> ½ farm chicken
> ½ farm duck
> ½ kg (1 lb) loin of venison, or saddle of hare
> ½ kg (1 lb) salt pork loin
> lard
> 150 ml (¼ pt) dry white wine
> 250 g (8 oz) gammon or ham
> 6 long frankfurters
> pepper, salt, bouquet garni
> 1 tablespoon flour

Put the *sauerkraut* on to stew gently with the stock, apples and half the onions. Brown the chicken, duck, venison or hare and salt pork in lard, transferring each piece to a large pan as it is ready. Put the remaining onion into the frying pan last of all and when it is lightly coloured pour in the wine and an equal quantity of water. Stir vigorously to dislodge any nice bits in the pan and when the mixture bubbles pour it over the meat; cover the pan tightly and leave to simmer for half an hour. Add the gammon or ham and the frankfurters, cover again and simmer for ten minutes. Remove the meat from the pan and cut it off the bones, which can be kept for stock-making. Cut the meat into chunks and the frankfurters into 2 cm (¾ in) lengths.

Transfer about half the *sauerkraut* to the meat pan, which still contains the cooking juices. Put the meat and sausages on top. Cover with the remaining *sauerkraut* and pour over the liquid in which it cooked. Add plenty of pepper and a little salt if you like. Go carefully, and taste the *sauerkraut* first, as you do not want to over-salt the dish. Add the bouquet garni, cover the pan tightly again and leave to simmer for a further hour or an hour and a half. Check the liquid level from time to time, and add more wine or stock – or tomato juice – if the mixture looks at all dry.

25 minutes before the end of cooking time, make a *roux* with a tablespoon of lard and the flour. Remove a couple of ladles of stock from the cooking pan, stir the mixture into a sauce and tip it back into the pan.

Cover again and leave for 20 minutes. Correct the seasoning, remove the bouquet and serve very hot, in the cooking pan if it is respectable looking. Enamelled iron pots are ideal for *bigos*; of you haven't got one, carry out the second long cooking in an earthenware or stoneware casserole in a slow oven. Allow a couple of hours at least. You can always reheat *bigos* if it is cooked too soon.

In Poland this is very much a winter dish for a hunting lunch. It would be equally good for a shooting party, or for bonfire night, as the second part of the cooking can continue at a slow bubble on the edge of the fire. Be careful not to cook it too fast, that is all. Remember that the ingredients should melt together, ending up as a juicy stew rather than a thick soup.

CALABRESE see BROCCOLI

CAPSICUM see PEPPERS

CARDOON

Like the globe artichoke, the
cardoon is an edible thistle and a
native of the Mediterranean.
Unlike the artichoke, the cardoon is cultivated for its leaf-
stalks. They are blanched in
banks of earth, like celery, and
are eaten in similar ways. It has
never caught on in this country, and I notice that it has disappeared from
the catalogues of our leading nurserymen in the last few years. You would
have to buy the seeds abroad. And you are not likely so see it on sale
unless you are housekeeping in Spain, southern France or Italy, where it
is much appreciated as a vehicle for delicious flavours such as anchovy
and Parmesan cheese.

You could say that cardoons follow the anchovy belt of the Mediterranean. In Liguria the stalks are eaten raw, dipped into a rich *bagna cauda*
sauce made of anchovies, olive oil and butter; on special occasions white
truffles will be stirred into the sauce as well, for an extra fragrance. In
Provence, cardoons are a favourite winter and Christmas dish with simpler versions of *bagna cauda*. The stalks are also delicious when cooked
and finished with butter, cream, Parmesan, and taste better than celery
which has too pronounced a flavour for such mild-tasting richness and
small piquancy.

The flower heads of cardoon before they open can be eaten like
artichokes. Once they burst into a splendour of thistle purple you can dry
them and use them as a rennet substitute. This was much done in the past,
according to regional cookery books of the Charentes, and the practice
lingers on in a few old-fashioned farms. Occasionally the soft cheeses, the
caillebottes, made in this way can be bought in Charentais markets (see
page 158). I should not be surprised if *caillebottes* were revived as a
regional speciality one day soon, just as the making of goat cheese has
been encouraged in the Deux-Sèvres in the last few years.

HOW TO PREPARE CARDOONS

If you grow cardoons you will have to cut away the roots, prickly outer
stems and leaves yourself. Only the inner stalks and firm hearts are eaten.

When cardoons are sold in markets and shops this initial trimming has been done. In the recipes following, any weights mentioned refer to the trimmed heads. As a general guide, allow 2–2½ kg (4–5 lb) total weight of trimmed cardoons for six people.

If you are eating the cardoons raw, separate the stalks, remove the obvious strings and inner white skin, but go carefully. If you are too enthusiastic, there will not be much cardoon left. Cut the stalks into suitable lengths, slice the heart across thinly, putting the pieces into water acidulated with lemon juice to prevent discoloration. An alternative is to rub each piece with a cut lemon. Dry the pieces just before serving them.

If you intend to cook the cardoons, separate stalk and heart and cut them up without stripping away string and skin. Drop the pieces into acidulated and salted water for 15 minutes; if the cardoons need to be completely cooked, give them a little longer until they are just tender. Drain and refresh them under the cold tap. Now the string and stalk can be peeled away fairly easily. Complete the cooking according to the recipe. A total cooking time of 30–40 minutes is enough. Do not be put off by some French cookery books which tell you to give them three hours. This is nonsense.

If you wish to give cooked cardoons a little extra flavour, blanch them for 15 minutes as above, then pour off the water and peel them. Return them to the pan with just enough good stock to cover and complete the cooking say 15–20 minutes.

A number of celery and fennel recipes may also be adapted for cardoons: it is worth turning to those sections for further ideas.

BAGNA CAUDA

In the Piedmont dialect of Italy, *bagna cauda* means hot bath, the bath being a pot of anchovy and garlic sauce kept hot on the table over a small burner. Pieces of one or more vegetables are served with it. Everyone takes their choice and dips it into the pot with a fork, in the style of *fondue*. Prudence dictates that you keep a bit of bread in the other hand to catch any sauce that drips on its way to your mouth. Another essential is plenty of rough red wine. In Piedmont, it would be Barbera. *Bagna cauda* is a country dish, with a lusty flavour; to avert indigestion, you need that sort of heavy red wine. It is no good sipping genteelly at a tumbler of water or a glass of fine claret.

Bagna cauda may come at any time of the day, with a meal, on its own, whenever somebody begins to feel peckish: a sociable dish. In fact it is mainly a winter pleasure as the main vegetable, at least for a Piemontese, is cardoons. It is the kind of thing that he might eat when trying out the new wine, the way a Frenchman in Touraine will eat hot new bread and fresh walnuts, or bread and oysters, with the still cloudy *bernache*.

Do not let the lack of cardoons stop you sampling *bagna cauda*. Celery is the obvious substitute, with pieces of sweet pepper that have been grilled and skinned, heads of chicory separated into leaves, sticks of carrot. Prepare them all in advance and arrange them boldly, like a painting, on a big dish or in baskets. Cut up plenty of good white bread, preferably home-made.

Last of all make the sauce. Proportions vary widely, but the basic ingredients are always the same. Melt a packet of butter in a wide stoneware or enamelled iron pot, stirring in 150 ml ($\frac{1}{4}$ pt) olive oil. As they heat together, chop four or five cloves of garlic finely and tip them into the pot. They should dissolve and soften rather than fry. Next chop up half a tin of anchovies, or six soaked and filleted anchovies from the salt. Stir them into the sauce with a wooden spoon, until they have dissolved as well. Taste the sauce and make any adjustments, e.g. more anchovies. If someone has brought you a white truffle from a winter business visit to Milan – unlikely perhaps, but I offer it as a suggestion and in the hope that it may happen to me one day – now is the time to use it. Shave it in thin slivers into the sauce, using the cucumber blade of the grater, just before you transfer the pot to its table-burner. White truffles, unlike the black kind, should never be cooked in the true sense of the word, just heated through by the food they adorn.

For provençal versions of *bagna cauda*, turn to the recipes for celery with anchovy sauce, substituting cardoons for the celery.

FRENCH CARDOON SALAD

Blanch and peel cardoon heads, then complete the cooking in stock. Drain the pieces well, mix them with olive oil vinaigrette and leave to cool. Scatter generously with chopped parsley, chervil and chives and serve chilled.

BUTTERED CARDOONS
CARDONS AU BEURRE

Melt a good large knob of butter in a shallow pan, and stew the blanched cardoons in it for 15 or 20 minutes, turning them frequently. Finish with chopped parsley.

CARDOONS WITH ANCHOVIES
CARDONS AUX ANCHOIS

Cook half a tin of anchovies, chopped, in 125 ml (4 fl oz) olive oil. When they have melted into the sauce, add 100 g (3 oz) butter, cut in bits. When the mixture returns to boiling point, put in the blanched cardoons and turn them in the sauce over a low to moderate heat for ten or 15 minutes.

CHRISTMAS GRATIN OF CARDOONS
CARDONS DE NOEL

Cardoons are a favourite vegetable of the *gros souper* on Christmas Eve in Provence.

Blanch the cardoons until they are cooked. Make a béchamel sauce using olive oil rather than butter. Season it with grated nutmeg. Lay the drained and peeled pieces of cardoon into a *tian* or gratin dish that has been lightly oiled. Pour the sauce over the top. Scatter with grated Parmesan and Gruyère. If you have some meaty juices left over from roasting a turkey or chicken or pheasant, pour them over the top as well. Bake in a hot oven until the sides are bubbling and the top lightly browned.

CARDOONS IN THE LYONS STYLE
CARDONS À LA LYONNAISE

Lyonnaise usually means 'with onions'. This robust sauce is full of flavour and piquancy. Do not strain out the onions.

> *2 kg (4 lb) cardoons, prepared, cooked*
> *1 medium to large onion, chopped*
> *2 tablespoons butter*
> *1 heaped tablespoon flour*
> *5 tablespoons dry white wine*
> *5 tablespoons white wine vinegar*
> *250 ml (8 oz) good beef stock*
> *1 heaped teaspoon tomato concentrate (optional)*
> *breadcrumbs*
> *melted butter*

Lay the cardoon pieces in a buttered gratin dish, close together. Soften the onion in the butter, then raise the heat and, as it begins to colour, stir in the flour and cook to a pale brown *roux*. Moisten with the wine, vinegar and stock. Simmer for five minutes, correct the seasoning and add the tomato concentrate if you like. If the sauce is very thick, add a little more stock. Pour it over the cardoons. Scatter an even layer of breadcrumbs on the top and pour a little melted butter over them. Put into a moderate to hot oven until the gratin bubbles at the side and the crumbs have formed a golden-brown crust.

CARDOONS WITH BEEF MARROW SAUCE
CARDONS À LA MOELLE

Ask the butcher to let you have enough beef shin bones to provide 100 g (3 oz) of marrow. You will have to scrape it out with a thin knife; or you could stand the bones upright in a large pan of water and simmer them until the marrow can be removed more easily.

> *100 g (3–4 oz) beef marrow*
> *2 tablespoons butter*
> *1 heaped tablespoon flour*
> *4 tablespoons meat jelly, glaze, or juice from a roast*
> *up to ½ litre (¾ pt) good beef stock*
> *1 tablespoon lemon juice*
> *salt, pepper, parsley*

If the marrow is uncooked, cut it across into slices and poach it for about ten minutes in a little water, then drain. If the marrow has been cooked in the bones, all you need to do is to cut it up and set it aside. Meanwhile make a sauce by melting the butter, stirring in the flour, then the meat jelly or glaze or juice, and enough stock to make a light sauce. Start on the liquid side, then boil it down steadily. Add the lemon juice, then the marrow to reheat. Check the seasoning and stir in a little chopped parsley.

Put the cardoons into a hot serving dish, pour over the sauce and serve immediately.

LA CAILLEBOTTE À LA CHARDONNETTE, OR LES CAILLEBOTTES

A type of soft creamy cheese which was once made more frequently than it is now in farms of the Poitou and Vendée. The milk, preferably goat's milk for the good flavour, is turned with dried cardoon flowers, *la chardonnette*, rather than rennet. Sometimes the first cow's milk of the morning was rapidly curdled this way, so that it would be ready for breakfast.

All recipes stress that new milk should be used, before it is cooled if possible. Do not expect to be successful with pasteurized milk; even if you can get it to form a curd, it will not taste very nice.

First prepare a little package of dried cardoon flowers. Put enough in a square of muslin to make a tight bundle the size of a large walnut – in terms of weight, 10 g or about a third of an ounce. Tie it and put it into a pan with a litre of milk (1¾ pt). Add a little sugar if you intend to eat the cheese as a dessert. Bring slowly to about 60°C (140°F). On no account allow the milk to come close to boiling point. If you are nervous of your stove overheating, put the milk and *chardonette* into a double saucepan; this way it will take about half an hour to get to the right temperature. As the milk warms up, stir the bundle about, pressing it against the side of the pan. It will probably show signs of curdling before you reach the right temperature and should curdle completely soon after you remove the pan from the stove. Pour the milk into a shallow bowl, discarding the cardoon flowers. Cut the curd into cubes, and drain them of whey. Some recipes instruct you to put the cubes back over the heat, until they are swimming in whey.

Serve the *caillebottes* with salt and radishes as a first course. For a dessert, pile them on a dish and pour over some sweetened, lightly whipped cream, flavoured with brandy.

CARROTS

When you are turning over a box of carrots in the winter, or buying a collection of pot herbs for stew or stock, it is difficult to realize that carrots were once as exotic as artichokes or Chinese leaf or avocados seemed in the 1960s. Ladies of the Stuart court pinned the young feathery plumage of young carrots to their heads and on their splendid hats. The leaves drooped down from exquisite brooches on sleeves, instead of the more usual feathers. I had always thought that the growing of carrot tops in a small dish of water was part of the nursery curriculum of my lifetime. Here is the famous garden designer J. C. Loudun most solemnly recommending the idea in his enormous and serious *Encyclopaedia of Gardening* of 1827, as 'an elegant chimney ornament' for winter, 'Young and delicate leaves unfold themselves, forming a radiated tuft, of a very handsome appearance, and heightened by contrast with the season of the year.' One may imagine the chimney pieces of grand houses or *cottages ornées*, sprouting with carrot tops set on pretty New Hall saucers.

Why so much enthusiasm for so antique a vegetable? Carrots were found in the excavations of Swiss lake villages at Robenhausen. They were eaten, without great enthusiasm, by the Romans, who did not even think they were good for the bowels. Carrots were not, it seems, forced down the infant Nero with the purpose of making his hair curl, or giving him cat's eyes to see in the dark. Perhaps he would have grown up more satisfactorily under such nannyish attentions. Though to be fair to the Romans, they at least dignified vegetables and plants with more serious virtues than nannies were accustomed to do.

The thing was that these early carrots were not the bright orange, even red ones we eat to-day. Ours have come from a purple carrot grown first in Afghanistan at the beginning of the 7th century A.D. Seeds of this purple carrot and a yellow mutant came with the Moors to Western Europe, along the coast of North Africa. Like spinach and aubergines, they spread from Spain into Holland, France and finally England. The Dutch seem to have produced the modern bright orange carrot in the Middle Ages. About 1390 they were still strange to many of the French and in particular to the young girl who married the Goodman of Paris: he wrote a book for her on running the house and looking after him (everything from dealing with fleas, to how to welcome him home – slippers by the fire and a welcoming bed in every sense of the word). An important part was the cookery. He has plenty to say about vegetables, but there is only one that he has to explain to her, 'Carrots are red roots sold by the handful in the markets: in each bunch there is a white one'. Nowadays we have long slender carrots, shorter stumpy carrots, even round carrots like ping-pong balls, but to be successful they must be as bright a colour as possible.

As far as Europe is concerned, carrots became a common standby in the last century. Vichy apart, where carrots were once eaten daily in many hotels as part of the cure (for overloaded digestions), we do not seem to have relied on them medicinally. The gentlemen of Teheran in the 1870s took carrots stewed with sugar as an aphrodisiac, to increase both the quality and the quantity of sperm; they were taken for dropsy, too, and if cooked and pickled in vinegar, they were said to reduce a swollen spleen (this comes closer to Vichy). For the wealthy of the Belle Epoque in Paris, oysters and caviare and champagne were regarded as more helpful in matters of love; carrots belonged to bourgeois cookery and the penitential weeks at Vichy.

Nonetheless cooks in France, Italy and England were producing excellent sweet carrot cakes and tarts, in particular an angel's hair charlotte that would have delighted the weariest Persian with its rich beauty. A pity we have dropped such recipes from our general European repertoire. The nearest we come to such things here is at Christmastime, when shredded carrots are sometimes substituted – for economy – for part of the expensive quantity of dried fruit.

HOW TO PREPARE CARROTS

Small, young carrots, straight from the garden or local nursery, need only to be rinsed and brushed gently under the tap to remove soil. Top and tail, leave them whole, and put them into a pan with as little water as possible, just enough to keep them from burning. Add a light sprinkling of sugar, a tablespoon of butter, a pinch of salt – then wedge the lid tightly in place with foil. Cook over a moderate heat until just tender. Correct the seasoning.

Another good way of cooking them, is to wrap them with a few tablespoons of water and the other ingredients I mentioned, in a baggy piece of foil. They can either be put into the oven, or added to a pan in which something else is cooking, so that they steam in their own juices. When the carrots are cooked, drain off the liquid, taste them for seasoning and serve with a large tablespoon of butter, into which you have forked plenty of either mint or parsley. This is carrots in the English style, *à l'anglaise*. The simplest style for something of quality.

Carrots which are a little older, but still young, can be scraped, and cut up in various styles and cooked by the same methods. The Chinese way of slicing them is a good one. Make a cut on the cross near the tail end of the carrot. Give it a one-third turn and slice again in the same direction. It's a rolling movement, and quickly done once you get the way of it. These are *kun tao k'uai*, rolling knife pieces, but it seems to me that it is the carrot that rolls.

Middling sized carrots, still fairly young but with a more pronounced flavour, are ideal for glazing, for salads, for serving with sauces. If they are not too tender, they can be used to make angel's hair charlotte, p. 173, and other sweet cakes. Sometimes they can be scraped, sometimes they need to be thinly peeled.

As carrots become older and turn into a standby winter vegetable, they have the advantages of maturity – these are flavour, and the ability to stand up to strong treatment. This means delicious combinations with beef – braised beef, salt beef, spiced beef, and stewed beef – favourite dishes at the time of year when one needs comfort from food. They make good soup, too. If you wish to cook these carrots to serve on their own, either glazed or in sauces, it is wise to blanch them for five or ten minutes first in boiling salted water. Undeniably these carrots need peeling, but do it as thinly as possible with a potato peeler.

When carrots are old and wood-centred at the end of winter, their flavour goes and they are not much good as a solo turn. Grate the outer part and use it with plenty of other vegetables and seasonings in soups and stews; throw away the hard core.

CRÉCY SOUP
POTAGE CRÉCY

In France, carrot soups always seem to be called *Potage Crécy* – in deference to the chalky area of the country carrots are supposed to like best. Or perhaps one should say areas, because two places dispute the name, Crécy of battle fame in the north, where our king Edward III sat in a windmill and watched his army fight to victory, and Crécy in the Brie country which we associate more with cheese.

> *1 onion, chopped*
> *60 g (2 oz) butter*
> *½ kg (1 lb) carrots, cut in lengths*

    125 g (4 oz) diced potato
    1 heaped tablespoon rice
    1 litre (2 pt) water
    125 ml (4 fl oz) milk
    1 large egg yolk
    2 tablespoons cream
    salt, pepper, chopped fresh parsley or chervil

Soften the onion in butter. Add the carrots, potato, rice and seasoning, stirring them about to mix them well together. Then pour in the water. Simmer, covered, until the carrots are well cooked. Then put them through the fine plate of the *mouli-légumes*, or purée in a blender. Reheat with the milk and more water if required. Beat the yolk with the cream, add some of the soup, then return the mixture to the pan. Stir for a few minutes without allowing the soup to boil, and season well. Serve scattered with herbs and with cubes of bread fried in butter.

PINKS CARROT AND ORANGE SOUP

Fairford in Gloucestershire is worth visiting for two things. In the church, in red and purple glass, the monstrous figure of Satan swallows the damned. On the edge of the town, is Pinks restaurant, where the food is well cooked and lively in choice and presentation. Here is their carrot and orange soup which can be served hot in winter, chilled in summer. They say firmly that frozen concentrated orange juice gives a better result than fresh. In January and February, Seville orange juice can be added to give a slight bitterness to the flavour.

    1 onion chopped
    60 g (2 oz) butter
    ½ kg (1 lb) diced carrots
    1 litre (1¾ pt) light chicken or veal stock
    1 carton frozen concentrated orange juice
    salt, pepper

    EITHER
    100 ml (3–4 fl oz) single or whipping cream
    grated rind of an orange

    OR
    150 ml (¼ pt) natural yoghurt
    chopped mint

Make a smooth soup with the first four ingredients, following the method for Crécy soup (page 162). Season to taste with the thawed orange juice, mixing it in gradually so that it never becomes too strong. Correct the seasoning.

FOR HOT SOUP Reheat with the cream and extra stock or water if you like a thinner soup. Add the rind just before serving. Small croûtons of bread fried in clarified butter go well with the hot soup.

FOR CHILLED SOUP Mix in the yoghurt, dilute with water and correct the seasoning. More yoghurt can be put in if you like. Serve with two ice cubes in each bowl and a scatter of chopped mint.

CARROT SALAD

Grated raw carrots, dressed with an olive oil and lemon juice vinaigrette and plenty of chopped herbs – either green fennel or tarragon, or chives and parsley mixed. Chill well and drain off any surplus liquid before serving.

CARROT SALAD WITH RAISINS

As above, but instead of herbs, use seedless raisins and split almonds. The flavour is improved if you toast the almonds.

GREEK CARROT SALAD

Cook the carrots until just tender, but with a little bite left in them. Drain and mix in the Greek oil and lemon dressing on p. 559. Serve warm with grilled meat.

CARROT GRATIN

Try this recipe for middle-aged carrots.

Slice cooked carrots into a gratin dish that has been lightly buttered. Pour over a velouté sauce, a mustard sauce, or a creamy béchamel sauce. Mix together a tablespoon each of grated Parmesan and Gruyère or Cheddar, and two or three tablespoons white breadcrumbs. Scatter over the top; dribble on melted butter – not too much – and brown to a rich golden colour in the oven or under the grill.

GLAZED CARROTS VICHY STYLE

Ideally the carrots should be cooked in Vichy water, from the springs of this prosperous spa in the Bourbonnais. A pinch of bicarbonate of soda with tap water is the accepted substitute, but I should be surprised if you would know the difference between glazed carrots cooked with Vichy water, or tap water plus bicarb, or tap water on its own. The thing is the sugar and butter and the reduction of the liquid to a shiny glaze. The carrots matter, too. In 1789, the great agriculturist Arthur Young described the Bourbonnais as one of the finest provinces of France – for him that meant good turnips and root vegetables, as well as 'The finest climate in France, perhaps in Europe; a beautiful and healthy country; excellent roads; a navigation' – down the Loire to Orléans, then overland – 'to Paris; wine, game, fish, and everything that ever appears on a table, except the produce of the tropics'. He longed to buy an estate there, but felt too old. So, mindful of Arthur Young, buy the best flavoured carrots for this recipe and for the next one from Nevers, in the rich lands on the opposite bank of the Loire.

Slice the carrots diagonally. Put them into a pan and cover them with water. Add a tablespoon of butter and a teaspoon of sugar, with just a pinch of salt. Bring to the boil and boil steadily at first, then more vigorously until the carrots are cooked and the liquid reduced to a small amount of shiny colourless glaze. These two moments should coincide, which means that you must watch the last part of the cooking to see that the glaze does not caramelize. Correct the seasoning and serve sprinkled with parsley.

CARROTS IN THE NEVERS STYLE

In England by the 1720s whole herds of cattle were being kept in reasonable health on a winter diet of roots. This cut out the autumn slaughter and salting of beef, and made it possible to improve breeds. When Arthur Young was travelling in Burgundy in the 1770s, he was appalled to see the backward folly of farmers who had not yet taken to roots and clover. 'All from Autun to the Loire is a noble field for improvement, not by expensive operations of manuring and draining, but merely by substituting crops adapted to the soil.'

Matters did not improve until after the Revolution, when the farmers became so successful in root growing and beef production, that braised or roasted beef with glazed carrots, turnips and onions came to be described in classic cookery as *à la nivernaise*, in the style of Nevers.

To link the vegetables with the beef, they are cooked in beef stock and not water, by the Vichy method. Of course the stock has to be barely salted, on account of the reduction, otherwise the method is the same.

The French are skilful in tying dishes together in this way, to a delicious harmony of flavours.

CARROTS WITH SPINACH OR LEEKS

Two veg. take on a more elegant appearance if you combine them on one dish, in this way.

Grate ½ kg (1 lb) young carrots finely into a saucepan with enough melted butter in it to cover the base. Cover tightly and simmer, shaking the pan occasionally, until the shreds are cooked – about five minutes.

Have ready 1½ kg (3 lb) fresh spinach, cooked and buttered, or 1 kg (2 lb) thin leeks cooked whole. Put the spinach in a circle on a round heated dish, season the carrots with salt, pepper, lemon juice and green herbs and put them into the centre. With the leeks, arrange them side by side on a long dish, with the seasoned carrots at head and foot. Or else arrange them into a bed of carrots.

If you serve the vegetables on their own, provide some small embellishment such as bread fried in butter, or fingers of toasted cheese. They should be light and savoury, not strong in flavour.

Carrots and peas, or broad beans, or small new potatoes, can all be combined in the same way. Medium carrots can also be used, though they may need a little more cooking and extra attention to prevent burning; the idea is that the carrot cooks in the butter and its own juices.

FORESTER'S CARROTS
CAROTTES À LA FORESTIÈRE

Forestière seems an odd way of describing a dish made with cultivated mushrooms grown in Nissen huts and of the same genus as field mushrooms. I take it that the chefs who created the classic repertoire and needed names for their recipes, succumbed on occasion to romantic fancy. If you want to introduce an air of the forest, to enhance the flavour, substitute a small packet of dried ceps – soaked – for 60 g (2 oz) of cultivated mushrooms. Or use 125 g (4 oz) fresh ceps.

> 750 g (1½ lb) carrots, sliced
> barely salted light beef stock
> ½ teaspoon sugar

butter
grated nutmeg
250 g (½ lb) mushrooms, sliced
150 ml (¼ pt) whipping or double cream
chopped parsley, chives
salt, pepper, lemon juice
6 baps

Cook the carrots Nevers style, see p. 165, with the addition of a little grated nutmeg, until the liquid is reduced to a small syrupy quantity. If the carrots are elderly, blanch the slices first for five to ten minutes in boiling salted water.

Meanwhile fry the mushrooms in a little butter, until lightly browned. Add with the cream and herbs to the cooked carrots. Stir gently for a few minutes over the heat, so that the sauce is well amalgamated and thick enough to coat the vegetables. Correct the seasoning with salt, pepper and lemon juice.

The baps should be split across into two, and hollowed out (be careful not to break through the crust). Either brush them with melted butter and put them into the oven to crisp up – 10–15 minutes in a hot oven. Or fry them all over in butter. The oven method is best, because it gives an even colour and crispness, but if you are cooking nothing else in the oven at the time, it comes expensive.

Pile the carrots and mushrooms into the bases of the baps and perch their lids on top. Serve immediately.

PANCAKES WITH CARROT FILLING
BLINCHIKI

The Russians are splendid pancake makers, and they love tiny pastries filled with savory mixtures – *pelmeni* from Siberia, *vareniki* from the Ukraine and more southerly Russia, *pirozhki* everywhere made from a yeast dough and often deep-fried. The famous *blini* pancakes, eaten with caviare, and sometimes made with a proportion of buckwheat flour, we know about. But thin *blinchiki* are less familiar.

BATTER
250 g (8 oz) flour
3 eggs
2 tablespoons melted butter
¼ teaspoon salt
about 400 ml (¾ pt) milk

FILLING
1 kg (2 lb) diced carrots
150 ml (¼ pt) water
3 heaped tablespoons butter
salt, pepper, sugar
4 hard-boiled eggs, crumbled

Make a pancake batter, adding milk gradually to make the mixture fairly thin. Leave to stand.

For the filling, put carrots, water and a tablespoon of butter into a pan. Cover it closely and leave over a moderate heat, until the carrots are tender. Drain the carrots, mash them with the remaining butter and seasonings to taste. Mix in the egg.

Try out a small quantity of batter, to make sure that it makes thin but not too fragile pancakes to wrap round the filling. Adjust the batter if necessary, then cook as usual *but on one side only*, to make 24 15 cm (6 in) pancakes. Put a tablespoon of cooled filling in the centre of the cooked, brown side of each pancake. Fold over the free edges, envelope style, to make neat packages: Fry in butter until golden brown and slightly crisp; do not hurry this, or the butter will burn and spoil the flavours.

For other *blinchiki* fillings, see under *Cabbage*, p. 144 and 145.

CARROTS WITH APPLE

A Rhineland recipe gives a lift to carrots.

¾ kg (1½ lb) carrots
375 g (12 oz) sliced onions
60 g (2 oz) butter
375 g (12 oz) Cox's Orange Pippins, peeled, sliced
salt, sugar, lemon juice, pepper

Cut the carrots into sticks and cook them in very lightly salted water, with 1½ teaspoons of sugar, until they are almost tender. As they cook, fry the onions in the butter until they are soft and yellowish, then raise the heat so that they brown a little. Add them to the almost-cooked carrot sticks with the apples and complete the cooking. Strain off the juices, which should be used up in another dish, then season the carrot mixture with a little more salt to bring out the flavours – not enough to fight the natural sweetness. Add lemon juice to taste and pepper. Serve with pork, salt pork or ham, duck, or on its own with wholemeal bread.

CHICKEN FRICASSEE FROM BERRY
FRICASSÉE DE POULET À LA BERRICHONNE

As you will see from the Nevers recipe, carrots are an important item in several classic garnishes, especially in garnishes for beef. A very similar one is a collection of thickly sliced, glazed carrots, small whole glazed onions and lightly browned strips of bacon: this is for braised beef, veal or liver, which is then described as *boeuf braisé à la bourgeoise*, *rôti de veau à la bourgeoise* and so on – the housewife's garnish, because the ingredients are always to be found in reasonably prosperous homes.

Here is a farmer's dish from George Sand's province of Berry. This and the following English recipes are given briefly, as they have appeared in earlier books of mine in some detail.

> 1 kg (2 lb) middle-aged carrots
> 2 tablespoons butter
> 2 kg (4 lb) chicken, jointed
> 3 shallots or medium onion, chopped
> 1 heaped tablespoon flour
> ½ litre (¾ pt) chicken giblet stock
> 125 ml (4 fl oz) double cream
> 1 tablespoon wine vinegar
> 2 egg yolks
> salt, pepper

Shred the carrots into matchstick strips on a mandolin slicer, or cut them with a sharp knife. Blanch them for five minutes, then drain them well. Brown them lightly in the butter in a large heavy frying or sauté pan. Remove with a slotted spoon. Put in the chicken pieces, with a little more butter if necessary, and the shallot or onion. When they are a golden colour, sprinkle over the flour and mix it in well, turning the chicken. Return the carrots to the pan. Pour in the stock. Simmer uncovered until the chicken is done – about three quarters of an hour. If the sauce reduces too much, add a little water. Turn the chicken at least once. Remove the cooked bird and carrots to a dish. Beat the cream, vinegar and yolks together and use them to thicken the sauce. Check the seasoning and pour the sauce over the chicken.

BOILED SALT BEEF AND CARROTS

This is a favourite dish of mine, though I think it is best eaten in a Jewish restaurant rather than at home. There they cook the salt beef in vast joints, which means it is far more succulent than the small pieces sold by family butchers and supermarkets. If you are dealing with a large party of people, and can therefore order a piece of 2–3 kg (4–6 lb), you can make a good dish of it all the same.

> 2–3 kg (4–6 lb) piece of salt beef
> 2 large onions, each stuck with 3 cloves
> 2 blades of mace
> 1 teaspoon lightly crushed allspice berries
> small piece of nutmeg
> 1 heaped teaspoon lightly crushed peppercorns
> 1½ kg (3 lb) carrots, left whole or split in half lengthways
> 60 g (2oz) butter
> pinch sugar, salt
> parsley

Ask the butcher if the beef needs any soaking and follow his instructions. Put the beef into a large deep pan and pour in enough tepid water to cover it by 2–3 cm (a good inch). Bring slowly to the boil and remove the scum. Put in the onions and spices. Cover closely and simmer for three and a half to four hours. This can be done either on top of the stove or in the oven. The thing is to keep the water at a desultory bubble.

For the last hour, add the carrots (and any other vegetables you like, such as potatoes in their jackets, a bundle of leeks tied together, smallish whole onions). Towards the end of the cooking time, fish out the carrots; cut them into convenient pieces and cook them in butter with a good pinch of sugar to glaze them and with seasoning if necessary.

Remove the cooked beef to a hot serving dish and arrange the carrots and any other vegetables around it. Sprinkle with parsley. Serve with some of the beef liquor in a sauceboat and with pickled cucumbers and rye bread (unless you decide on potatoes, of course).

If you like dumplings, poach them in the beef liquor if there is room in the pan. Otherwise pour off the amount necessary into a separate pan 30 minutes before the end of the cooking time.

FRENCH BRAISED BEEF WITH CARROTS
BOEUF À LA MODE

The best variation on the theme of beef and carrots, topside being a finer cut than shin or brisket.

> 1½–2½ kg (3–5 lb) topside, nicely tied
> 125 g (4 oz) fat bacon
> 2 medium onions, chopped
> a little oil
> 4 tablespoons brandy
> 1 calf's foot or 2 large pig's trotters
> ¾ kg (1½ lb) sliced carrots
> 300–500 ml (½–¾ pt) red wine
> beef stock
> 2 cloves garlic, finely chopped
> double bouquet garni
> chopped parsley
> salt, pepper, sugar

Chill the bacon until firm, then cut it into strips and lard the beef. This is optional, but a great improver of leanness.

Brown the beef with the onion in a little oil. Transfer to a deep casserole or pan. Warm the brandy and set it alight, pouring it over the beef. When the flames have died down, tuck in the calf's foot or trotters and a third of the carrots. Pour in the wine and enough beef stock to come two-thirds of the way up the beef. Add the garlic and bouquet but no seasoning. Bring to a feeble simmer, skim if necessary and cover tightly. Simmer for four hours. This is easiest done on top of the stove, so that you can hear whether the liquid is boiling too hard and make adjustments accordingly. Test the beef; a large joint may need longer.

Towards the end, cook the remaining carrots in another pan in beef stock, until they are tender. Add a good pinch of sugar.

To serve hot, put the beef on a dish surrounded by the separately cooked carrots and sprinkle with parsley. Serve some reduced cooking liquor as sauce.

To serve cold, cool the beef in its own liquor for four to six hours. Remove it to a close-fitting dish, lined with the separately cooked carrots. Taste, then boil down the cooking liquor to a good flavour and strain over the whole thing. Leave to set in the refrigerator.

ITALIAN CARROT AND ALMOND CAKE
TORTA DI CAROTE E MANDORLE

The almonds for this cake can be used blanched, or washed and unblanched. Grind them coarsely in an electric chopper or with a nut mill, but whichever way you choose the nuts must not end up in a powder like ground almonds.

> 4 egg yolks
> 250 g (8 oz) sugar
> grated rind of a lemon
> 250 g (8 oz) finely grated carrots
> 250 g (8 oz) grated almonds
> 1 heaped tablespoon self-raising flour
> 4 egg whites, stiffly whipped
> pine kernels (optional)
> icing sugar

Beat the yolks, sugar and rind thoroughly together. This will take five minutes by hand, about half that with an electric beater. Mix in the carrots, almonds and flour. Fold in the whites gently with a metal spoon. Line a 5 cm (2 in) pan, about 20 cm (8 in) in diameter, with vegetable silicone parchment. Put in the cake mixture. If you have some pine kernels, sprinkle a heaped tablespoon over the top of the mixture before you put it into the oven. Pine kernels are always an improvement – nearly always, anyway – but the cake is really good without them. Bake at 180°C/350°F/Gas Mark 4 for about 45 minutes. Sprinkle with icing sugar before serving.

Ask people if they can guess what the main ingredients are. No one will think of carrots.

ENGLISH CARROT PUDDING

We have always been enamoured of sugar, though it was mainly a court and aristocratic vice until the 18th century and the sugar plantations of the West Indies. When new vegetables came in, they were viewed without strong savoury prejudice – sweet spinach tart, sweet puddings made from potatoes, artichoke bottoms in pies flavoured with sugar, egg yolks and wine – and carrots fitted well into this pattern. Not only were they sweet in themselves, but the glorious colour gave a most appetizing look to an open tart.

flaky or shortcrust pastry
60 g (2 oz) finely grated carrot
60 g (2 oz) white breadcrumbs
1 large egg
1 large egg yolk
150 ml (¼ pt) whipping cream
60 g (2 oz) butter, melted, cooled
2 tablespoons brandy
1 tablespoon orange-flower water (optional)
3–4 tablespoons sugar
grated nutmeg

Switch the oven on to 180°C/350°F/Gas Mark 4 and put a baking sheet on the centre shelf. Line a 23–25 cm (9–10 in) tart tin, with a removable base, with pastry. Mix remaining ingredients together in the order given, beating each one in well. Orange-flower water need not be added, but it gives an authentic breath of 18th-century cookery. Taste the mixture and put in a little more sugar if you like. Pour into the pastry crust and slide the tart on to the baking sheet – its immediate warmth helps cook the bottom crust more rapidly, so preventing sogginess. Bake for 35–45 minutes until the filling is slightly risen and crusted a beautiful golden colour. Serve warm with cream.

ANGEL'S HAIR CHARLOTTE
CHARLOTTE AUX CHEVEUX D'ANGE

250 g (8 oz) medium grated carrots
250 g (8 oz) sugar
peel and juice of a large lemon
32 boudoir biscuits
juice of a large orange
125 ml (4 fl oz) each double and single cream, or 250 ml (8 fl oz) whipping cream
60 g (2 oz) split, toasted almonds

Put the carrots, sugar, lemon peel cut in strips, lemon juice and 150 ml (¼ pt) water into a pan. Heat slowly until the sugar is dissolved, then bring to the boil and cook hard until the mixture is jammy and thick.

Line a charlotte mould with foil. Dip the biscuits briefly into orange juice and arrange them, sugar side out, round the walls of the mould. Fit more biscuits, cut to shape, in the bottom.

Whip the cream(s) together until stiff. Fold in the cooled angel's hair and almonds. Pile into the mould. If the biscuit palissade comes up too high, trim it off neatly, putting the cut ends over the top of the filling. Cover and chill at least five hours. Turn out to serve.

CAULIFLOWER

The largest cauliflower I have ever seen, a great curdled depth of white cupped in a green fringe of leaves, was about 45 cm across. It was so large that the elderly Turk who was carrying it, in the outskirts of Nicosia, could not get his arm right round.
Only enough to clamp it to his side, as he shuffled along in his droopy black clothes. A good omen. We were on our way to Kythrea, said to be the home of the cauliflower.

In fact this is not true. Cauliflowers may have flourished in the fields around Kythrea, with its never failing Kephalovryso spring, but they were developed by the Arabs in the Middle Ages (even as late as 1699, the best seed came from Aleppo according to John Evelyn). They were being grown in Europe certainly by the 16th century, even in this country which is on the end of the vegetable chain – Gerard mentions them in his *Herball* of 1597. A Jewish-Italian traveller, Elias of Pesaro, wrote back home in 1563 from Cyprus that cabbages and cauliflowers were to be found there in abundance, 'for a quattrino one can get more almost than one can carry' – perhaps he, too, had seen elderly Turks struggling home with vast cauliflowers. As he mentions cauliflowers in the same breath as cabbages, without explaining them, his friends in Pesaro must have been as familiar with them as with the other vegetables he lists – beetroots, spinach, carrot, peas, lentils, white kidney beans, beans.

A confusing thing – winter cauliflower is sometimes called broccoli, both by the Italians and ourselves. I stick to cauliflower all the year round, keeping broccoli for the sprouting varieties with many stalks or shoots and flowering heads. As a cook, never be put off by colour. Sometimes one may see cauliflowers with deep purple heads; they should be cooked like a cauliflower rather than purple-sprouting broccoli. Be guided by size and formation. If you are in real doubt, choose a recipe that is appropriate to the flavour of either.

CHOOSING AND COOKING CAULIFLOWER

A fresh cauliflower of quality looks and smells beautiful. The ones I like best are covered with a mass of creamy pointed whorls, gently studding the whole surface; often they are shaded, round the edges at least, by leaves of long and tender greenness. Other people will prefer the curded kind that looks as white as a cloud on a blue, propitious day. Whichever you happen to like, the general rule for choosing vegetables prevails – if the cauliflower looks back at you with a vigorous air, buy it; if it looks in need of a good night's sleep, with dry, curling, cut down leaves and brownish edges, if it has a slightly rank air that would give the game away to a blind man, leave it where it is.

The problem of cooking cauliflower is how to preserve its beautiful look and flavour, without overcooking the top and without breaking it up. First trim any tiny blemishes away and slice neatly across the stalk so that it can easily be kept upright. If the stalk is thick, cut a cross into it or core it slightly so that the boiling water can get to the centre. Then turn it upside down into a bowl of salted water – this discourages any insect that may be lurking in the flowery depths.

Drain the cauliflower (if you like, rub the white part over with the cut side of a piece of lemon; I don't do this, as I like the creamy colour of cooked cauliflower, but it is a matter of taste). Stand it in a metal basket and lower it into a pan which contains 5 cm (2 in) salted water at a rolling boil; once the water returns to the boil, lower the heat so that it bubbles steadily rather than violently. Cover the pan with its lid, or a piece of foil if the cauliflower stands up above the edge. Leave until the stalk is just tender, sprinkle the flower part with salt and replace the lid for a few moments until the stalk is tender but not collapsing.

Remove the basket from the pan. Let it drain off briefly. Have a large sieve handy, lined with a piece of butter paper or foil and quickly turn the cauliflower into it, head down. As quickly, turn the cauliflower into a hot, deep serving dish, so that it is at last right way up and in reasonably good shape. With a spoon you can push away unsightly pieces underneath the general domed effect.

MICHAEL SMITH'S CAULIFLOWER SOUP

I loathed cauliflower soup, all the crèmes Dubarry of the French repertoire failed to move me, until I tried this recipe of Michael Smith's from his *Fine English Cookery* (Faber paperback, 1977). It is a winner, because the cauliflower is not sieved or liquidized into the soup, but left in tiny florets. Use a reasonably good home-made chicken stock, that is not too strong in flavour.

> *1 medium-sized very white, firm cauliflower*
> *600 ml (1 pt) chicken stock*

600 ml (1 pt) milk
60 g (2 oz) butter
30 g (1 oz) plain flour
salt, grated nutmeg
yellow colouring (optional)
1 egg per person (optional)

Cut the stalk from the cauliflower and divide the head into small 'finger-tip-sized florets. Patience at this stage will be very rewarding.' Poach them in boiling chicken stock for about seven minutes, until cooked but still firm. Remove them with a slotted spoon. Pour the milk into the pan and bring to boiling point.

In a second pan melt the butter, stir in the flour and cook for two minutes. Gradually whisk in the liquid and simmer for five minutes. Strain it back into the rinsed-out first pan, add the cauliflower and reheat gently. Correct the seasoning with salt, and add nutmeg to taste.

Colour the soup with a drop or two of egg yellow colouring, enough to make it creamy but in no way yellow. Avoid the lemon colouring as it sometimes has a nasty flavour. (I do not do this always, as the soup usually has a good colour anyway.)

Put a poached or lightly boiled egg into each plate and pour on the soup. (I often boil the eggs until just hard, chop them with the egg slicer, and serve them in a separate bowl, with another bowl of fried bread cubes for crispness.)

CAULIFLOWER AND GREEN BEAN SALAD
INSALATA DI CAVOLFIORE E FAGIOLINI

A good salad from the Abruzzi, given by Wilma Reiva La Sasso in her *Regional Italian Cooking* (Collier Books, 1958).

Cook a cauliflower, divided into florets, and a roughly equal weight of small green beans in two separate pans of water. They must not be overcooked, but still slightly crisp. Drain them well and arrange them on a shallow dish. Add extra seasoning if necessary and sprinkle over with finely chopped mint and garlic. Mix olive oil and a little lemon juice. Pour evenly over the salad and leave it in a cool place – not in the refrigerator – for the flavours to blend. 30 minutes to an hour is the ideal time.

GREEK CAULIFLOWER

The Greeks have a way of serving vegetables tepid with the delicious lemon and olive oil dressing on p. 559. It sounds unpleasant, but tastes good. Do not be niggardly with the herbs.

CAULIFLOWER WITH PARSLEY BUTTER
CHOUFLEUR MAÎTRE D'HÔTEL

Make up *Maître d'hôtel* butter, p. 551, using half a packet of butter for six people. Dab it over the hot cauliflower in its dish. Scatter on a little more parsley and serve.

CAULIFLOWER PARMESAN

Put the cooked cauliflower into a heatproof dish. Sprinkle the top with a heaped tablespoon of grated Parmesan cheese and dribble over some melted butter. Place under a very hot grill until the cheese melts and turns golden-brown. This gives a light cheese flavour; if you want to increase it, add some grated Gruyère or hard Farmhouse Cheddar and see that at least some of it goes down into the cracks of the cauliflower.

For cauliflower with a cheese sauce see page 179.

RED HOT CAULIFLOWER

Cook the cauliflower and divide it into florets, then turn to the red hot broccoli recipe on p. 115. This Roman way with winter greenery is particularly successful with cauliflower. It is my own favourite recipe for cauliflower.

CAULIFLOWER IN THE SICILIAN STYLE
BROCCOLI ALLA SICILIANA

The Italians certainly grew cauliflowers as magnificent as the one we saw in Nicosia. Perhaps they still do. In 1824 Thomas Appleton, the United States consul at Leghorn, sent a big package of seeds to Jefferson at Monticello, with a letter in which he said 'Their cauliflowers which I have seen at Naples would not enter into a peck measure'. Now a peck is a measure of two gallons, or quarter of a bushel, the capacity of the average bucket.

Oddly enough Italian cookery books do not give many recipes for cauliflower, but here is a good one from Ada Boni.

1 kg (2 lb) cauliflower
250 ml (8 fl oz) olive oil
1 large onion, chopped
12 stoned, sliced olives
½ tin anchovies, chopped
60 g (2 oz) Cacciocavallo cheese, sliced, or Provolone, or strong
  Cheshire
salt
250 ml (8 fl oz) red wine

Separate the cauliflower into florets as evenly as possible. Pour a little olive oil into a heavy non-stick pan. Add some of the onion, olives and anchovies. Then put in a layer of cauliflower, with some cheese, a sprinkling more of oil and a very little salt. Repeat the layers. Finish by pouring on any oil remaining, with the red wine. Cover and stew gently until the cauliflower is tender. Do not muddle up the dish by stirring. The liquid should have evaporated more or less by the time the cauliflower is done, but be prepared to raise the heat and boil it away. Turn on to a hot serving dish and sprinkle generously with little cubes of fried bread.

CAULIFLOWER AU GRATIN

With a very little extra trouble, the old standby of cauliflower cheese can be turned into something really delicious. Although Cheddar can be used, and other hard cheeses, none of them have the rightness of Parmesan and Gruyère, with their sweet and piquant flavours.

Cook the cauliflower, then drain it well, taking care to keep 150 ml (¼ pt) of the cooking liquid to make a mornay sauce (p. 555).

While the sauce is simmering down, break up the cauliflower into pieces and fry them lightly in 60 g (2 oz) butter, stirring them about so that they are evenly coated, but not brown. Now mix in a few tablespoons of the sauce.

Pour a layer of sauce into a gratin dish, arrange the cauliflower on top, and pour over the rest of the sauce. If the cauliflower was a small one, you may not need the whole quantity.

Sprinkle the top with two tablespoons of breadcrumbs and one tablespoon each of grated Parmesan and Gruyère. Drip a tablespoon of melted butter as evenly as possible over the crumbs and cheese and bake in a moderate or fairly hot oven until the top is brown. If the oven has to be at a lower temperature for other things, brown the top under the grill and finish the reheating in the oven – or vice versa.

DUBARRY GARNISH FOR VEAL AND LAMB

Break the cauliflower into large florets and chop the stalk parts. Fry them gently in a knob of butter so that all the pieces are coated with it; then mix in a good 300 ml ($\frac{1}{2}$ pt) mornay sauce, turning the cauliflower over and over in it. Make sure finally that the flower parts are upright; sprinkle them with grated Parmesan and Gruyère and slip the frying pan under the grill for the top to brown. Do this after the meat is cooked and arranged on its serving dish, so that as soon as the cauliflower is coloured and delicious, you can arrange the pieces round the meat.

CAULIFLOWER WITH CREAM SAUCE
CHOUFLEUR SAUCE CRÈME

Break the cooked cauliflower into nice pieces and cut the stalk left over into 1 cm ($\frac{1}{2}$ in) dice. Arrange them on a dish, in a mound so that the less glorious parts are concealed.

> *béchamel sauce made with 150 ml ($\frac{1}{4}$ pt) each milk, cauliflower*
>   *water and cream*
> *salt, pepper, grated nutmeg*
> *3 hard-boiled eggs, shelled*
> *12 triangles of bread fried in butter*
> *chopped parsley*

Flavour the béchamel sauce with salt, pepper and a hint of nutmeg. Pour some of it over the cauliflower and the rest into a sauceboat. Quarter the eggs and arrange them with the fried bread around and on the cauliflower. Scatter with parsley and serve.

CHOUFLEUR À L'AURORE

The above recipe made with *sauce aurore*, p. 553, rather than béchamel, is even more delicious in appearance as well as flavour.

CAULIFLOWER WITH ALMONDS
CHOUFLEUR AUX AMANDES

If you are serving cauliflower with poultry, veal, sweetbreads or any light delicate meat, break it up when it is cooked and turn it over gently in butter over a low heat. Dish it up and scatter it with almonds, blanched and split, or flaked, and browned either in butter or under the grill. They should only be lightly browned, crisp yet still sweet and juicy.

CAULIFLOWER WITH PRAWN SAUCE

Sherry vinegar, the essential flavouring here, was first introduced into this country in 1976 by William Tullberg of Calne, the maker of Urchfont mustards. He gave me a bottle in March, at a time when gardeners had a surplus of tiny cauliflowers. I put the two together. The result was unexpectedly successful.

> *6 small cauliflowers, or 1–2 larger cauliflowers*
> *6 triangles bread fried in butter*
>
> SAUCE
> *600 ml (1 pint) prawns in their shells*
> *about ½ kg (1 lb) fish trimmings*
> *1 onion stuck with 3 cloves*
> *1 carrot, quartered*
> *bouquet garni*
> *1 litre (1¾ pt) water*
> *2 tablespoons sherry vinegar*
>   *or 4 tablespoons dry white wine*
>   *or 2–3 tablespoons dry white vermouth*
>   *or 2 tablespoons each white wine and wine vinegar*
> *1 heaped tablespoon butter*
> *1 heaped tablespoon flour*
> *125 ml (4 fl oz) cream*
> *1 tablespoon grated Parmesan cheese*
> *salt, pepper*
> *chopped parsley*

First make the sauce. Shell the prawns. Set their meat aside for later, put their debris into a pan with the fish trimmings, vegetables, herbs, water, vinegar and wine if used. Boil steadily, but not too hard, with a lid on the pan, for 30 minutes. Strain the liquor into a measuring jug, pressing to extract as much as possible. If necessary, reduce by boiling to something under half a litre (¾ pt). Melt the butter, stir in the flour and cook for two minutes. Moisten with the fish stock and cream and cook to a rich creamy consistency. Trim the cauliflowers so that the white heads are free of all but a few tender green leaves. Cook them in the usual way. Add the cheese and prawns to the sauce, with seasoning to taste, and leave for a further five minutes to heat the prawns through, without allowing the sauce to boil again. If you think a little extra sherry vinegar would be a good idea, add it as well, but go carefully as it should not be overdone.

Put each small cauliflower into an individual pot or arrange the larger cauliflower(s) on a serving dish. Pour over the sauce and scatter with parsley. Tuck the croûtons round the side and serve immediately, very hot.

CAULIFLOWER IN THE BOURBONNAIS STYLE
CHOUFLEUR À LA BOURBONNAISE

The Bourbonnais is centred on Moulins and the spa town of Vichy. To the east the river Loire forms one boundary; to the south the mountains of the Auvergne form another.

> 2–3 shallots, chopped
> 60 g (2 oz) butter
> 60 ml (2 fl oz) wine vinegar
> 250 ml (8 fl oz) double cream
> salt, pepper
> 1 cauliflower, cooked
> chopped parsley

Cook the shallots until soft but not brown in the butter. Add the vinegar and boil hard until it is reduced to about one tablespoon of liquid. Stir in the cream, until the sauce is rich and thick. Correct the seasoning. Pour the sauce over a cooked, drained cauliflower and scatter it with parsley.

CAULIFLOWER IN THE POLISH STYLE
CHOUFLEUR À LA POLONAISE

This way of serving vegetables turns them into a special and delicious course on their own – it has contrast of texture, crispness, butteriness, softness and freshness, it looks beautiful and elegant. Try it too for chicory, leeks, green and purple-sprouting broccoli, boiled fennel and celery. Make sure that the vegetables still have a little bite to them and that they are well drained. Arrange them in a round shallow dish in a circle with a small hole in the middle.

Cauliflower should be cooked in the usual way, lightly. Drain it well and break it into florets. Fry the florets gently in 60 g (2 oz) butter; they should not brown properly, though a few touches of golden colour make them look more appetizing. Arrange them, stalk inwards, on a warm dish. Have ready the following:

> 2 hard-boiled eggs, shelled
> 2 heaped tablespoons chopped parsley
> 30 g (1 oz) white breadcrumbs
> 125 g (4 oz) butter

Fork the eggs to crumbs and mix them with about two-thirds of the parsley. Fry the crumbs in the butter until both are rich golden-brown. Put most of the egg and parsley in the centre of the cauliflower, with the rest scattered generally over it. Pour sizzling butter and crumbs over the whole thing, then sprinkle on the remaining parsley.

CAULIFLOWER À LA GRECQUE
CHOUFLEUR À LA GRECQUE

In spite of the name, this is a favourite French dish. The method can be adapted to other vegetables – mushrooms, leeks, courgettes, young carrots, even parsnips – with suitable adjustment of cooking time. The tenderest vegetables do not need the preliminary blanching. The problem of this dish is to catch the cauliflower while it is just a little crisp, but not undercooked. Hover around the pan and keep tasting.

> 1 cauliflower, broken up
> 250 ml (8 fl oz) dry white wine
> juice of 2 lemons
> sprig thyme
> bay leaf
> small branch green fennel
> ½ teaspoon each coriander and peppercorns, lightly crushed
> 5 tablespoons olive oil
> salt, pepper

Put the cauliflower into a metal basket and place in a large pan of boiling salted water to blanch for five minutes. Count from when the water returns to the boil. Remove the basket and run the cold water tap over the contents. Drain.

Put the remaining ingredients into a pan, and add 150 ml (¼ pt) water. Simmer for eight minutes, covered. Add the drained cauliflower and cook for 10–15 minutes. This time the cauliflower should not be in the basket, but allowed to bathe in the liquid. When it seems just cooked, but still with some bite, remove with a perforated spoon to a salad dish. Strain the liquor into a clean pan and reduce it by hard boiling to a concentrated flavour. Add a little more olive oil, if you like, once you have removed the pan from the heat. Pour over the cauliflower and leave it to cool. Serve chilled with a scatter of parsley.

CAULIFLOWER MOUSSAKA
MUSACA DE CONOPIDA

In Rumania, cauliflower is sometimes used to make a dish very close to aubergine *moussaka*. Turn to the recipe on page 57 and substitute one large lightly cooked cauliflower for the aubergine. Divide it into small florets and use in layers in exactly the same way. Do not fry the cauliflower.

CELERY AND CELERIAC

There are three sources of celery flavour – the pale greenish heads of celery on sale most of the year, usually in long plastic bags, the white winter celery encrusted with black soil sold by good greengrocers rather than in supermarkets, and the round turnip-looking root known as celeriac that appears in the autumn and lasts through until the spring with careful storage.

I do not deny the usefulness of modern varieties of green celery. Splendid to have that kind of flavour in the summer time and to be able to rely on it as a standard kitchen item. It is, I suppose, the crown of that long gardening effort which began in Italy in the late 15th and early 16th century, the effort to turn smallage, a plant useful in medicine but too bitter to be served with meat or as a salad, into a vegetable that people could eat for their own good but with enjoyment. On the other hand, white celery, at its best after a few early frosts, is much crisper and better flavoured. It is also stringier, and has more waste. Nonetheless, buy it. Enjoy it in its season. This is the Christmas celery, the real thing, what John Evelyn called 'the grace of the whole board'. And if you like to be reminded from time to time that vegetables come from the earth, growing things, you will not mind the few seconds scrubbing required, to clean the heads.

I do not understand why celeriac, so convenient and storable a form of celery flavouring, has never become as popular here as it has, say, in Germany. It seems that the garden designer and seedsman Stephen Switzer introduced it in the 1720s (the seeds came from Alexandria). He may even have invented the name, certainly he was the first to use it in print, in a pamphlet on growing 'foreign kitchen vegetables' that he offered with his seeds. This makes it almost as old a plant in our gardens as celery. Its lack of success is odd.

HOW TO PREPARE CELERY AND CELERIAC

For eating raw, trim off any roots still attached to the head of celery and separate it into stalks. Scrub away soil from the outer stalks under the cold tap and rinse the cleaner stalks lightly. Set aside the tough pieces and the bright green leaves to use as flavouring for stock, soups and stews; leave the pale yellowish leaves, as they are good to eat, either with the stalk, or chopped up with other herbs in a celery salad.

Some of the worst string will come away as you break off the green leaves: the rest can be pulled and sliced away with a sharp knife. With green varieties of celery, this stringing is often unnecessary.

If you want to keep celery, wash it as little as possible and store it in the refrigerator in an air-tight box. Celery with the earth still on can be kept in the cool, on a larder floor for instance, well wrapped in newspaper. But it is best to eat celery the day you buy it.

Celery is a tricky vegetable to cook. The safest method is to slice the stalks across into pieces and to blanch them briefly in boiling salted water. They can then be finished in various ways. Or you can cook it Chinese style. I have eaten so many watery, stringy, braised celery hearts in hotels and dining cars, that cooking celery whole is something I avoid. The only time I have eaten it with pleasure, was in a French restaurant in Tel Aviv. The combination of Israeli horticultural skill and French culinary style produced a wonderful result, tender, delicious the whole way through. The heads had been trimmed down to a length of about 16 cm (6 in), and then split in half before being cooked. This you can only do with modern green celery that grows close and tight and clean, with no particles of earth between the stalks.

Celeriac, being a root, needs careful looking over before you buy. Unlike a turnip or swede, it has a convoluted and bossed exterior which means that you can get quite a lot of waste when you remove the outer part. Choose the smoothest looking celeriac you can find, for this reason, and cut it into pieces before you start peeling so that you can see properly what you are doing.

Celeriac discolours when it is cut and although this doesn't matter from a taste point of view it gives a greyish, unsightly tinge that can put people off. Perhaps this is why it has never become popular here. To prevent discolouring, drop the pieces, as you cut them, into a basin of water acidulated with a tablespoon of vinegar to a litre (2 pt). Another way is to rub the cut surfaces over with a piece of lemon.

To maintain as much whiteness as possible, cooks often blanch celeriac not in salted water, but in a *blanc à légumes*, p. 549: 30 g of flour gradually diluted with a litre of water (1 oz to 2 pt), and flavoured with salt, lemon juice and aromatics.

If you need celery to flavour stocks or stew, and have none, use a good pinch of lovage instead.

DANISH CELERY AND CHEESE SOUP

A number of good soups can be made with celery, but this is the most delicious of them all. The creamed blue cheese gives a savoury yet tactful richness to the light flavour of celery. If you are using a milder blue cheese than the Danish kind, be prepared to add a little extra.

> *1 medium head celery, chopped fairly small*
> *2 medium onions, chopped*
> *50 g (scant 2 oz) lightly salted butter*
> *30 g (1 oz) plain flour*
> *1 litre (2 pt) chicken stock*
> *salt, pepper*
> *60 g (2 oz) Danish blue cheese*
> *chopped parsley*

Cook the vegetables in the butter for ten minutes in a covered pan without browning them. Stir in the flour thoroughly, then moisten with the stock. Season. Cover and simmer for 40 minutes, until the celery is really tender. Mash the cheese to a cream and whisk it gradually into the soup just before serving it; once you start doing this, lower the heat to make sure the soup remains well below boiling point. Correct the seasoning, add parsley, and serve with croûtons.

CELERY AND DILL WEED SOUP

Make the soup as above, but omit the cheese. Flavour it finally with dried dill weed, adding it gradually to taste and giving the flavour time to develop in the soup between doses. Stir in a couple of tablespoons of double cream to finish.

CELERY ENGLISH STYLE

Serve the best inner stalks standing in a glass jug or jar, with a large piece of Farmhouse Cheddar, plus biscuits, bread and salty farm butter.

When a Stilton is coming to an end, mash the last pieces with butter and serve with the celery.

CELERY FRENCH STYLE

Instead of serving celery with the cheese, the French put it on the table for a first course, together with a bowl of *gros sel* or sea salt crystals and a pat of unsalted butter. The idea is to fill the channels of the celery with butter, sprinkle the butter with salt, and eat it with bread. Much to be recommended for the best quality winter celery.

CELERY WITH CREAM

We may deserve our reputation for badly cooked vegetables in hotels, inns and restaurants, a reputation that is three centuries old. At home, though, things must have been different, to judge by the cookery books of the 18th century which are full of good vegetable dishes like this one.

Wash, clean and trim three heads of celery. Cut across so that they fall into pieces of 7–8 cm (3 in) length. Boil them in salted water until just tender. Pour off the water (keeping it for soup). Have ready 150 ml (¼ pt) whipping cream beaten with two egg yolks and seasoning. Stir this into the hot celery, over a low heat, until the sauce thickens without boiling. Serve at once as a vegetable, or better still on its own with toast.

CELERY AND MUSSEL SALAD
CÉLERI EN SALADE AUX MOULES

Mussels go well with salads; their small piquant richness enhances both the crispness of celery and the softness of potato. If you want to serve the salad as a first course, reduce the quantities of mussels and potatoes by one-third, or more depending on the rest of the meal. I find this dish, and the simpler potato and mussel salad on p. 407, ideal for Christmas and New Year meals; the fresh flavour cuts into the heavy eating of that time of the year in a vigorous way.

> ½ kg (1 lb) potatoes, preferably Desirée
> 3 litres (good 5 pt) mussels
> 4 shallots, chopped, or 4 heaped tablespoons chopped onion
> 100 ml (3–4 fl oz) dry white wine
> 1 head of celery
> 300 ml (½ pt) mayonnaise
> Dijon mustard
> 3 hard-boiled eggs
> chopped parsley

Scrub and boil the potatoes in their skins. Peel and dice them into a bowl while still warm. Meanwhile scrub the mussels, discarding any that remain open when tapped smartly with a knife. Put the mussels into a pan with a quarter of the shallot or onion, and the wine. Cover, set over a high heat, and leave for five minutes until the mussels open. Discard the shells and strain the very hot liquor over the diced potatoes. Leave the mussels and potatoes to cool.

Cut the celery into fine slices. Flavour half the mayonnaise with mustard to taste, starting with a teaspoonful. Mix in the celery. Mix the remaining mayonnaise with the cold mussels, drained potatoes and the rest of the shallot or onion. Shell and quarter the eggs.

Put the mussel salad in the centre of a large plate. Surround it with celery salad and arrange the eggs in a circle between the two. Scatter with parsley. Serve well chilled.

SOME CELERY SALADS

WALDORF Slice equal quantities of celery and unpeeled eating apples (crisp and reddish-skinned for preference). Mix with mayonnaise, adding some chopped walnuts. Decorate with walnut halves.

BOULESTIN'S SALAD Mix equal quantities of diced cold turkey or chicken, celery and unpeeled eating apples. For each 250 g (8 oz) poultry, add 100 g (3½ oz) diced Gruyère cheese. Dress with vinaigrette or mayonnaise.

BEAUCAIRE Set aside the heart from a head of celery. From the rest slice 125 g (4 oz) celery strips. Mix with 125 g (4 oz) blanched celeriac julienne strips, 250 g (8 oz) sliced tomatoes, 200 g (scant 7 oz) sliced beetroot, 12 hazelnuts, 125 g (4 oz) ham cut in squares. Moisten with vinaigrette, then mix with 250 ml (8 fl oz) mayonnaise. Decorate with extra beetroot and tomato slices and push the celery heart into the centre.

CELERY CURLED SALAD Cut the stalks into lengths of 4–5 cm (about 2 in), then slice them down into long matchstick shreds. Put them into a covered plastic box of water with ice cubes in it. Leave them for at least two hours. The shreds will curl slightly, and be marvellously crisp.

CELERY WITH ANCHOVY SAUCE
CÉLERI AUX ANCHOIS
L'ANCHOUÏADO AU CÉLERI

Ideally for this dish you need to buy a dozen salted anchovies from the delicatessen. They should then be soaked in cold water for 15 minutes.

Rinse them under the cold tap and carefully remove the fillets, throwing away the rest.

1) For a simple, cold anchovy sauce, crush or pound the fillets with enough olive oil to make a rich cream. Add a little wine vinegar to taste. Divide this mixture between small individual bowls, so that each person can dip his celery stalks into the sauce at each bite.

2) For a more elaborate and festive dish, make a hot sauce in a presentable double saucepan. Or stand a wide stoneware pot over an ordinary saucepan half full of boiling water. Pour a third of a litre of good olive oil into the top of the pan or the pot (about 12 fl oz). Add the anchovy fillets and with a spoon or fork crush them so that they melt into the oil. Now add up to 1 cup of chopped shallots and chopped garlic, in the proportion of three to one and a teaspoonful of thyme. Stir everything well together over the heat until it is warmed through and aromatic. Just before serving add a little wine vinegar. Leave the sauce over its hot water so that it keeps warm as people dip their celery into it.

You can serve the celery, divided into stalks, in a glass jug. Or cut each head in half, and put one half on each plate, together with a thick slice of French bread with an anchovy fillet laid across it. This way, the prettier of the two, comes from *La Cuisine Rustique: Provence*, by Maguelonne Toussaint-Samat, Robert Morel Editeur, 1970.

CELERY STUFFING AND SAUCE FOR POULTRY

These recipes go particularly well with boiled turkey, even a small battery turkey, but can as well be used with boiled or roast farm chicken. If you want to use a boiling fowl, i.e. a mature creature that will require a couple of hours boiling at least, perhaps longer, omit the stuffing as it can become too soggy for pleasure. Never attempt to boil a battery chicken; they have too little flavour.

Halve the quantities below if they are intended for chicken.

STUFFING

        250 g (8 oz) chopped onion
        250 g (8 oz) chopped celery
        125 g (4 oz) butter
        350 g (12 oz) white breadcrumbs
        rind and juice of 1 large lemon
        4 tablespoons chopped parsley
        2 small eggs
        salt, pepper

Soften the onion and celery in butter without browning them, for 20 minutes. Cool slightly, then mix with the remaining ingredients. Stuff the bird lightly, leaving room for the mixture to swell. If the bird is to be boiled, sew up the vents firmly to keep the stuffing in place.

SAUCE

>*1 head celery, chopped*
>*80 g (scant 3 oz) butter*
>*½ litre (good ¾ pt) béchamel sauce*
>*150 ml (¼ pt) double or whipping cream*
>*salt, pepper, lemon juice*

Blanch the sliced celery for ten minutes, drain and complete the cooking in the butter until very soft. Pour in the béchamel and continue to cook slowly for another ten minutes. Purée in a blender or put through a *mouli-légumes*, then sieve to get rid of any remaining stringiness. Reheat, add cream and check the seasoning.

GAME BIRDS BRAISED WITH CELERY

Adjust this recipe either to a roasting pheasant or to pigeons and grouse and so on that are only suitable for the casserole. The point of the dish is the way the celery takes up the rich flavours of the game and at the same time gentles their strength and provides a complementary flavour of its own.

>*gamebirds for 4–6 people*
>*butter*
>*stock made from giblets boiled in chicken stock*
>*2–3 diced rashers of green back bacon*
>*150 ml (¼ pt) port*
>*1 large head celery, cleaned, sliced*
>*good pinch lovage*
>*salt, pepper*
>*200 ml (7 fl oz) whipping cream, or half single, half double cream*
>*1 large egg yolk*
>*chopped parsley*
>*salt, pepper, lemon juice*

Brown the birds all over in a little butter. Tuck them into a close-fitting pot, breasts down. Pour in 300 ml (½ pt) stock if you have roasting birds;

otherwise pour in enough stock barely to cover them. Add the bacon and port. Close the pot tightly with foil and a lid.

FOR A ROASTING BIRD Put it into the oven at 180°C/350°F/Gas Mark 4 for half an hour. Then turn it breast side up and pack in the celery all round, with lovage and seasoning. Cover closely again and return the pot to the oven for a further half hour, or until the bird, or birds, are cooked. Put on to a serving dish, surrounded by the drained celery. Beat the cream and egg with the cooking juices (reduce them first if they taste watery). Stir over a gentle heat until the sauce thickens, without boiling. Correct the seasoning, adding parsley and a little lemon juice if necessary. Pour a little sauce over the celery, serve the rest in a sauceboat.

FOR CASSEROLE BIRDS Put the pot into a lower oven, 150°C/300°F/Gas Mark 2, or simmer on top of the stove, until the birds are almost done. You can judge this by looking at the breastbones: if the meat is just beginning to come away at one end, or can easily be raised with the point of a knife, you have got to the right stage. Meanwhile slice up and blanch the celery in boiling salted water for five minutes, then drain it well.

Turn the nearly cooked birds on to their backs, and put in the celery and lovage with extra seasoning if required. Push it down into the juices. Cover the pot and complete the cooking.

Take the birds one by one from the pot, and carve away the large breast pieces; keep the legs and carcases for soup. Strain the liquor from the celery into a wide pan, and boil it rapidly down to 250 ml (8 fl oz) of concentrated flavour – be guided by flavour rather than by precise quantity. Finish the sauce with the cream and egg yolk as above.

Put the cooked celery round the hot serving dish, with the breast pieces down the middle. Pour over a little of the sauce and serve the rest separately. Game chips, matchstick potatoes or fried breadcrumbs will complete the dish with their crispness.

Note If you like the sauce a little thicker, use an extra yolk, or a teaspoonful of cornflour slaked in a little stock.

CELERIAC SOUP

The unusual idea of combining celeriac and dried ceps is East European. If you cannot buy dried Polish ceps at the delicatessen, Italian *funghi porcini* will do just as well as they are the same mushroom (though they tend to be more expensive). If you dry your own, you will have no trouble.

15 g (½ oz) dried ceps
100 g (3–4 oz) prepared celeriac, chopped
100 g (3–4 oz) chopped onion
60 g (2 oz) butter
150 ml (¼ pt) each soured and single cream
1 level tablespoon flour
chopped dill weed or parsley
salt, pepper

Pour a ladleful of very hot water over the ceps, and leave them to soak for 20–30 minutes. Sweat the celeriac and onion meanwhile in the butter, in a covered pan, until they begin to soften, then add the mushrooms and their liquor. Simmer, covered, until the vegetables are tender. Purée in a blender, or put through the *mouli-légumes*, and return the purée to the pan. Mix the creams with the flour to make a smooth paste and stir into the soup as it reheats. Cook slowly for about five minutes, until the taste of flour has gone. If the soup is too thick for you – it may well be if it was blended – dilute with more water. Stir in chopped dill weed to taste, or parsley, and seasoning. Serve with croûtons.

CELERIAC CHIPS
CÉLERI-RAVE SAUTÉ

The day before you want to eat the celeriac, peel a fine large root and cook it whole for about 20–30 minutes. Pierce it with a skewer or larding needle to make sure it is just tender. Leave it to cool. Next day cut it into ½ cm (¼ in) chips, or into dice. Cook them gently in butter until they are nicely golden. If you are serving them with roast poultry or meat, pour in two or three tablespoons of the rich juices, and turn the celeriac over gently. Serve scattered with parsley.

If you wish to serve the celeriac on its own, don't worry about the meat juices. Put the browned chips or dice into a gratin dish, scatter the whole thing with 100 g (3–4 oz) of grated Gruyère and put under a hot grill for a few moments.

GRATIN OF CELERIAC ALLA MILANESE

For general remarks about making gratins, see page 411. If they are made with hot ingredients for immediate eating, they can be completed under a hot grill to give the browned and bubbling surface. If they are made in advance, for eating later the same day, or for freezing, they will need to bake through in the oven. The temperature does not much matter: be guided by whatever else you may be cooking at the same time. The browning can always be completed under the grill.

Cut one large or two smaller celeriac into slices rubbing them with lemon juice to prevent discoloration. Alternatively, peel and slice them into acidulated water.

Cook them in boiling, salted water until just tender. Arrange in a buttered gratin dish, scattering each layer with grated Parmesan cheese and dabs of butter. Pour tomato sauce over the whole thing. Top with cheese, breadcrumbs and butter. When making the tomato sauce, recipe on p. 511, include a hot chilli or some cayenne pepper, if you like.

CELERIAC SALADS

When using celeriac in any salad, slice it into a bowl of acidulated water, then bring the pan with more acidulated water to a hard boil. Drain the celeriac and plunge it into the boiling water. When the water returns to a vigorous boil, tip the whole thing into a large colander set in the sink. Run under the cold tap, dry on a cloth. This brief blanching makes the celeriac much nicer to eat, without destroying its crisp chewiness.

CÉLERI-RÉMOULADE Cut one celeriac in julienne strips and mix with $\frac{1}{4}$ litre ($\frac{1}{2}$ pt) mustard-flavoured mayonnaise (Dijon mustard). Decorate with chopped parsley.

SICILIENNE Mix equal quantities of celeriac julienne strips, diced cooked artichoke bottoms, sliced russet apples and sliced tomatoes. Use an oil and lemon dressing.

OTHER SUGGESTIONS

Some cardoon and fennel recipes can be adapted successfully for celery. Turn to the sections starting on pages 154 and 254.

CELTUCE, ASPARAGUS LETTUCE

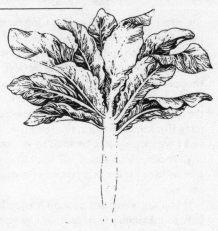

Celtuce has been grown for centuries in China, particularly in the western region near Tibet, where it is called *woo chu*. The main point, the virtue of the vegetable, is its stem which is long and swelling, with a celery kind of crunch when you eat it. The leaves can be eaten raw like lettuce, or cooked like greens. Hence the name of celtuce. It has a lush, tall growth, rather like a bolted lettuce. In Chinese markets all the tough lower leaves are cut away and the stem is sold with only its top plume of leaves. The general opinion of western gardeners is that the leaves are a poor substitute for spinach and lettuce, the stalk is the thing, with a delicate gentle flavour and agreeable texture.

Celtuce is a latecomer to our gardens. You are never likely to see it on sale in a greengrocer's shop. The seeds were sent from China in 1938 by a missionary, the Reverend Carter D. Holton, to W. Atlee Burpee Company, the great seedsmen of the United States. Four years later they offered 'celtuce' in their catalogue.

It seems that this was a reintroduction, and that seeds had been on sale in America in the 1890s, as asparagus lettuce. Again a reference to the delicious stalk.

HOW TO PREPARE AND COOK CELTUCE

Like Swiss chard, celtuce is two vegetables in one. They should be treated separately. Cut off the large leaves growing up most of the stem, and throw them away. Keep the top ones, the small younger leaves, for a salad, or cook them spinach style.

The stem should be cut into handy pieces for peeling, then it can be sliced across, or lengthways into strips, which is better if you intend to serve it raw (with salt like celery, or with vinaigrette or mayonnaise). If you want to cook the stalk, turn to recipes for celery, celeriac and cardoon, but remember to reduce the cooking time to four or five minutes. Celtuce is not stringy or hard.

Celtuce is not mentioned in Chinese cookery books published in England, but is an obvious candidate, both leaves and stalk, for Chinese dishes in which spinach or celery figure.

To stir-fry celtuce stalks, heat through two slices of fresh ginger, a small chopped onion and one chopped clove of garlic in three or four spoons of oil. Use a large light frying pan. When it is very hot, after a minute, add the celtuce and stir about until the pieces are coated with oil. Pour in 125 ml (4 fl oz) chicken stock, two tablespoons soya sauce, a teaspoon of sugar and ¼ teaspoon salt. Continue to stir until the celtuce is tender but still slightly crisp. You can, if you want to make a change, cream the sauce by stirring in at the end a teaspoon of cornflour slaked with two tablespoons of water.

If you are using this method for the leaves you may have to raise the heat to maximum to evaporate any wateriness.

CHARD see SWISS CHARD

CHAYOTE

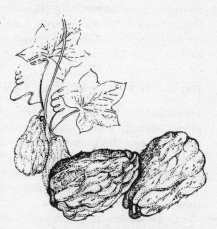

Although the chayote grows on a vine, it does not look much like the gourds we know well, such as courgette or pumpkin. It is more of a ridged green pear in shape, a pear-cucumber to look at, and the vine that seems so tender does not lie submissively on the ground, but scales roofs and trees. The harvesting of chayotes can involve incredible gymnastics, unless you take care to plant them at the foot of a trellis and train the ramping growth sedately on strings across a courtyard or terrace. Then the fruit becomes a deliciously precise crop, hanging down evenly spaced like Christmas decorations. You have only to reach up to pick them and can see why the Chinese call them Buddha's hand gourd.

A friend of mine in Australia, where chokos as they are known there now run wild, will preserve the last fruit of the season so that she can plant a more manageable crop. One year she took into her kitchen an enormous fruit weighing well over a kilo (the ideal eating size is from about 175 g or 6 oz). It soon started to wander. A long pale stem with rudimentary leaves and clinging tendrils burst through the choko from the single flat seed, and explored every cranny of the room, a triffid of a plant, until it found the door. Then she disentangled it carefully and cradled the shrunken parent to a hollow she had made by the trellis, where it could take root and rampage fruitfully. She mostly picks her chokos when they are only a few days old and about 8 cm (3½ in) long. Then they can be cooked, skin and all, like courgettes, and served as a salad with an olive oil and lemon dressing. Some are left to grow larger, up to about 20 cm (8 in); they go into chutneys, or they are simmered in a red wine syrup like pears and eaten for pudding. You will judge from this that the taste is not pronounced, delicate according to some people, insipid according to others. It is more of a texture and substance than a flavour, though it has its devotees, especially in Mexico and the Caribbean islands. This you might expect, as the chayote is a native of tropical and sub-tropical America (the name comes from the Aztec *chayotl*).

In Central America and in the other hot parts of the world where it flourishes, the young shoots and leaves of the chayote are eaten as well as the fruit. Unlike the other gourds, it is a perennial, and after a year or two the root-tubers swell to the size and appearance of yams. They can be cooked in the same way, and have a similar taste.

HOW TO BUY AND PREPARE CHAYOTE

Do not be put off by different names – choko, chaco, xuxu, christophene – nor by odd appearances, as there are a number of varieties of chayote – spiny and smooth-skinned, white and green, rounded and ridged, more and less pear-shaped. In this country, you will see them in West Indian shops and some supermarkets. If you are house-keeping in France, they are occasionally sold in shops that specialize in Algerian items such as *couscous*, *harissa*, *ras el hanout*.

You are unlikely to have much choice in the matter of size. Recipes in Caribbean and Mexican cookery books stipulate fruit weighing between 250 and 375 g each (8–12 oz), up to 20 cm (8 in) long. As with courgettes, unless you grow them, you are unlikely to be able to eat them at an ideal 6–8 cm (2–3 in) long. At this stage, follow recipes for courgettes, as a general guide.

Although they are sometimes peeled before cooking, larger chayotes can be boiled in their skins. This is desirable with so pale-flavoured a vegetable, as a way of keeping in the goodness. It is essential of course if you intend to stuff and then bake it. If on the other hand you want the chayote to absorb flavour from the cooking liquid, as with some of the sweet recipes, you should peel it first.

It is a vegetable that invites spicy vigorous treatment, though you can choose to play up its delicacy with a cream sauce. This latter way is to be avoided unless you are prepared to lavish butter, cream and care, otherwise the precious chayote will taste as dull as institutional marrow in the dreaded 'white' sauce, made with marge and thin milk.

CHAYOTE SALAD
CHAYOTTE EN SALADE

The particular pleasure of this salad is provided by the contrast of smooth chayote and crisp fried garlic bread. For success, use olive oil in the dressing and plenty of chopped herbs.

> *3 pear-sized chayotes, or 2 large ones*
> *3 tablespoons lemon juice*

1 teaspoon Dijon mustard
good pinch sugar
6 tablespoons olive oil
salt, pepper
chopped parsley and chervil
large slice white bread
olive oil for frying
1 large clove garlic, crushed

Boil the chayote in their skins until tender – pierce one with a larding needle after 30 minutes to see if it is cooked; if not, leave a little longer. Run the chayote under the cold tap until you can remove the skins. Cut each one into four pieces and arrange them on a large round dish in a flower shape. Include the large seeds which are delicious.

While the chayotes are cooking, make the dressing: beat together lemon juice, mustard and sugar, then add the oil gradually. Season and mix in the herbs. Pour this over the chayote pieces while they are still slightly warm if possible. Leave in a cool place for an hour or two.

Half an hour before serving the salad, remove the crust from the bread and cut it into small cubes. Heat four or five tablespoons of olive oil with the garlic, slowly to extract the flavour, then fry the bread to a golden-brown. Drain on kitchen paper and scatter over the salad just before serving.

CHAYOTE CREOLE
CHAYOTTES À LA CRÉOLE

Creole cookery – or criollo cookery as it is called in the Spanish-speaking countries of Latin America – is the regional cooking of Central and South America as it was influenced by the invaders from Europe. In other words it is a blend of European and Indian ideas and foodstuffs. To us the most noticeable difference is the increased pepperiness, the abundant use of tomatoes with chillis and peppers in variety, of sweet-corn both as vegetable and cereal, and of beans and squashes that we cannot always identify.

3 large chayote
3 large onions, chopped
1 clove garlic, chopped
6 tablespoons olive oil
½ kg (1 lb) well flavoured tomatoes, peeled, chopped
tomato concentrate

bouquet garni
1 heaped teaspoon sugar
salt, pepper, cayenne pepper

Boil the chayote for 15 minutes, then peel and quarter them. Meanwhile cook the onions and garlic in the oil, slowly for five minutes, then faster so that they begin to brown very lightly. Put in the tomatoes and a tablespoon of concentrate, the bouquet, half the sugar, salt, pepper and a generous pinch of cayenne.

Add the quartered chayote to this sauce, and cover the pan. Simmer for an hour. If the sauce becomes very oily, or too dry, add three or four tablespoons of hot water: it should reduce to a savoury pulp. Correct the seasoning, and serve with plainly boiled rice (what the French call *riz créole*).

STUFFED CHAYOTE

For six people, allow three chayote, each weighing 300–375 g (10–12 oz). Boil them unpeeled in salted water until tender, testing them after 30 minutes. Run them under the tap to cool them, then halve and scoop out the insides, leaving enough to make a good firm shell. Mash the insides, including the seeds, to a pulp. Make up one of the stuffings below, and fill the chayote shells. Sprinkle them with breadcrumbs and melted butter, or with breadcrumbs and Parmesan cheese and melted butter, and bake at 200°C/400°F/Gas Mark 6 for 15–20 minutes. The tops should be golden-brown; they can be finished under the grill, if need be.

NEW ORLEANS SHRIMP STUFFING Put the mashed chayote into a frying pan with a good knob of butter, and cook quickly to evaporate excess moisture. As the mixture begins to dry out, mix in 50 g (scant 2 oz) breadcrumbs and fry a little longer. Stir in four tablespoons of basic tomato sauce and 200–250 g (6–8 oz) shelled shrimps or prawns. Season with salt, pepper and cayenne.

If you have no tomato sauce to hand, brown lightly a chopped medium onion and a chopped clove of garlic in butter. Add a huge tomato, skinned and chopped, with parsley, thyme and quarter of a bay leaf. Season and bubble vigorously to make a sauce-like purée. Add a little sugar and vinegar if necessary, to emphasize the flavours. Do not overdo them.

MEAT STUFFING Make up a basic minced meat with 375 g (¾ lb) raw beef and other ingredients in proportion – see aubergine moussaka page 57.

Add the chayote pulp and one fresh red chopped chilli, minus seeds. Instead of the chilli you could use *harissa* or *ras el hanout*, or cayenne. Boil hard to evaporate moisture, the mixture should be juicy, but not wet.

You could also adapt this recipe to make individual chayote *moussakas*, by topping the meat stuffing with an egg-enriched béchamel sauce flavoured with Parmesan.

CHEESE STUFFING Soften a large, chopped onion in a generous tablespoon of butter. Add the mashed chayote and seasoning and cook more vigorously to dry out the mixture. Stir in 60 g (2 oz) breadcrumbs and cook for a minute or two. Stir in 75 g (2½ oz) Cheddar cheese, coarsely grated, and 50 g (1½ oz) grated Parmesan, then take the pan off the heat before the cheese has a chance to get stringy.

This mixture can be varied by adding some fried mushrooms. Allow 125–175 g (4–6 oz), and slice them small or chop them coarsely. Cook them in butter with a chopped onion and a small chopped clove of garlic. Add the cheese to the mixture after stirring in the mushrooms, adding it slowly, tasting as you go. You may not need the full quantity. Aim to balance and bring out the mushroom flavour, rather than to have a dominant flavour of cheese.

CHAYOTE IN RED WINE

A sweet dish to be served as a pudding, with cream.

> 6 pear-sized chayote, peeled and left whole, or 3 larger ones,
>   peeled and halved or quartered
> 150 g (5 oz) sugar
> 300 ml (½ pt) water
> 150 ml (¼ pt) red wine
> 5 cm (2 in) piece cinnamon stick
> 4 cloves
> lemon juice

As you prepare the chayote, put the sugar and water on to dissolve and simmer for two minutes. Use a pan that will take the chayote fitting closely together in a single layer. When the syrup is ready, put in the chayote carefully and add the red wine and spices. Cover and simmer until tender. Remove the chayote to a bowl, arranging them upright, like pears, if they are small and whole. Taste the cooking liquor and reduce it by boiling if the flavour needs to be concentrated (this will depend on how much evaporation took place during cooking). Add a little lemon juice to bring out the flavour.

Strain the juice over the chayote, to come no more than half-way up. Serve with whipped cream, lightly sweetened with icing sugar.

OTHER SUGGESTIONS

Chayote chutney See the recipe for pumpkin chutney, and substitute peeled chayote for pumpkin.

Chayote à la grecque Peel and quarter the chayote and blanch for about 15 minutes, until half-cooked. Then follow the recipe for cauliflower *à la grecque*, page 183.

CHICK PEAS

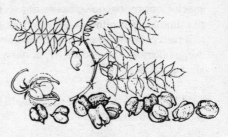

Chick peas are the oddest, most recognizable of the dried pulses, with a plump hazelnut shape ending in a fine point. The second word of their botanical name *Cicer arietinum*, describes their resemblance also to the skull of a ram – *aries*. *Cicer* gave the Italians *ceci* (often eaten with noodles, butter and Parmesan and in soups), the French *pois chiches*, and ourselves chick peas. Nothing to do with poultry. It gave Cicero his family name – an ancestor had a wart on his face like a chick pea. So, as often happens, what started as a nickname was passed on to sons and grandsons, becoming a surname.

Chick peas are so old a basic food of India and south-western Asia that they are unknown in wild form (like onions and leeks). They spread to the Middle East, then round the Mediterranean with the Arabs to Algeria, Morocco, Spain and Portugal, where they are added to stews of mixed meats such as *dfeena*, *couscous*, and *cocido*, and to soups like *harira*.

Unless you have a solid-fuel or oil-fired cooker that is perpetually burning, you might as well buy canned chick peas. They are certainly far more expensive. The drained weight of 250 g (8 oz) – the equivalent to 125 g (4 oz) dried chick peas – costs almost as much as 500 g (1 lb) of the finest Italian *ceci*. But you save cooking heat and they stand up well to canning.

HOW TO CHOOSE AND PREPARE CHICK PEAS

At good Italian delicatessens there will usually be a choice of chick peas. Go for the larger ones. They are a little more expensive, but they cook tender more quickly.

Soak them for 48 hours, in cold water. If you are in a hurry, cover them generously with boiling water and leave for 12 hours. As with all dried vegetables, when they have doubled their weight they have soaked long enough for you to begin cooking them. In the old days, soda was often recommended as a softening aid in the soaking water; modern chick peas do not need this.

Drain and put them into a pan with plenty of water, a bouquet garni and an onion, with a clove of garlic, but no salt. Bring to simmering point and cover. Cook for two to six hours, as necessary. A pressure cooker reduces this time to a few minutes. However long you cook chick peas they never lose their charming shape, or collapse into a mush. Add the salt about five minutes before the end of the cooking time, when the chick peas are already tender.

CHICK PEA AND SESAME SEED PASTE
HUMMUS BI TAHINA

A blender is essential for making *hummus bi tahina*, for it will take you a long time and much effort to *mouli* the mixture to the correct consistency. The point is to reduce the chick peas (*hummus* or *hommos* in Arabic) to a richly flavoured cream, with sesame seed paste or *tahina*, which can be bought from health food stores and delicatessens. It is seasoned with garlic and lemon and served on its own, or with other dishes as a first course. Pitta – a small pouch of Greek or Arab bread – is used to scoop up the cream; pitta is easy to make, the recipe is on p. 568.

> *150 g (5 oz) chick peas*
> *juice of 2 lemons*
> *2 cloves garlic, halved*
> *100 g (3–4 oz) tahina*
> *olive oil, salt, cayenne, chopped parsley*

Cook the chick peas, then drain them, saving their liquor. Put four tablespoons of this liquor into the blender goblet, with the lemon juice and garlic. Set the blades whirling at top speed, then add chick peas and about half the *tahina* alternately, through the hole of the lid. If the mixture clogs, add a little more liquor and some olive oil. You should end up with a thick, coherent, creamy purée. Season with salt and add more *tahina* if you like, but do not overdo it. Scrape out into a bowl, or bowls, and cover with a thin layer of oil and a sprinkling of cayenne pepper and parsley.

PURÉE OF CHICK PEAS WITH GARLIC AND ONION

Here is another crushed chick pea recipe. Claudia Roden says that although it is a 'peasant' dish, it is extremely popular in the Lebanon.

> *375 g (¾ lb) chick peas, soaked*
> *2–4 cloves garlic, crushed*
> *salt*

> olive oil
> chopped onion, or thin onion slivers
> 250 g (8 oz) pitta, toasted, or other white bread, sliced and toasted
> paprika

Boil the chick peas in a litre (1¾ pt) of water until tender. Drain off the liquid and set it aside while you crush the chick peas with a potato masher or electric chopper. Flavour them with garlic, salt, olive oil to taste and some chopped or slivered onion.

Cover the base of a shallow dish with broken pieces of toasted bread. Pour on enough of the chick pea cooking water to soak them. Spread the chick pea purée on top, sprinkle with a little more oil and a dusting of paprika. Heat through in the oven.

VARIATION 1 Mix up some yoghurt with crushed garlic and dried mint and pour over the hot chick peas just before serving.

VARIATION 2 Fry a couple of tablespoons of pine kernels in butter and mix them in with the puréed peas, along with the garlic, oil and salt.

ISRAELI FALAFEL

With *hummus bi tahina*, tomato and pepper or chilli salad, cucumber and yoghurt salad, all stuffed inside pitta bread, *falafel* make a sustaining lunch. The problem is to get the *falafel* right, to reduce the uncooked chick peas to a gritty paste. If you do not get it fine enough, the *falafel* will disintegrate in the frying pan. I use an electric chopper, putting in the onion, garlic and parsley at the same time. This makes a successful and reasonably coherent paste. Claudia Roden suggests mincing the chick peas mixture twice, through the fine blade of the mincer. You could also cheat by cooking the chick peas slightly, or by changing the soaking water with freshly boiling water every so often to make them more yielding; the risk then is that the paste will be too soft. Perhaps the first time you would be wise to use half quantities, to see how you get along.

> 125 g (4 oz) chick peas, soaked 24 hours
> 1 medium onion, sliced
> 1 large clove garlic, halved
> 25 g (scant oz) chopped parsley leaves
> ground cumin and coriander seed
> salt, pepper, cayenne
> ¼ level teaspoon baking powder

Reduce the chick peas to a paste with the onion, garlic and parsley, as I have indicated above. The pulped onion and parsley helps to bind the chick peas. Season to taste with cumin and coriander, starting with a level half teaspoon of each. Add salt, pepper and cayenne. Finally mix in the baking powder. Leave for 30–60 minutes in the refrigerator. Then form the paste into little cakes the size of a walnut and flatten them slightly. Fry them in oil until they are brown and crunchy outside; once you have confidence that the paste you make will hold together, you can deep-fry them instead.

AMERICAN THREE-BEAN SALAD

Serve as part of a salad meal, or on its own or with a hot spicy *chilli con carne* (see page 389). You can vary this salad as you like; I often put in some black beans as well, or as a substitute if I am out of haricot beans. The salad can be made the day before you need it. If you do this, do not put in the parsley and chives until the day you eat it. An important point is to put the beans hot into the vinaigrette dressing, so that they absorb the flavour.

> *100 g (3–4 oz) each chick peas, haricot beans and dark red kidney*
>   *beans, all soaked separately*
> *8 tablespoons chopped spring onion or onion*
> *2 cloves garlic, finely chopped*
> *plenty of chopped parsley and chives*
> *5–6 tablespoons olive oil*
> *1 tablespoon wine vinegar or lemon juice*
> *salt, pepper, sugar*

Cook the chick peas for two hours, adding the haricot beans after one hour. Keep an eye on the pot, so that the haricot beans are not over-cooked to bursting point. Cook the kidney beans separately for an hour and a half, as they dye the water and anything cooked with them (or use canned red kidney beans instead). Mix the remaining ingredients in a bowl, adjusting the seasonings to your taste. Put in the drained hot vegetables, turning them over well. Leave to cool down, then put in the refrigerator to chill. Scatter some extra parsley and chives over the top before serving.

This is a good and most beautiful dish, the colours and shapes of the beans are like a painting.

THUNDER AND LIGHTNING
TUONI E LAMPO

A favourite dish with the Italian members of our family is the homely mixture of chick peas and pasta known as thunder and lightning – I am not sure why, or even which is which, though chick peas have the more thunderous form. It is a good way of using up the left-over chick peas. Boil a roughly equal weight of pasta and add them to the chick peas (reheated if necessary). Add a spoonful or two of Parmesan and some butter or olive oil. Serve with more Parmesan and butter or olive oil on the table, or with a home-made tomato sauce.

DFEENA OR DAPHINA

This is a Jewish speciality from North Africa for all night cooking and Sabbath eating. To the normal beef stew ingredients – ¾ kg (1½ lb) stewing beef browned with an onion, carrot and garlic and covered with stock – add 1 split calf's foot, 250 g (8 oz) soaked chick peas, 6 peeled potatoes, and 6 well washed, uncooked eggs in their shells. Add extra water or beef stock to cover the whole thing, and keep the eggs on top. Bring to the boil, cover and simmer slowly; according to your convenience, and the temperature of your oven, it can take from four to twelve hours. The longer it cooks the better, so long as the liquid is kept just below the simmer which makes it a good recipe for the new electric slow cooking casseroles or for solid-fuel cookers.

When everything is tender, remove the eggs to a plate and allow them to cool. Extract as many bones as you can. Taste the stock, and see how watery it is. Either reduce it hard by straining it off into a frying pan and boiling it down, or add salt and pepper if the strength of the flavour is more or less right. Return the eggs to the top of the pot, reheat and serve.

Do not take fright at the idea of eggs cooking for hours. They become creamy-hard, and taste very good mashed down with the meat and potatoes. In fact they make a good addition to any beef stew, if you are short on meat.

If you cannot get a calf's foot, and are not bound by Jewish food laws, you could substitute pig's trotters or a decent piece of belly of pork, complete with rind and not cut up. If you have the kind of butcher who sells pork rind for a few coppers, buy several sheets, cut them into convenient pieces and roll them up. Tie or spear them with cocktail sticks and put them into the stew.

CHICORY – WHITE, GREEN and RED

The winter I was finishing this book, we had much entertainment growing chicory in the tank cupboard in the bathroom. The temperature was gently warm, far less than the airing cupboard's. The white and yellow leaves grew fast. Every other morning we had the ritual inspection, with offerings of water, and often I would sneak up with the scissors to cut a few leaves for the salad, to try a little before the chicons were properly formed. In fact even the chicons I left alone never formed as perfectly as the ones we could not afford to buy. But they tasted just as good, even better, as their freshness gave the flavour an added zest.

One thing has to be sorted out. That is the name. The kind I am talking about is a cultivated variety of *Cichorium intybus*, the wild familiar European chicory with the blue flowers tight to the stalk. The French call it *endive*, or sometimes *witloof* which is the correct name I suppose, the name given by the Belgians who developed it around Brussels in the 1840s by cutting off the natural foliage and forcing the roots in darkness to produce the white, yellow-tipped chicons that we hoped to find in our tank cupboard.

To the French *chicorée* signifies the blanched and curly leaved endive (q.v.), or the batavian endive or batavia (q.v.) that looks more like an extra crisp Webb's Wonder lettuce. These were developed from *Cichorium endivia*. You will understand from the botanical names why I think our solution is more logical.

Do not imagine that the matter is now cleared up. There are other chicories about these days. They are developments of wild *Cichorium intybus*, what the French call *chicorée sauvage*. They have long been growing *barbe de Capucin* (Friar's beard) in France, *pain de sucre* (sugar loaf) and other improved blonde varieties for crisp winter salads. They can be greenish white to yellow, and have something of the shape of a cos lettuce, or the baggier variety of Chinese leaf. They can be forced to produce chicons, but to a gardener they are most useful for salads at a time when greengrocers can only provide those floppy tasteless greenhouse lettuces. Have you spotted that these wild chicories should, botanically speaking, be called *endives* by the French? They are after all related to the forced *endive witloof*. I suppose they call them *chicorées* because to the cook at least they are more like the *chicorées-frisées* and *chicorées-scaroles-batavia*.

Our adventurous seedsmen and importers have lately introduced yet another chicory, this time from Italy (at least the Italians have the decency to call them *cicorie* – *cicoria* in the singular). They are a beautiful red, the colour of some doge's or bishop's robe in a Titian painting. You can grow them and you can buy them, sometimes under the name of red salad, though they do not seem to have got far beyond central London. Take a walk down Rupert Street market in Soho, in October or November, and your eye is sure to be caught by them. There are three kinds – one like a deep Burgundy-red giant Brussels sprout, the *rossa di Verona*, another a flaunting deep red and creamy streaked tulip, the *rossa di Treviso*, and a third shaped like our familiar Belgian chicory, but dark red – this is the forced kind. Sadly their colour fades in cooking to a brownish tone. One is tempted to keep them entirely for salad. But they do cook well, especially the first two.

Is your head spinning? I hope not, for there is yet another chicory now being grown by private gardeners. This is the Magdeburg chicory with huge roots that you can dry, roast and grind up for adding to coffee.

It is worth considering the perfection of the Belgian chicory, as it lies neatly meshed in a nest of deep blue paper. Gibault, in his *Histoire des Légumes* of 1912, tells the story of the accidental discovery of the technique that produces the *witloof* chicons. The head gardener of the Brussels botanical garden wanted to bring on some chicory for the winter in frames. He took some roots from the salad bed, lopped off most of the foliage and planted them. To his surprise small tight shoots began to pierce through the soil, rather than the loose salad leaves he had expected. Next season he did the same thing deliberately, but kept the process secret. When he died, his widow told her gardener what to do, so that if deprived of her husband she need not be deprived of chicory. He passed it on and the result is an enormous family industry in Brabant to-day. The forcing beds are piped so that warmth can be provided if the weather turns cold. They are further protected and kept in darkness by corrugated iron sheets bent over Nissen-hut style. Bales of straw are built up to cover the whole thing, so that the temperature inside remains constant while the thermometer of a Belgian winter goes up and down.

HOW TO CHOOSE AND PREPARE CHICORY

In style of flavour there is little difference between red and white chicory. Something of the natural bitterness always remains in cultivated chicory to give a characteristic edge to the crispness, though I have found this to be slightly more pronounced in the red kinds making them particularly good for salads. The style of texture, too, is the same in both, particularly after cooking, but uncooked the red leaves are a more supple, less crackling version of the white *witloof* chicory. This difference makes the red ideal for a plain salad, what one would normally describe as a green salad, whether on its own or mixed with other fresh leaves. For mixed salads of the Waldorf kind, in which chicory might be seen as an alternative to celery, the white is better on account of its firmer crispness; it partners apple, walnuts, cold chicken, ham, Gruyère cheese and so on with an equal vigour.

An advantage of all chicory, white or red, is that it is easy to see its condition. The blue paper that wraps and interleaves white chicory is not just a pretty marketing idea; it is there to be folded over between sales, so that the light cannot get to the chicory and turn it green and bitter. Always go for the whitest chicons. Reject any that are damaged or turning brown at the cut edge or at the sides of the leaves. Chicory is expensive, so be more determined than usual to insist on perfect quality. If you get it in the right condition, you should be able to keep it yourself for a couple of days or even longer; wrap it up to exclude the light and store in the bottom of the refrigerator.

Red chicory should also have a clean and vigorous look. The outer leaves should be neither withered nor brown at the edges. Again it will keep quite well for a couple of days, though not as well as white chicory.

Quantity is a little difficult. In the end you must choose by eye, allowing so many pieces per head. With a salad for instance one head between two people will often be quite adequate. In the cooked recipes, there is more of a problem, depending on whether the chicory is only part of a much larger meal, or whether it is the main feature. If you are in real doubt, go for a safe quantity such as two smallish heads per person, or one huge one (the huge ones can always be split lengthways at some stage in the cooking, so that they make more of a showing).

To prepare white chicory

Remove a thin slice from the root end. Some people remove the inner cone – it is said to taste bitter but I have never found this. If any of the leaves have developed a brown edge, cut it away. Then rinse the heads quickly, never leave them soaking. If the chicory is to be cooked, it will often require blanching, i.e. cooking in boiling salted water, with a teaspoon of sugar and the juice of half a lemon, until half or almost completely tender, depending on the final stages. Or you can blanch it in a *blanc à légumes*, p. 549.

To prepare red chicory

According to the requirements of the recipe, twist off or slice away the miniature parsnip-like root that is sometimes still attached to the head. Cut away withered or brown leaves. Rinse briefly. If you blanch red chicory, give it less time than the white, barely five minutes, as it cooks more quickly with its loosely packed leaves.

Most of the recipes following can be used for both white and red chicory, though one kind or the other may be preferable for the best results. When the word chicory occurs on its own, the more familiar white chicory is intended or the forced red chicons.

SOME CHICORY SALADS

1) Mix a more or less equal quantity of sliced chicory and cooked cut up potatoes with double cream seasoned with wine vinegar, salt, pepper and a little sugar. Criss-cross with strips of smoked German or Polish ham, unless you can afford French *jambon de campagne*, Bayonne or Parma ham which are even better.

2) CHRISTMAS SALAD Mix equal quantities of diced cold poultry or gamebird and diced, unpeeled eating apples and sliced chicory. Dress with vinaigrette alone for guineafowl or pheasant. Mayonnaise can be added for light poultry, together with some cubed Gruyère cheese.

3) To make the most of two large chicons, cut off the base and separate out the leaves. Keep the tiny centre cones intact. Fill the leaf hollows with curd cheese or ricotta mixed with a little cream and seasoning, or with crumbled hard-boiled eggs bound into a light paste with some melted butter and cream. Stud the filled leaves with bits of black olives, strips of

anchovy fillet or capers. Arrange the leaves on a dish with the tiny cones in the centre. Scatter with parsley.

CHICORY WITH CREAM

The best way of cooking chicory that I know, especially if it is to be served as a vegetable with a joint of meat or a bird. If you want to serve it on its own, add triangles of toast to the dish, or lightly fried bread.

Trim and cook the chicory in the usual way until it is just tender. Drain it well. Melt a large knob of butter in a pan large enough to hold the chicory in a single layer, and turn the chicory in the butter until it begins to turn golden. Do not hurry this part of the cooking; there should be no hint of dark brown.

Finally pour in some cream – allow a tablespoon per head of chicory – raise the heat and bubble the whole thing so that the liquids combine into a small amount of savoury rich sauce. Keep turning the chicory. The outer part should be meltingly succulent, contrasting with the inner core that still retains a certain bite.

Note Concentrated juice from roasting meat, game or poultry can be substituted for cream, for *endives au jus*.

CHICORY POLONAISE

This is a favourite recipe of mine for the first course of a winter meal. You can use red or white chicory, whether it is the firm chicons or the floppier style. Remember that the latter will shrink in cooking, so allow for this.

FOR THE FIRM CHICONS Blanch the heads until just tender, then finish them in butter so that they are streaked with golden-brown. Arrange them on a big round dish like the spokes of a wheel. For six people, crumble three hard-boiled eggs with a fork and mix in a couple of tablespoons of chopped parsley. Scatter this over the centre of the chicory. Brown three tablespoons of crumbs in a little knob of butter in the breadcrumbs pan and pour it over the whole thing.

FOR THE LOOSE-LEAFED WHITE OR RED CHICORY Cook the heads and drain them really well in a colander, pressing the moisture out gently as they cool. This can be done in advance of the meal. Now turn to the beetroot recipe on p. 100, and prepare Polish stuffed eggs. Just before serving the meal, heat the chicory through in butter, browning it very slightly, and fry the crumbed side of the eggs. Arrange the chicory on a dish, with the eggs in the centre.

Of course you can follow the first method above for the floppy-leaved chicories as well, but as they do not have the trim look and texture of chicons, this second method with the halved eggs gives the dish a livelier flavour.

CHICORY WITH HAM AND CHEESE SAUCE
ENDIVE AU JAMBON SAUCE MORNAY

Whether the meal is a grand dinner or family supper, the combination of the watery, faintly bitter chicory, with the mild smokiness of good ham and the rich flavour of mornay sauce, makes an irresistible dish.

> *6 fat heads of chicory, trimmed*
> *6 thin slices of ham*
> *Dijon mustard*
> *butter*
> *mornay sauce, made with meat stock and milk, p. 555*
> *extra Parmesan cheese and breadcrumbs*

Blanch the chicory for ten minutes in boiling salted water with a squeeze of lemon juice and a good pinch of sugar. Drain thoroughly. Spread each slice of ham with a little mustard and wrap it round a head of chicory.

Butter a gratin dish and fit in the swaddled chicory, flap sides down so that it stays neat. Pour over the sauce, scatter with cheese and bread-crumbs, and bake for about 20 minutes at 190–220°C/375–425°F/Gas Mark 5–7. Remove when it looks as good as it smells, with a golden-brown top.

Note Nutmeg is usually added to mornay sauce but you may like to increase the quantity slightly. It goes especially well with chicory dishes of this kind.

STUFFED RED TREVISO CHICORY
CICORIE ROSSE DI TREVISO RIPIENE

This is an unusual dish for flavour, a typical southern Italian mixture full of zest. It is a favourite recipe of mine which can be made with a variety of vegetables, though I think that red Treviso chicory is best. Elizabeth David gives the recipe using batavian endives which are grown widely in southern Italy, but it is not easy here to buy them small enough (though obviously the dish could be made with one large batavia). As an alternative she suggests small round lettuces: I have not tried this, but would think that a crisp variety should be used, not the floppy kind. With red chicory it is marvellous.

Go for the larger tulip-shaped heads, that are elongated well beyond the Brussels sprout stage to a length of 15 cm (6 in). If they are really large, three should be enough for six people as a first course. If you intend the dish as a main course – and it comes well after egg noodles, butter and Parmesan – buy six heads, and increase the other ingredients by about two-thirds.

If red chicory, crisp lettuce or batavia are all unobtainable, buy large heads of white *witloof* chicory. Hollow them out from the root end (keep the part you cut away for salad), as the leaves are too crisp, too tightly curled to be opened out for stuffing.

Unfortunately there is no good substitute for pine kernels; almonds can be used of course, but they do not have the right waxy texture. Pine kernels, or pine nuts as they are often called, are usually sold in health food shops and Italian delicatessens. They are expensive, but one uses them in small quantity. The other essential is olive oil of decisive flavour.

    3 large heads red Treviso chicory

    STUFFING
    75 g (2½ oz) breadcrumbs
    12 stoned, chopped, black olives
    6 chopped anchovy fillets
    1 heaped teaspoon capers
    30 g (1 oz) pine kernels
    30 g (1 oz) sultanas
    2–3 cloves garlic, chopped
    2 heaped tablespoons chopey
    5 tablespoons olive oil
    pepper, salt

    PLUS
    olive oil
    2 anchovy fillets
    1 tablespoon capers
    1 large clove garlic
    1 teaspoon sugar
    6 tablespoons dry white wine

Slice the root ends from the chicory (do not twist them out, or the leaves will fall apart). Carefully open out the heads to make a cavity. Mix the stuffing ingredients – you will not need much salt on account of the pickled items – and divide between the chicory, pushing it down into the cavities. Mould the heads back into shape with your hands, then tie them together with string.

Put a little olive oil into a heavy pan to heat – the pan should be large enough to take the stuffed chicory in a single layer – and chop the anchovy fillets, capers and garlic together quickly to make what the Italians call a *battuto*, a mixture of things chopped so finely together that they become indistinguishable. Put this *battuto* into the oil to heat through, then place the chicory on top, putting the top of the heads closely against the side of the pan (this helps to ensure the stuffing stays in place). Add sugar, and enough water to give a centimetre's ($\frac{1}{2}$ in) depth or a little less. Cover tightly and simmer for half an hour. Turn the chicory over and give it another half an hour, covered as before. Pour in the wine, and leave a further 10–20 minutes. Transfer the chicory to a hot dish and cut away the string. Taste the cooking juices and see if they need boiling down. The flavour should be lusty and appetizing. Pour the juices over the chicory. Serve with plenty of bread, and white wine to drink.

CHINESE ARTICHOKES

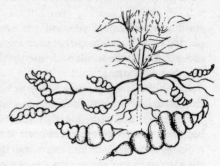

Chinese artichokes are the most elusive of vegetables. They are easy to grow, but no one grows them commercially – in England, at any rate. They are easy to cook. They are delicious to eat, exceptionally so. But unless you are prepared to badger Louis Roche Ltd, of Old Compton Street in Soho, or send to Vilmorin in Paris for tubers for the garden, you may live a lifetime without tasting them.

They look a little strange, these small and finely shaped tubers that grow like potatoes. If you went and dug some up, never having seen them before, you might think you were raising a crop of healthy maggots. Then if you rubbed one or two clean in your hand, you would soon notice the beauty of form and pearly translucency that have made Chinese poets compare them to jade beads and give them poetic names such as *kan lu*, meaning sweet dew. Dr Bretschneider, who first sent Chinese artichokes to Europe for acclimatization from the gardens of Peking, described them as looking like turreted shells. He also commented on their nutritional quality, though this is the least interesting aspect of so rare and exquisitely flavoured a vegetable.

When at last a few of Dr Bretschneider's tubers reached the Societé d'Acclimatation in Paris, in 1882, in reasonable condition – most earlier consignments had rotted – they were planted at Crosne, in the garden of the Vice-President, Monsieur Pailleux. They flourished, and have been known in France ever since as *crosnes*. The first person in England to grow them, seems to have been 'Mr Haskins, gardener to Sir H. Thompson, of Hurstside, West Moulsey'; he exhibited some at the December meeting of the Royal Horticultural Society in 1887. Here they have not caught on, unfortunately, as they have in France.

If you succeed in buying some, set aside a few in peat and sand to sow the following March. They can be dug up in late November or December. A day or two out of the ground the white glow turns a patchy yellowish fawn, perfectly appetizing, though less virginal to look at. Naturally their subtleties of flavour diminish with keeping, though even after a month they may be still delicious and worth eating on a special occasion.

HOW TO PREPARE CHINESE ARTICHOKES

Preparation is simple. Top and tail the artichokes, then rinse them clean as quickly as possible. French instructions tell you to put them into a cloth with some coarse or sea salt, then rub them to remove the outer skins. I find this makes no difference at all and can well be cut out, as the skin is so delicate that it is quite imperceptible to the tongue. It does make yellowing artichokes look a little whiter, that is all.

The usual way of cooking them, the first blanching, can be done either with water or stock. With water, put the artichokes into a pan with just enough to prevent them burning. Add a pinch of salt, and a lump of butter or a tablespoon of oil according to your plans for finishing them. Bring to the boil, cover and cook them for five to 15 minutes.

If you intend to make a sauce for the artichokes, you can cook them slightly differently. Substitute veal or chicken stock for water, and use enough to cover them completely. You will end up with the most deliciously flavoured liquor that can then be used in making a velouté sauce or sauce for a gratin.

Whichever method you choose, the point is not to waste a scintilla of their flavour. To tip any of their cooking liquor down the sink would be a crime.

CHINESE ARTICHOKES WITH MEAT JELLY
CROSNES DU JAPON AU JUS

If you have any good jellied essence from roasting beef, game or poultry, just a few tablespoons, stir it into the Chinese artichokes when they are almost cooked, and their liquid practically gone. Add chopped parsley, chives or tarragon as appropriate.

CHINESE ARTICHOKE SALAD
CROSNES EN SALADE

Cook the artichokes by the first method, with a little water and oil rather than butter. When they are just tender, the water gone, mix them while still hot with an olive oil vinaigrette. Leave to cool. Serve slightly chilled, but not too cold or you will loose the finest edge of flavour. Scatter with a little parsley. Wholemeal, rye or granary bread and butter go well with the salad.

If you want to add an embellishment, mix in a few shelled prawns, but not too many. They must be subordinate to the artichokes.

CREAMED CHINESE ARTICHOKES
CROSNES À LA CRÈME

A recipe from the most meticulous of French cookery books, *La Cuisine de Madame Saint-Ange*. It was first published in 1927. If you really want to know about French cooking, from a household point of view, this is the book to buy. Every detail of the recipes is explained.

Boil the artichokes for about five minutes. Drain them and give them a further five to seven minutes in the saucepan, over a low heat, with a good knob of butter and half a teaspoon of lemon juice. For ½ kg (1 lb) artichokes, bring 300 ml (½ pt) double cream to the boil, then add it to the pan stirring it in well. Season with a pinch of salt, a little white pepper and a very little nutmeg. Cover and leave a further 15 minutes over a moderate heat, so that the cream reduces by a quarter. Taste and adjust the seasoning.

Just before serving, stir in three spoonfuls of fresh double cream, then quickly remove the pan from the heat so that its natural flavour is not lost.

Madame Saint-Ange describes the taste of Chinese artichokes as something between the base of a globe artichoke and salsify. This is as close as you can get.

CHINESE ARTICHOKE AND MUSHROOM BAPS

Turn to Forester's carrots on page 166, and substitute Chinese artichokes. It may seem wrong to put so delicate a vegetable with any other, but this is a marvellous combination, subtle and harmonious.

Serve the vegetables, minus the lightly fried baps, with veal or chicken. If you add a little extra cream, no other sauce will be required to go with the meat.

SALADE MOUSMÉ

Invented by the chef Henri-Paul Pellaprat. This is a delicious way of serving Chinese artichokes if you have only a few.

> 250 g (8 oz) Chinese artichokes
> lemon juice
> vinaigrette
> 6 large tomatoes

inner celery stalks
4 tablespoons mayonnaise
salad greenery

Simmer the Chinese artichokes until tender in salted water, with a good squeeze of lemon juice. Drain them when cooked and put immediately into a basin containing vinaigrette. Turn them over in it and allow them to cool.

Slice a lid from the tomatoes. Hollow them out. Pour a little vinaigrette into each one and swill it round so that the interior of each tomato is nicely seasoned. Slice up the celery, enough to fill the tomatoes just over half full, and mix it with mayonnaise. Put this celery mixture into the tomatoes and arrange a few Chinese artichokes on top as garnish.

Place the tomatoes on some salad greenery, and arrange any Chinese artichokes left over around the tomatoes.

CHINESE LEAF

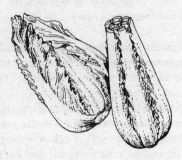

Chinese leaf, Chinese cabbage, Shantung cabbage, wong bok, Chihli cabbage, pe-tsai (which means white vegetable), celery cabbage, *Brassica pekinensis*, are all names for one special cabbage from Eastern Asia and northern China in particular – hence the botanical name. It is long, crisp and heavy, weighing from 1 to 2 kg (2–4 lb). It is often compared for shape to a cos lettuce, but this is misleading because it is solidly packed by comparison, such a fine column of pale crinkled greenness, striped with smooth white ribs that taste delicately of cabbage and celery combined. The texture is not tough, but lightly crunchy. Raw in salad, cooked in Chinese or Western style, it will delight the cabbage-hater. Before 1970 most of us in Britain had never seen a Chinese leaf, many of us had never heard of it. The Israelis started to experiment with it about 1966 and in 1970 they were able to export 11 tons, ten of them to this country. In 1975 we took 350 tons and in 1976 out of an export of 2,600 tons, we ate our way through 850 (the rest went to Scandinavia, Germany and the Benelux countries).

Turning new foods to our own way is what the history of eating is all about. No need to limit Chinese leaf to Chinese meals, any more than we limit tomatoes and potatoes to recipes from their native Andes. On the other hand, the Chinese way of stir-frying could be adapted with advantage to our western meals: stir-fried Chinese leaf goes deliciously with roast beef or pork or duck.

HOW TO CHOOSE AND PREPARE CHINESE LEAF

The tall, stiff-looking Chinese leaf is far superior to the rounder, more wrinkled, cabbage-looking kind. In density and length, it looks far more like a huge head of celery.

Whichever kind you buy, it needs to be really fresh. No good choosing one with the edges of the leaves looking wilted or fawn-streaked. No good keeping it lying about in a warm kitchen, either. Remember that it is a delicate relation of the familiar cabbage and treat it tenderly. Never wash it before you need it. Keep it wrapped in plastic in the lowest drawer of the refrigerator, slice off what you need and then wash it. Put the rest back into store immediately. A large Chinese cabbage will last for a week at least.

The general recommendation is to slice off the end part for salad, and cook the thicker stalks. If your Chinese leaf is only likely to do two meals, then use the whole of the inner part for salad, and the outer leaves for cooking.

I follow another method. I separate the whole cabbage. This leaves me with a pile of thick stalked leaves, a second pile of tiny cabbages that sprout from the centre stem, and the thick stem itself. I cut the leafy part from the thick stalks of the first pile with scissors for a mixed green salad. The stalks themselves are sliced into convenient pieces and put with the main stem, which is peeled and cut across; these are cooked and served with creamy sauce, or a hollandaise, as if they were seakale or asparagus. The pile of miniature Chinese leaves can be served in a jug like sticks of celery; or they can be braised and treated like *witloof* chicory; or they can be cut up and used to make a crisper kind of salad than the leafy part I trimmed from the outer stalks. This kind of treatment is particularly suitable if you are feeding a small family.

Some people grate the thick end of the whole head of Chinese leaf on the coarse blade of the grater. This is cooked and served with a mornay sauce – a good method, though I cannot agree that Chinese leaf treated this way tastes like asparagus. It has a freshness that is pleasing, as the stalks of some other plants have when they are crisp and healthily fresh, but you would never mistake it for asparagus.

Chinese leaf can be cooked like other, lustier cabbages, but remember to reduce the time. Hover round the pan, especially if you are preparing stuffed Chinese leaves, or you will end up with an expensive disaster. It is not a vegetable for careless cooking.

STIR-FRIED CHINESE LEAF

First weigh out the Chinese leaf. Use the heart of a large head, cutting away ruthlessly any tough-looking stalk, and shredding the green leaves until you have between ¾ and 1 kg (about 2 lb). You will also need

4 tablespoons oil
1 heaped tablespoon chopped onion
2 cloves garlic, finely chopped
2 slices green ginger, peeled
salt

1½ tablespoons dry sherry
1½ tablespoons soya sauce
1 teaspoon sugar
¼ chicken stock cube
1 tablespoon melted lard or chicken fat or sesame oil

Have the heat turned high before you begin. Put the oil into a large shallow pan; as it warms up and spreads, put in the onion, garlic and ginger. Stir-fry with a spoon or chopsticks for one minute, to flavour the oil. Rapidly put in the Chinese leaf, sprinkle it with about a teaspoon of salt, and keep it turning and moving in the pan for about three minutes; all of it should be filmed with the oil. Now put in the sherry, soya sauce, sugar and chicken cube. Lower the heat to moderate, but continue to turn and stir for another two or three minutes. If the leaves begin to look soupy, raise the heat. If the leaves seem to risk being browned, add a little water or stock *by the spoonful*. You should end up with a tiny amount of juice and crisp yet tender cabbage. Just before serving, add the final spoonful of fat to give an appetizing gloss. Taste for seasoning.

Spinach, celery, leeks, celtuce, Swiss chard (keep leaves and stalk separate), cabbage hearts, peas with or without their pods, can all be cooked in the same way, with slight adjustments of timing.

CREAMED CHINESE LEAF HEARTS
NAI YU TS'AI HSIN

Dairy products and white meats have never been important in Chinese or Japanese cookery. This recipe shows how sparingly milk is used; much of the creaminess comes from the thickening cornflour.

3½ teaspoons cornflour
3 tablespoons milk
2 tablespoons melted chicken fat
½–¾ kg (1–1½ lb) shredded Chinese leaf
1 scant teaspoon salt
¼ teaspoon sugar
up to 150 ml (¼ pt) chicken stock
2 thin rashers canned Yunnan ham, or lean Danish smoked
    bacon, or Westphalian ham, or prosciutto crudo, chopped

Mix the cornflour to a cream with the milk. Heat up your frying pan or *guo*, add the chicken fat and when it is hot, the Chinese leaf. Keep it moving and turning. After a minute, add the salt, sugar and half the stock. Go on stirring for about five minutes, until the leaf is tender but still crisp. Be ready to add the remaining stock if the cooking juices become too evaporated. Remove the leaf to a hot dish. Stir up the cornflour cream, pour it into the pan and mix with a wooden spoon to make a smooth sauce. Again add a little stock if needed at this point, but there should not be much sauce. Check the seasoning. Pour the sauce over the leaves, scatter ham or bacon on top, and serve.

Note The bacon is not cooked – it is there to add a little meaty flavouring to the dish.

CHINESE LEAF SALAD

Mix ½ kg (1 lb) sliced Chinese leaf with 1 sliced red pepper, a small chopped onion and half a cucumber, sliced. Dress with an olive oil vinaigrette, or with the Chinese dressing of two tablespoons each of soya sauce and wine vinegar, one tablespoon of sesame oil and sugar to taste, up to one tablespoon.

CHINESE LEAF AND PINEAPPLE SALAD

To ½–¾ kg (1–1½ lb) sliced Chinese leaf, add half a small chopped onion and three cubed slices of pineapple. Bind lightly with mayonnaise. Put into a bowl and scatter with walnuts (pecans in the original American version), chopped red pepper and a few capers. Surround with a ring of watercress.

CHINESE LEAF IN TOMATO SAUCE

A variation of one of the recipes put out by Carmel. If you have a good ¼ litre (½ pt) of the basic tomato sauce given on page 511, you can produce this dish quickly; if not, follow the instructions below which will take a little longer, about half an hour.

3 tablespoons butter
1 medium onion, chopped
1 small clove garlic, chopped
1 small can tomatoes, or 250 g (8 oz) chopped, skinned fresh tomatoes
pinch of dried oregano
bouquet garni

100–125 ml (3–4 fl oz) stock
salt, pepper, sugar, wine or sherry vinegar
¾ kg (1½ lb) sliced Chinese leaf
extra butter
chopped parsley

Melt the butter and simmer the onion and garlic in it, without browning them. Add the tomatoes, herbs and 100 ml (3 fl oz) stock. Simmer for 15 minutes, covered, then taste and bring out the flavour with seasonings. A dash of vinegar helps particularly with fresh tomatoes. The sauce should not have reduced much, hardly at all.

Add the Chinese leaf, raise the heat a little to keep the whole thing boiling, and cover. Stir from time to time, adding extra stock if the Chinese leaf appears to be sticking; or remove the lid and boil hard if the sauce turns watery. Taste a shred occasionally and remove from the heat the moment it is cooked, still a little crisp but not raw. Mix in a knob of extra butter, turn into a hot dish and scatter with parsley. Serve on its own with triangles of fried bread, or with rice, or as a vegetable with meat.

MONGOLIAN FIRE POT
SHUA YANG JOU

Mongolian fire pot is a one-pot dish of a kind to be found all over the world from Japanese *nabé-mono* to French, Swiss and Dutch *fondues* (or Lancashire hot-pot though it does not need a table burner). They make relaxing and sociable meals, an entertainment as well as food. They all require the same equipment, one large pot and a source of heat. The Chinese have a specially designed metal pot and burner in one that can be bought in this country, but the dish can be perfectly well cooked with fondue equipment, and fondue forks may well be easier for your visitors than chopsticks.

The meat must be cut into thin slices, paper thin, or 'flying thin' as Mrs Chao calls them in *How to cook and eat in Chinese* (Faber). For this you do not need Peking skill with the cleaver. Put the lamb to chill in the coldest part of the refrigerator until it is firm. Then you can easily cut fine, thin, nicely shaped slices. As with many other Chinese dishes, everything happens at the last minute, so it is wise to have someone to help.

1½ kg (3 lb) boned lamb from leg or shoulder
125 g (4 oz) transparent Chinese noodles
250 g (½ lb) fresh leaf spinach

½ kg (1 lb) Chinese leaf
12 part-cooked rolls, or home-made bread dough
2 litres (3½ pt) stock made from chicken and lamb bones
2 large spring onions or small leeks
1 large clove garlic, chopped
2 good teaspoons chopped fresh ginger
60 g (2 oz) green coriander or parsley chopped

SAUCE
12 tablespoons soya sauce
2 tablespoons pale dry sherry
2 tablespoons sesame oil
1 rounded tablespoon soft brown sugar mixed with 1 tablespoon
 very hot water
½ teaspoon cayenne
5 teaspoons canned fermented red bean curd

Cut the chilled meat into thin slices. Divide between six plates and put into the refrigerator (cover them with film wrap if you need to stack plates). Pour ½ litre (good ¾ pt) very hot water over the noodles, leave for half an hour, drain and put to one side of a large serving dish. Wash, trim and dry the spinach leaves. Slice and blanch the Chinese leaf for three minutes in boiling salted water, drain and put with the spinach on the noodle dish. Put the rolls in a steamer, set over boiling water, cover and leave for 20 minutes; if you are using home-made dough, leave it to rise for an hour in a warm place while you are coping with the above preparations, then form it into rolls, prove for 15 minutes, then steam for up to 30 minutes – they should end up white and puffy and light. Mix the remaining ingredients together and bring them slowly to the boil. Mix the sauce ingredients; put some into six bowls and the rest into one bowl to be kept as a seasoning for the final stew.

Set the table by putting a plate of meat, a small bowl of sauce, chopsticks or fondue forks, and a porcelain Chinese soup spoon or dessertspoon in each place, with a napkin. Put the table burner in the centre, so that everyone can reach it. Place the noodles and vegetables on one side, the hot rolls on the other and keep the extra bowl of sauce somewhere near you so that it doesn't get used too soon. Also provide yourself with a soup ladle.

When all your visitors are seated, bring in the boiling stock and set it over the burner. Aim to keep it at a steady boil. Each person picks up a slice or two of meat with chopsticks or fork, and 'rinses' it in the stock.

After a few seconds, the meat will have changed colour and be ready to eat; different people may like it cooked for different lengths of time. Dip the meat in the sauce before eating it, and take an occasional mouthful of hot roll.

When the meat is all finished up, ladle a little stock into each person's sauce bowl to make some soup for them to drink, while you put the noodles and vegetables into the stock to cook for a few minutes. The stock by this time will have a delicious flavour from the meat. Share this vegetable stew round the table, adding extra sauce from the spare bowl as seasoning.

Finish the meal with a huge basket of tangerines or grapes or both. Or with peeled oranges, the traditional end to many Peking meals.

CHOKO, see CHAYOTE

CHRISTOPHENE, see CHAYOTE

CORN SALAD, see LAMB'S LETTUCE

COURGETTES

The first encounter with courgettes for anyone over 45 is likely to have been in Italy, either before the war or when travel became possible again afterwards. I remember eating this delicious vegetable at a *pensione* in Florence in 1949. We had it several times a week. My Italian was too Dantesque to ask the cook, but a fellow student put me right. He had first sampled it in less happy circumstances when the army was fighting its way up the Italian peninsula at the end of the war. 'It's zucchini,' he said, 'the Italian answer to cabbage.' They were lightly cooked and dressed with olive oil; I never got tired of them in two months. Zucchini are courgettes; in American cookery books, you will find them under that name, as they came in with Italian immigrants.

The following year, 1950, Elizabeth David published her first book, *A Book of Mediterranean Food*. I bought it and re-lived my first journey to Italy. Ostensibly she wrote about Mediterranean and French food, in the days when we were still caught up in rationing, and memories of snoek and the Woolton Pie. If you look at it superficially, she was writing cookery escapism. In reality, she was doing far more. Her recipes were authentic and encouraging; her information was full and evocative. She asserted what all food should or could be, clear principles of simplicity, flavour and freshness that could be practised wherever the reader lived, with whatever materials could be found. In doing this she also upturned a century of poor cookery, when plain living was so often mean living as far as food was concerned, good living so often pretentiousness.

In bringing us back firmly to the source of good things and civilized eating, Mrs David introduced us to vegetables and foods that had previously been enjoyed only by the wealthy. Courgettes in particular, courgettes in this country, are her creation. True a couple of writers in 1931 gave recipes for *courgettes* (their italics). One of them was Marcel Boulestin, who translated the word as 'baby marrows'. The Oxford English Dictionary in its A–G supplement, gives the first use of *courgettes* to E. Lucas in the same year, in *Vegetable Cookery*. Surely the honour should go to Mrs David? She was the first to relieve courgettes of their italics, taking them into our language completely, naturalizing them, when she wrote in her Mediterranean book, 'Slice the courgettes (very young marrows)' in the courgettes and tomato recipe that you will find on page 229. By the time she finished her master book, *French Provincial Cooking*, published in 1960, she was able to write 'Enterprising growers are supplying us with little courgettes as an alternative to gigantic vegetable marrows'.

HOW TO CHOOSE AND PREPARE COURGETTES

Even if you haven't a garden, you can buy a couple of courgette plants from a nurseryman and grow them in the backyard or on a balcony in a tub. Until you have grown them yourself and picked them at 5–7 cm (2–3 in) length, you cannot imagine how delicious they can be. Put them straight into the pan, after cutting off any stalk, with a knob of butter and seasoning. Jam on the lid with foil and stew gently for about five minutes. That is all you need, or should do with so perfect and fresh a vegetable.

Shop courgettes will be larger. Reject them if the skin is so tough that it cannot be wrinkled easily with your fingernail; once it hardens, the courgette is on the decline into vegetable marrow (q.v.). Never peel young courgettes, but check the rounded end. It sometimes has a hardened patch, brownish in colour, that should be sliced away. Again it is a question of age and size.

If you want to fry courgettes, or make fritters, salt them as you do aubergines. Seasoning apart, they soak up less fat if you do this first. Slice them across or diagonally, spread them round a colander, or on a pierced strainer plate, and sprinkle on a tablespoon of salt for 1kg (2 lb) courgettes. Leave them for an hour, then eat a small piece to see if they are too salty and need rinsing. Dry them in a clean cloth before frying, or they will spit and burn you.

Sometimes courgette and cucumber recipes can be used interchangeably. Very young courgettes, for instance, can be eaten raw in salad, though they are normally cooked.

Courgette Flowers

Fragile, rich, yellow flowers of courgette, and of squashes and pumpkins and marrows, can be dipped in batter and fried to light golden-brown fritters. Or they can be stuffed. Whichever style you choose, make all the preparations before you pick the flowers, as they wilt rapidly. Remove the stem and calyx.

The Greeks pack a little rice stuffing (p. 565) into the cup of the flowers and gently fold the petals round it. Pack these brilliant dumplings close together in a saucepan. Add water barely to cover them, and simmer for 20 minutes. Serve with the egg and lemon sauce on p. 558, made with some of the cooking liquid. A meat and rice stuffing is sometimes used, but it is, I think, a little heavy; the whole thing needs twice as long to cook.

In Central America, squash flowers are sometimes chopped and added to scrambled egg. For an omelette, cut them in slightly larger pieces, fry them in butter for a few moments, then beat them with some parsley into the eggs and seasoning.

COURGETTE AND DILL SOUP

Follow the recipe on p. 238 for French cucumber soup, substituting dill weed for parsley. Be discreet with quantities – dill weed, particularly in its dried form, is a strong herb. If you have no dill weed, try using the feathery leaves of garden fennel instead.

COURGETTES IN 1931

It seems that the two recipes for courgettes given by Marcel Boulestin, in *What Shall We Have To-day?* were among the first in an English cookery book.

The first is for *courgettes au beurre*. Boil the 'little vegetable marrows', when they are no longer than 7½ cm (3 in), in salted water. Do not peel them, do not cut them up. Drain and serve with fresh butter and chopped parsley.

The second is for a simple *moussaka*, *courgettes à l'orientale*. Peel, slice, salt and fry some courgettes, cutting the pieces lengthways. Mince finely some left-over roast lamb; mix it with a little of its stock, season with nutmeg, salt and pepper. Butter a fireproof dish. Put in a layer of lamb, then courgettes and a little grated cheese. Repeat until the lamb and courgettes are finished. Top with cheese and breadcrumbs. Bake in a moderate oven for half an hour.

If you wish to turn the courgettes and lamb out, make up the dish in a charlotte mould, which you have buttered and crumbed.

COURGETTES IN 1950

In *Mediterranean Food*, Elizabeth David gives a recipe for *courgettes aux tomates*, one of the two best recipes for them. (The second we also owe to her, the sweet-sour recipe from *Italian Food* of 1954, later in this section.)

'Slice the courgettes (very young marrows), peeling them unless they are very small. Salt them and leave to stand for 30 minutes. Put them in a fireproof dish with plenty of butter and 2 sliced and peeled tomatoes. Cook for about 10 minutes on a very low flame.'

For four or six people I use 1 kg (2 lb) courgettes, 125g (4 oz) butter and ½ kg (1 lb) tomatoes. This large quantity takes longer to cook, so I gives the courgettes about ten minutes on their own first, or 15 minutes, then add the tomatoes for the final ten minutes only so as not to lose their fresh flavour in the dish.

COURGETTE SALAD

Cook the courgettes whole in boiling, salted water. Drain them, then halve and quarter them. Mix with an olive oil and lemon juice vinaigrette (*lemonolatho* p. 559) while still hot. Just before serving, drain off any surplus liquid and turn the courgettes with plenty of parsley and chives. The secret is plenty of herbs and not just a niggardly sprinkling.

If you grow basil, use that with a smaller amount of parsley and just a few chives; let the basil predominate.

See also the recipe for *chayote salad*, p. 198, which can be used for courgettes.

COURGETTES FRIED AND FRITTERED

Cut small unpeeled courgettes into ½ cm (¼ in) sticks. Salt and dry them. Shake them in seasoned flour and deep-fry in oil very briefly. They should be crisp and light.

For fritters, salt the courgettes after they have been cut into long, ½ cm (¼ in) thick slices. Dry and dip into a light fritter batter (p. 568) and shallow or deep-fry in oil.

COURGETTES WITH CREAM AND ROSEMARY

A dish that can be served on its own, or with veal, chicken and lamb. Use fresh rosemary from the garden. Other herbs can be substituted – parsley, chives, tarragon, fennel – but rosemary gives the best flavour of all.

> ¾ kg (1½ lb) courgettes
> 75 g (2–3 oz) butter
> 100 g (3–4 oz) double or whipping cream
> 10 cm (4 in) sprig rosemary
> salt, pepper

Cut the unpeeled courgettes into thick diagonal slices. Blanch them in boiling salted water for about five minutes. Drain them well and put them back into the pan with the butter. Cover closely and leave them to finish cooking over a gentle heat. The courgettes must not brown or stick to the pan, so shake them from time to time. When they are tender, stir in the cream and rosemary, and leave to simmer for another five minutes. Turn the courgettes fairly often, so that they are coated with the sauce and delicately flavoured with the rosemary. Fish out the sprig of rosemary before serving.

COURGETTE TART

The mild flavour and gentleness of this tart could be insipid if the filling were not properly seasoned. At the same time, resist the temptation to overdo the herbs, especially if you choose rosemary. When using single cream, sharpen the tone with a little lemon juice, or add a splash of wine vinegar to the courgettes as they cook.

> *375 g (12 oz) courgettes, unpeeled, cubed*
> *butter*
> *1 small onion, chopped*
> *sprig rosemary or 1 teaspoon chopped tarragon*
> *salt, pepper*
> *2 large eggs, or 4 yolks*
> *150 ml ($\frac{1}{4}$ pt) soured or single cream*
> *about 3 heaped tablespoons grated Parmesan cheese*
> *22 cm (8–9 in) pastry case, baked blind*

Cook the courgettes with a good lump of butter, a few tablespoons of water, the onion, herb and seasoning, in a tightly covered pan. Keep the heat moderate so that they cook without burning. Remove from the heat when the courgettes are barely done. Allow to cool to tepid and fish out the rosemary sprig if used. Beat together the eggs or egg yolks and cream, with the Parmesan. Stir into the courgettes and correct the seasoning. Add a little more Parmesan if you like. Pour into the pastry case. Bake for 30–40 minutes at 180°C/350° F/Gas Mark 4. The top should not brown much, if at all. Aim to keep the fresh yellow and green look, with only the lightest patch of deeper colour. For the best flavour serve warm, rather than cold or straight from the oven.

COURGETTES WITH SHRIMP SAUCE

Cook 1 kg (2 lb) sliced courgettes in boiling water until just tender. Drain them well and arrange them on a dish. Pour over them the following sauce:

> *1 rounded tablespoon butter*
> *1 rounded tablespoon flour*
> *150 ml (¼ pt) fish or chicken stock, heated*
> *150 ml (¼ pt) single cream*
> *small carton Young's potted shrimps*
> *salt, pepper, cayenne, mace*
> *lemon juice*

Melt the butter, stir in the flour, then the stock and cream to make a fairly thick sauce. Just before serving, add the potted shrimps with their butter. When everything is smoothly mixed, add seasonings to taste and lemon juice.

Serve with boiled and buttered rice, or toast, or lightly fried bread, or wholemeal bread.

SWEET-SOUR COURGETTES SICILIAN STYLE
ZUCCHINI IN AGRODOLCE

The Sicilian mixture of pine kernels, sultanas, anchovies and vinegar or lemon juice is a most sympathetic one to northerners. Even used with fish or, as in this recipe, with courgettes, the taste is not alien to a nation of mint sauce and chutney eaters. Such invigorating blends, which we owe ultimately to the Middle East, can be seen in refined descendants of medieval food, in which the borders between sweet and sour, sweet and peppery, even between meat and fish, were not as clearly defined as they are to-day.

> *1 kg (2 lb) courgettes*
> *1 large clove garlic, crushed*
> *2 tablespoons olive oil*
> *2 tablespoons wine vinegar*
> *2 tablespoons water*
> *30 g (1 oz) pine kernels*
> *30 g (1 oz) sultanas*
> *8 anchovy fillets, chopped small*
> *salt, pepper*

Cut the courgettes into strips about 5–8 cm (2–3 in) long. Salt them. Cook the garlic slowly in the oil and, after a couple of minutes, add the drained and dried courgette strips. Keep them moving until they are coated in oil and just beginning to colour. Pour in vinegar and water, cover with foil and simmer for 15 minutes. Remove the foil, add the nuts, sultanas and anchovies and cook more rapidly until the liquid is reduced to about two tablespoons. Keep stirring, so that the courgettes are bathed in the flavours of the sauce. Correct the seasoning and serve with bread.

SWEET-SOUR COURGETTES (2)
ZUCCHINI IN AGRODOLCE

A simpler, even more successful recipe than the Sicilian one is another favourite of mine, particularly when I want to serve courgettes dry with meat, grills and so on, rather than on their own in a sauce. This version comes from Elizabeth David's *Italian Food*; the splendid thing is the hint of cinnamon.

Salt 1 kg (2 lb) sliced, unpeeled courgettes for an hour in a colander. Pat them dry and cook them gently, in two pans if necessary, in olive oil. When they are nearly tender, season with plenty of black pepper, a little powdered cinnamon, four tablespoons of wine vinegar and two table-spoons of sugar. Turn them over in the juices, which should evaporate into a small quantity of sauce. The dish is unlikely to need more salt, but taste it to make sure. The courgettes shrink and brown slightly, but remain soft. They become impregnated with the flavours of the other ingredients. Keep an eye on them towards the end, otherwise this is a simple recipe.

STUFFED COURGETTES, MARROWS AND OTHER SQUASHES

Although it is not essential to blanch courgettes as a preliminary to stuffing and baking them, it does save on oven heat and time. Courgettes that are still tender, even though they may be quite large, should be blanched whole in boiling salted water; the time will naturally depend on their thickness – 10–15 minutes is about right.

Once the courgettes become quite large and weighty, it is a good idea to slice them in two before blanching them. Remove the seeds first. The English style is to stuff them and tie the two halves together again for baking. This takes a long time to cook. I find it more practical to place the halves side by side in a pan and pile the stuffing into the centre. Instead of dabbing bits of butter over the top to keep the stuffing moist, try using chopped tomato instead, if there is no tomato in the stuffing.

For a small family faced with a huge marrow, it is best to slice it down into rings about 2–3 cm (1 in) thick, allowing one or two per person. This frees the rest of the marrow for other recipes. Do not peel the rings, but blanch them for five minutes only, after removing the seeds in the middle. Place the rings in a buttered baking dish and fill the centres with stuffing, allowing it to spread a little over the marrow in a domed shape.

Allow 45–60 minutes at 190°C/375°F/Gas Mark 5 for these dishes, or longer at a lower temperature if this is more convenient.

For stuffings, see pp. 58, 200–1, and 562–7.

STUFFED COURGETTES IN YOGHURT SAUCE

A recipe from Claudia Roden's *Middle Eastern Food*.

> *18 medium courgettes*
> *salt*
>
> FILLING
> *250 g (8 oz) lean lamb or beef, minced*
> *100 g (good 3 oz) long grain rice*
> *1 tomato, seeded, chopped*
> *2 tablespoons parsley*
> *ι tablespoon each raisins and pine kernels (optional)*
> *salt, pepper*
> *½ teaspoon ground cinnamon or allspice*
>
> PLUS
> *¾ litre (1½ pt) yoghurt, plus 1 eggwhite*
> *3 cloves garlic, chopped finely*
> *salt*
> *1 teaspoon dried mint*
> *1 tablespoon butter*

Hollow and stuff the courgettes as above. To make the filling, mix the ingredients together and check the seasoning. Beat yoghurt and eggwhite together; simmer and stir for 10 minutes, uncovered.

Range the courgettes in the pan and simmer them in water for 35 minutes. Towards the end of this time, the water should almost have disappeared. Add the yoghurt – hot or cold – and bring back to boiling point. Cover the pan and simmer for a further 20–25 minutes. Just before the end of the cooking time, crush the garlic with a little salt, mix in the mint and fry in the butter in a small pan. Scrape this mixture into the yoghurt sauce. Add more salt if necessary and serve hot with boiled rice.

STUFFED COURGETTES IN TOMATO SAUCE
For the yoghurt in the recipe above, substitute the basic tomato sauce, p. 511. Omit the garlic and mint finish.

OTHER SUGGESTIONS

Courgette parmigiana Turn to the recipe for aubergine parmigiana, on p. 49, and use sliced, salted but unpeeled courgettes instead. Top with split anchovy fillets and only the lightest sprinkling of Parmesan cheese. If you are lucky enough to grow basil successfully, it can be chopped over the layers of tomato sauce.

Ratatouille Courgettes are a main ingredient in *ratatouille*, the provençale vegetable stew. For the recipe, turn to p. 52.

Courgette soufflé A purée of courgettes makes a delicate soufflé, see p. 468.

Kolokythia moussaka (*Courgette moussaka*) Substitute 1½ kg (3 lb) medium-sized courgettes for the aubergines in the *moussaka* recipe on p. 57.

Kolokythia papoutsakia Substitute 1½ kg (3 lb) medium-sized courgettes for the aubergines in the recipe on p. 59. There is no need to salt them, but blanch them whole in boiling, salted water until they are just tender. Drain them and split them in half from the stalk end. Scoop out the pulp very carefully as courgette skins are even more tender than aubergine skins. In *The Home Book of Greek Cookery*, Joyce M. Stubbs recommends cinnamon as a flavouring for the meat stew, and not oregano. It makes a change, but I prefer the oregano, particularly if it comes from Greece.

CRESS, see MUSTARD AND CRESS. Also LAND or AMERICAN CRESS. Also WATERCRESS.

CUCUMBER

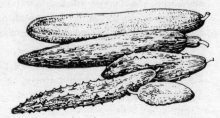

The cucumbers of Ur-Nammu, who lived in Mesopotamia 4000 years ago, refreshed him so well that he built a temple to the god Nanna, to save the garden where they grew from destruc-
tion. In the end Nanna could not save it, this earliest known of the world's vegetable gardens. Even the most speculative of archaeologists, with all the scientific aids available, could not place it exactly, though the site of the temple is known. Ur-Nammu paid a high price to guard his cucumbers (and his onions, leeks, lettuce and melons), though anyone eating chilled cucumber soup at the end of a hot day, may think he spent his money wisely. They will certainly think sympathetically of the Children of Israel, two years in the desert, who sighed wearily for the cucumbers of Egypt.

This ancient taste reached another peak of fashion with the Roman Emperor Tiberius, who had a remarkable partiality for cucumbers, one of his enthusiasms that we can, I think, feel happy to share. There is not a day in the year when our tables, like his, are not supplied with cucumbers, though from rather different sources. The imperial cucumbers were raised in beds mounted on wheels. Like hospital patients they were rolled out into the sun. When the day turned chilly, they were moved back under frames glazed with transparent stone.

Although neither Ur-Nammu nor the Children of Israel seem to have worried about it, cucumbers were reckoned to be decidedly indigestible by Tiberius's day. Early garden varieties were windy food, with a bitter comeback. Even with our improved cucumbers, smooth, ridged, slim, squat, green or yellow, Japanese or Chinese, short or long, one may occasionally have a bitter one from the garden. Nurserymen have long been busy on these shortcomings and now we can grow what are coyly known as Burpless Cucumbers. It seems to have been that polite race the Japanese who achieved this small miracle.

It would certainly have appealed to my grandmother, who told me never to peel a cucumber – for the sake of the family's digestion (I think she really meant wind). I still don't, unless the skin is notably coarse, because I prefer cucumber that way, and cannot bear to waste anything edible. There seems to be plenty of people whose grandmothers told them the opposite, who never eat a cucumber without peeling it first for precisely the same reason that I was told not to peel mine. With a burpless cucumber, I suppose it doesn't matter. One may be guided entirely by the recipe and the occasion. Cucumber sandwiches for Lady Bracknell, obviously must contain peeled slices of cucumber. A dish making use of boiled cucumber 'boats' requires that the skins should be left in place. Another point of argument is whether to salt cucumber slices in advance, in the French manner (again I think the idea is to extract indigestible juices), or to leave them as they are.

You could follow the half-way practice of peeling them down in strips. Or you could take a heavy-pronged fork and pull it down the cucumber before slicing it thinly. This makes an attractive edging, less lumpy than the plain peel, less wasteful than the denuded style. Then to salt, or not to salt, is up to you.

Cucumbers have come to a special glory in recent years. Partly it is the result of being able to buy them round the 12 months. Even more it is the experience of travel and immigration that has opened our eyes to the cucumber's possibilities. Dill-pickled, sweet-sour cucumber from Poland, *gazpacho* from package holidays in Spain, cucumber and yoghurt salads and soups from Cyprus and the Middle East, lovely creamy dishes of cooked cucumber from France, as well as stuffed cucumber and Doria cucumber garnish for sole or light meats, seem a long way from cucumber sliced thickly into rather boring salads and cucumber with salmon that was all that many people knew of in the Thirties.

HOW TO CHOOSE AND PREPARE CUCUMBER

Always press the stalk end of a cucumber, gently, to see if it is soft. It should be firm at every point, and have the appearance of bursting with juiciness. Naturally smooth-skinned cucumbers will have a glossier, fuller appearance, than the other kinds. Ridged or warty-looking skins cannot look so tight, but they are none the worse for that. It is the feel that matters, the look of good health.

If you need to keep a cucumber for a little while and have no refrigerator, stand it stalk end down in a tall jug with a little water in the bottom. Cucumbers do not like intense cold, so if you do not have a cool larder always keep them in the vegetable drawer of the refrigerator, either in their plastic skin, or well wrapped in film. Ideally they should be used quickly, if you want them at their best.

If you must peel cucumbers, do it as thinly as possible. To salt them, slice to the thickness you require either with a sharp knife or, for very thin slices, on the cucumber blade of the grater, and spread them all over a colander. Sprinkle with tablespoon of salt (and a tablespoon of vinegar if you like, mixing the two together first) and leave for an hour. Pat dry in a cloth; rinse them first if they taste a little on the salty side.

I have seen people conscientiously removing the seeds from cucumber. This is not necessary, unless they are hard, or the recipe demands it for constructional reasons.

Always taste a piece of cucumber before making a dish, especially if you grow your own. Occasionally one is bitter.

CUCUMBER SOUPS

The cooling delicacy of cucumber is emphasized further by mixing it with liquids into a chilled soup, and by giving it a slight edge with yoghurt or soured cream. Hot countries all have their chilled soup – okroschka and cucumber and sorrel soup from Russia, for instance, or the Spanish tomato and cucumber soup known as *gazpacho* (p. 510), or the Turkish cucumber and yoghurt soup which is *cacik*, p. 239, diluted further with yoghurt.

Surprisingly heat seems to emphasize the virtues of cucumber too, as you will see from the first recipes in this section. Be careful, though, to cook it gently and as briefly as possible. This will preserve all the cucumber delicacy to make a fresh-tasting soup for winter days.

FRENCH CUCUMBER SOUP

> 1 medium onion, chopped
> 60 g (2 oz) butter
> ½ large cucumber, peeled, diced
> 1 heaped teaspoon flour
> 1¾ litres (3 pt) water or light stock
> salt, pepper
> 2 egg yolks
> 2 tablespoons double cream
> extra lump of butter
> chopped parsley

Cook the onion gently in the butter in a covered pan for five minutes. Stir in the cucumber, cover again and leave for another five minutes. The vegetables must not brown, but become impregnated with butteriness.

Sprinkle in the flour, then add about half the liquid. Simmer until the cucumber is just cooked. Then purée in a blender, adding the remaining water or stock to obtain a fairly thin consistency. Correct the seasoning. Reheat and pour a ladleful of the soup into a basin in which the yolks and cream have been beaten together. Pour this mixture back into the pan, and stir over a low heat for a few minutes to bind the soup smoothly. Add the lump of butter. Scatter with parsley and serve.

CUCUMBER AND DILL SOUP

Follow the recipe above, using dill weed instead of parsley. Remember that dried dill weed is much stronger than fresh, so start with ½ teaspoon or even less, adding more to taste as the flavour develops.

RUSSIAN CUCUMBER AND SORREL SOUP

> 250 g (½ lb) sorrel*
> 200 ml (7 fl oz) double and single cream mixed
> 300 ml (½ pt) natural yoghurt
> ½ litre (¾ pt) cold consommé or beef stock
> 250 g (½ lb) cucumber, chopped
> 3 hard-boiled eggs, chopped
> chopped chives
> chopped green fennel or tarragon
> salt, pepper
> lemon juice

Remove the thickest sorrel stems, rinse and cut up the sorrel with scissors. Put it into a pan with no extra liquid and stir it over a moderate heat until it is reduced to a dark green purée. Put into a big soup tureen and leave to cool. Add the remaining ingredients in the order given, mixing them in well and adjusting the quantities of herbs and seasonings to taste. Serve well chilled.

CACIK
TURKISH CUCUMBER AND YOGHURT SALAD

This refreshing salad, Turkish by origin, is popular all over the Near and

---

* Spinach can be used instead of sorrel, but increase the amount of lemon juice to get the right sharpness.

Middle East. It can be eaten by itself, diluted with more yoghurt to make soup, or served as a vegetable-cum-sauce with grilled lamb. Dried, crushed wild mint is a favourite herb in that part of the world, particularly in the Lebanon. Indeed the use of mint in Middle-Eastern – and Indian – cookery is far more resourceful than it is in our English dishes.

> *1 large cucumber*
> *1 heaped teaspoon sea salt, finely ground*
> *4–5 cloves garlic, crushed with salt*
> *½ litre (¾ pt) yoghurt, preferably from goat's milk*
> *plenty of black pepper*
> *1 tablespoon dried mint, or 4 tablespoons fresh chopped*

Dice the cucumber, peeled or unpeeled according to your preference. Put the pieces into a colander, sprinkling them with the sea salt. Leave to drain for at least an hour. Press the cucumber with your hands to extract the largest possible amount of liquid, then dry it in a clean cloth. Meanwhile crush the garlic to a cream with a little salt, using a heavy knife blade or cleaver rather than a garlic press. Add it to the yoghurt, with plenty of pepper and nearly all the mint. Half an hour before serving the salad, mix in the cucumber and chill it well. Sprinkle with the remaining mint before serving it with wholemeal bread and butter, or with grilled chops.

Note Bulgarian walnut and garlic sauce (p. 562) and white bread can be served with this salad or the soup; omit garlic from the ingredients above.

DANISH CUCUMBER SALAD
AGURKESALAT

In Denmark, as in Cyprus, a cucumber salad will be served often with hot meat, grills, roast lamb or pork, chicken, or with hamburgers.

> *1 large cucumber*
> *1 level tablespoon salt*
> *1 rounded tablespoon sugar*
> *150 ml (¼ pt) wine or cider vinegar*
> *pepper*
> *chopped parsley or dill leaves*

Score the cucumber down with a fork, slice thinly on the wide grater blade and salt the slices. Then squeeze them dry in a clean cloth and spread them out in a dish. Mix the sugar and vinegar and pour it over them. Pepper well. Leave for an hour or two in the refrigerator. Taste some and see if the seasonings need adjustment. Drain off the liquor and serve the cucumber sprinkled with parsley or dill.

SALAD ELONA

Alternate circles of thinly sliced peeled cucumber with circles of hulled and halved large strawberries. Season with salt, plenty of black pepper, a pinch of sugar. Pour over two or three tablespoons of dry white wine, or white wine vinegar. Serve with cold salmon, salmon trout, or chicken.

MRS STEPHENSON'S CUCUMBER MOUSSE

This recipe came from Mrs Victoria Stephenson, who makes a variety of mousses and salads for Justin de Blank's provision shops in London. Sometimes we eat it on rye bread, with a bottle of white wine, at five or six in the evening, sometimes as a first course. It goes with cold salmon trout and wholemeal bread and butter, as the main course of Sunday lunch in June.

>	½ large cucumber, peeled, diced small
>	1 heaped tablespoon salt
>	3 tablespoons tarragon or wine vinegar
>	15 g (½ oz) gelatine
>	6 tablespoons hot water
>	300 ml (½ pt) whipping cream, or 150 ml (¼ pt) each single and double cream
>	½ kg (1 lb) curd cheese or ¼ kg (½ lb) each sieved cottage cheese and cream cheese
>	black pepper
>	chopped chives, parsley, spring onion

Mix the cucumber, salt and vinegar in a bowl thoroughly. Then turn it into a colander, put a heavy plate on top and leave for an hour at least. Then remove the plate and press the cucumber with a clean cloth to get the last of the liquid away.

Dissolve the gelatine in hot water and whisk in the cream(s) gradually until the mixture is smooth and very thick, but not stiff. Break up the cheese and add it to the cream. Mix in the cucumber.

Taste and add a little more vinegar and salt if necessary, but be careful to overdo neither. Sometimes a couple of pinches of caster sugar will help to bring out the flavour; this depends on how good the cucumber was to start with. Grind in plenty of black pepper and add an abundance of chopped chives, parsley and spring onion – enough to make a strongly speckled effect. Tip the mixture into an oiled decorative mould and leave overnight in the refrigerator to set. Turn out on to a serving dish.

Always provide wholemeal or rye bread with cucumber mousse, whether it is the only dish or one of several; the rich cool mixture needs that kind of flavour to set it off.

CUCUMBER TARTARE

Concombres à la tartare comes from *Le Pot-au-feu*, a cookery and household magazine which first appeared in Paris in 1893 and continued at least until 1913.

The introduction to the recipe, in an 1899 issue, observes that *concombres à la tartare* has been popular for some time in the best Paris restaurants. Although it is 'extremely simple', two whole pages are devoted to it, including foolproof instructions for mayonnaise.

> ½ *large cucumber*
> 75 g (2½ oz) *canned tunnyfish in oil*
> 50 g (scant 2 oz) *butter*
> *salt, pepper, cayenne*
> 100 ml (3 fl oz) *olive oil*
> 1 *teaspoon wine vinegar*
> ½ *teaspoon Dijon mustard*
> 1 *raw egg yolk*
> 3 *tablespoons tomato purée*

First cut six slices from the cucumber half, each one about 2 cm (¾ in) wide. Remove the peel, going round each slice as evenly as possible: this is why you do not peel the cucumber first – it is easier to avoid grooves and bumps by peeling round, rather than down the cucumber. Scoop out the centre seeds, to make the slices into rings. Poach the rings in boiling, salted water for 15–20 minutes until tender, then run them under the cold tap and leave on a cloth to drain.

Mash the tunny and butter together to a paste, seasoning it with salt, pepper and cayenne. Make a mayonnaise with the oil, vinegar, mustard and yolk (method p. 560). Flavour it with tomato purée; this can be a cooked tomato purée flavoured with a little concentrate, or a large, firm, fresh tomato that has been skinned and seeded and put through the *mouli-légumes*. Another way is to chop a large tomato and cook it very briefly and quickly in butter to evaporate the juice; good seasoning is required.

To complete the dish, fill the rings with the tunnyfish butter, place them on a dish and pour the tomato mayonnaise around them.

VARIATION Substitute prawns or shrimps for tunny. Shell them before weighing out 75 g (2½ oz). Keep half a dozen in their shells to decorate the cucumber rings.

This light dish was to be followed by breast of mutton with peas, then cold stuffed chicken, and *coeur à la crème* for pudding. With our shorter meals, you might like to augment the cucumber slices above with some hard-boiled eggs, quartered. Or you could double the quantity, making twelve rings for six people.

CUCUMBER SAUCE

Slice half a large cucumber thinly. Stew it in a heavy pan with a good tablespoon of butter and about 60 g (2 oz) cooked spinach, or twice that amount of raw spinach. When cooked drain and purée in the blender. Add to about 300 ml (½ pt) béchamel or velouté sauce, and finish with a spoonful or two of double cream. Reheat gently.

Serve with hot salmon, salmon trout, chicken and veal. It is also good with fat fish such as herring and mackerel, which have been stuffed and baked in a hot oven.

The idea of the spinach is mainly to add colour, but the flavour is pleasant too. It can be omitted, and you could in that case complete the sauce with some chopped cucumber, lightly cooked in the sauce before you add the cream.

CUCUMBER RAGOÛT

Hannah Glasse's version, from *The Art of Cookery*, 1747. Very good. Courgettes could be substituted for cucumbers.

> 2 cucumbers, sliced, unpeeled
> 2 medium to large onions, sliced
> clarified butter
> 6–8 tablespoons chicken stock
> 2–3 tablespoons dry white wine
> salt, pepper, ground mace
> 1 dessertspoon flour mashed with 1 dessertspoon butter

Brown the cucumbers and onions lightly in clarified butter, in separate pans. When they have collapsed nicely, put the onions in with the cucumber. Add the stock and wine, adjusting the quantity to the original sizes of the cucumber and onions. Season. Cover and simmer until cooked. Add the flour and butter paste in little pieces gradually, stopping when the sauce is nicely thickened without turning into glue. Keep the whole panful at simmering point or just below, while you complete this operation. Check the seasoning again and serve.

The liquid levels are the slightly tricky part. The amount of stock and wine you add will partly depend on the amount of juice left from frying the cucumber; and as the vegetables stew together, they will produce more liquid. Aim to end up with the vegetables lightly bound together in a little sauce, not swimming in a bowlful.

SOLE, WHITING, SALMON TROUT OR TROUT DORIA

In other words, fish and cucumber. First prepare the garnish. Peel the cucumber or cucumbers as thinly as possible, then with a Parisienne cutter make as many olive shapes or balls as possible. Keep all the bits and pieces for soup. Melt a little butter in a large shallow pan and put in the turned cucumber, preferably in a single layer. Season with salt and a little sugar. Cover closely and leave to stew gently until tender. Shake the pan from time to time and keep an eye on the contents to make sure they do not brown. Check the seasoning at the end.

Flour the fish, either whole or in fillets, very lightly, then cook it in clarified butter. Put on to a hot dish and surround with the cucumber, or put it at each end. Rapidly melt some fresh butter and when it is slightly browned, pour it over the fish. Sprinkle over a little parsley, tuck in some lemon quarters and serve immediately.

Cucumber cooked this way can be served with thin escalopes of veal, chicken or turkey, dipped in egg, rolled in breadcrumbs and fried in butter.

CREAMED CUCUMBER
CONCOMBRES À LA CRÈME

Prepare and turn the cucumber as above. Cook the balls in a little butter in a covered pan until they are almost done, but not quite. Pour on boiling cream, enough to bathe them nicely, but not to cover completely. Keep the lid off the pan while you complete the cooking. The cream should be much reduced with the cucumber liquid. If the sauce seems too much on the thin side, add a spoonful or two of béchamel sauce. Serve with fish and light meats.

CHICKEN WITH CUCUMBER CREAM SAUCE
POULET AUX CONCOMBRES

Here is one of those convenient, pan-fried French dishes, that they often make to avoid switching on the oven in summer. Meat, vegetable and sauce cook together gently. Nothing disastrous can happen. Anything that is not quite right can easily be adjusted. The snag is that the ingredients have to be good; it is no use chopping up a poor battery hen and a couple of yellowing cucumbers.

> 2 large cucumbers, peeled
> 1 chicken, jointed
> seasoned flour
> clarified butter or oil
> 3 tablespoons brandy
> about 600 ml (1 pt) chicken stock
> 175 ml (6 fl oz) whipping or double cream
> salt, pepper
> 1 tablespoon each flour and butter (optional)

Cut the cucumbers into sturdy matchstick pieces, about 3 cm (1½ in) long. Blanch them for three minutes in boiling salted water, then drain them well. This is not strictly necessary, but it does seem to make for a finer flavour.

Turn the chicken in the flour and fry it until golden-brown, not deep brown, in butter or oil. Keep the heat moderate to avoid any risk of burning. Pour over the brandy, set it alight and turn the pieces over in the flames. Stir in the cucumber and enough chicken stock to bring the level up almost to the top of the chicken pieces. Leave to bubble steadily, not fiercely, until the meat is just done. Raise the heat, stir in the cream and correct the seasoning. You can boil the sauce down a little to thicken it, or mash flour and butter together and add little bits to the sauce until it is the thickness you require. If the sauce shows any tendency to oiliness, stir in a little water or chicken stock. Serve very hot, on very hot plates.

STUFFED CUCUMBERS
CONCOMBRES FARCIS

One of the best cucumber dishes I know.

> 3 cucumbers
> 175 g (6 oz) rice
> 1 large onion, chopped
> 1 clove garlic, chopped
> 100–125g (3–4 oz) butter
> 250 g (8 oz) mushrooms, sliced
> 5 rashers bacon
> 3 eggs, beaten together
> chopped parsley, salt, pepper

Cut each cucumber in two through the middle, then slice each piece longways to make two 'boats'. Scrape out the seeds with a teaspoon, and blanch the pieces for 15 minutes until they are cooked. Drain well.

Meanwhile boil the rice. Cook the onion and garlic gently in half the butter, until they are soft. Add the the mushrooms, raise the heat and fry until they are cooked. Season. Cut the bacon into strips, discarding any rind, and stir them into the mushroom mixture. After two or three more minutes remove the pan from the heat. Drain the cooked rice and mix into the mushrooms. Pile into the hot cucumber pieces, putting any left over round them. Last of all cook the eggs into one or two thin omelettes, roll up and cut like a jam roll so that they fall into strips. Reheat for a few seconds only in the remaining butter, sprinkle with parsley and seasoning and scatter over the rice. Serve very hot.

CUSTARD MARROW or SQUASH

The custard marrow is a charming vegetable to look at. The taste is pleasant, too, though not up to the appearance. The skin has a matt cream or crocus-yellow or green or striped colour, with the green stalk popping out of the top as if it were a beret. The edges are curved into thick scallops; the whole shape is something like a fluted brioche, a squashed home-made one, without a pronounced top-knot. In its native America, the custard squash is also called the scallop or scollop squash, elector's cap, patty pan squash – the French have picked up the last one as *pâtisson*. Another pastrycook name is cymling, spelt in various ways though all are versions of our simnel, as in simnel cake for Mothering Sunday. And simnel comes from the Latin *simila*, fine flour. Why custard? Perhaps because the dryish quality of the cut flesh looks something like a baked custard when you first put in the spoon. Remember that the word custard developed from *croustade*, the crust or pastry case in which eggs and milk were often baked, and you are back again with cakes and buns and pastry shops, with the pleasure its shape has given.

Custard marrows are not a standard item of greengrocery. One is more likely to see them in some odd place. A butcher's window in Oxford market usually has two or three in the autumn, a sign of the owner's pride in his garden. I would never have the cheek to ask if they were for sale. The first ones I ever bought took some courage to acquire. There was a couple of them, placed in a particularly decorative manner amongst the summer vegetables of a greengrocer's in Tours. They were the focus of the display; they were set apart, to be looked at. I took refuge in my bad French, and had the feeling they were put into my basket with reluctance, as a courtesy due to foreigners. Were they going to a good home? Shan't we miss them? Such thoughts flickered over the pleasant face of the woman who served me. The amiable courtesy of that part of the world prevailed. I was grateful.

HOW TO CHOOSE AND PREPARE CUSTARD MARROW

The ideal is to have them really small, like courgettes, no more than 6–8 cm long (2½–3½ in), about the size of *petits fours*. As with all young squashes, the skin should be easily pierced by your finger nail. It does not need to be removed. Indeed it would be impossible to remove it without spoiling the shape. These small custard marrows can be thinly sliced and stewed in butter in a tightly covered pan or eaten raw in salads. Or they can be blanched whole and finished in ways appropriate to courgettes – pp. 229–33.

When custard marrows do appear for sale they are more likely to be 11–18 cm (4½–7 in) across in diameter, with a deeper, more bunlike shape. Larger than that they are best avoided. These custard marrows can be cut across into slices, blanched and stewed in butter, again like courgettes. Or they can be steamed and stuffed. This last method has the advantage of preserving the attractive shape. Cooked in this way, they make an unusual vegetable dish of some beauty. Allow one per person in the smaller size, one for every two people as you go over 14 or 15 cm (6 in).

As with all tender squashes, there is no need to remove the seeds or the centre flesh. It is only when the skin gets hard, as with pumpkins and marrows and winter squashes generally, that the seeds become unpleasant and the flesh between them fibrous and cottony.

Although custard marrows can be kept for a few days in the refrigerator, like courgettes, they are best cooked as quickly as possible while their delicate flavour is at its height.

For soups, soufflés, fritters, gratins of custard marrow, turn to the section on courgettes.

BAKED STUFFED CUSTARD MARROW

Pick out evenly matched custard marrows, one per person if possible, otherwise three larger specimens for six people. Trim off the stalks. Put 1 cm (½ in) water, with salt, into a large shallow pan. When it boils, fit in the custard marrows whole. Cover tightly and leave them to steam until tender. Allow 20 minutes, then test them. There should be enough water to steam the marrows without them boiling dry, but not a lot of water to ruin their flavour. Drain the marrows; let them cool a little, or run them under the cold tap.

With a small sharp knife, cut out a lid from the stalk end of each marrow

and set it aside. Hollow out the centres, leaving a solid enough crust to
contain the filling. Chop this inner pulp, allowing any moisture to drain
away, and put it into a basin. Add the following:

> 250 g (8 oz) curd cheese, or medium fat soft cheese
> 1 tablespoon chopped parsley
> 1 teaspoon chopped fresh tarragon
> large spring onion, chopped
> 2 tablespoons double cream
> salt, pepper, a pinch of cayenne

Put this mixture into the cavities of the custard marrows, piling it up. Press
the lids down gently on top (or, if the lids are raggy, scatter with breadcrumbs and melted butter). Range the marrows in a buttered oven proof
dish and pour in a scant 1 cm (½ in) of milk. Bake at 180°C/350°F/Gas
Mark 4 for about 20 minutes, or a little longer.

Note: If you cannot buy a medium fat soft cheese, buy cottage rather than
full cream cheese which will melt away in the oven leaving no curd
behind. Enrich the cottage cheese with extra cream or melted butter.

VARIATIONS This dish can obviously be made with a number of different
stuffings: turn to pages 562–7 or to the sections on peppers or aubergines.
Remember that custard marrow is a gently flavoured vegetable, respond-
ing best to the milder tones of summer. A better alternative to spicy
stuffing would be a mixture of some other vegetable with a light cheese.
American cooks often combine green beans with a mild American Cheddar and mix them in with the pulp and some butter. Caerphilly would be
the English cheese to try.

DANDELION LEAVES

We have used dandelions for wine and beer, but rarely for salad. Bitterness has been the problem, as it was once with many other kinds of salad greens, celery, lettuce, endive, chicory. The solution is the same, the plants must be blanched. I suppose that dandelions may have been unpopular, too, for their diuretic nature; we do not seem to have the same affection for our old name of piss-a-bed as they do in France, where *pissenlit* is a far commoner name than *dent-de-lion*. In France, too, strains have been developed for forced cultivation in the manner of chicory and seakale. English nurserymen can supply them. If you prefer to buy them on holiday in France, look out for *pissenlit à coeur plein amélioré*. On the whole it is easier to buy these seeds, than to scurry over the lawn with an armful of plant pots looking for suitable wild dandelions to blanch. Jason Hill took a more enterprising attitude in his *Wild Foods of Britain*, 1939 (he was, I think, the first and the best with this kind of writing). He says firmly that the blanched leaves of wild dandelion are as good as curly endive, 'They may be had in winter by lifting large roots from September onwards and packing them, after cutting off the large leaves, in a box of light soil, sand or peat, and putting this under the scullery table or in a warm cellar; the soil should be kept just moist and the leaves will be ready in about a month.' Darkness is essential. Other gardening writers claim they do better without warmth.

As well as putting them into salads, like Jason Hill's curly endive, or in the manner of the favourite French recipe following, you may also cook the blanched leaves 'like spinach'. Do not expect the same quality of flavour and, if you use wild leaves, be prepared to change the cooking water and put in a splash of vinegar as well as salt. My advice is to stick to salads.

DANDELION SALAD WITH BACON
SALADE DE PISSENLITS AU LARD

A popular hot salad in several parts of France, and one of the best ways of

enjoying dandelion leaves, blanched, from the garden. If you use wild
dandelion leaves, pick young ones and mix them with a crisp lettuce.

> *350 g (¾ lb) dandelion leaves*
> *2 tablespoons vinaigrette*
> *salt, pepper, hint of sugar*
> *125 g (4 oz) piece smoked streaky bacon, diced*
> *1 cm (½ in) slice white bread, cubed*
> *oil*
> *2 tablespoons wine vinegar*
> *1 hard-boiled egg, crumbled*

Tear the washed greenery into pieces and place it with the vinaigrette in a
slightly warmed salad bowl. Sprinkle on a little seasoning. Fry the bacon
and bread in oil until golden-brown and tip the panful, complete with
sizzling fat, on to the salad. Turn it well. Rapidly heat the vinegar in the
frying pan and when it is bubbling hard add that to the salad, turning it
again. Serve immediately with a sprinkling of egg over the top.

This salad is delicious on its own but it also goes well with omelettes,
e.g. a ham omelette – omit the hard-boiled egg.

The hot oil and bacon fat and vinegar wilt the salad leaves slightly.
Crispness is provided by the bread. Altogether a fine combination of
flavours.

Lamb's lettuce and a beetroot, diced, can be used instead of dandelion
leaves; or they can be mixed with them. For another hot salad, using a few
dandelion leaves with lamb's lettuce and walnut oil, see p. 284, Winegrower's Salad.

EARTHNUTS
or PIGNUTS

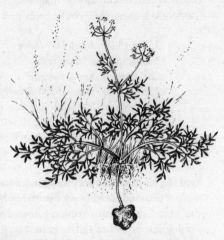

Children who are brought up in towns, as I was, are likely to be unfamiliar with earthnuts, one of the few wild foods that are worth eating. My introduction to them was therefore late – and startling. The three of us were ambling down a sunken lane through woods, to a large dolmen beside a stream near Ernée, in Mayenne. The day was hot and dragging, the picnic basket heavy, our steps slow. Suddenly my husband moved smartly to the side of the lane and scrabbled at the earth bank. His fingers moved fast, then with increasing gentleness as they worked their way through into the soil. 'What are you doing, Daddy? What is it?' Silence, puffs, then a discreet shout of triumph as he brought up a muddy knob attached to the root of a thin umbellifer. He rubbed off the muck, scraped it a bit with a knife, then handed it to our young daughter to eat. 'Pignuts.' Soon we were busy down the lane, digging out a salad of these crisp, clean-tasting tubers for our picnic. He told us that earthnuts, pignuts, cuckoo potatoes, did not care for cultivated land. No one had been able to turn them into a good vegetable. They remain the occasional delight of country children, as they had been for him and his brothers in Cornwall.

Rustics, too, as John Evelyn observed in *Acetaria*, like them 'crude . . . with a little pepper', but they are best 'boil'd like other roots, or in pottage rather, and are sweet and nourishing'. A herbalist of the same period remarked that the earthnut 'stimulates venery'. I have not noticed this. Earthnuts are their own pleasure. That is all.

ENDIVE

Cichorium endivia, endive, is the Struwwelpeter of vegetables, a frizzy headed salad looking like a wide-open lettuce. The kinky, shredded leaves are yellow to white in the centre from blanching, then they shade out to vivid green. Although endive can be cooked, it is better kept for salads. The chewy crispness and agreeable bitterness taste good after dishes such as *lasagne, moussaka* and stews. There are no special salad recipes for endive, as there are, say for lamb's lettuce, but bear its strong texture in mind when you mix it with other ingredients. It works well, for instance, with apple and a creamy dressing (see the Batavian salad Normandy style, p. 73).

I have tried to disentangle the chicory/endive names in the chicory section. Let me add to the confusion in your mind by saying that the French call endive *chicorée frisée*, curly chicory. You may decide to give the whole thing up, and take to pointing when you get to the greengrocery. This is an old dilemma. Three American scholars, who have recently published two volumes on Ancient Egyptian food, are as confused as Pliny was in the 1st century A.D. They retreat before such an intricate botanical knot. Language, paintings, tomb sculpture, excavations give no help. It seems as if the wife of the man who built the pyramids was reduced to pointing, too.

At least we do know that the names go back to Egypt, where both plants are thought to be indigenous, by way of French, medieval Latin and Greek. Chicory and *cichorium* to the Egyptian word for the plant, *kehsher*: endive and *endivia* and *intybus* to *tybi* meaning January. In other words chicories are the winter salad as opposed to lettuce – another plant prized by the Egyptians – which is eaten in summer. Though we are no longer stuck with seasonal limitations, the crisp endives and chicories of wintertime make a far better salad than greenhouse lettuces which make an expensive, floppy mouthful.

See also Batavia, and Chicory.

ESCAROLE, see BATAVIA

FENNEL

The garden I should most like to
have visited for its vegetables
was made at Monticello in Virginia by Thomas Jefferson.
When he returned there after
his second term as President of
the United States, he was free to
concentrate on it and grow the
seeds that were sent him from all
parts of Europe. In his *Garden
Book*, over 90 different kinds
and varieties of vegetables are

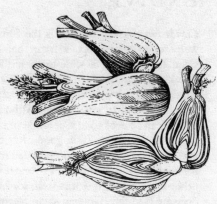

listed. Some came via an Italian friend, Philip Mazzei, who lived nearby.
Some, like his fennel seeds, arrived from Italy direct. In 1824 the American consul at Leghorn, Thomas Appleton, sent him five varieties of
cabbage, eleven of broccoli, three of cauliflower, and two of fennel, with
an enthusiastic letter describing the superb vegetables of Italy: 'the fennel
is beyond every other vegetable, delicious. It greatly resembles in
appearance the largest size celery, perfectly white, and there is no vegetable equals it in flavour. It is eaten at dessert, crude, and with, or without
dry salt, indeed I preferred it to every other vegetable, or to any fruit.'

Fennel, sometimes called Florentine fennel to distinguish it from the
feathery herb, is eaten still in Italy as a dessert, rather as we might serve
celery with cheese, except that fennel is put on as a companion – or
alternative – to fruit rather than cheese. The 3rd Earl of Peterborough,
one of the first people to grow fennel in this country, ate it in the same
way. Presumably he brought the seeds back from his adventures in Italy,
to grow them in his kitchen garden at Parsons Green in Fulham. Although
his unorthodox diplomatic and military escapades dismayed William of
Orange and Queen Anne, they kept sending him out on missions which
invariably turned to disaster: he must have been a quick-witted and
charming man. In London between-whiles he invited the liveliest people
of the day to Parsons Green (Locke and Newton were two of his friends).
Everyone enjoyed the Earl's conversation, and his delicious food. His
fruit was extra good as it was grown against the sun-warmed garden walls
– a new idea from Italy – so that one may imagine fine desserts at his table
including fennel in the Italian style that Thomas Appleton described.

Fennel is often eaten with vinaigrette, or as part of a salad, a dish of *crudités*, or an *ailloli garni*. It can be cooked and served cold *à la grecque* or *à la niçoise*, i.e. in tomato flavoured sauces. You can try it with cheese instead of celery, but choose goat cheese, or a fresh Parmesan at the *grana* stage, or a creamy French cheese: Cheddars and Cheshires are not quite right. Many celery and cardoon recipes are suitable for fennel, with small adjustments: often any sauce that accompanies it can be improved with a teaspoon of Pernod, which underlines the anise flavour. And this applies, too, to recipes in which fish or chicken are being cooked with fennel; I have tried using Chinese five-spice instead, which contains a proportion of star anise, but Pernod or some other anise preparation is more successful by far.

HOW TO CHOOSE AND PREPARE FENNEL

Recognizing fennel is easy enough: it looks like celery that has been pressed down into bulbousness by a giant hand, but instead of fretted yellowish leaves, it has feathery vivid green ones (fennel comes from the Latin for hay, fresh-cut hay I would imagine). Choosing fennel is less easy. Often it is sold in a tight wrap of plastic, but even when it isn't appearance is no guide to flavour. In this it resembles apple or celery, but as the price is high one minds more; perfect-looking fennel can be pale in flavour, battered-looking fennel can be superb. I suspect freshness and soil are the explanations. Or perhaps the commercial variety grown in Italy for export is like some of the celery we import, beautiful but dumb.

In spite of this, I buy fennel whenever I see it – not so often in Wiltshire, more often in France where it is grown locally – because its flavour is unique in vegetables, its texture clean and crisp.

When cooking fennel, allow one head per person, as the outer layer can be stringy and has to be cut away with the hard round stalks. As the trimmings can be used for soups, fish and chicken stocks (so can any cooking water), they are not wasted. The leaves can be chopped and scattered over the dish, or kept to add to a mixture of *fines herbes* for omelettes and stuffings. For most cooked recipes, the heads are halved, then quartered.

When fennel is to be eaten raw like celery, or in salads, four heads or less for six people should be plenty. After trimming the heads of the tough parts, slice them across thinly. The slices can be separated into thin strips if you like. If sliced fennel has to wait around, or seems less crisp than it should be, put it into a covered bowl of water with ice cubes, and leave in the fridge for an hour or a little longer.

FENNEL AVGOLEMONO SOUP

The soup recipes in the celery section can be adapted to fennel successfully, but as a change try this version of Greek *avgolemono* devised by
Margaret Costa. The familiar *avgolemono* of Greek holidays is made with
chicken stock, although fish stock is occasionally used instead; the
unusual part is the addition of fennel.

STOCK
¾–1 kg (1½–2 lb) bones, heads, skin of white fish
1 onion, sliced
1 leek or small bunch spring onions, sliced
1 small carrot, sliced
1 large head fennel, sliced (or you could use the trimmings from 6
    fennel heads)
1 small slice celery
½ large clove garlic
bouquet garni, including sprig of tarragon if possible
12 fennel seeds
12 peppercorns
8 coriander seeds, lightly crushed
1 glass dry white wine or dry cider (optional)
salt, pepper

SOUP
3 large egg yolks
juice of 2 lemons
1 teaspoon Pernod or anisette (optional)
6 thin lemon slices
chopped fennel leaves

Put all the stock ingredients into a large pan. If you have any water in
which fennel has been cooked for another dish, now is the time to use it.
Add water so that the ingredients are well covered. Bring to the boil and
simmer for 40 minutes. Strain into a measuring jug, pressing the juices
through slightly. Bring up to a litre (1¾ pt) with water, or reduce to the
same measure by boiling (this depends on your interpretation of the word
'simmer'). Correct the seasoning.

Just before the meal, bring the stock to just below boiling point. Whisk
the yolks and lemon juice together, then add a ladleful of the stock slowly
from a reasonable height, so that by the time it encounters the egg yolks it

is well below boiling point. Keep whisking. When the mixture is amalgamated, add it slowly to the soup which should be off the heat so that it, too, is well below boiling point. Taste and add more lemon juice if you like, and bring out the fennel taste with the Pernod or anisette. Divide between six warmed soup bowls and float a lemon slice on top of each one, with a few fennel leaves. Alternatively cool the soup quickly in a bowl of iced water, then pour it into a tureen and leave in the fridge to chill.

These somewhat pernickety finishing processes are to prevent the egg curdling.

FENNEL SOUFFLÉ
SOUFFLÉ DE FENOUIL

First prepare a fennel purée: quarter half a kilo (1 lb) of prepared fennel, blanch it for 15 minutes in boiling salted water, then drain and stew it for ten minutes in butter until it is really soft. Purée in a blender, or pass twice through the *mouli-légumes*. Season, adding some Pernod if you like, and a tablespoon of grated Parmesan cheese.

Now turn to the soufflé recipe on p. 277 and complete in the usual way.

FENNEL SAUCE

With one large head of fennel, prepare a fennel purée as above, then add to approximately 300 ml (½ pt) thick béchamel sauce. Stir in whipping cream to dilute slightly, or more milk. Heat through and stir in chopped fennel leaves. Serve with poultry or fish.

If you are making the béchamel sauce specially for this occasion, put in a teaspoon of fennel seeds when you are flavouring the milk, along with the other aromatics.

UMBRIAN FENNEL STUFFING

Fish from Lake Trasimene in Umbria are sometimes filled with ham, garlic and fennel and then spit roasted. You can do the same thing with our fresh-water fish, and then bake them in the oven or grill them over charcoal. There is no reason, either, why the mixture should not be used for salt water fish – it is particularly good with mackerel, for instance, as well as the more usual red mullet or bass.

2–3 *slices* prosciutto crudo di Parma, *or other raw ham from
Germany or France, or lean bacon as a last resort*
1 *large clove garlic*
1 *head of fennel, trimmed, quartered*
olive oil or butter
60–90 *g (2–3 oz) breadcrumbs (optional)*
1 *egg (optional)*
salt, pepper

Take a large chopping board and put on it the first three ingredients.
Chop them all together with a heavy knife, until they are reduced to a
crumbly mass: this is what the Italians call a *battuto*. Heat enough olive oil
in a frying pan to cover the base and stew the *battuto* in it, until the fennel
softens. It should not brown at all, so cover the pan if you cannot keep the
heat really low.

The mixture is now ready for use. Or it can be augmented with breadcrumbs and bound with an egg. Season, using plenty of pepper.

FENNEL BAKED WITH PARMESAN CHEESE
FINOCCHIO ALLA PARMIGIANA

My favourite fennel dish, the best one of all by far. The simple additions
of butter and Parmesan – no other cheese will do – show off the fennel
flavour perfectly. The point to watch, when the dish is in the oven, is the
browning of the cheese. Do not let it go beyond a rich golden-brown.

6 *heads fennel, trimmed, quartered*
butter, pepper
3 *tablespoons grated Parmesan*

Cook the fennel in salted water until it is tender. It is important to get this
right: the fennel should not still be crisp, on the other hand it should not
be floppy either. Drain it well and arrange in a generously buttered gratin
dish. Be generous, too, with the pepper mill. Sprinkle on the cheese. Put
into the oven at 200°C/400°F/Gas Mark 6, until the cheese is golden-
brown and the fennel is bubbling vigorously in buttery juices.

TAGLIATELLE WITH FENNEL

2 *heads fennel, trimmed, quartered*
salt
½ *kg (1 lb) tagliatelle*

olive oil or butter
plenty of grated Parmesan or Pecorino cheese
fennel leaves, parsley

Cook the fennel in plenty of salted water. When the pieces are tender, fish them out and slice them up; keep the slices hot. In the fennel water, cook the pasta until soft. Drain, and mix in the fennel and two or three tablespoons of oil or butter. Sprinkle on some cheese and the chopped fennel leaves if you like, with parsley, and serve with a bowl of grated cheese.

GOLDEN FENNEL AND FENNEL FRITTERS

Fennel can be eked out most successfully by turning it into fritters or frying it. Allow three large heads for six people. Trim them and cook them whole in boiling salted water for about 20 minutes. Drain and cool them to tepid, then slice them about ½ cm (¼ in) thick.

For golden fennel, dip the slices in egg and breadcrumbs, then fry them in butter or oil. Serve them with tomato sauce, or a mayonnaise flavoured with Pernod and chopped fennel leaves.

For fennel fritters, dip the slices into fritter batter, and fry them in a good depth of oil – or deep-fry them in a wire basket – until crisp and brown. Serve with lemon quarters, or Pernod-flavoured mayonnaise as above.

Note By blanching the fennel whole, you give the tougher outside layers a chance to cook until they are almost tender, while the inside remains fairly crisp; the slices should not disintegrate, so if you have only been able to get hold of small heads, reduce the cooking time by about five minutes.

FENNEL À LA GRECQUE

In this version of the French *à la grecque* method, tomatoes are included in the sauce, as well as the usual white wine, lemon and olive oil.

2 medium onions, chopped
1 large clove garlic, crushed
150 ml (¼ pt) olive oil
½ kg (1 lb) tomatoes, peeled, chopped
bouquet garni

1 level teaspoon coriander seeds, lightly crushed
100 ml (3–4 fl oz) dry white wine
juice of 2 lemons
salt, pepper, sugar
3 heads fennel, sliced, blanched 5 minutes
chopped fennel leaves

First make the sauce. Stew the onion and garlic in the oil until they are soft. Add the tomatoes, bouquet and coriander, then the wine, lemon and seasonings. Simmer for 15 minutes, then put in the fennel, with water if necessary to cover it; if you do this, add a little extra seasoning but not much, as you will later be reducing the sauce. Cover the pan and stew for about 30 minutes.

Remove the fennel to a serving dish. Taste the sauce and boil it down if it is copious and watery. Otherwise, correct the seasoning and leave it as it is. Pour hot over the still warm fennel. Cool, then put into the refrigerator to chill thoroughly. Serve scattered with fennel leaves.

OTHER FENNEL SALADS

You can make another cooked fennel salad by boiling the slices in salted water, then draining them well and pouring on an olive oil and lemon vinaigrette while they are still warm. Chill and serve sprinkled with chopped fennel leaves and parsley. The dish can be embellished with hard-boiled egg, mixed in with the herbs.

Sliced fennel can also be mixed with halved spring onions and topped with slivers of anchovy fillets. Use a vinaigrette dressing. You can pad the mixture out with cooked rice, hard-boiled egg and tomatoes, to make a bowl of mixed salad. Or you can arrange slices of raw fennel, celery, carrot, and so on, plus cooked potatoes, globe artichokes, eggs, haricot and green beans, salt cod, on a huge dish with ailloli, the garlic mayonnaise on p. 561.

If you can get hold of some really good tomatoes of the Marmande type, add them to a plain fennel salad with an olive oil and lemon dressing. Chop three or four thin slivers of lemon peel with parsley and scatter over the top. There is no point in making this salad with poor tomatoes.

FENNEL À LA NIÇOISE

The Nice style usually means tomatoes and tiny black olives. This is a good recipe, too, for celery or cardoons.

     6 fennel, quartered
     3 medium onions, chopped
     1 large clove garlic, finely chopped
     olive oil
     ½ kg (1 lb) tomatoes, skinned, chopped, or medium can tomatoes
     bouquet garni
     100 ml (3–4 fl oz) dry white wine
     salt, pepper, sugar
     100 g (3–4 oz) black olives, stoned if they are large
     60 g (2 oz) grated Gruyère

Cook the fennel in salted water. Meanwhile make a tomato sauce by softening the onion and garlic in a little olive oil for a few minutes, then raising the heat to brown them slightly. Add the tomatoes and bouquet. When the mixture stews to a pulp, after about 15 minutes, pour in the white wine. Season with salt, plenty of pepper, and sugar to taste.

    Put the drained fennel into a gratin dish and dot the olives about. Pour over the tomato sauce, and sprinkle with the cheese. Bake in the oven, or put under the grill, until the cheese browns slightly.

BUTTERED FENNEL

A good way of cooking fennel if it is to accompany fish or chicken (see recipes following as well) though it can also be served on its own.

     6 heads fennel, trimmed
     1 large onion, chopped
     1 large clove garlic, finely chopped, or crushed
     100 g (3–4 oz) butter
     salt, pepper, chopped fennel leaves

Slice the heads not too thinly. Simmer them in salted water until just tender. Meanwhile soften the onion and garlic in half the butter in a covered frying pan. Add the drained fennel and the rest of the butter. Continue to simmer uncovered, until the fennel is meltingly tender. Season and serve sprinkled with the chopped leaves. Cream may be added for the last few minutes of cooking time, if you like.

CHICKEN WITH FENNEL
POULET AU FENOUIL

There are various versions of this dish in French cookery.

1) Prepare buttered fennel as above. Roast and cut up the chicken, and arrange it on the fennel, on a hot serving dish. Skim the fat from the roasting juices, and pour them over the whole thing.

2) Quarter the heads of fennel, and simmer them in salted water until they are just tender. Drain off the water, and put a good knob of butter into the pan. Lower the heat and leave the fennel to stew gently for half an hour, turning it from time to time. Aim to brown the fennel lightly, to an appetizing golden-brown colour. Put in the centre of a serving dish, with the carved chicken round it. Skim the surplus from the chicken juices, boil them up with two or three tablespoons of fortified wine (Madeira or Marsala) and pour over the fennel and chicken. Scatter with fennel leaves.

3)   *6 small heads fennel, sliced*
        1 chicken
        butter
        3 shallots, or 100 g (3 oz) onion, chopped
        4 tablespoons dry white wine
        ¼ litre (8 fl oz) whipping cream
        1 level tablespoon flour
        2–3 tablespoons grated Parmesan
        salt, pepper

Blanch fennel for seven minutes in enough boiling, salted water to cover. Cut the chicken in half, from the breast, but do not entirely separate it at the base; press it open into a frog shape. Butter a large roasting pan well, or oven dish. Put fennel and cooking liquor over the base with shallot or onion and wine. Place chicken on top, dab with butter and roast at 200°C/400°F/Gas Mark 6, until cooked. Baste from time to time. Strain off liquid into a pan; mix cream with flour and add to chicken liquor. Stir over moderate heat until thickened. Correct seasoning. Pour round and over the chicken, sprinkle with Parmesan and return to the oven for a few moments.

HAMBURG PARSLEY or PARSLEY ROOT

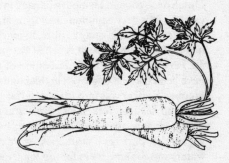

Hamburg parsley is a large-rooted variety of the familiar herb. It looks like an under-privileged parsnip. The parsnip is rounded and full-shouldered; Hamburg parsley is a leaner carrot shape. The parsnip washes to a glowing yellowish-cream skin; Hamburg parsley emerges from its muck a dispiriting shade of greyish-white. Do not let this lack of appearance put you off. Hamburg parsley has a far more tactful and delicate flavour than parsnip, which can dominate any dish unpleasantly to which it may have been added with too much enthusiasm. Hamburg parsley has been compared in flavour to celeriac. They are often suggested as alternatives to each other. I do not see why. They have a similar kind of texture, but they do not taste the same.

Hamburg parsley, as our name for it suggests, is popular in northern Germany, but it goes farther than that. Look up a borshch recipe in a Russian or Polish cookery book and you will see parsley root listed with the standard pot herbs. The same thing with a Croatian soup of pig's trotters, or a hare soup from Bulgaria. There, below onion and carrot, you will see parsley root listed. Some translations will even abbreviate the name misleadingly to parsley, but you can tell that the large-rooted kind is meant because of its position in the list and because it will be given as '1 parsley' rather than as a tablespoon of chopped parsley. Politics apart, you could not expect so familiar a pot-herb to be called 'Hamburg' parsley.

One thing you are unlikely to find, however hard you search, is a recipe for parsley root on its own as a vegetable. I wonder why? Gardeners say it is easy to grow. It is larger than many young carrots, and it tastes well when prepared in similar ways to carrots. The only author I know of who treats Hamburg parsley as a vegetable in its own right is John Organ in his book, *Rare Vegetables*, published by Faber in 1960. Mr Organ suggests turning the roots into chips, just like potatoes. He also gives a recipe for fritters, using a Balkan inspired yoghurt batter and mentions that the young leaves are good in salads.

In his *Gardener's Dictionary* of 1741, Philip Miller implies that the Dutch once cooked Hamburg parsley in much the same manner as young carrots. 'It is very common in all their markets. They bring these roots in bunches, as we do young carrots, to market in summer; and the roots are much of the same size: it is called Petroseline Wortle by the Dutch who are very fond of it.' In a later edition of the dictionary, Miller claimed to have been the first to grow it in this country, about 1727, the same time as Stephen Switzer was importing celeriac seeds. The turnip-rooted parsley seems to have been even less successful than the turnip-rooted celery. A few gardeners apart, we do seem to have been incurious about many vegetables. Perhaps if most cooks were drowning them all in panloads of water, and cooking them far too long, only the strongest flavoured could hope to survive and meet with approval. Sad to think of the fine delicacies of salsify, celeriac, Hamburg parsley, custard marrows lost in a watery grave.

HOW TO PREPARE HAMBURG PARSLEY

Rinse any earth from the roots, helping it away with a small brush (I keep a nailbrush specially for root vegetables). Never peel them as the skin is fine. If they are to be sliced or cubed, cut the pieces rapidly into acidulated water to prevent them discolouring. The alternative is to sprinkle them with lemon juice.

Before cutting them up, remove any stalks and whiskery roots.

HAMBURG EEL SOUP
HAMBURGER AALSUPPE

One of the best soups of European cookery, but only for the open-minded. More timorous cooks may be shaken by the idea of mixing dried pears and eel, beef and eel, cauliflower and sage and eel – if they are prepared to contemplate eel in the first place. I gave one version of the recipe in *Fish Cookery*. Here is a better one, improved partly by the use of dried rather then fresh pears, and partly by the method of making beef stock and soup all in one process (though if you found it more convenient, you could carry out the first two hours cooking in advance and complete the rest just before the meal).

> ¼ kg (½ lb) dried pears, cut in strips
> ¼ litre (½ pt) dry white wine
> ½ kg (1 lb) cheapest stewing beef, cut in 6 pieces
> 2 leeks, sliced thinly

3 carrots, sliced thinly
4 stalks celery, chopped
2 parsley roots, sliced thinly
bouquet garni
3 sage leaves
1 medium onion, sliced
salt, pepper
¼ kg (½ lb) shelled peas
½ cauliflower, broken into florets
1 kg (2 lb) eel, cut in 5 cm (2 in) lengths
white wine vinegar
3 shallots, each stuck with a clove
thinly cut peel of a lemon, in wide curling pieces
2 tablespoons sugar
4 egg yolks, or 2 yolks and a good tablespoon of beurre manié *(p. 552)*

Soak the pears in wine overnight, adding a little water if need be to cover them easily. Next day, put the eight ingredients following into a huge pan, with 2½ litres (generous 4 pt) water. Bring to the boil, skim and simmer for two hours, covered. Season. Remove the bouquet and beef (use up in another dish), add the peas and cauliflower and complete the cooking.

Three-quarters of an hour before the basic soup above is likely to be finished, put the eel into a pan. Cover it barely with water, add a good splash of wine vinegar and the shallots. Bring to the boil and simmer for 20 minutes, until the eel is tender. At the same time put on the pears to cook in their soaking liquor with the lemon peel; add more water or wine if necessary.

Here are the last operations of all – the grand combination of the three panfuls. First remove the eel from its liquor, cut away the fillets from the bone and skin which should be thrown away. Strain the liquor into the basic soup and put in the eel pieces. Lower the heat. *From this point on the soup should not return to the boil.*

Tip the cooked pears and their juices into the soup as well. Taste and adjust the seasoning, adding the sugar gradually to your liking. If you think a little more sharpness is in order, pour in a cautious quantity of wine vinegar.

Beat the egg yolks, then pour in a ladleful of the hot soup. Return the whole thing to the pan and stir for a couple of minutes. If you use only two yolks and would like a thicker soup, add the *beurre manié* bit by bit, stirring all the time.

Pour the whole thing into a large tureen. I doubt your visitors or family will want anything much afterwards, especially if they eat bread with the soup. Follow up with fruit and cheese.

POLISH BARLEY SOUP
KRUPNIK

The dominant flavour of this good and comforting soup comes from the tiny quantity of dried ceps, the soft texture from pearl barley. Nonetheless the deliciousness of the soup also depends on a good beef stock, just as the Hamburg eel soup does, and this means a stock that includes parsley root. There are two ways of tackling this soup, depending on whether you have some beef stock to hand already, or not. If you are starting from scratch, follow the first part; if not turn to paragraph (2).

1) Make a beef stock, using the following ingredients:

> 200 g (6 oz) shin of beef, minced
> 300 g (10 oz) beef bones, chopped into large pieces
> 1 large carrot, chopped
> 1 large onion, chopped
> 1 large parsley root, chopped
> 125 g (4 oz) celeriac, chopped, or 2 stalks celery
> butter
> sugar, pepper, salt
> bouquet garni

Put meat and bones into a pan with 2½ litres (good 4 pt) water. Mix the vegetables together and add half to the pan. Brown the rest lightly in butter, sprinkling on a teaspoon of sugar as they begin to colour. This produces a richer tone and flavour. Add to the pan of meat, swilling out the brown parts if necessary with some of the water from the pan. Season and put in the bouquet. Bring to the boil, skim and simmer, covered, for one to two hours. Strain. (A second weaker stock can be made by pouring more water on the debris and simmering it again.)

As well as stock, you will also need:

> 15 g (½ oz) dried ceps
> 60 g (2 oz) pearl barley
> 30 g (1 oz) butter
> 2 potatoes, peeled, cubed (optional)
> salt, pepper, chopped parsley

Heat up your stock in a pan – there should be at least 1½ litres (2½ pt) – and pour a ladleful on to the dried mushrooms in a small bowl. Leave them to soak for 20–30 minutes. Add the pearl barley to the panful of stock, with half the butter. Simmer, covered, for at least an hour until the barley is very tender. Add the mushrooms to the pan with their liquor as soon as they have finished soaking, so that they cook in with the barley. When the barley is ready add the potatoes if you are using them and continue to simmer until they are done. Taste and add extra seasoning. If the flavour is on the strong side, add water to dilute it. Tip into a tureen, scatter with parsley and serve.

2) If you are already supplied with beef stock, and like to have a jumble of vegetables in your soup along with the mushrooms and barley, make the recipe this way:

Sweat the vegetables on the stock list of ingredients gently in butter. Do not allow them to brown. Soak the mushrooms as above. Put the barley in with the softened vegetables, plus 1½ litres (2½ pt) beef stock, and complete the recipe as above.

If you do not want to add potatoes to the soup, you could stir in a little *beurre manié* to thicken it slightly. Be careful not to overdo this, as you already have barley in the soup which has a thickening effect.

HOP SHOOTS

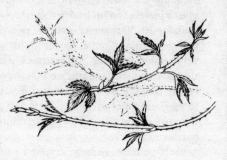

Food from the hedgerows seems a pretty, Arcadian idea. It has even supported a bestseller or two, though I wonder if most of the readers had not gone back in a couple of years to blackberries, hazel-nuts, field mushrooms, woodland strawberries, the conventional wild harvest. The truth is that much wild food, including escapes from gardens in the past, is not very nice. Who wants to eat fat hen when spinach is easy to buy? To be honest, who wants hop shoots in preference to more delicate matters, such as asparagus and seakale? The answer to the second question is, of course, that asparagus and seakale, the choicest of shoots, are not around in spinach abundance or at spinach price. And hop shoots do have a pleasant taste, even if it's not of the finest quality. The tiny sheaves of green drooping heads look delicious, too, on the plate. Moreover they are there for picking at a time when little else is around to bring home from a walk.

If you live in Kent or Worcestershire, near hop-gardens, you may be able to come to an arrangement with a grower and serve hop shoots as a regular early vegetable. They are much prized from February or March until May in other hop-growing areas of Europe, in South Germany and Belgium. They even have a small corner of their own in classical cookery, being the main item of the Antwerp garnish. In other parts of the country, one needs to track down the odd plant or two of hops in late summer, when their cone-like green flower heads are so easy to recognize in the hedges, then you will know where to look in the following year. These hops have escaped from cultivation in the days when many houses brewed their own beer, so you can find them all over the country, not just in what we now think of as hop-growing neighbourhoods. Geoffrey Grigson remarks, in his *Dictionary of Plant Names*, that 'hop' may come from a Germanic word meaning 'that which gropes and fumbles around'.

I have always picked the tips of the young shoots when they are already quite long, before the leaves unfurl, cutting a hand's length down. At this stage the shoots look like miniature asparagus. Apparently in Belgium, Germany and France the shoots are cut when they first appear from the soil in March and are sold in the markets. I have never eaten them at this stage, but can believe that they are better and plumper than the kind I have picked. Unfortunately the hops I know are so entangled with hedgerows, that one cannot get at the first shoots, but I pass on the idea for those who have access to hop gardens. It is these shoots rather than the thin tendrils of the later plant that are intended in a number of Belgian recipes, though of course either can be used.

HOW TO PREPARE HOP SHOOTS

With the early March shoots, cut from the ground, all you have to do is remove the sandy earthy part and give them a quick rinse. For the green tips, tie them in loose bundles before washing them rapidly. Never soak hop shoots of either kind in water for any length of time.

Drain them well and cook them in salted water, acidulated with the juice of half a lemon.

Serve the cooked shoots with melted butter, or hollandaise or tomato sauce, or with one of the sauces in the recipes following.

HOP SHOOTS ALOSTOISE

Alost, or Aalst to give the more usual Flemish name, is a brewing town in East Flanders, between Ghent and Brussels. The town is also famous for onions, so take care and make enquiries first if you ever have occasion to order an *alostoise* dish in a restaurant; you could be disappointed.

Cook ½ kg (1 lb) hop shoots in acidulated water. Make a ¼ litre (scant ½ pt) of béchamel sauce. Off the heat, at the last moment, beat in a large egg yolk and a good knob of butter. Mix in the hop shoots. Serve with lamb, pork, poached eggs or as an omelette filling.

HOP SHOOTS WITH MUSHROOMS
HOPFENKEIME

A recipe from Robin Howe's *German Cooking*, Deutsch 1953. She remarks that hop shoots are frequently eaten in Southern Germany, cooked like asparagus and served with melted butter, or with hollandaise sauce. Either of these ways are good. This recipe I find useful because we rarely have the chance to gather more than a few bundles of hop shoots in our part of Wiltshire; the mushrooms stretch them nicely.

> *about 250 g (8 oz) hop shoots*
> *250 g (8 oz) mushrooms*
> *butter, salt, pepper*

Prepare and cook the hop shoots in acidulated water, then drain them. Slice and fry the mushrooms in butter. Turn the hop shoots in a separate pan in more butter; put on the lid and leave them to steam gently for 15 minutes. Serve the vegetables together on the same dish. Toast or bread goes with the dish, to mop up the delicious juices.

JERUSALEM ARTICHOKES

These are none the worse for not being artichokes and having nothing at all to do with Jerusalem. They came from North America. Europeans – French explorers – saw them first in 1605 in Massachusetts, a crop grown by the Indians at Nausett Harbour, Cape Cod. One of the party was the great Samuel Champlain, founder of Quebec, who described them, in writing of his journeys, as roots 'with the taste of artichokes', i.e. of the bottoms of the globe artichoke of Europe.

That accounts for artichoke in the name, though I wouldn't say the likeness in taste was strong, unless they are dug and eaten immediately.

This new vegetable, held by some to be more suitable for pigs than men, was soon grown abundantly in France, and was recognized as a relative of the sunflower (which had been introduced from the New World some thirty years before), it was another of those plants whose flowerheads twist round with the sun. In 1617 a French merchant in London, John de Franqueville, who was greatly interested in plants and gardens, gave 'two small rootes' of artichoke to the young English botanist John Goodyer. Goodyer planted them in his garden (at Buriton, under the high Hampshire downs), and the two tubers flourished and gave him a peck of roots, 'wherewith I stored Hampshire', Goodyer said.

Can't you imagine de Franqueville telling Goodyer that of course the plant was a *girasol* (a French as well as an English word), a heliotropium, a turnsol, whose flowers, if it produced any, would turn with the sun; and can't you imagine Goodyer's Hampshire neighbours discovering from him in turn that it was a 'girasol artichoke' they were planting, a name they quickly changed to 'Jerusalem artichoke', which at least sounded satisfactory and intriguing in a vague way, heliotropism and Indians across the water not meaning much to them?

Jerusalem artichokes have such a special flavour that they never could have become a kind of neutral basic food like the New World's potatoes. They are for eating now and then, not every day. Goodyer, as it happens, did not like them, they upset him, 'which way so ever they be dressed and eaten, they stir and cause a filthy loathsome stinking wind, thereby causing the belly to be pained and tormented'. His contemporary Dr Tobias Venner, physician to all the ailing at Bath, tried this new vegetable and found it too strong to recommend to his patients in search of health and long life: it was no good for weak digestions, in fact it was 'somewhat nauseous and fulsome to the stomach'.

That might not be your experience. It is not mine. I love Jerusalem artichokes. But certainly they are provokers of wind in some insides. So be careful. A little and not too much, too often.

HOW TO CHOOSE AND PREPARE JERUSALEM ARTICHOKES

Try to pick out the smoothest, most evenly sized Jerusalem artichokes you can find. The knobblier they are the more waste you will have. This is why it is difficult to be dogmatic about the weights required. Always buy far more than you need – at least they are fairly cheap for such a fine vegetable – as any left over can go into soups and purées.

When you get them home, scrub them well with plenty of water. Cut away the dry stringy roots and little dark tips. If the artichokes are smooth, you can peel them thinly with a minimum of waste, and use them straightaway in the recipes following. Drop them into acidulated water as you peel them, or they will discolour.

Knobbly artichokes are best boiled in their skins for about ten minutes until they are half-cooked. Run them under the cold tap, then peel away the skins. Now you are ready to cook them according to the recipes, though you may have to make a reduction in cooking time.

PALESTINE SOUP

I make no apology for repeating this soup, that I gave in *English Food*, as it is so good. However I do make some changes in the basic recipe from time to time, for variety, so they are included as well.

> ½ kg (1 lb) large Jerusalem artichokes, or ¼ kg (½ lb)
>   each artichokes and potatoes
> 1 large onion, chopped
> 1 clove garlic, chopped
> ½ stick celery, chopped
> 125 g (4 oz) butter
> 2 rashers green back bacon, or 60 g (2 oz) ham
> 1 litre (1¾ pt) light chicken or turkey stock, or water
> ¼ litre (8 fl oz) milk (optional)
> salt, pepper

6 tablespoons cream
chopped parsley and chives

Scrub and boil the artichokes in salted water for five to ten minutes, then run them under the tap and remove the peel. I think this is the easiest thing to do, though if you prefer to peel them raw, it doesn't matter. It entirely depends on the shape of the artichokes.

Cut up the artichokes. If you are using potatoes, peel and dice them.

Put them with the onion, garlic and celery in a large pan with half the butter. Cover tightly, and stew over a low heat for ten minutes, giving the pan an occasional shake or stir. Now add the bacon or ham, and cook a moment or two longer. Pour in the stock or water, and leave to simmer until all the vegetables are soft. Blend if you want a smooth soup, or put through the *mouli-légumes* if you like a knobbly texture. Reheat, adding more water, or the milk, to dilute to taste. Correct the seasoning. Finally stir in the last of the butter, the cream and herbs. Serve with croûtons of bread fried in butter.

ESCOFFIER'S PALESTINE SOUP
PURÉE DE TOPINAMBOURS

Escoffier's recipe is a whiter, more refined version of the one above. The interesting thing is the inclusion of toasted hazelnuts, which blend beautifully with the artichokes and enhance their fine subtlety.

1 kg (2 lb) Jerusalem artichokes, peeled
1 large tablespoon butter
5 large filberts, toasted, crushed
light veal or chicken stock
150 ml (¼ pt) milk
1 tablespoon potato flour (fécule)
salt, pepper
a good knob of butter

Chop the artichokes, then sweat them in the butter in a covered pan for about ten minutes. Add the nuts and ½ litre (good ¾ pt) stock. When the artichokes are tender, rub through the sieve (it is quicker to put the soup into a blender, then through the sieve). Mix the milk and flour, add to the soup and reheat gently, putting in more stock for a thinner soup if you like; I find this purée a little too thick without extra dilution, but it is a matter of taste. Check the seasoning and add the knob of butter just before serving. Provide a bowl of cubes of bread fried in butter.

JERUSALEM ARTICHOKES WITH VARIOUS SAUCES

If you have a generous crop of artichokes, you can afford to cut them to one size, turning them into small potato shapes. Keep the trimmings for soup or a beef stock. As you prepare them, put them into a bowl of acidulated water. For six people allow ¾–1 kg (1½–2 lb) or a little less if they are to be served as a vegetable with meat.

You can boil the artichokes, but they will keep their shàpe better if you steam them tender. It takes longer but it is worth it. Season them and put them into a dish.

While the artichokes cook, prepare a very thick parsley or béchamel sauce, or a tomato sauce. When the artichokes are done, add enough of their cooking water to the sauce you have chosen, to dilute it to a creamy consistency. Correct the seasoning and pour over the artichokes. Scatter them with parsley.

Steamed artichokes are also delicious with an hollandaise or béarnaise sauce: keep the cooking water, and use it to make up some velouté or parsley sauce to store in the freezer for the next time.

JERUSALEM ARTICHOKES EN DAUBE
ADOBO DE PATASSOÙN

A pleasant recipe from *La Cuisine Rustique: Provence*, by Maguelonne Toussaint-Samat, which is rather different from our more northerly recipes. The stew is likely to turn into something of a purée, as artichokes collapse easily, so serve it with plenty of bread or fried croûtons to mop it up with and provide a contrast of texture.

> *1 kg (2 lb) Jerusalem artichokes, scrubbed*
> *3 tablespoons olive oil*
> *60 g (2 oz) green bacon, cut in strips*
> *1 clove garlic, left whole*
> *salt, pepper*
> *4 cloves*
> *pinch of cayenne or nutmeg*
> *250 g (8 oz) tomatoes, peeled, chopped*
> *sugar (optional)*

1 tablespoon flour
1 glass red or white wine
1 glass water

Peel and slice the artichokes fairly thickly, then cook them for five minutes in a *blanc* (p. 549). Drain the slices, then put them into a fireproof pot with the oil and the chopped bacon. Cook them moderately quickly for a few minutes, then add the garlic, seasoning and spices and the tomatoes with a teaspoon of sugar or less, if they are on the tasteless side. Cover the pot and leave to stew gently for 15 minutes. Then stir in the flour, bubble for a minute or two, and then add the wine and water. Leave to stew again, until the Jerusalem artichokes are thoroughly tender and the sauce nicely matured. If the sauce seems watery leave the lid off the pot; otherwise cover the pot for the last part of the cooking.

JERUSALEM ARTICHOKE PURÉE

Sieve together equal quantities of cooked artichokes and potatoes – or slightly less potato than artichoke. Reheat with a good knob of butter and a spoonful or two of cream. Mix in chopped parsley. Good with game, beef or veal, and with poultry.

JERUSALEM ARTICHOKE CHIPS

Really large artichokes can be deep-fried like potatoes, either as chips, or in slices or matchstick shreds. Chips are the most satisfactory, as one has the contrast of the soft delicious inside and the crisp outside. With the thinnner slices, the flavour is rather lost in the crispness.

STOVED ARTICHOKES

The very best way of cooking artichokes: it emphasizes their exquisite flavour and the golden-brown crust holds their tenderness in shape so that there is no squash of texture to spoil the dish. Serve them on their own in small dishes, or as a vegetable with veal, lamb, chicken, beef, so long as you can leave the meat to keep warm while you spend time just before the meal attending to their cooking only. The method is a combination of frying and steaming: once you grow used to it, there is no need to concentrate quite so hard. It is a superb way of cooking root vegetables of quality – new potatoes, for instance, and very young turnips – as it concentrates the flavour. For this reason it is no good for older roots; they would taste too strong for pleasure and need blanching before being finished in an open pan in butter.

> *about 2 kg (4 lb) Jerusalem artichokes, scrubbed*
> *olive oil*
> *butter*
> *salt, pepper*
> *1 large or 2 small cloves garlic*
> *small bunch parsley*

Pick out the smoothest, least knobbly artichokes you can find. Cut off any oddities, then peel them into fairly even pieces the size of Queen olives; the larger artichokes can be cut to this kind of shape and then halved. (Keep peelings and cuttings to flavour stock for soups or stews.) Aim to end up with at least 1½ kilos of prepared pieces. Put them into acidulated water as you go to prevent discoloration.

As the artichokes need to be cooked in a single layer, you may need to use a couple of pans. Heat them over a moderate flame, putting in enough oil to cover the bases comfortably. Put a heaped tablespoon of butter into each pan, or two tablespoons if you are using one large pan. Drain and dry the artichokes and put them into the sizzling fat. Cover them for the first ten minutes, but not too tightly, so that they partly fry and partly steam. Turn them over after five minutes, and keep the heat steady so that the fat does not burn although the artichokes begin to turn a nice golden-brown. Remove the covers from the pans, and give the artichokes a further ten minutes, turning them over from time to time to colour evenly. Remember that they are far more tender than, for instance, new potatoes, and may collapse on you rather suddenly. The idea is to keep the softness inside the skins; you can always remove them gradually from the pans.

Put the artichokes into small individual pots, or round the meat, sprinkling them with salt and pepper. Have ready the garlic and parsley chopped finely together and scatter this evenly over the top. Serve straightaway if possible, though the artichokes can be kept warm for a while, so long as the oven is not too hot, which is helpful if they are being served with the main course.

WINTER ARTICHOKE SALAD

I discovered by accident how well the earthy flavour of Jerusalem artichoke combines with the sweetness of prawns. Christmas came round one year, and there was no possibility of having the usual smoked eel or smoked salmon. I bought some prawns in their shells instead. Wondering exactly how I was going to serve them, I drove home and came in through the back kitchen past a basket of Jerusalem artichokes. The dish was a great success. It looked beautiful with the contrast of grey, pink and green: the flavour was fresh and unusual, an appetizing start for the lengthy Christmas dinner. It is also good before strong and substantial game dishes.

The quantity of prawns can be varied; you want enough to provide the contrast, but not so many that they become too dominant. I sometimes buy a pint and a half of prawns in their shells (for six people), set a few aside for decoration, then shell the rest; the debris goes into a pot with fish bones to make the next day's soup, or stock for a prawn sauce to serve with cauliflower.

> *1 kg (2 lb) Jerusalem artichokes*
> *about 125 g (4 oz) shelled prawns, or shrimps*
> *olive oil vinaigrette*
> *plenty of chopped parsley and chives or spring onion*
> *6 prawns in their shells*

Cook the artichokes in their skins. Peel and slice them neatly. This means discarding the squashy parts, but do not throw them away – they can be used to flavour mashed potato, or potato soup. Put the slices in a shallow dish and dot the prawns or shrimps over the top. Pour on enough vinaigrette to moisten the salad and add a good scattering of herbs. Arrange the prawns in their shells on top. Serve well chilled.

Note The artichokes can be cooked and sliced well in advance. Cover them with vinaigrette. Add the shellfish an hour before serving, and chill.

JERUSALEM ARTICHOKE SOUFFLÉ

Although a soufflé sounds grand, it is an inexpensive dish – particularly when root vegetables provide the flavouring. If you cannot buy Jerusalem artichokes, use parsnip or turnip instead. With parsnips, add 100–125 g (3–4 oz) chopped mushrooms fried in butter, or chopped walnuts, when you blend the purée into the soufflé mixture. Turnips benefit from a flavouring of cinnamon, a quarter of a teaspoon is enough, rather than the hazelnuts.

375 g (¾ lb) large Jerusalem artichokes, or ½ kg (1 lb) small ones
60 g (2 oz) butter
50 g (1½ oz) flour
150 ml (¼ pt) artichoke cooking liquor
125 ml (4 fl oz) milk
4 egg yolks
salt, pepper, cayenne
1 tablespoon chopped hazelnuts
1 tablespoon chopped parsley
5 egg whites
1 heaped tablespoon breadcrumbs
1 rounded tablespoon grated Parmesan cheese

Scrub the artichokes and boil them in salted water. Peel off the skins, and sieve the inside part. You should end up with about 200 g (6–7 oz) of dryish purée. Be careful not to throw away the cooking liquor.

Melt the butter in a large pan, stir in the flour and leave it to cook for a few moments. Moisten with the 150 ml (¼ pt) of cooking liquor and the milk. Mix in the purée. Remove the pan from the heat and whisk in the egg yolks one by one. Correct the seasoning, adding a pinch of cayenne, the hazelnuts and parsley. The flavour should be on the strong side, as the egg whites will have the effect of toning it down. Fold in the stiffly beaten egg whites carefully, raising the mass with a metal spoon. Do not worry if all the white is not mixed smoothly in, a few smallish lumps do not matter.

Grease a 1½ litre (2½ pt) soufflé dish with a butter paper. Tip in crumbs and cheese and turn the dish about so that the sides and base are evenly coated; keep the surplus to one side. Spoon in the soufflé mixture. Sprinkle the surplus crumbs and cheese on top. Bake for 30 minutes at 200°C/400°F/ Gas Mark 6. With soufflés, it is a good idea to put a metal tray into the oven when you switch it on: it makes a hot base for the dish.

OTHER SUGGESTIONS

Clear artichoke soup See Soyer's clear turnip soup, p. 525.
Curried artichoke soup Turn to the curried parsnip soup recipe on p. 363, and substitute Jerusalem artichokes.

KOHLRABI or CABBAGE-TURNIP

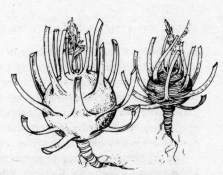

There are better vegetables than kohlrabi. And worse. I am thinking in particular of winter turnip and swede; certainly kohlrabi is a pleasant alternative to that grim pair. It is not a true turnip, but a cabbage with its stem swollen into a turnip shape, a cabbage-turnip, by analogy with the French *céleri-rave*, celery-turnip, our celeriac. We have adopted rather than translated the German name, which indicates a lack of warmth on our part. No doubt we thought there was enough turnip flavour in our winter diet already, for when it was first grown on any scale in this country, in the early 19th century, it was fed to cattle. There were garden varieties. They do not seem to have caught on. Mrs Beeton observes, without giving a special recipe for it, that kohlrabi 'although not generally grown as a garden vegetable', is 'wholesome, nutritious and very palatable', if used when young and tender. This was in 1861. Over a century later both gardening and cookery writers are still obliged to treat it as a novelty because their readers are unfamiliar with it.

The sphere of enthusiasm for kohlrabi extends from Germany eastwards as far as Europe is concerned, and includes Israel (the first time I saw kohlrabi in quantity was in the market at Tel-Aviv, beautiful green and purple shapes, with trimmed stalks bristling upwards from the swelling). In *The Golden Peaches of Samarkand*, Professor Edward Schafer remarks that kohlrabi was eaten in T'ang China, an import from Europe that had travelled 'by way of the Serindians, the Tibetans and the Kansu corridor'. A great pharmacologist of the time called it 'indigo from Western lands' because the leaves reminded him of the Chinese indigo plant; and he recommended it as a tonic. To the Chinese as to Westerners until recently, the primary purpose of plants was medicinal. They did one good. They cured ailments. They prevented sickness. They kept your body in balance. Some of them one could enjoy (the Romans loved asparagus, though they esteemed it primarily as a diuretic, an aphrodisiac, a soother of the stomach and as the main ingredient of a pomade that prevented bees stinging). Others one did not. Happily this tyranny of goodness has now ceased.

HOW TO CHOOSE AND PREPARE KOHLRABI

Buy kohlrabi when it is between a golf ball and a tennis ball in size. Never larger, or it will be as tough as any ancient turnip.

Preparation is simple. Just cut off the stalk – most kohlrabi in shops has been trimmed of its leaves – and rinse it. If your kohlrabi comes from the garden, cut away the stalks and leaves, keeping the young, fresh-looking ones. They can either be cooked with the round part, or on their own in the same ways as spinach or turnip tops.

The simplest way of cooking kohlrabi is to steam it (you can also boil it, but steaming is better for flavour). When it is small, there is no need to remove the skin, but most kohlrabi benefits from peeling after cooking unless the recipe makes this impossible. It can then be sliced and finished in butter with parsley and a few drops of lemon juice. Or you can mix it with a béchamel-based sauce – such as cream or mornay sauce, sauce aurore – and serve it as a vegetable. Another way is to put it into a dish, cover it with one of these sauces and top it with crumbs, grated cheese and melted butter to make a gratin.

Some people are attached to kohlrabi grated over salads. Another way is to treat it like celeriac in the recipe for *céleri-rémoulade* (page 194); grate it into julienne shreds, on a mandolin slicer if possible, and then blanch them briefly, cool and mix with mayonnaise.

Other celeriac and a number of turnip recipes are suitable for kohlrabi. It can also be used instead of turnip when a mixture of root vegetables is required.

ELISABETH BOND'S KOHLRABI

An Austrian friend turned my dislike of kohlrabi into a muted enthusiasm, when she gave me this recipe of hers which brings out the best of its natural flavour. It is a version of the caramelizing methods that many Europeans use for turnips and the Danes use for new potatoes.

When you prepare the kohlrabi, include any young leaves with their stalks; chop them and put them with the slices.

> ½ kg (1 lb) kohlrabi, peeled
> 30 g (1 oz) butter
> 2 tablespoons oil

1 level tablespoon sugar
1 level teaspoon flour
light stock or water
salt, pepper, chopped parsley

Halve and slice the kohlrabi thinly, to make half-moon pieces. Heat the
fats in a heavy frying pan, stir in the sugar and keep stirring until you have
a golden-brown caramel; the heat should be high, but take care the sugar
does not blacken. Put in the kohlrabi, with any leaves, and stir it about –
the heat should be high still – until it is well coated and a nice brown
colour. Then lower the heat and cover the pan until the kohlrabi has
cooked and most of the liquid evaporated. Sprinkle on the flour, then
enough stock or water to make a binding sauce. Bring to the boil and cook
a minute or two longer. Add salt, pepper and plenty of parsley.

KOHLRABI SOUP

Follow the recipe above, but add a larger quantity of liquid – about a litre
(1¾ pt) – then blend or put through the *mouli-légumes*. Dilute further if
you like, correct the seasoning and reheat. Stir in the parsley last of all.

STUFFED KOHLRABI

A good dish for kohlrabi that is reaching the danger limit of tennis ball
size. Depending on the rest of the meal, allow one or two for each person.
The pulp scooped out can be chopped with any young leaves and put into
the base of the baking or stewing dish for the second stage of cooking. Or
it can be used separately.

12 kohlrabi, trimmed
about 300 ml (½ pt) soured cream
butter

STUFFING
½ kg (1 lb) minced veal or brains
1 heaped tablespoon chopped onion
100 g (3–4 oz) butter
salt, pepper
plenty of chopped parsley
grated rind of half a lemon
1 egg

Hollow out the kohlrabi to make cups. Simmer for 15 minutes in salted water, until half-cooked. Rinse under the cold tap and remove the peel if it is tough.

Make the stuffing. If you are using brains rather than veal, soak them in salt water for at least an hour, then remove the membranes and poach them until they are just firm in salted water. Chop them roughly. Soften the onion in half the butter. Raise the heat, stir in the veal or brains and cook for about ten minutes until lightly browned. Add the rest of the butter gradually, if necessary. Cool, then put in more salt to taste, plenty of pepper and plenty of parsley. Stir in the lemon rind. Bind the mixture with an egg. Fill the kohlrabi cups with the stuffing.

There are two ways you can complete the cooking. The nicest is to put the cups into a buttered baking dish and mix the cream with half its volume of water before pouring it over the kohlrabi; dab butter on top of the stuffing. Set in the oven pre-heated to 180°C/350°F/Gas Mark 4 for about half an hour, or a little longer, until the kohlrabi are cooked properly. The other way of finishing them off is to put them into a saucepan or sauté pan, with a good 1 cm ($\frac{1}{2}$ in) depth of boiling water. Cover and simmer slowly until they are done. Keep an eye on the liquid level. Use the soured cream to make a sauce – mix two level teaspoons of flour with half the cream, then gradually add the rest. Heat in a small pan, stirring all the time, for about five minutes, then add salt to taste.

For other ideas for stuffed kohlrabi, turn to the recipe for Escoffier's stuffed turnips on page 529.

LADY'S FINGERS see OKRA

LAMB'S
LETTUCE or
CORN SALAD

We were never allowed into the
large grey-walled vegetable
garden of our boarding school in
Westmorland (the school motto
was 'One heart, one way', but it
should have been 'No'). This
blank Africa of our familiar
world was tucked away behind
the playing fields on a south-facing slope. We knew it concealed delights
that we would not enjoy – raspberries, black currants and so on. Once,
though, my mother was invited to see it, and I was tolerated. We were
introduced to sugar peas (staff consumption only) and to a long row of
soft leaves in a bed against the north wall. I was told they were called
'lamb's tongues', and was permitted to chew one. It was easy to see
how the shape and soft texture of the leaves justified the name –
but the colour? Lambs with green tongues? That worried my literal mind.
Those were the days when the colour of one's tongue was of perpetual
interest to adults. 'Put out your tongue! Ahhhhhh. You need a dose of
rhubarb.'

It was a relief to discover later that the proper name is lamb's lettuce,
translating the *Lactuca agnina* of the herbalists. Gerard was the first to
use the name, in 1597, in his *Herball*. In the wild form, it is known as corn
salad. The French, who grow it a great deal more than we do, call it
mâche. Oddly enough the first record of the name occurs in a French-
English dictionary, compiled and published in England in 1611, by Ran-
dle Cotgrave. From homely use, it has acquired the affectionate names of
doucette, little soft one, from the velvety leaves, and *bourcette*, little
purse, from their shape.

In France, *mâche* is grown and sold as a useful salad to bridge the
difficult months of winter, just before the spring. It seems to me a much
better solution than the floppy, expensive greenhouse lettuces we have
to put up with. Incidentally, *mâche* can be cooked and eaten in the
same way as spinach, but its slightly astringent flavour comes over best
in salads.

LAMB'S LETTUCE AND BEETROOT SALAD
SALADE DE MÂCHE ET DE BETTERAVE

As you will see from this and the following recipe, beetroot is a favourite
companion of lamb's lettuce.

> *1 tablespoon wine vinegar*
> *4 tablespoons olive oil*
> *salt, pepper, sugar*
> *350–500 g (¾–1 lb) lamb's lettuce*
> *1 fine beetroot, boiled*
> *1 hard-boiled egg*

Mix the vinegar, oil and seasonings in a salad bowl. Lay the servers, one
across the other, over this dressing. Wash the lettuce very well (it can be a
little gritty), dry it and put it on top of the servers. Cut the beetroot into
strips, sprinkle them with seasonings and a few drops of vinegar and place
them in the centre of the lamb's lettuce. Crush the egg with a fork and
scatter it over the top. Turn the salad at table.

VARIATION Instead of the egg, put in two apples, peeled, cored and
quartered. Sprinkle them with lemon juice to prevent discoloration
before adding them to the bowl. Cox's Orange Pippins make the best
substitute for French Reinettes.

WINEGROWER'S SALAD
SALADE DE VIGNERONS

A version of the hot salads which one finds in several parts of France. This
recipe, though, belongs to the vine and walnut growing districts in particular, where one can buy walnut oil. A few good groceries sell it in England,
but be on the look out for mixtures of the real thing with tasteless oils. The
price – high – is one indication of authenticity, though details should be
spelt out on the label. The flavour is delicious and unmistakable.

> *250 g (8 oz) lamb's lettuce*
> *125–150 g (4–5 oz) blanched dandelion leaves*
> *1 fine beetroot, boiled*
> *salt, pepper, sugar*
> *5 tablespoons walnut oil*

125 g (4 oz) fat belly of pork, diced small, or smoked streaky
  bacon
3 tablespoons wine vinegar

Wash the salads well, dry them and place them in a slightly warmed salad bowl. Cut the beetroot into strips and put it in the centre. Season. Heat the oil and cook the pork or bacon dice until they are nicely browned. Pour the whole thing over the salad. Quickly tip the vinegar into the hot pan, let it boil up to a good foaming frizzle and pour it over the salad. Serve immediately.

WILD CORN SALAD

In summer in France we pick wild lamb's lettuce, wild corn salad (*Valerianella locusta*), from the hot flint wall that buttresses our garden. It takes quite a time and the lizards don't like it, vanishing rapidly into the holes between the stones. Then the leaves have to be stripped gently, so as not to tear their softness, from the clusters of tiny white and bluish-white flowers. A holiday occupation, when the hours of the day stretch out slowly in sunshine. We sit round the table under the lime tree, with a bottle of wine, and throw the leaves into a bowl of water, fingers busy as we chatter and look at the enormous view. Then the leaves have to be swished about thoroughly to remove the dry grit of their preferred habitat.

This simple salad has no need of beetroot, or anything else, only a straightforward olive oil or walnut dressing. Don't spoil its flavours, and the family's work, with one of the tasteless oils or with malt vinegar.

OTHER SUGGESTIONS

See Dandelion salad with bacon, page 250.

LAND CRESS

Writers on growing rare vegetables and our best seedsmen encourage gardeners to grow 'American' or land cress, as a useful winter salad and a substitute for watercress, which it resembles in leaf and peppery flavour. The convenience is obvious. Winter is not an easy time for salading and watercress has to be left to commercial growers who can be sure of the purity of their boreholes and spring water.

I do not understand why it should be called 'American' (land cress is an apt name – like watercress). The plant is a native of Europe as well as America, and it has been eaten as a salad here for a long time, if not on any great scale. In 1741, Philip Miller had this to say about it in his *Gardener's Dictionary*: it grows wild on dry banks 'in several parts of England' and was 'formerly used as a Winter Salad'. He calls it Winter Cress. When Evelyn listed 'cresses' in the salads grown at Sayes Court, he must have meant winter or land cress, because they were included with greenery for the winter months.

The decline in its use that Philip Miller recorded presumably continued, so that it was reintroduced as a garden plant from America, with the name of American cress. It can be prepared and used in all the ways appropriate to watercress, though the flavour is not quite as good. Try it in salads of mixed vegetables, endive, chicory, spinach and celery for instance.

LAVER

Laver is the one seaweed we can decently count in English or Welsh cooking as a vegetable. The coasts of the Bristol Channel are the modern laver world. Places to buy it, in the form of laver-bread, a black, almost viscous pulp, are Cardiff, Swansea, Port Talbot, Newport, also Bristol, Barnstaple and Ilfracombe (I get laver-bread by post from Howells of St Mary Street, Cardiff).

It was more widely sold and eaten in the past, in Scotland and Ireland for instance until recently. In 18th-century Bath, according to Christopher Anstey's *New Bath Guide*, 'fine potted laver' used to be cried in the streets, along with oysters and pies. It is now sold in markets and by fishmongers.

Once you have learnt to recognize laver (*Porphyra leucosticta* and *Porphyra umbilicalis*) it is more bothersome than difficult to prepare laver-bread for yourself. The seaweed is common on rocks between high and low tide, the fronds, purple-pink, wavy and fine, have to be washed free of sand and salt (with a little bicarbonate of soda to take away the bitterness) and then stewed in fresh water until they become tender 'and can be worked like spinach with broth or with milk or a pat of butter and a squeeze or two of lemon juice'.

I have lifted that quotation from *Kettner's Book of the Table* (1877), which was written not by Kettner the restaurateur but by the celebrated Victorian critic Eneas Dallas. Dallas complained that laver had lost its popularity and was no longer to be met so frequently in London clubs. If only French cooks had ruled England, he wrote, 'they would have made it as famous as the truffles of Perigord'.

Before I tasted laver – with oatmeal, bacon, roast lamb – I was intrigued by the name. Welsh? No, it was not Welsh, it was straightforward Latin, the name Pliny used for a water plant which certainly wasn't a seaweed. Our seaweed was called laver first by 17th-century botanists. The older name, as in Scotland and Ireland, was slawk or sloke.

Laver grows round the world, and is one of the favourite seaweeds in Japanese cuisine. The Japanese improve it by cultivation. It is dried in sheets – they can be bought in Japanese shops in London and elsewhere – and is used especially in combinations with rice.

LAVER SOUP

I devised this soup to use up the remains of a tin of laver, which we had mainly eaten with roast lamb, and it turned out well. One friend told me she had seen it quoted as a genuine Welsh recipe in one of the inaccurate books of Welsh cookery now on the market to trap tourists. I used the bones from the lamb to make some stock, but chicken or fish stock do well instead.

> 125 g (4 oz) chopped onion
> 125 g (4 oz) chopped carrot
> 125 g (4 oz) diced potato
> 2 heaped tablespoons butter
> generous litre (2 pt) stock
> 2–3 tablespoons laver
> salt, pepper, sugar

Soften the onion, carrot and potato for 15 minutes in the butter in a covered pan. Add the stock and laver. Simmer until the vegetables are really tender. Purée in the blender or put twice through the *mouli-légumes*. Taste and correct the seasoning if necessary, adding a pinch of sugar to bring out the flavour. Add more stock or water if the soup is too thick. Reheat and serve with toast fingers.

LAVER-BREAD WITH BACON

The Welsh and Irish way of eating laver. I remember years ago, having it for breakfast in one of Cardiff's main hotels. Very good.

Take about three heaped tablespoons of laver and mix them with enough oatmeal to be able to form small coherent cakes. Turn them in oatmeal, then fry them in bacon fat and serve with bacon, or with bacon, sausage and lamb chop as part of a mixed grill.

LAVER WITH LOBSTER

This unusual combination is served by Mrs Bobby Freeman, who used to have an hotel at Fishguard. She now writes about Welsh food.

Mix the laver with orange juice and the grated rind of half an orange, and serve hot or cold with hot or cold lobster.

LAVER SAUCE

A Welsh recipe sent me by Mrs Freeman – stir laver-bread into béchamel sauce, to taste. Adjust the seasoning and serve with lamb.

SOYER'S LAVER RAMIFOLLES

London clubs are not famous these days for their food. I am told that they aim to comfort their members by reproducing the culinary traditions of nursery and prep-school. It is difficult to imagine anyone being comforted by reminders of those unhappy institutions, but there is no accounting for the sad way in which some of our countrymen enjoy themselves. Club food was another matter in the middle of the last century. At least it was another matter in the 1840s at the Reform Club, where the kitchens were designed and ruled by the French chef, Alexis Soyer. One thing he tried to do, both at the club and in his books, was to get us to appreciate our own good things, including laver. For a time he succeeded. Other clubs followed. But by 1877, when Eneas Dallas was writing, laver had almost disappeared from their menus.

Here is the recipe Soyer invented for winning over the reluctant to laver. The name of *ramifolle* was an invention of his own, without any particular meaning. He used it for a pancake sandwich, stuffed with chicken forcemeat, dipped in egg and breadcrumbs, and fried in pieces like cutlets. When he came to this recipe, he said that the pieces should be cooked like *ramifolles*, so I have transferred the name.

Steam ten fine potatoes, then mash them with 60 g (2 oz) butter, a teaspoon of salt, pepper, and about 300 ml (½ pt) milk. Spread out half the mixture on a flat sheet or dish, 1 cm (½ in) thick. Cover with a layer of laver-bread. Put the rest of the potato on top. Chill until firm. Cut in squares, dip in beaten egg, then in breadcrumbs and fry them in lard until they are brown.

WELSH LAMB WITH LAVER

Whenever we go to Cardiff to look at the paintings in the art gallery, we always visit the market to buy Welsh lamb and laver-bread. Butchers in our part of south-west England claim to sell Welsh lamb, and no doubt it is, but it does not have the fine flavour of the lamb one can buy in Cardiff market which tastes as if it came from the hills. Highland lamb in Scotland must be as good, but I have never had the good fortune to eat it.

1 best end of neck (both sides), split down the backbone and
   chined
salt, pepper
glass of port, Madeira, or brown sherry
250 g (8 oz) laver-bread
orange juice
lemon juice
new potatoes

Cut the top 5 cm (2 in) of fatty skin from the bone ends of the two pieces of meat. Season them and stand them upright on a roasting pan with the bones criss-crossing together. Cover the bones with foil to prevent charring. Into the pan, pour a glass of water. Put into the oven at 200°C/400°F/Gas Mark 6 for about 45 minutes. Remove the lamb to a serving dish; pour off the fat from the meat juices, then tip in the wine and bubble the mixture into a sauce. Add a little lamb stock if you like, but not more than a couple of tablespoonfuls.

While the lamb cooks, heat the laver with the juice of an orange. Taste it and add a little more juice if you like, and a couple of teaspoons of lemon juice to sharpen the flavour. Scrape and cook the new potatoes.

To serve, put the potatoes into the cavity of the lamb, allowing them to spill out at each end. Arrange the laver down each side. Serve the thin gravy in a small jug.

LEEKS

Everyone knows about leeks
and the Welsh – Fluellen and St
David's Day. A lot of people
know about leeks and the Scots
– cockie-leekie and Mussel-
burgh seeds for the garden. A fair number will have read about Northum-
brian gardeners and the giant leeks they grow for competitions in the
autumn. But how many know about the social collapse of this ancient
vegetable, and its disappearance in the 16th and 17th century from the
tables of the gentry?

Their disappearance, except as a pot-herb, may seem extraordinary
now, but I would say that leeks have returned to a modest popularity only
in the last 25 years, stimulated by our new interest in European cookery
and the dishes of our past that survive in the regions. Even so we cannot
buy young leeks in the summer, which is maddening when you want to
make chilled vichyssoise soup on a hot day, or a leek salad; commerce and
habit keep them firmly as a winter vegetable.

The reasons for the leek's fall from grace are understandable, once
there were attractive new vegetables coming into the country; overcooking makes them slimy and they can be gritty, if they have not been
patiently cleaned. Overcooking was a major fault in the past and a servant
girl in a dark kitchen without running water would not have had much
success in cleaning leeks properly.

There was a third reason, too, which had never occurred to me until I
was looking up Mrs Beeton on leeks. She gives two soups only, one of
them cockie-leekie, but adds a note that leeks should be 'well-boiled' –
overcooked – or they will 'taint the breath'. This fear was also behind
many fussy 19th-century recipes for onions. It seems to have been a major
nightmare of the time, not just a silly refinement. I remember my grandmother's obsession with her digestive system, her purges and peppermint
tablets; I remember, too, how constipation hung over some families like a
mushroom cloud. If digestions were as bad as all this suggests, and they
probably were when diets were stodgy without fruit or many vegetables,
then breath must often have been bad. Anything that could have added to
this social fear – leek, onion, above all garlic – was prudently avoided, or
subdued by strong-arm water treatment.

So we had three centuries without leeks, more or less. Sad when one thinks of the 10th-century Irish hermit who boasted of the 'fresh leeks, green purity' that grew by his cell. Or of the medieval lepers who grew leeks in the garden of Sherburn hospital in Co. Durham, under the Prince Bishop's instructions. Gilbert White attributed the decline of leprosy in this country – it was still rampant in Scandinavia – to the general increase in vegetable eating, as well as to more cleanliness. He was not much of an enthusiast for leeks himself: he must have shared the general polite repugnance. Occasionally he notes their planting in his Journal, but the cucumbers are far more important to him. So are the beans, carrots and turnips. It is their progress that he records. The leeks, once planted, were ignored.

HOW TO CHOOSE AND PREPARE LEEKS

Large leeks can be useful for stuffing – see the unusual recipe on page 301 – and for making soup. On the whole – I risk north-eastern wrath in saying so – medium or small leeks have a much better texture and taste. A thing to watch is leeks at the end of the season in March and April; they may look fine, but the centres can be a hard yellow stalk. Buy them ready lopped if you can, as this exposes the centre core and you can see and feel what you are buying. The rest of the winter, leeks with earth, roots and a bit of reality to them are likely to be fresher than the trimmed, washed neat cylinders in a plastic pack. Of course this means waste and extra cleaning, but it is worth while. You will notice that most recipes are for prepared leeks, so always buy more than you think you will need.

When you start cleaning, do not slice the top green leaves off recklessly with one stroke. Slit round the outside layer beneath the first green leaf, and you will often find that the white part goes up inside further than you had thought. By judicious cutting away of the layers, you waste less. When the leeks are free of the coarse green part – it can be used in soups and stews – chop off the roots. Then stand the leeks, root end up, in a large jug of water to soak for a while, so that the grit has a chance to float out. If any remains as a dark shadow under the white skin, make small slits so that it can be rinsed away without spoiling the shape of the leek.

If the recipe demands sliced leeks, this soaking can be dispensed with. Just trim and rinse them, then cut them into slices and give them another rinse in a colander. You can soon see any earth remaining, and separate the layers to release it.

Avoid overcooking. Leeks can still be slightly resistant in the middle, but only slightly. Watch them.

COCKIE-LEEKIE

No vegetable book would be complete without cockie-leekie. As I have
given a full account of it elsewhere, in *Good Things*, here is a summary.

> *1 kg (2 lb) piece stewing beef, or good beef stock*
> *1 boiling fowl*
> *salt, pepper*
> *1½ kg (3 lb) leeks*
> *½ kg (1 lb) prunes, soaked*

Put the beef into a large pot of a size to hold the chicken as well later on.
Add plenty of water. Bring to simmering point, skim and cook for 30
minutes. Add the boiling fowl, seasoning and half the leeks, which should
be trimmed and cleaned but left whole and tied in a bundle. Make sure
everything is covered with the water. Cover and simmer until cooked –
about two and a half hours, though this will depend on the antiquity of the
chicken and your interpretation of simmering.

Instead of beef, you can use good beef stock, and cook the chicken and
leeks in this.

20 minutes before serving time, add the prunes. Five minutes before
serving time, add the remaining leeks, trimmed and sliced. They should
remain slightly crisp and fresh of flavour. Discard the bundle of leeks; it
will be tasteless and too limp for pleasure by this time.

To serve, slice the meats and put a little of everything into each bowl.
Or else strain off the liquid and serve it with chopped parsley first; the
meat, prunes and leeks can follow as the main course.

LAMB AND LEEK BROTH

> *125 g (4 oz) pearl barley, soaked 4 hours*
> *1 kg (2 lb) scrag end of neck of lamb, sliced*
> *200 g (7 oz) diced carrot*
> *150 g (5 oz) diced turnip*
> *1 small stalk celery, chopped*
> *salt, pepper, sugar*
> *2 leeks, trimmed, chopped*
> *chopped parsley*

Drain the barley and put it with the lamb into a large pot, with 2 litres (3½ pt) water. Simmer for one hour. Add the next three items with seasoning and a pinch of sugar. Simmer for one to one and a half hours until the meat is cooked. Take out the pieces, discard the bones, and, if you like, the fattiest parts. Skim the surface fat from the soup. Return the pieces of meat to the soup, reheat and add the leeks. Serve five minutes later, sprinkled with parsley. Serve with wholemeal bread and butter as a meal in itself.

GERMAN LEEK AND BARLEY SOUP
POTAGE D'ORGE À L'ALLEMANDE

Not long ago I was given a manuscript receipt book kept by Emily, wife of Lord Shaftesbury, the great social reformer. Her apparently straight-forward collection of recipes, adjuncts to the normal kitchen repertoire, special small delights for her husband and many children, becomes a moving comment on their lives once one begins to work out the implications of the book. The nervous draughts, for instance: they were not for feminine vapours, but to settle Lord Shaftesbury's nerves, that had been bruised first by the sufferings and deprivations of his own appalling childhood, then by the sufferings of so many of the children of England. 'They come between me and my rest,' he once said. Early in the book she notes that 'Sassafras shavings will prevent bugs'. Her husband often came home with a 'household of vermin' on his back. Judging by a later sentence, 'to get rid of Bugs . . . Sassafras Shavings' they cannot have been much good as a preventive.

This soup recipe comes from the gayer side of Lady Shaftesbury's existence, from her mother who late in life had married her first lover, Lord Palmerston. He may have been Lady Shaftesbury's father. Certainly no father could have been more loving, or have rescued her and her husband more delicately from their financial problems. 'Excellent', she writes, 'from Mama's cook.' I have reduced the quantities, otherwise the recipe is the same:

> 60 g (good 2 oz) pearl barley
> 2 litres (3½ pt) veal or chicken stock
> 60 g (good 2 oz) butter
> white part of 4 fat leeks, sliced
> 375 g (¾ lb) sliced celery
> grated nutmeg, salt, pepper
> 125 ml (¼ pt) cream
> 1 egg yolk

Blanch the barley for five minutes in boiling salted water. Drain off the water and add three-quarters of the stock. Cover tightly and simmer for an hour, or until soft – some barley may require an extra half an hour. Meanwhile melt the butter in another pan and stew the leek and celery in it without allowing them to brown. Pour in half the remaining stock, cover and simmer until the vegetables are soft enough to be sieved into the barley pan. Flavour with nutmeg, salt and pepper to taste and add the remaining stock, plus extra water if you think the soup is on the thick side. Beat the cream and egg yolk together in a well heated tureen and then pour in the boiling soup, mixing it thoroughly to blend in the cream and yolk. Serve immediately.

POOR MAN'S ASPARAGUS

Allow two small leeks per person, or one medium sized leek. Trim off all the green part. Split the larger leeks almost in half down their length. Blanch them in boiling salted water until they are tender, but not over-cooked to sliminess, about ten minutes. Drain rapidly, then run them under the cold tap until they are cool enough to handle. Press out the surplus water gently.

Spread the small leeks side by side in a shallow dish. Completely halve the larger leeks, and arrange them cut side down, side by side. Pour over a good olive oil vinaigrette, with plenty of chopped parsley and some chives, spring onion, or chopped onion. If the leeks are still warm when you do this, so much the better. Finally, just before serving the salad, scatter over a crumbled hard-boiled egg.

Note These leeks only look like asparagus. They have their own virtues of flavour. One of the best winter salads.

LEEK AND PRAWN SALAD

To the poor man's asparagus salad, of the previous recipe, add 250–275 g (8–10 oz) shelled prawns or shrimps. Serve with wholemeal or rye bread and butter.

A DISH OF LEEKS

From Nancy Shaw's *Food for the Greedy*, published in 1936. Simple but luxurious looking. Choose leeks of a similar size and cut off nearly all the green. Leave just enough to add to their appearance.

6–12 leeks, according to size
butter
4 tablespoons grated Parmesan, or 2 of Parmesan and 5 of
   Cheshire
cream

Blanch the leeks for about 20 minutes if they are medium size, ten minutes if they are small (this dish is not suitable for giant leeks). They should be just cooked.

Butter a gratin dish, then scatter on it about half the cheese. Lay the leeks on the cheese. Pour cream over the leeks to come almost to the top and scatter with the remaining cheese. Bake in a fairly hot to hot oven for about ten minutes, until the top is a nice golden-brown.

GERMAN LEEKS AND WINE

Stew the prepared leeks slowly in butter, in a covered pan, for about five minutes, turning them once or twice, so that they are all buttery. Pour in 175–200 ml (6–7 oz) dry white wine and leave to finish cooking. If there is rather a lot of liquid at the end, remove the leeks to their serving dish and boil the juices down. Whisk in a final knob of butter and pour over the leeks. Serve with triangles of fried bread.

MUSHROOM AND LEEKS
FUNGES

The mark of the medieval court feast in England was roast meat and a variety of roast game far beyond anything we eat now. It was an orgy of protein, slightly muted by soups and sweet tarts and helped down by sauces that were sharp, spiced and thick with breadcrumbs. Vegetables do not appear on the feast menus. Perhaps they were slipped in as side-dishes. Perhaps they were too ordinary for grand occasions. Certainly few recipes appear in the manuscripts, as Lorna Sass remarks in her *To the King's Taste*, an adaptation of dishes from the *Forme of Cury* of 1390 to modern taste. This is her good version of a laconically described vegetable mixture, entitled *funges*: 'take funges, and pare hem clene and dyce hem. Take leke, and shred hym small and do hym to seeth in gode broth. Color it with saffron, and do there-inne powder-fort.'

8 small leeks, topped and tailed
3 tablespoons butter

¾ kg (1¾ lb) mushrooms, quartered*
250 ml (8 fl oz) vegetable or chicken stock
½ teaspoon brown sugar
pinch saffron
½ teaspoon fresh ginger root, chopped small
2 tablespoons butter mashed with 2 tablespoons flour
salt, pepper

Slice the leeks and fry them in the butter until they collapse. Then add the mushrooms and stir them about thoroughly. Add stock, sugar, saffron and ginger. Cover and leave a few minutes until the vegetables are cooked. Add the butter and flour in little bits to thicken the juices, keeping the liquid under boiling point. Stir all the time. Season to taste and serve.

25/09/05

BETTY BOLGAR'S LEEK AND MUSHROOM TART

Not long ago after I had been trying the medieval leek and mushroom recipe above, we were invited to a lunch party at a friend's house near Cambridge. The centre piece of the big salad main course on a very hot day was this tart.

shortcrust pastry made with 175 g (6 oz) flour
375 g (12 oz) prepared leeks, halved, sliced
1 medium onion, chopped
butter and oil
salt, pepper
125 g (4 oz) mushrooms, sliced
1 large egg
1 egg yolk
150 ml (¼ pt) milk and single cream mixed

Line a 20–22 cm (8–9 in) tart tin with a removable base with the pastry. Stew the leeks and onion in a little butter and oil for about ten minutes, or longer, until they are soft. Do not allow them to colour. Put a lid on the

* There were no cultivated mushrooms in the 14th century. Probably *funges* means field mushrooms, or horse mushrooms. If you do use field mushrooms, add the stock little by little, or the dish will be swamped: you may need hardly any as the mushrooms will exude their own juices in abundance.

pan at first if you like, though you should not allow them to become very wet, moist rather than liquid. Season them. In a separate pan, cook the mushrooms in a little butter rather more rapidly for about four minutes. Season them too and add to the leek pan when the leeks are cooked. Remove from the heat and cool, to tepid. Beat together the egg, yolk, milk and cream. Mix with the vegetables and spread over the pastry case.

Have the oven pre-heated to 190°C/375°F/Gas Mark 5, with a baking sheet on the centre shelf when you switch on. Slide the tart tin on to the hot baking sheet; this helps the pastry underneath to cook more crisply. Leave for 30 minutes. Take a look and see how the tart is getting along. If the filling is lightly browned on top and puffed up, it can come out. Otherwise leave it a further five or ten minutes. Serve warm, preferably soon after it is cooked, rather than reheated from an earlier baking.

FLAMICHE OR FLEMISH LEEK PIE
FLAMICHE AUX POIREAUX, FLAMICHE À PORIONS, FLAMIQUE À PORGEONS

In *Good Things* I gave a recipe for a Cornish leek pie, with several variations of a dish that is also popular in Wales, Burgundy and Central France. Here is the finest version of them all, from Picardie in the very north of France on the Flemish border with Belgium. The name hints at the Flemish origin of the dish, though these days it has come to mean no more than pie. The pastry can be shortcrust or puff, or even a bread dough enriched with butter, 50 g (1½ oz to 1lb) to half a kilo of dough. The filling can be leeks alone, leeks and onion, leeks and bacon, with or without cream and eggs. Instead of the flour in my recipe, you could use a couple of egg yolks, but I find that they overcook in the great heat of the oven and the filling has too curdled a look. Sometimes at the end, I pour a little extra cream into the pie, through the central hole, and put it back into the oven for five minutes.

> *125 g (4 oz) butter*
> *1 medium onion, sliced*
> *375 g (12 oz) prepared leeks, sliced*
> *125 g double or whipping cream*
> *1 rounded teaspoon flour*
> *salt, pepper*
> *puff pastry made with ½ kg (1 lb) flour*
> *beaten egg to glaze*

In half the butter, cook the onion slowly to soften it. Add the rest of the butter and put in the leeks. Cover the pan and leave the vegetables to stew for five minutes so that the leeks wilt to a moist rather than a wet mass. If there is much liquid raise the heat to evaporate it, taking care not to brown the vegetables. This is better than draining off the liquid and losing its flavour. Beat the cream gradually into the flour to make a smooth paste and stir it into the leeks. Cook for one minute, then remove from the heat altogether, add seasoning and leave to cool down to tepid. (If you use egg yolks instead of flour, mix them with the cream. Take the leek and onion off the fire, and mix in the cream and eggs without further cooking. Leave to cool. The heat of the vegetables will thicken the custard, so that the mass is coherent and not too sloppy.)

Roll out the pastry and cut two large circles, one slightly bigger than the other. Put the smaller of the two on a moistened baking sheet. Spread the leek filling in the middle, leaving a 1½–2 cm (½–¾ in) rim. Brush the rim with egg. Place the larger circle over the top. Press the edges firmly together so that the filling cannot escape. Knock up and scallop the edges. Make a hole in the centre of the lid and score the pastry lightly with the tip of the knife if you like. Brush the whole thing over with egg. Bake for 15 minutes at 220–230°C/425–450°F/Gas Mark 7–8 to set the pastry and brown it slightly, then lower the heat to 180–190°C/350–375°F/Gas Mark 4–5 for a further 20 minutes.

The pie should be well risen and a beautiful golden-brown. Keep an eye on it towards the end, in case you need to lay a protective butter paper over the top to prevent it catching.

If you are nervous about baking a pie flat like this, put the lower circle of pastry into a 22–23 cm (9 in) flan tin, or on to a metal pie plate with a slightly raised rim. This is certainly a good idea if your filling seems to be on the wet side.

This recipe is sometimes made with a filling of onions alone, or onions, potatoes and chopped bacon.

NORTH COUNTRY LEEK PUDDING

In the north of England, in particular the north-east, the use of pastry with leeks gives way to a boiled suet crust. These boiled puddings are the kind of dish I grew up with and love for their winter comfort, though I cannot claim them as the height of fine cookery. The intention was to make the most of the small amount of tasty ingredients that was the most some families could afford. Even in better off homes, their frugality was appreciated. They could be eaten on their own, or with boiled meat, though I never recall eating them before meat in the Yorkshire style.

Leek puddings can be made in two ways. The old-fashioned roly poly, wrapped in a floured cloth, was boiled in a big iron pot with meat, or in a pan on its own if there was no meat; nowadays it is easier to use foil – few people keep a boiling cloth for puddings. The newer style basin pudding is steamed rather than boiled, which improves both flavour and lightness and the filling can be much more lavish.

This kind of food very much belongs to this country, even if we turn our backs on it in these prosperous and calorie-conscious days. We make it shamefacedly, even, as a special treat, waistlines having become a higher principle of morality than purses – at least in the south of England. Every time I put a suet pudding on the table, one or other of us remembers the French émigré who came here at the end of the 17th century and exclaimed 'BLESSED BE HE THAT INVENTED PUDDING, for it is a manna that hits the palate of all sorts of people'.

SUET CRUST Sift 250 g (8 oz) self-raising flour, or plain flour and two level teaspoons of baking powder, with half a level teaspoon of salt. Mix in 125 g (4 oz) grated beef suet, which is better than the packaged kind. Bind with enough water or milk and water to make a soft, unsticky dough.

ROLY POLY LEEK PUDDING Roll the pastry out into a large oblong. Dot it over with one or two prepared, chopped leeks, excluding the coarse green part, and a chopped medium onion. Add if you like 200–250 g (6–8 oz) chopped bacon pieces, cooked bacon or bacon rashers – these are a great improvement if you are eating the pudding on its own. If you are serving it with a piece of boiled gammon or bacon, there would be no point in putting it into the pudding. Sprinkle the whole thing with salt and pepper and approximately one teaspoon of dried sage, or rather more fresh chopped sage.

Brush round the rim of the pastry with water. Roll the whole thing up, pressing gently on the edges to seal them. Do not overdo this, after all the contents are not liquid. Place the roly poly, seam side down, on a piece of foil that is longer than the pudding at each end by a hand's length. Bring the long sides together and seal them with a double fold; enclose the pudding loosely so that it has room to swell, but make the seal tight. Seal each end as well, then bend the ends up to form lugs by which you can move the pudding about.

Lower the whole thing into a large pan of water at a rolling boil, or into an even larger pan in which the bacon or gammon or other meat may be boiling. Leave for two hours.

If there is no extra meat, you can serve the roly poly with un-Northumbrian tomato or mornay sauce, or with butter. This may be unorthodox, but it is a lot nicer than some of the horrible gravy I have seen poured over suet puddings in my time.

SUZY BENGHIAT'S STUFFED LEEKS

> *2 fine fat leeks*
> *250 g (8 oz) minced lean beef or lamb*
> *1 large onion, chopped*
> *1 large tomato, skinned, chopped*
> *salt, pepper*
> *2 tablespoons finely chopped parsley*
> *2 teaspoons dried mint*
> *½ teaspoon ground cinnamon or allspice*
> *100 g (3 oz) long grain rice*
> *375 g (¾ lb) dried apricots*
> *6 tablespoons olive oil*
> *juice of 2 lemons*

Top and tail the leeks, so that you end up with two white cylinders. Slit them along, but only half-way through, so that you end up with a number of curling, oblong pieces. Wash well. Mix the remaining ingredients down to and including the rice. Pour boiling water over the apricots and leave them to soak.

Now separate out the unblemished pieces of leek. You may get 24, or you may not. Any broken bits and the small inside pieces should be chopped and added to the stuffing. Now put some stuffing into each leek roll – when you get to the smaller ones, place a second piece over the gap. Heat half the oil and fry the leek rolls until lightly coloured. You may need to do this in batches.

If the oil has burned, wipe out the pan. If it still looks appetizing, don't bother. Put in half the apricots, then the rolled leeks, fitting them closely together, seam sides down. Dispose the rest of the apricots on top and between them. Pour in the remaining olive oil, the lemon juice and about 250 ml (8 fl oz) of the apricot liquid. Cover and simmer for about an hour. Check the liquid from time to time and top it up with the apricot water first, then with plain water. The rice will absorb some of the liquid, so it does need watching.

The juices are delicious. Taste them for seasoning before you serve the dish, boil them down a little if they seem too copious. Like the stuffed courgettes on page 234, the stuffed leeks can be eaten hot or cold.

Note Recipes for stuffed courgettes can be adapted to leek leaves. See pp. 233–5.

OTHER SUGGESTIONS

Gratin of leeks with ham or bacon See Chicory with ham and cheese sauce, page 213. Use the cooking water from the leeks to make the sauce, in place of stock.

Leeks à la grecque A dish for small, slender leeks. Cut away the coarse greenery and follow the recipe for Cauliflower à la grecque, page 183. There is no need for preliminary blanching if the leeks are the size they should be. Allow two per person.

Leek and potato soup Make up the potato soup on p. 396, adding 300 g (10 oz) chopped white leek to the onion, and using butter and water rather than lard and stock. Chill and dilute the soup with cream for *crème vichyssoise glacée*, finishing it with chopped chives rather than parsley.

Leek roulade Substitute cooked leek for spinach in recipe on p. 469.

LENTILS

Lentils have been around a long time. The Lady of
Han, wife of the Marquis of Tai, took a bag of red
lentils with her to the tomb. Not only did she adorn
China in the 2nd century B.C., but she still exists
with muscle and skin and red silk wrappings (un-
less the scientists of Ch'ang-sha have entirely dismembered her in their
search for information). I would think that red lentils were an odd taste
for eternity, but as she took an example of every food it was possible
to eat at the time – if you were rich – red lentils cannot be claimed as
her special favourite, neither do they figure much in modern Chinese
cookery.

The occasional salad of slate-green lentils can surprise one by its
pleasantness. Lentils and pig's ears or red lentil soup with bacon after a
winter walk has its virtues. Indian cooks do interesting things with lentils,
but I cannot regard them with deep enthusiasm.

The nicest way of eating lentils is with game – venison, hare, pheasant,
partridge. Their earthy flavour sets off the rich juices perfectly.

HOW TO CHOOSE AND PREPARE LENTILS

There are three kinds of lentil on sale in this country. The commonest, the
yellowish red, split Egyptian lentil, can be bought at any grocery; it
disintegrates rapidly when cooked, so that it is only suitable for soups and
purées such as the Indian dhall dishes (it benefits from the spices and
aromatics of Indian cookery). The two others, the brown German lentils –
which have probably been grown in Canada – and the tiny slate-green
French *lentilles de Puy*, from Velay in the Auvergne, can only be found in
delicatessens and good groceries; they remain intact and have an earthy
flavour which many people prefer. The Puy lentil is finer than the German, but they can be used interchangeably. Few people at a blind tasting
could tell the difference.

Lentils need no soaking. The Egyptian ones are soft in about 20
minutes, which means they can be used for last minute dishes – unlike
most other dried vegetables. German and Puy lentils take a little longer;
30 to 40 minutes is often enough, though the time can vary with the
brand.

A point to watch – when you have weighed out brown or green lentils, put them on to a plate and see if there are any pieces of grit that should be removed. They can jar unpleasantly on one's teeth.

Basic recipe for German or French lentils

As a general guide, allow ½ kg (1 lb) lentils for six people. A little less will not matter. On the other hand, any left over can be used up in various ways.

> ½ kg (1 lb) lentils, picked over
> 1 large onion
> 1 large clove garlic
> bouquet garni
> 2 tablespoons butter
> salt, pepper
> chopped parsley, lemon juice

Put the first four ingredients into a pan. Cover generously with cold water. Cover and bring slowly to the boil. Simmer for half an hour, then try the lentils to see how they are. If necessary continue to simmer until they are tender, testing them at regular intervals and replenishing the liquid with boiling water if it disappears from view.

Tip the whole thing into a sieve. Pick out and discard the onion, garlic and bouquet. Return the lentils to the rinsed out pan with the butter. Season to taste – be generous with the pepper. Just before serving stir in the parsley and emphasize the flavour if you like with a squeeze of lemon.

Serve with game, salt pork, roast pork or sausages.

LENTIL SALAD

Follow the basic recipe, but with half quantities, up to draining point. Discard the flavourings. Tip the hot lentils into a bowl containing 150 ml (¼ pt) olive oil vinaigrette. Leave to cool – until the next day if you like. Put in plenty of chopped herbs, parsley, chives or spring onions, tarragon or chervil, and serve chilled. Strips of grilled, skinned red pepper, olives, hard-boiled eggs and anchovy fillets, can be added to make a good first-course dish.

CREAMED LENTILS

This a good dish with hare, on account of the sweet wine and cream.

Prepare and drain the lentils, following the basic method. In the rinsed out pan melt the butter with two or three tablespoons of sweet white or fortified wine. Stir in the lentils and add seasoning. Heat for five minutes. Mix in six tablespoons of cream and some chopped parsley.

LENTILS WITH FEATHERED GAME

Follow the recipe above for creamed lentils, but for cream substitute the juices from the roasting pan in which the game has cooked.

LENTILLES À LA PONOTE

Ponot and the feminine *ponote* are the adjectives from Puy, and they are used as nouns to designate the inhabitants.

When the lentils are cooked and drained according to the basic method, have ready a sauce made by cooking a chopped onion and a chopped clove of garlic in some butter. When they are soft, raise the heat so that they brown very lightly. Stir in a level tablespoon of flour. Moisten with beef stock and several skinned, chopped tomatoes. Reheat the lentils in this sauce and serve them all together in a dish, sprinkled with parsley.

LENTILLES AU POULET FUMÉ OR CONFIT D'OIE

Lentils are often used as background for smoked poultry or potted goose in France. Smokey meat flavours do go well with their earthiness.

Prepare the lentils according to the basic recipe. Towards the end of the cooking time, reheat the poultry in a little stock, or the goose in its own fat (this is a suitable recipe for making the most of canned *confit d'oie*).

Put the hot meat with the lentils in the rinsed out pan, with any juices, and cook gently for a further ten minutes, covered. Remove any large bits of poultry and cut them up. Serve the lentils with the poultry on top, or showing through the surface, scattered with chopped parsley.

Small amounts of left-over game can be used instead of smoked chicken or potted goose, but be careful not to overcook it to rags.

LENTIL AND PORK STEWS

Several countries of northern Europe, especially Germany and France, are fond of cured pork cooked with lentils. They can be delicious, especially if you provide a mixture of meats. One does better to buy a small piece of smoked bacon, a couple of salted pig's ears or trotters or tails, and a boiling ring of some kind, than one big joint of gammon.

Put the first four ingredients of the basic lentil recipe into a roomy pan. Push in the meats you have chosen (soak the salty ones first if necessary). Add a carrot and leek if you like. Cover very generously with water. Bring to the boil, cover and simmer for one or two hours, according to the meat you have chosen.

If you have overdone the water, remove the lid from the pan towards the end of the cooking time and boil it away so that the dish is moist rather than sloppy. At the very end you can put in grilled fresh sausages, or slices of garlic sausage.

Finally taste the lentils. Add salt if necessary, and pepper. Extract the bouquet garni. Cut up the largest pieces of meat. Judge whether the dish would benefit from butter – it is likely to be fat enough if you used a piece of streaky bacon, for instance. Mix in a good chopping of parsley and chives and serve.

Note We have a great weakness for pig's ears – especially as local butchers give them to me free or for a few coppers. They are delicate and should be place on top of the lentils to cook; allow 45–60 minutes. Butter and parsley are needed with them. Serve with German or Dijon mustard.

POTAGE ESAU

You will remember Esau the hunter and Jacob the farmer, the twins born to Isaac and Rebekah? Esau, the firstborn 'came out red, all over like an hairy garment'. Jacob pressed close behind, with his hand on Esau's heel. When the two were grown up, Esau came back from hunting deer one day, faint with exhaustion and hunger. Jacob was stirring a soup of the lentils he had grown. He was an unpleasant, wily man: Esau was an impetuous fool and ended by selling his birthright for Jacob's red mess of pottage. The farmer tricked the hunter, just as farmers gradually superseded the huntsmen in our progress towards civilization. In fact lentils have been found in some of the earliest Near Eastern sites, at Jarmo in Iraq, dating from the sixth or seventh millennium B.C.

This soup, in spite if its name, is very much a product of European cookery, as you can see from the role that bacon plays in its flavouring.

>   *1 large onion, chopped*
>   *1 medium carrot, chopped*

2 heaped tablespoons bacon fat, or lard or butter
250 g (8 oz) red Egyptian lentils
1½ litres (2½ pt) ham or gammon stock, or water plus a ham bone,
*   or water plus 250 g (½ lb) bacon pieces, preferably from smoked*
*   meat*
salt, pepper
150 ml (¼ pt) cream
chopped parsley
fried bacon bits, or bread croûtons

Brown the onion and carrot lightly in bacon fat if possible, otherwise lard or butter. Stir in the lentils, then the stock, or water and bone or bacon pieces. Simmer until the vegetables are tender. Purée three-quarters of the soup in the blender, or a little more. Or put it twice through the *mouli–légumes*. Return it to the pan. In this way you combine smoothness with the attractive texture of the puréed lentils. Bring back to the boil and season. Stir in cream and parsley. Serve with the bacon bits or croûtons.

PURÉE CONTI

Conti is the name for lentil soup in classic French cookery. I have based this recipe on one in *La Cuisine de Madame Saint-Ange*. The book came out in 1927, but happily Larousse brought out a reprint in 1958. It's a carefully written set of instructions, ideal for foreigners like ourselves who may not be familiar with the dishes already and find themselves lost when they try to use modern French cookery books. Madame Saint-Ange never leaves you in doubt.

Bring 375 g (12 oz) French lentils slowly to the boil in 1½ litres (2½ pt) water. Meanwhile chop a *mirepoix* (p. 547) of 50 g (scant 2 oz) each carrot, onion and streaky bacon, then brown it in 50 g (scant 2 oz) butter. Add to the lentils with a bouquet garni. Simmer for two hours. Remove the bouquet and sieve the soup (or put it through a *mouli-légumes*) into a clean pan. Add 250 ml (8 fl oz) water and a pinch of sugar to enhance the flavour. Bring to the boil and simmer for a further half hour, skimming the soup carefully from time to time. Add an occasional spoonful or two of water, to help the scum to rise. Just before serving, mix in 60 g (2 oz) butter and pour through a sieve into the tureen. Serve with croûtons of bread fried in butter.

A delicious edge can be given to the soup by cooking 150 g (5 oz) of sorrel, cut into strips, in 40 g (1½ oz) butter, better still in bacon fat, for 15 minutes. Add this to the skimmed soup ten minutes before it has finished cooking. Beat two egg yolks with a ladleful of the soup and pour it into the warmed tureen just before you pour in the soup. Stir well together. Serve with croûtons.

GERMAN LENTIL SOUP
LINSENSUPPE

You might think that all smoked bacon is smoked, all green bacon unsmoked. Things, though, are not always what they seem: 'smoked' these days can mean 'flavoured approximately with chemicals'. The difference is obvious once you become familiar with the real, right thing. You soon realize which hams and bacons have been painted and injected to simulate cured meat. A friend who runs the delicatessen in Oxford market insisted that I buy the real thing to make her recipe for *linsensuppe*, proper German smoked belly bacon, *geräuchter bauchspeck*. The flavour is rich and deep without harshness, without cosmetic tinge. Ask for it at your local delicatessen and buy a piece of cured pork fat, *fetterspeck*, to use rather than lard.

> MEAT LIST
> 250 g (8 oz) lentils, unsoaked
> 250 g (8 oz) smoked belly bacon
> 1 large leek, chopped
> 1 small onion, chopped
> 2 tablespoons chopped celery
> bouquet garni
>
> VEGETABLE LIST
> 375 g (12 oz) potatoes, diced
> 60 g (2 oz) fetterspeck *or* lard
> 1 medium onion, chopped
> 1 tablespoon flour
> salt, pepper
> 1–2 small leeks, chopped
> chopped parsley
>
> OPTIONAL
> 2–6 frankfurters

Put the meat list ingredients into a roomy pot with 1¾ litres (3 pt) water. Cover and simmer for about two hours until the bacon is just cooked. Now add the potatoes from the vegetable list and leave to simmer.

In a small pan heat the *fetterspeck* until the fat runs, or melt the lard. Add the onion and brown it slightly. Stir in the flour and moisten with some of the liquid from the lentil and meat pot. Simmer for ten minutes, then stir into the soup to thicken it slightly. Season with salt and pepper to taste. Add the remaining ingredients, and give them five minutes' simmering to heat them through.

LENTILS WITH SPINACH
SHULA KALAMBAR

In her *Middle Eastern Food*, Claudia Roden writes that *shula kalambar*, this lovely earthy tasting dish, was given to sick people in medieval Persia. There was a snag – to be effective as a cure the ingredients had to be paid for by begging.

> 250 g (8 oz) lentils, cooked
> ½ kg (1 lb) spinach, cooked, drained, chopped
> ½ teaspoon each ground coriander and cumin
> salt, pepper
> large knob of butter

Mix the lentils with the spinach and add the seasonings. Reheat and add more spices and salt and pepper to taste. Stir in the butter and serve immediately.

APICIUS' LENTIL AND MUSSEL DISH
LENTICULA EX SFONDYLIS

This earthy lentil and mussel dish comes from Apicius' *De re coquinaria*, which is a 4th-century compilation, not by Apicius at all, though it does contain much of two books he did write, including this recipe. It seems he had a weakness for mussels. He was the great gourmet of the reign of Tiberius. His dinners were magnificent whether for friends, or for himself alone. No poached egg on a tray for Apicius, when he felt like a solitary evening. One day he realized he had got through 100,000,000 sesterces. Only 10,000,000 were left. Rather than face starvation, or a permanent retreat to the country, he took poison.

> 250 g (8 oz) lentils
> 250 ml (8 fl oz) canned or bottled grape juice, or grapes
> 8 tablespoons olive oil vinaigrette

1 teaspoon honey
½ teaspoon each ground coriander seed, cumin and dried mint
1 kg (2 lb) mussels
anchovy essence

Put the lentils on to boil. Either boil down the grape juice by two-thirds or crush the grapes in a blender or *mouli-légumes*, sieve them and reduce by two-thirds. Set this aside for later seasoning of the dish: it is an equivalent to the Roman seasoning *defrutum*, made by boiling down the grape must at the vintage.

When the lentils are done, drain them and mix immediately with the vinaigrette, honey and seasonings. Scrub and scrape the mussels, throwing out any that are damaged or that will not close when sharply tapped with a knife. Put them into a heavy pan, cover, and set over a fierce heat for five minutes. Shake the pan occasionally. Remove the opened mussels and discard their shells. Cut them in half and add them to the lentils. (Any mussels that refuse to open should be rejected.)

Flavour the dish with the *defrutum* and anchovy essence using enough to enhance the flavours of the main ingredients without bossiness. Bring to boiling point, taste again for the seasoning and serve straightaway. Mussels should never be overcooked or they turn rubbery.

DHALL CURRY

I have always found lentil purées insipid and heavy until I tried them in the Indian style. This recipe is adapted from one of 15 in an excellent chapter on cooking lentils, or dhall, in E. P. Veeraswamy's *Indian Cookery* (Mayflower paperback). You can use it for any other of the varieties of lentil that you may find in an Indian grocery shop.

250 g (8 oz) red Egyptian lentils
600 ml (1 pt) water
60 g (2 oz) clarified butter
1 medium onion, finely chopped
2 cloves garlic, finely chopped
1 tablespoon coriander seeds, ground
1 teaspoon ground turmeric
½ teaspoon ground cumin seed
½ teaspoon ground chillies or cayenne
¼ teaspoon ground fenugreek
2 tablespoons wine vinegar

*150 ml (¼ pt) tomato sauce, or 2 tablespoons tomato concentrate
    dissolved in 2 tablespoons water*
salt

Cook the lentils in the water in a covered pan, until they are soft and the consistency of a thick porridge. Heat the butter in a large frying pan and cook the onion and garlic in it, slowly, for five minutes. If you have the spices in whole form, put them all into a small electric mill to grind them together. If they are already ground, mix them together in a small bowl. Mix them to a paste with the vinegar and add to the onion. Fry for another two or three minutes, then put in the lentils and the tomato sauce or diluted concentrate. Simmer, stirring occasionally, for another ten minutes, with a lid on the pan. If you like the mixture on the liquid side, add more water.

Serve with boiled rice, poppadums and any other side dishes you like to eat with curry.

CURRIED LENTIL SOUP

Dilute the dhall curry with water to a soup consistency. Purée or sieve most of it – see *Potage Esau*, p. 307 – to make a smooth texture with bits in.

A good lentil soup for people who do not eat meat at all.

LETTUCE and LETTUCE SALADS

In his carved chapel at Karnak, the pharaoh Senusret I, of the Middle Kingdom, offers the god Min two flasks of milk. Min raises a smooth arm against the background of three tall lettuces in detailed relief. They stand on a grid that is thought to represent the field they are growing in. For the Egyptians, Min was the ithyphallic god of increase. Lettuces were sacred to him, perhaps because of the 'straight vertical surge' of their growth, perhaps because of the milky juice they exude which could be taken as a symbol of mother's milk or semen. Juicy plants like this are rare in dry Egypt.

Lettuce milk was valued for medicinal purposes by Greek physicians later on though the Greeks valued lettuce for a different reason – because it helped people to sleep. That tradition ends, with us at any rate, in the Flopsy Bunnies eating too many 'shot' lettuces, falling asleep because 'the lettuces had been so soporific', and nearly ending up in Mrs McGregor's rabbit pie.

The tall lettuces in the Egyptian relief inevitably put one in mind of our cos lettuce, which originated – presumably – in the fruitful island of Kos, in the Dodecanese, near what is now the Turkish coast. But when? The Ptolemies held the island for 300 years, until the Romans conquered Cleopatra in 30 B.C. The fame of the island was the healing sanctuary of Asklepios, the god of medicine; Hippocrates was born there about 460 B.C., the greatest of all physicians. Was the cos lettuce developed in those days from the tall lettuces of Egypt, as one of the medicinal plants that doctors relied on? My guess is that whatever its origin this variety came west, to Rome first of all, when Kos and Rhodes were held by the Knights of St John (they lost the two islands to the Turks in 1522, by which date the lettuce was well established in Italy). It came to France via the gardens of the Papal court at Avignon. At first it was called the Roman lettuce in this country, and it remains the *lattuga romana* to the Italians and *laitue romaine*, or just *romaine*, to the French.

In 1629 Parkinson, in *Paradisus in Sole*, wrote of the superb red Roman lettuce that John Tradescant, the great gardener and collector, had introduced: when stripped of its outside leaves, it weighed 17 oz. He mentioned other kinds as well. John Evelyn seems to have been the first to use the word cos, in 1664, in his salad calendar (page 322). Again in 1699, in *Acetaria*, he refers to the 'Cos lettuce from Turkey', as if in his time there had been a direct reintroduction from Kos. It must have been a recent arrival, to add to the other varieties he mentions – Lop, Silesian and Roman were all similar. For a while the names existed side by side. Not until this century did cos prevail as an umbrella word sheltering all the long leaved lettuces.

The other kind of lettuce is the round cabbage lettuce, which in its varieties goes from soft, sometimes regrettably a sad floppiness, through the firmer large-hearted butterhead cabbages, to the crispness of Iceberg and Webb's Wonder which have as much bite as cos.

A new kind of lettuce has been introduced quite recently. It is called a loosehead lettuce, as it has no heart, but grows in a great circular patch. From it, you cut just the amount of leaves you need. One American variety, Salad Bowl, is on sale from seedsmen here. In France, where it is known as a cutting lettuce, *laitue à couper*, you will find several varieties, including the delicious tawny Oak Leaf with appropriately jagged leaves, the *feuille de chêne*. This kind is practical for a family vegetable patch. It survives the heat well, and will provide a supply of lettuce over a long period.

Although lettuces can be used for soups and braised dishes, their prime purpose is not in dispute. The lettuce 'ever was and still continues the principal foundation of the universal Tribe of Salads; which is to cool and refresh'. None of us would disagree with John Evelyn about that, though we might demure at his insistence, following the opinion of Ancient Rome, that lettuce upholds morals, temperance and chastity. In his little book of instructions for the gardener at Sayes Court, he gives a scheme for salads throughout the year – see p. 322 – and in *Acetaria* of 1699, he propounded the theory of good salad-making. 'In the composure of a salad, every plant should come in to bear its part, without being overpowered by some herb of a stronger taste . . . but fall into their places, like the notes in music, in which there should be nothing harsh or grating.' The perfect, polyphonic salad.

When the 'herby ingredients', which had been gathered into a special withy basket with compartments for the different leaves, were brought into the kitchen, they were to be 'sprinkled' with spring water, then drained in a colander. 'Lastly, swing them all together gently in a clean coarse napkin' said Evelyn. For the dressing the oil should be of a 'pallid olive green . . . smooth, light and pleasant upon the tongue; such as the genuine omphacine and native Lucca olives afford'. The vinegar must be the best wine vinegar, infused with flowers or herbs. The salt should be 'the brightest Bay grey-salt', clean, dry, without clamminess. Sugar must be perfectly refined, the mustard from Tewkesbury (the best mustard of the day). The pepper should not be 'bruised to too small a dust'. Instead of vinegar, try lemon or Seville orange juice for a change.

HOW TO CHOOSE AND PREPARE LETTUCE

Do not be put off by earthy roots and damaged outside leaves. They may be a better indication of local fresh origin and quality than a tidier lettuce in a plastic bag. Take a look at the centre part before you buy. A lettuce that looks droopy is past its best.

If you need to keep lettuce for two or three days, take off the worst of the soil, roots and damaged leaves, then put the whole thing unwashed into a plastic bag. Squeeze out the air gently, fasten it tightly and put into the refrigerator. Washed lettuce can be kept in the same way, or in a plastic box, but the stalk begins to turn brown after 24 hours and premature washing does no good to the flavour.

When it comes to preparing a lettuce, start as early as you can and hang it up in a wire basket to drip dry in peace. This saves you time and clean cloths, though you will still have to finish it with a cloth, either by dabbing the leaves or by swinging the whole thing (outside the back door).

Bolted or shot lettuce, or the undamaged but coarse leaves of large lettuces, can be used to make a good soup. Lettuce stalks can be peeled, sliced and steamed. Serve them with a creamy sauce as if they were cucumber or seakale or Swiss chard stems.

BASIC GREEN SALAD

After you have washed and dried the lettuce, tear the larger leaves into pieces. Never shred a lettuce with a knife or scissors, unless you are going to cook it, or make a last minute Greek salad (p. 316). Cutting bruises the lettuce and darkens the edges; if it has to wait for any length of time it becomes a sad looking mass.

In the salad bowl, mix the following:

> *1 small clove garlic, crushed with salt*
> *¼ teaspoon sugar*
> *a large pinch of sea salt*
> *1 scant tablespoon Orléans wine vinegar*
> *4 tablespoons olive oil*
> *plenty of chopped parsley, chives and tarragon if possible*

If you mix them in the order given, you will find that the garlic dissolves into the dressing. Crush it with a wooden spoon in the bowl, or a mortar; garlic crushing machines are so difficult to clean properly that they give an unpleasant air of stale garlic – I have thrown mine away. If you split the clove first with the knife, so that the flat side is downwards, the garlic can be crushed easily. Whisk the dressing well together with a fork or small whisk, then taste it and adjust the seasonings.

Lay the salad servers across each other over the dressing, then rest the lettuce on top. The salad can now be left in the larder or fridge until you are ready to bring it to the table and turn the lettuce over in the dressing. If you do this too soon the lettuce becomes slimy.

Some people turn the leaves first in oil, then add the other dressing ingredients. I find that this is not satisfactory. The flavours do not seem to blend well together into the salad.

ANJOU SALAD
SALADE ANGEVINE

A vinaigrette is a poor thing when made with the tasteless oils that are so much sold these days. They have to be livened with many seasonings to give them at least a semblance of virtue. My favourite oils are the Tuscan green oil – difficult to find these days, alas – and the golden oil of Beaumes-de-Venise in Provence, which is fruity but light at the same time. An olive oil which has a thick olive flavour is the Greek Minerva oil. Then there is a fourth oil, the oil needed for this particular salad, walnut oil. The flavour comes through well, it has not been refined out of recognition. One can buy it at Roche's in Old Compton Street, Soho, and other high-class groceries. If you make a summer journey through the west of France, you should be able to find it quite easily. A vinegar I can recommend – sold by Woolworth in this country – is the Martin-Pouret brand from Orléans: it is matured slowly in barrels in the old-fashioned way, and it shows in the flavour.

> lettuce, batavia, endive, lamb's lettuce or dandelions – or a mixture
> 4 tablespoons walnut oil
> 1 tablespoon wine vinegar
> salt, pepper, pinch of sugar
> parsley, tarragon, chives, chopped
> 2 slices white bread
> cut clove of garlic
> extra walnut oil

Separate and wash the salad leaves. Mix the oil, vinegar, seasonings and herbs in the bottom of a salad bowl. Dry the salad and place it in the bowl, resting on the crossed salad servers. Cut the bread free of crusts, rub it with garlic, then cut into cubes and fry them in walnut oil, with a piece of garlic. Remove the garlic if it begins to darken. When the bread is golden-brown, drain, cool slightly and add to the salad. Turn it at table. These cubes are known as 'chapons'. They improve green salads with their crunch.

GREEK SALAD

From about 10th May in Greece, people go out to pick the leafy tips and buds of wild marjoram to dry and store for year-long use. The flavour is characteristic of much Greek cooking and rather different from that of our northern marjoram. It is sold in Greek and Cypriot delicatessens under the name of rigani. Its main use is with meat, particularly with kebabs, but it also adds an unmistakable note to the typical Greek salad that one gets in every *taverna*. The different items are built up in layers on side plates, but it would be more sensible for a family to use one large dish.

> *12–15 lettuce leaves, cut in ribbons*
> *6 fine large tomatoes, peeled, sliced*
> *10 cm (4 in) piece cucumber, peeled, sliced thinly*
> *salt, pepper*
> *10–12 spring onions, chopped, or slices of sweet raw onion*
> *12 sprigs fresh mint, chopped*
> *1 heaped teaspoon dried rigani*
> *4 tablespoons olive oil*
> *2 tablespoons lemon juice*
> *175 g (6 oz) feta cheese, crumbled or sliced or cut in cubes*
> *18 black olives*

Put the shredded lettuce all over the dish. Then arrange the tomato and cucumber on top in layers, seasoning as you go. Scatter with onion and herbs. Mix the oil and lemon juice and pour over the whole thing. Dot the cheese over the top and surround with the black olives. Serve immediately. This is not a salad to be kept waiting around or the lettuce will begin to look tired and the cucumber will exude moisture.

Often rings of green pepper are added as well, or a few capers.

This is an ideal salad to serve after some plainly grilled fish, such as herrings or sardines, especially if you are eating out of doors, and can cook the fish over charcoal. Serve it with lemon quarters and plenty of bread. Retsina wine and this fresh-tasting salad complete a perfect summer meal.

SALADE NIÇOISE

Until I really went into the matter, I had always thought that *salade niçoise* was characterized by long, cooked string beans, but really they are no more than one of the seasonal extras that can be added to the base of lettuce, tomatoes, eggs, anchovies and olives. You choose your extras, too, according to what you are eating afterwards – *salad niçoise* is always the first course. It is a brilliant dish, or should be, so keep the various items chunky and arrange them boldly like a Matisse painting.

THE BASE
1 crisp, hearted lettuce cut in 6 wedges
3 hard-boiled eggs, halved or quartered
3 large firm tomatoes, peeled, quartered
9 rinsed anchovy fillets, halved lengthways
12 black olives, stoned
1 heaped teaspoon capers

EXTRAS
tunnyfish, divided into chunks
cooked whole string beans
a pepper, red or green, cut into diamonds
small whole new potatoes, cooked

The most effective arrangements I have seen have usually been in a deep bowl. The lettuce goes in first, as a cradle for the other things; if you cannot get hold of a really firm lettuce, make do with the best leaves and use them to line the bowl, then arrange the rest of the ingredients, topping the whole thing with the anchovies, olives and capers. Pour over a vinaigrette of really good olive oil, tarragon vinegar, garlic, salt and pepper. You can sprinkle over chopped chervil, parsley, tarragon and chives.

Note for gardeners – young artichoke hearts, so small they can be eaten raw, are often added; they should be quartered.

SALADE DE ROMAINE AUX CAPUCINES
NASTURTIUM SALAD

When large bright nasturtium flowers appeared in English gardens at the end of the 17th century – they were introduced from Peru – they were often used to garnish food. Nasturtium (the botanical name is *Nasturtium indicum*, Indian cress) means literally nose-twister, referring to the peppery strength of its flavour which permeates leaves, flowers and seed. Hannah Glasse, in the 18th century, suggests putting 'stertion' flowers round her magnificent dishes of potted tongue, or with the piled up salad called Salmagundy. Some people declare that nasturtium seeds can be pickled and used instead of capers, but they are a poor substitute for the real thing.

Wash and separate the leaves of a cos lettuce, then tear them in convenient sized pieces and slice the firmer parts of the white stems into thin strips. Make a good vinaigrette in a salad bowl, lay the lettuce and the strips of stem on top and decorate the whole surface with nasturtium flowers which have been carefully washed and allowed to dry. Vary the colours as much as possible, and do not turn the salad until it is ready to be eaten, as it makes a glorious addition to a table of cold food.

CAESAR SALAD

So many stories and versions of America's Caesar Salad seem to be in circulation, that one might think it was a dish of Roman antiquity. In fact it was invented half a century ago, in 1924, by Caesar Cardini, who ran a popular restaurant at Tijuana, just south of the American-Mexico border. The following ingredients and method came from his daughter, Rosa, and were given in *Julia Child's Kitchen*.

>   2 large cos lettuces
>   2 large cloves garlic
>   salt
>   175 ml (6 fl oz) olive oil
>   3 slices white bread, cubed
>   2 large fresh eggs
>   the juice of a lemon
>   30 g (1 oz) freshly grated Parmesan
>   pepper
>   Worcester sauce

The basis of the salad is prepared in advance, but the dressing and mixing are done at the table. If you can think of it in time, crush the garlic and salt into the oil for the croûtons, several days in advance. It is important that all the ingredients should be of first class quality.

Separate the lettuces. Pick out six to eight leaves for each person, nicely shaped, unblemished leaves 8–18 cm (3–7 in) long. Wash and dry them tenderly so that they do not break, then roll them in a cloth and put them in the refrigerator until you have completed the other preparations.

Next crush the garlic with a quarter of a teaspoon of salt and mix it with three tablespoons of oil. Put the bread cubes on a baking sheet and put them into a cool to warm oven to dry out; baste them from time to time with the garlic flavoured oil, so that they become crisp all through but nicely browned. When they are ready, put them in a bowl on a tray. Julia Child gives a quicker method of preparing the croûtons. First toast the bread, then cut the croûtons and fry them for a minute in the strained garlic oil. So long as they end up rich and crisp all through and a nice brown, it doesn't matter which method you choose.

Next put the eggs into quite a large pan of vigorously boiling water – it should not go off the boil so take the eggs from the fridge, if you keep them there, in good time. Boil the eggs for one minute only, then remove them to a bowl and put that on the tray with the croûtons.

Pour the lemon juice into a little jug. Put the cheese into a small bowl. Arrange them on the tray as well, along with the pepper mill, the salt cellar, the Worcester sauce and the rest of the oil in a second jug. You will need spoons, too.

Lastly unwrap the cos leaves and put them in a really large bowl so that there is plenty of room for turning them. You are also supposed to chill the plates for the salad.

TO PREPARE THE SALAD AT TABLE If you are given to baroque flourishes in the dining room and take chafing and flaming in your stride, this is your moment. If you find such performances nauseating rather than fun, just go at it quietly while everyone is talking and with luck they will not notice what you are up to. The salad will taste the same.

First scoop under the salad with the salad servers. Then pour four tablespoons of oil over it. Move the servers to the back of the bowl, opposite to you, then bring all the salad over and up in a wave. Be careful you do not misjudge the movement and end up with the wave in your lap. This is why you need a big bowl and why you may find it easier to stand, rather than sit. Sprinkle on a quarter of a teaspoon of salt. Grind the pepper mill over it eight times, pour on two tablespoons of oil and turn the salad over again in the same way. Julia Child uses the word 'toss' – this puts me in mind of hay-making and cabers; I think it is more prudent to turn.

Pour on the lemon juice and six drops of Worcester sauce. Now break in the eggs, praying the thin shells do not crush to pieces in your hands or over the salad. Turn again twice, so that the lettuce is covered in the creamy egg and oil mixture. Sprinkle on the cheese. Turn again. Scatter the croûtons over all and turn twice.

Do not sit down yet. You have not finished. You now have to arrange the salad 'rapidly but stylishly' leaf by leaf on the chilled plates (which by now are unchilled, so there is little point in bothering in the first place, unless you are a speedy operator). At the side of the leaves put a few croûtons.

The approved manner of eating Caesar salad is to pick up the leaves with your fingers, asparagus style, then eat the croûtons with a knife and fork. Cloth napkins and small bowls of water for the fingers are essential, as the dressed stems of the lettuce are far more messy than asparagus. Serve it on its own as a first course.

LETTUCE SOUP

A good recipe for outside lettuce leaves – the ones left over from a Caesar salad, for instance – or for bolted lettuces. If they are very coarse in flavour, blanch them for three minutes in boiling salted water, run them under the cold tap and then continue with the recipe.

> 2 tablespoons butter
> 1 large onion, chopped
> 1 clove garlic, chopped
> about 250 g (8 oz) lettuce leaves and stalk
> 1 tablespoon plain flour
> 1 litre (1¾ pt) light chicken stock or water
> salt, pepper
> 150 ml (¼ pt) double or whipping cream
> 1 large egg yolk
> chopped parsley or, preferably, chervil

Melt the butter and cook the onion and garlic slowly for five minutes, then stir in the lettuce, which you have sliced up. Mix everything round for a minute or two, then sprinkle on the flour. Stir again and cook for a minute, then gradually add the stock or water. Simmer for five minutes (or longer if you include tough lettuce stalk). Purée the soup in a blender, or put it twice through the *mouli-légumes*; pour it into the rinsed out pan through a strainer. If puréeing has made the soup too thick, dilute it with water. Correct the seasoning.

Bring the soup to boiling point. Whisk the cream and egg together in a soup tureen, pour on a ladle of boiling soup and whisk again, then add the rest of the soup. Stir in chopped parsley if you have no chervil.

Serve with croûtons of bread fried in butter.

BRAISED LETTUCE

It is worth braising lettuce if you have a glut of firm, well-flavoured cabbage or cos lettuces. Choose heads that are closely packed, and cut one for each person.

Trim off the outer leaves, then blanch the lettuces whole for five minutes in boiling salted water. Drain them well, and cut them in half, lengthways. Unless they are tightly and roundly formed, fold them in half so that the thin part at the top is bent over on to the stalk. Tie them neatly. Into a shallow pan that will hold the lettuces in a single layer, make a bed of pieces of very thin bacon, with chopped carrot and onion on top. Tuck in a bouquet garni. Lay the lettuces closely together on top. Cover and sweat over a moderate heat or in a hot oven for ten minutes. Pour in enough light stock barely to cover the lettuces. Put the lid on again and simmer until the lettuces are tender. Drain them, remove them to a hot dish and keep them warm – if you do not intend to move them again, cut off the string. Otherwise leave it, so that the lettuces do not flop about when they are transferred to their serving dish, or put around the meat.

Strain off the braising liquor into a clean pan. Skim off the fat and boil it down to a concentrated flavour. Correct the seasoning. Off the heat, whisk in a good knob of butter and squeeze in a little lemon juice. Pour over the lettuces to serve.

NIVERNAISE GARNISH Braised lettuce, glazed carrot, onions, turnip – boiled potatoes if you like. Above all for beef, ham.

CHOISY GARNISH Braised lettuce and potatoes cut into large olive shapes, then stoved in clarified butter (see page 402). Sprinkle with chopped parsley. Serve with grilled beef steak, noisettes of lamb.

JUSSIÈRE GARNISH Braised lettuce, glazed carrots (optional), braised onions, stoved potatoes as above.

JOHN EVELYN'S

	SPECIES	ORDERING AND CULTURE	MONTH	ORDERING AND CULTURE	SPECIES
IX. Blanched.	1. *Endive*	Tied up to blanch			Rampions
	2. *Chicory*		January	Blanched as before	Endive
	3. *Celery*	Earthed-up			Succory
	4. *Sweet Fennel*				Fennel, sweet
	5. *Rampions*	Tied up to blanch			Celery
	6. *Roman**				*Lamb-Lettuce*
	7. *Cos*	Lettuce			Lob-Lettuce*
	8. *Silesian**				Radish
	9. *Cabbage*	Tied close up / Pome and blanch themselves			Cresses
	10. *Lob-lettuce**		February		Turnips
	11. *Corn-Sallet*	Leaves, all of a middling size			Mustard seedlings
	12. *Purslane*				Scurvy-grass
	13. *Cresses*, broad	Seed-leaves, and the next to them.			Spinach
	14. *Spinach*, curled	The fine young leaves only, with the first shoots		Green and unblanched	Sorrel, Greenland
	15. *Sorrel*, French		and		Sorrel, French
	16. *Sorrel*, Greenland				Chervil, sweet
	17. *Radish*	Only the tender young leaves			Burnet
	18. *Cresses*	The seed-leaves, and those only next to them			Rocket
	19. *Turnip*				Tarragon
	20. *Mustard*	The seed-leaves only			Balm
XXVI. Green: Unblanched.	21. *Scurvy-grass*				Mint
	22. *Chervil*				Sampier
	23. *Burnet*	The young leaves immediately after the seedlings	March		Shallots
	24. *Rocket*, Spanish				Chives
	25. *Parsley*				Cabbage-Winter
	26. *Tarragon*	The tender shoots and tops			
	27. *Mints*				Lop*
	28. *Sampier†*	The young tender leaves and shoots	April	Blanched	Silesian* Winter — Lettuce
	29. *Balm*				Roman* Winter
	30. *Sage*, Red				Radishes
	31. *Shallots*	The tender young leaves			Cresses
	32. *Chives and Onion*			Green herbs unblanched	Purselan
	33. *Nasturtium*, Indian	The flowers and flower-buds			Sorrel, French
	34. *Rampion*, Belgrade	The seed-leaves and young tops	May		Sampier
	35. *Trip-Madame‡*				

* Lop or Lob, Silesian, Roman are varieties of Cos lettuce.
† Sampier is probably marsh samphire, q.v.
‡ Trip-Madame is Yellow Stonecrop, *Sedum reflexum*, used in the 16th and 17th century as a salad.
'Note that by parts is to be understood a pugil; which is no more than one does usually take up between the thumb and the two next fingers, by fascicule a reasonable full grip, or handful.'

SALAD CALENDAR

PROPORTION	MONTH	ORDERING AND CULTURE	SPECIES	PROPORTION
10 2 5 ⎬ Roots in number 10 4	and June	Note, *that the young seedling leaves of orange and lemon may all these months be mingled with the Sallet*	*Onions*, young *Sage*-tops, the Red *Parsley* *Cresses*, the Indian *Lettuce*, Belgrade *Trip-Madame*† *Chervil*, sweet *Burnet*	Six parts Two parts One of each part Two parts
A pugil of each Three parts each One of each part Two parts	*July*	Blanched, *and may be eaten by themselves with some* Nasturtium *flowers*	Silesian *Lettuce* Roman *Lettuce* *Cress* *Cabbage*	One whole *lettuce* Two parts Four parts
One part of each Twenty large leaves One small part of each Very few	*August* and *September*	Green herbs *by themselves, or mingled with the* blanched	*Cresses* Nasturtium Purslane Lop-Lettuce Belgrade, or Crumpen-*Lettuce* Tarragon Sorrel, French Burnet Trip-Madame	Two parts One part Two parts One part Two parts of each One part
Two pugils or small handfuls A pugil of each Three parts Two parts 1 Fasciat, or pretty full gripe Two parts One part	*October* *November* and *December*	Blanched Green	Endive Celery Lop-Lettuce Lamb's-Lettuce Radish Cresses Turnips Mustard seedlings Cresses, broad Spinach	Two if large, four if small, stalks and part of the root and tenderest leaves A handful of each Three parts Two parts One part of each Two parts of each

THE POET'S SALAD

The Reverend Sydney Smith, greatest of English wits, perfected a special sauce for salads which became so popular that he turned the recipe into a poem.

> *To make this condiment your poet begs*
> *The pounded yellow of two hard-boil'd eggs;*
> *Two boiled potatoes, passed through kitchen sieve,*
> *Smoothness and softness to the salad give.*
> *Let onion atoms lurk within the bowl,*
> *And, half-suspected, animate the whole.*
> *Of mordant mustard add a single spoon,*
> *Distrust the condiment that bites so soon;*
> *But deem it not, thou man of herbs, a fault*
> *To add a double quantity of salt;*
> *Four times the spoon with oil of Lucca crown,*
> *And twice with vinegar procur'd from town;*
> *And lastly o'er the flavour'd compound toss*
> *A magic soupçon of anchovy sauce.*
> *Oh, green and glorious! Oh, herbaceous treat!*
> *Twould tempt the dying anchorite to eat;*
> *Back to the world he'd turn his fleeting soul,*
> *And plunge his fingers in the salad-bowl!*
> *Serenely full, the epicure would say,*
> *'Fate cannot harm me, I have dined today.'*

Use 125 g (4 oz) cooked potato, a teaspoon for the seasonings, a tablespoon for the olive oil and wine vinegar, with a scant teaspoon of anchovy sauce. Put into the bowl with items such as watercress, Cos lettuce or chicory, and mix at the table.

MANGE-TOUT PEAS, or SUGAR PEAS

This French name is a good one, because that is precisely what sugar peas are – eat-them-all peas.

This you can do because the pods have no tough inner skin. They lie close-fitting to the little peas, clearly showing their form like a skin-tight dress. These pods do not bulge into an enclosing shell in which the peas will develop, as garden pea pods do, a shell that is usually discarded except in a frugal moment.

Sugar peas have been grown by knowledgeable gardeners for at least a century and a half, a delicate and sweet variety that has gone quietly on its way. Now they come to us in the shops as a fine new vegetable, a true delicacy to be savoured, enjoyed as a treat. Their fresh, true flavour reminds us of what we have lost with the freezer-pack dominance of treated, shelled peas. When they first come in from Spain, in January, they convince me that the year really has turned towards spring, that winter is on the way out, slowly perhaps, but on the way. A sign of grace, and therefore not cheap. But why should they be? They are almost into the asparagus class. They deserve to be eaten on their own or as a single vegetable with a light meat dish, veal or chicken with a cream sauce for instance, or with the first English lamb of the season.

HOW TO CHOOSE AND PREPARE MANGE-TOUT PEAS

The flat pods should look bright green, fresh, juicy. The row of little peas down one side should undulate through the skin, rather than protrude in full-term manner. The stalk ends, with their moth-like traces of withered petals, will also give you an idea of how vigorous or stale they are. Some supermarkets sell their mange-tout peas in plastic bags, which helps to preserve the quality.

Preparation is simple. Just top and tail the pods, pulling the strings away from the edges as you go. It is difficult to get rid of all the string, but the delicious flavour is worth the small inconvenience. If you think this is likely to worry you, steam and eat the peas in your fingers like asparagus. If you are not worried by the thought of having to fish an occasional knobble of string from your mouth, try the other recipes as well.

As a general rule, allow ¾ kg (1½ lb) for six people.

MANGE-TOUT SOUP

A delicate soup of fresh colour and flavour. I make it in February from imported mange-tout peas, to give winter suppers a light clear beginning. At this time of the year, you have to use dried mint. Don't worry about this: it works beautifully. In the summer when home-grown mange-tout peas are about, fresh mint can be added. If you do not possess a blender, add a peeled and cubed potato to the vegetables, and put the soup twice through the *mouli-légumes*, first at medium, then at fine, to get the smoothest possible result. The soup should always be on the thin side.

> *1 medium onion, or 5 spring onions, chopped*
> *1 tiny clove garlic, chopped*
> *2 tablespoons butter*
> *250 gr (8 oz) mange-tout peas, prepared*
> *salt, pepper, sugar*
> *200–250 ml (6–8 fl oz) whipping or single cream*
> *about 1 teaspoon dried mint, or 1 tablespoon fresh chopped mint*

Stew the onion and garlic in the butter for about five minutes, with a lid on the pan. Cut the mange-tout peas into three pieces each, add them to the pan and turn them over in the butter for a further three minutes, still at a low heat. Pour in a generous litre (2 pt) water and add the seasoning with a quarter of a teaspoon of sugar. Cover the pan and leave to simmer until the peas are tender. Purée the soup in a blender, then pour it back into the pan through a sieve, so as to catch any remaining stringy threads. Push through any of the solider part with a wooden spoon. This is quick to do, as the blender has done all the work. Add the cream and bring the soup to just below boiling point, stir in the mint and leave for a minute. Taste again and add more seasoning if necessary. If the soup is too thick, add more water. Serve with a few very small cubes of bread fried in butter.

Note Do not keep the finished soup hanging around over a low heat or the freshness, both of colour and taste, will disappear. Better to let it cool, then reheat it as required, over a moderate heat.

MANGE-TOUT PEAS ASPARAGUS STYLE

The Chinese have a happy way of cooking garden peas in their pods and eating them with their fingers. I can recommend the same style, which is really the asparagus style, for sugar peas.

After preparing them, put them into a steamer over boiling salted water – ½ kg (1 lb) for four people, ¾ kg (1½ lb) for six. Sprinkle them with salt and cover the pan. Leave for at least ten minutes, then try them, and give them a little longer if necessary.

Mange-tout peas can also be boiled in the pan, but use only a little water and cover the pan tightly so that the peas are mainly steaming.

When they are tender, transfer them to a hot dish. If you have time, or a helper, arrange them in some kind of order, rather than in a jumble. They should have a slightly ceremonial look appropriate to such a delicious vegetable. Serve them with a jug of melted butter, or the delicious cream and butter sauce on p. 555, or with clotted cream heated gently and seasoned with a little lemon juice.

Be sure to provide cloth napkins and little bowls of water for people to rinse their fingers – unless you are eating in the kitchen, near a tap.

CREAMED MANGE-TOUT PEAS

If you have only ½ kg (1 lb) peas and six people to feed, your best plan is to follow this or the next recipe. Make your choice according to the rest of the meal.

Cook the peas in a good centimetre (½ in) salted water, with a tight-fitting lid on the pan to prevent evaporation. When the peas are tender, spread them out in a lightly buttered gratin dish and keep them warm. With the cooking water and some milk, butter and flour, make a velouté sauce the consistency of thickish cream. Improve the flavour with a tablespoon or two or three of double cream and enough Parmesan and Cheddar or Gruyère to liven the sauce without making it taste as strong as a proper cheese sauce. Pour this over the peas. Sprinkle the top with a little more cheese and put the whole thing under the grill for a few moments to turn a golden colour. Do not overdo this, or you will spoil the peas-and-cream taste.

STIR-FRIED MANGE-TOUT PEAS

After all I have said about the delicacy of these peas, the Chinese stir-fry method may seem a shocking way to cook them. Soya sauce? Sherry?

Strong chicken stock? Desecration. So it should be, but it isn't. Remember that this dish, like many other Chinese dishes, should be eaten with rice. The strong ingredients, used by the spoonful rather than in quantity, help keep the flavour of the peas on top, prevent it being drowned by the blandness of the rice.

Prepare the rice before you start cooking the peas. Allow 175–200·g (6–7 oz), if you are serving the two dishes together as a first course or a course on their own.

> ½ kg (1 lb) mange-tout peas, prepared
> 4 tablespoons corn or groundnut oil
> 5 tablespoons strong chicken stock, or clear vegetable broth
> 3 tablespoons butter
> 1 tablespoon soya sauce
> 1 rounded teaspoon cornflour
> 1 rounded teaspoon sugar
> 2 tablespoons sherry
> ½ tablespoon hoisin sauce (if possible)
> salt

Make sure the peas are dry or they will spit dangerously in the oil. Heat the oil in a large *guo* or frying pan (for this kind of cookery thin metal pans are better than thick ones). Add the peas and stir them over a medium heat for two minutes, or a shade longer. Quickly tip in the stock or broth, butter and soya sauce and continue to stir until the peas are just tender. Bring the smaller thinner ones up to the top, so that the larger ones cook more quickly. Mix the cornflour, sugar and sherry with three tablespoons of water and the hoisin sauce if used. Mix into the peas, stirring all the time, until the sauce is a shiny coating – about one minute. Taste and add salt.

MANGE-TOUT PEAS IN THE STYLE OF CORUNNA
TIRABEQUES A LA CORUÑESA

Corunna in Galicia, the north-west corner of Spain, may be famous to us for the burial of Sir John Moore – 'Not a drum was heard, not a funeral note, As his corse to the rampart we hurried' – but if you were to ask a Spaniard he would be more likely to say scallops and oysters, then he might remember the pork and all kinds of vegetables, including the tiny potatoes known as *cachelos*. Green Celtic Galicia is the wet part of Spain, where such things flourish. The Spanish *tirabeques* for mange-tout peas refers to the way you pull the stringy part out of your mouth – *tirar* meaning to draw out or pull – as you eat the tender pod and peas inside it. A good name. This recipe was sent in to the *Fruit Trades Journal*: it is a delicious way of cooking mange-tout peas.

> *good pinch saffron*
> *¾ kg (1½ lb) pork tenderloin*
> *salt, pepper*
> *1 onion, finely chopped*
> *5–6 tablespoons cooking oil or lard*
> *⅛ bunch parsley, finely chopped*
> *150 ml (¼ pt) dry white wine*
> *375 g (12 oz) tiny potatoes, scraped or peeled*
> *¾ kg (1½ lb) mange-tout peas, prepared*

First put the saffron filaments to infuse in 225 ml (7–8 fl oz) hot water. Next cut the pork into kebab-size chunks, season them and set aside for a while. In a deep sauté pan, cook the onion in the oil or fat until it is golden-brown, then add the pork and parsley and continue to fry slowly for a further ten minutes, turning everything about. Pour in the wine, leave for five or six minutes before putting in the potatoes. Pour in the saffron and its water. The potatoes do not need to be covered completely; a little more water may be added, but keep it to a minimum. Simmer for five minutes, then put the mange-tout peas on top. From now on, do not stir the pot. Put on the lid and leave for ten minutes. Take off the lid and leave for 10–15 minutes, by which time pork and potatoes and peas should all be tender. Correct the seasoning.

MARSH SAMPHIRE or GLASSWORT

Pickled samphire is still familiar, prepared either from the true samphire (*Crithmum mariti-mum*) of seaside rocks and cliffs or from the unrelated Marsh Samphire (various annual kinds of *Salicornia*) which makes a carpet on muddy salt marshes, and which it would be less confusing to call by its other name of glasswort (it used to be calcined to alkali for glass-makers). In Norfolk and Suffolk the marsh samphire is boiled and eaten as a vegetable and regarded as a summer delicacy. You may be tempted by the fact that it belongs to the same family as beet (another seasider by origin) and spinach, though you would never guess that from the narrow shiny succulence of its bright green jointed stems.

If you manage to buy marsh samphire from an East Anglian fishmonger, rinse it well, then trim off the small wiry root from the fleshy part. Boil it until just tender – better still, steam it – and serve it with melted butter or an hollandaise sauce. John Evelyn included samphire in his salad garden (see the salad calendar he made, p. 322), although it is not clear which kind he meant. He got his seeds from France, which also gave us the name of samphire, a corruption from Saint Pierre, i.e. the herb of St Peter, from its habit of striking deep into the crevices of the rocks.

JASON HILL'S SAMPHIRE PICKLE

In *Wild Foods of Britain*, which came out in 1939, well before the present mania for eating the hedgerows, Jason Hill gave a recipe for pickling the true samphire. He liked its 'aromatic, slightly resinous taste', which the marsh samphire does not have.

Fill a litre (1¾ pt) bottling jar with samphire, adding peppercorns and some grated horseradish. Bring equal quantities of vinegar and dry cider to the boil, adding salt, and pour over the samphire to cover it. Put the open jar into the oven to infuse for an hour (at about 150°C/300°F/Gas Mark 2). Remove it, cool the jar, then fix on the rubber ring and close.

MUNG BEANS, see SOYA BEANS

MUSHROOMS

Since cultivated mushrooms are available all the year round – at a price which remains fairly low and stable – we incline to ignore the wild kinds. The cultivated mush-room is in a way more a seasoning, more an addition or embellishment, than a vegetable on its own. In fact since mushrooms differ so from the green-leaved plants and the roots, I didn't think of them as 'vegetable' at all when I was planning the book. And I had already dealt with them – wild and tame – in a separate book, *The Mushroom Feast* (1975). But leaving them out won't do.

Mushrooms of several species have long been regarded as a delicacy in Europe. Roman gentlemen sometimes indulged in a little mushroom cookery during their dinner parties. At the courts of Theodoric, King of the Ostrogoths, and of our Richard II, mushrooms were on the menu. In the 18th century there were enough cultivated around Paris to be part of the enjoyable decadence of that wicked city. It was the French who at last produced a sterilized mushroom spawn at the end of the 19th century, which enabled them to be grown reliably as a commercial crop.

Grateful though we should be that a luxury has now come within everybody's range, most of us would admit to a preference for the wild field mushroom. Which makes it all the odder that we leave many more delicious mushrooms to go to waste every year. Puffballs and parasols, shaggy caps and horse mushrooms are often ignored. So, too, are the great mushrooms of the finest European cookery, cep, morel and girolle.

HOW TO CHOOSE AND PREPARE MUSHROOMS

CULTIVATED: Avoid brownish-looking mushrooms, they have lost their pinkish-white fresh texture and flavour, which is the reason for buying them. Use them quickly, and rub them over with lemon as you prepare them, important if they are to be eaten raw in a salad or sandwich.

CEP vary from the small tight *tête de négre* to huge plate-sized caps with a meaty texture. Examine any you find very closely, slicing those that seem less than perfect. There is no need to discard the spongy tubes underneath the cap, unless they seem soggy or damaged. The stalk is good too. Cut away the earthy base and any damaged part, peel if necessary, then chop: the best way with ceps is to fry them in olive oil, with chopped shallots and parsley. Add to stews of chicken, guineafowl etc.: a few dried slices – reconstitute in warm water for 20 minutes – can make a powerful difference.

MOREL: Fresh morels are a springtime delicacy. Split them so that any earthy particles can be rinsed from the hollow centre. Their ideal partner is cream. Cook them briefly in butter, then add cream or cream sauce and complete the cooking. Serve in pastry cases or as a sauce-cum-vegetable.

GIROLLE: Trim off the earthy base, remove bits of moss etc., rinse. Girolles exude a lot of moisture; drain off the liquid as it flows out, and start the girolles off again with fresh butter (keep the liquid for sauces, soups etc). Delicious with eggs on buttered toast. Recipes using mushrooms in conjunction with other vegetables occur throughout this book as you will see from the index. Mushrooms can also be stuffed and baked. Mushrooms make excellent fritters, their passage from pan to plate is rapid. The two recipes following are special favourites of mine.

MUSHROOM SOUP

Chop and sprinkle about half a kilo (1 lb) field mushrooms, or rather less cultivated mushrooms with lemon juice. Stew a heaped tablespoon of chopped onion or shallot in 60 g (2 oz) butter with a chopped clove of garlic. When they are soft, stir in the mushrooms, then a tablespoon of flour. Stir for a couple of minutes, and pour in gradually half a litre (1 pt) of light stock or water. Grated nutmeg can also be added. When cooked, purée in a blender or *mouli-légumes*. Correct seasoning and dilution. Add 150 ml ($\frac{1}{4}$ pt) cream.

MUSHROOM PIE

Grease a pie dish generously with butter (60 g or 2 oz). Lay in closely a layer of mushrooms, gill side up, with their stalks chopped and scattered over them, and seasoning. Cover with 175 g (6 oz) cooked long grain rice, then a large onion stewed to softness in butter with 2 leaves of sage, chopped. Finally put on a layer of three shelled and halved hard-boiled eggs and dot them with a further 60 g (2 oz) butter. Cover the whole thing with shortcrust pastry, brush it with beaten egg and bake at 220°C/425°F/Gas Mark 7 for at least 30 minutes, protecting the top from over-browning with paper when necessary.

MUSTARD
AND CRESS

In their combination of freshness, crispness, sharpness and
warm flavour, mustard and cress
still hold their own against all the other seed-leaf or sprouting salads now
in fashion – fenugreek, alfalfa, triticale and the various beans.

The two plants white mustard and garden cress, were very anciently
cultivated, both of them introduced probably from the old food centres of
Western Asia, via Greece. One traveller was pleased to find the humble
but once universally eaten garden cress growing round the monastery of
St James, under Mount Ararat. The monks said that their monastery
(which wasn't so far from the fabled site of the Garden of Eden in the
Aras valley) was built where Noah had his vineyard and made the first
wine. Perhaps Noah had his garden cress as well.

Growing mustard and cress together, in child-marriage or infant-
marriage, and then eating them in their seed-leaf state first became
popular in Victorian times. Queen Victoria and Tennyson, I am sure,
nibbled their thin, crustless, triangular sandwiches of mustard and cress.
Jane Austen who died in 1817, or Nelson's Lady Hamilton who died in
1815, were probably too early for the fashion, which had the appeal of
such quick and infallible cultivation without mess. It wasn't on damp
squares of flannel or blotting paper that the early Victorians liked to grow
their mustard and cress. Their fancy was for mustard and cress 'pyramids',
which were cones of earthenware grooved to hold the seeds and porous
enough to hold the necessary moisture.

One mid-Victorian cookery writer, Mary Jewry (*Warne's Model Cookery Book*, 1868) had an eye for arranging salads of radishes and mustard
and cress. The mustard and cress were piled high in the middle of a plate
and ringed with radishes, one red, one white, in turn. This salad was for
eating at breakfast – delicious with bread and salted farm butter.

NETTLES

Young nettles in the spring, when they shoot with fierce bright leaves, are good to eat. Not as good as spinach, whatever some people may claim, but not to be despised especially at a season of the year when greenery is scarce. There was a folk belief that nettles taken in April and May purified the blood. After a winter of stodgy storeable foods, one would have welcomed the lighter flavour, and felt better for the vitamins, less stuffy, and so in a sense purified.

Wear gloves and use scissors to snip off nettle leaves, and, as Florence Irwin remarks in *Irish Country Recipes*, choose them from a field in preference to a dusty road. Wash them well, cook them in their own juices and use the well drained purée like spinach. Naturally the sting goes with heat, so that is no worry. Delicious on fried bread with a topping of poached egg or egg mollet, or brains with a creamy sauce.

There is a story about St Columba and nettle soup, that Miss Irwin quotes from Joyce's *Lives of the Saints*. One day the saint met an old woman gathering nettles near his monastery in Ireland, and he asked her why. She replied that with her cow in calf she was deprived of her usual diet of milk and white meats and was making do with nettle soup. The ascetic Columba reflected that if she could keep going on such poor nourishment, with the uncertain prospect of the cow having a healthy calf, he should be able to survive on it even better, as his hopes were for the certain prospect of heaven. His monks were dismayed at Columba's decision, but he seemed to thrive on the soup which he had ordered to be made without milk or butter. After a while Columba too became puzzled at feeling so well. Being a sensible man, he didn't assume a miracle, but sent for the cook. 'There's nothing in your soup but what might have come from the iron pot, or the wooden pot-stick I stir it with.' An evasive answer. Columba went to look at the pot. That was all right. Then he picked up the pot-stick and found that the cook had hollowed it out like a reed, so that he could fill the channel with milk and enrich the soup without anyone noticing. If you have a mind to try St Columba's soup, in either version, here is the recipe:

IRISH NETTLE POTAGE OR SOUP

> ½ litre (1 pt) water, or milk, or milk and water, or meat or vegetable
>   stock
> 30 g (1 oz) butter
> 30 g (1 oz) rolled oats
> ¼ litre (½ pt) chopped young nettles
> pepper, salt
> 1 good teaspoon chopped parsley

Bring the liquid and butter to boiling point, then stir in the oats. When the
pan returns to the boil, add the nettles and seasoning. Cover and simmer
for 30 to 45 minutes, stirring occasionally. Taste and correct the seasoning, add the parsley, and leave for another two minutes. Then serve.

IRISH NETTLE BROTH

When the broth was half done, a separate pot of potatoes would be
prepared, to be ready at the same time. The meat was taken out and cut
up, with a piece or two put into each bowl along with some of the broth
and potatoes. Everyone mashed up his bowlful to his own liking, and ate it
with a spoon.

> 1 kg (2 lb) shin or any boiling beef or lamb tied in a piece
> 2¼ litres (4 pt) water
> 1 teacup pearl barley
> bunch of spring onions
> ¼ litre (½ pt) chopped young nettles
> pepper, salt, flour

Simmer the first three ingredients for two hours. Add the greenery and
give it another hour. Finally add seasoning – this should not be done until
the beef is tender, as salt can toughen it. If you like an even thicker soup,
mix some of the broth into a tablespoon of flour, then stir this mixture
when smooth into the soup, and leave to simmer for a further 15 minutes.

OTHER SUGGESTIONS

Nettle champ, see the recipe for potato champ on page 405, and substitute
a teacupful of young chopped nettles for the spring onions.

NEW ZEALAND SPINACH

New Zealand spinach has been the occasion of
an extraordinary culinary solecism. Extraordinary because it was committed by Alexandre
Dumas, who was a great authority on food. And
again extraordinary because in the *Grand Dictionnaire de Cuisine* that he finished at the end
of his life in 1870, he never – as far as I can discover – mentioned the vegetable he had made so much of in *The Three
Musketeers*.

You may remember the amusing episode at the inn, when D'Artagnan
discovers Aramis gloomily shut in a dark room with a Jesuit and the local
priest. He is about to relapse into holy orders again. The clerics depart.
D'Artagnan is starving. All I can offer, as it's Friday, says Aramis, is fruit
and an omelette of 'tetragon'. D'Artagnan accepts ruefully, then hands
over a letter he has brought. A sudden change comes over Aramis; the
lady who seemed to have rejected him, loves him after all. The servant
arrives with the omelette. Take that horrible vegetable back to the
kitchen, shouts Aramis, we'll have hare, a capon and a leg of mutton with
garlic, and four bottles of Burgundy.

Now *tetragon* is French for New Zealand spinach, from its botanical
name of *Tetragonia expansa*. And New Zealand spinach was unknown in
Europe until 1771, well over a hundred years after the adventures of the
three musketeers. It was discovered in New Zealand by Sir Joseph Banks
in 1770, who was one of the scientists on board Captain Cook's
Endeavour on its voyage round the world. It was first grown in Kew and
was making ground as a vegetable by 1820, though it has established itself
more successfully in France than here. This is because it flourishes in dry
summers, when it is too hot for spinach.

In August, I can always pick it out on the stalls in our local market at
Montoire, its characteristically shaped leaves are so bright a green. Of the
many vegetables described as being 'like spinach', New Zealand spinach
is the closest, and you cook it in exactly the same ways.

OKRA or LADY'S FINGERS

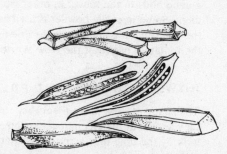

Lady's fingers, the five-sided pods of *Hibiscus esculentus*, are the most elegant of vegetables – as their name suggests. They could be classed with Chinese artichokes for beauty of form. In flavour and texture they do not come quite so high. The soft tapering green pods can be stringy if they are not as young, or as freshly gathered as they should be. Morever they exude a mucilaginous juice, like other members of the mallow family, which some people dislike, though American cooks of the south have turned this to culinary advantage.

I so not think it too far fetched to say that, as far as man is concerned, *Hibiscus esculentus* has a history of suffering and much anguish. No story here of gentlemanly plant hunters such as Dr Bretschneider who sent the Chinese artichoke from Peking to Paris, or Sir Joseph Banks who discovered New Zealand spinach on Cook's voyage in 1770 and brought it home to Kew. Forget the charming image of lady's fingers, forget the bindi of Indian curries, the bamies of East Mediterranean and Arab food. Think instead of the other names – okra and gumbo. These are American, as everyone knows who has ever looked at an American cookery book, or eaten in the restaurants of New Orleans; American now, certainly, American since the 18th and early 19th centuries, but not by origin any more than the plant itself. *Hibiscus esculentus* is a native of tropical Africa. Okra comes from *nkurama*, its name in the Twi language of the Gold Coast, gumbo from *ochinggombo* of Angola's Umbundu language. Like the plant and its first users, these names crossed the Atlantic with the slave trade, and became shortened in negro patois to more manageable forms. Indeed the negro patois itself is known as gumbo, though I do not know – dictionaries are silent on the matter – whether the two apparently identical words are connected.

At least in America the lady's finger found a perfect partner in the tomato. These two now form the basic elements of the chicken and seafood gumbos of the southern states, as they do also of Mediterrranean dishes. The word has been extended from the pod, to cover the whole stew because it is characterized by the jellied smoothness that the juice of gumbo adds to the sauce. In other words, gumbo is the defining ingre-dient. Sometimes filé powder is used as well as, or instead of gumbo, to the same end: it is made from the dried young leaves of the sassafras, like the tomato a native of the New World.

HOW TO CHOOSE AND PREPARE OKRA

The shape of okra make them easy to spot on the rare occasions when they appear at the greengrocer's. Sometimes they have a brownish stale look by the time they arrive here, that dims their natural green. In this state they should be avoided, but I can never resist them as we have so few chances of eating them fresh. Pick out the smallest, brightest looking ones.

Wash and dry the okra carefully. Cut away the conical cap from the stalk end, being sure not to expose the seeds and sticky juices inside or the okra will split and lose their shape as they cook. Some books recommend salting them and leaving them in the sun for an hour to bring out the mucilaginous juices, which can then be rinsed away (enough will be left to add the right texture to the final stew or sauce). Another way is to leave them for half or three-quarters of an hour in water acidulated with 125 ml (4 fl oz) vinegar to $\frac{1}{2}$ litre (generous $\frac{3}{4}$ pt). Whichever treatment you adopt, dry the okra before cooking them and always cook them gently so as to keep the shape of them intact.

Quantities are a little difficult. Cookery books of Mediterranean recipes suggest 1 kg (2 lb) for six people, American cookery books far less, as you will see from the gumbo recipes. I think okra are best enjoyed in smaller quantities; $\frac{1}{2}$–$\frac{3}{4}$ kg (1–$1\frac{1}{2}$ lb) is about right, according to the dish.

Canned Okra

Fresh okra are not sold in every market, nor are they commonly grown in vegetable gardens. At least they are imported from Greece in cans, as okra or bamia, so that no one need go without some experience of their unusual look and taste. The medium-sized can, 375 g or 13 oz, is about right for four people; and for the American gumbo recipes. For the Mediterranean stews, for vegetable dishes for more than four people, you will need to buy the larger size.

Canned okra can be used in all the recipes and will stand up to half an hour's cooking without losing their character, though obviously this depends on keeping the pot at simmering point rather than a rolling boil. The normal preparation consists in rinsing the okra under the tap to get rid of most of the mucilaginous juice; the amount of rinsing they require should be adjusted according to the recipe you intend them for. With a gumbo or other meat stew you need the juice for the correct texture. Of course a fresh vegetable is usually superior to the canned version, but okra do stand up to the processing well. They are an excellent standby for the store cupboard.

Here is a simple way of presenting them, using other store cupboard ingredients:

>*1 medium sized can okra*
>*1 medium onion, chopped*
>*1 large clove garlic, finely chopped*
>*olive oil or butter*
>*2 generous tablespoons tomato concentrate*
>*salt, pepper, sugar, pinch dried rigani or oregano*

Rinse the okra. Soften the onion and garlic in oil or butter in a frying pan. Put in the okra in a single layer. Dissolve the tomato concentrate in 300 ml (½ pt) water, and add. The okra should not be covered. Add seasoning and the herb. Simmer for 20 minutes, allowing the liquid to reduce to a sauce. Taste and correct the seasoning.

OKRA IN THE GREEK STYLE
BAMIES MEH DOMATES

>*250 g (8 oz) chopped onion, or a mixture of onion, spring onion*
>   *and leeks*
>*1 clove garlic, chopped*
>*3–4 tablespoons olive oil or butter*
>*½ kg (1 lb) okra, prepared*
>*½ kg (1 lb) tomatoes, skinned, chopped*
>*juice of 1 lemon*
>*salt, pepper, sugar*
>*chopped parsley*

Cook the onion and garlic slowly in oil or butter for about five minutes until they soften. Add the okra and raise the heat a little so that they brown to a light golden colour. Turn them about gently and do them in batches if you are pushed for space in the pan. When they are ready, spread them out in the pan in a close layer and put the tomatoes on top with the lemon juice and seasoning. Add a little water to prevent them catching before the tomatoes yield their moisture. Cook for about 20 minutes or longer, until the okra are tender and the sauce much reduced. Do not stir them about but shake the pan occasionally to prevent sticking (a non-stick pan is a good idea). If the whole thing becomes too liquid, as it can with English tomatoes, remove the cooked okra with a slotted spoon to a warm serving dish and boil the sauce down hard. Check the seasoning and serve hot, tepid or cold, scattered with parsley. Serve bread to mop up the juices.

CYPRIOT LAMB STEW
YAHNI

> 6 loin chops, or 12 cutlets, or 1 kg (2 lb) boned shoulder lamb, cubed
> 1 large onion, chopped
> olive oil
> ½ litre (¾ pt) tomato juice, or ½ litre (¾ pt) water with 2 good tablespoons tomato concentrate, or ½ kg (1 lb) skinned, chopped tomatoes
> salt, pepper, sugar, ½ teaspoon dried rigani or oregano
> up to 1 kg (2 lb) okra, prepared
> 6 medium potatoes
> juice of lemon

Fry the lamb and onion slowly in the oil, until the onion softens. The meat will not brown, but turn it over so that it is lightly sealed by the frying. Pour in the tomato juice, or water and concentrate, or chopped tomatoes with their juice. Add water if need be, to cover the meat. Season and add the herbs. Cover and simmer as slowly as possible, until the meat is almost tender. Time will vary a good deal, according to the cut of lamb you use. Test after one hour, allow up to three hours.

Meanwhile brown the okra lightly in oil. Peel and cube the potatoes, then sprinkle them with lemon juice to prevent discoloration. Add to the stew when the lamb is almost done and leave for another 45–60 minutes. If the sauce seems too watery, remove the lid, raise the heat and allow it to reduce. Skim any surface fat from the stew, or blot it away with kitchen paper. Check the seasoning. Serve with rice.

CHILLED GUMBO BISQUE

The gumbo stews of the southern states of America are often given their defining character by okra. Here is a delicious soup in which they may seem subdued: nonetheless they are essential to the satin smoothness and unusual flavour of the soup. This recipe is an anglicized version of a bisque devised by a New York friend to use up a can of crab claws. I never find canned shellfish satisfactory – it is almost as tasteless as the frozen kind – and prefer to use fresh prawns in their shells, or fresh crab claws which are sometimes sold separately. The important thing is to have some hard debris to flavour the stock, as well as soft meat for finishing the soup.

> 300–350 g (10–12 oz) prawns in their shells, or fresh boiled crab claws
> 1 litre (1¾ pt) fish or chicken stock, plus ½ litre (¾ pt) water or 1 kg (2 lb) fish trimmings plus 1½ litres (2½ pt) water and ¼ litre (8 fl oz) dry white wine or cider
> 100 g (3½ oz) chopped celery
> 100 g (3½ oz) chopped onion
> ½ green pepper, chopped
> 2–3 tablespoons butter
> 250 g (8 oz) okra
> 1 medium can tomatoes (approx. 400g, 14 oz)
> 100 g (3½ oz) rice
> salt, pepper, cayenne

Shell the prawns or crab claws. Put the debris into a large pan. Set aside the meat. To the pan, add the stock and water, or fish trimmings, water and wine or cider. Simmer for 45 minutes to extract the flavours, then strain into a measuring jug and add water to make 1½ litres (2½ pt). Meanwhile soften the celery, onion and pepper in the butter. Prepare and cut the okra into slices 1 cm (½ in) thick. Add them with the stock, tomatoes and rice to the vegetables. Season. Cover and simmer for an hour. Purée in the blender, dilute further if you like, then chill overnight or for at least four hours. Serve with some or all of the prawn or crab meat.

GUMBO

The difference between this recipe and the Mediterranean type of stew is the inclusion of peppers and chillis or cayenne, and a substantial piece of country ham (substitute smoked gammon or bacon, bought in a piece).

250 g (8 oz) country ham, cubed
175 g (6 oz) chopped onion
175 g (6 oz) chopped celery
1 red or green sweet pepper, chopped, minus seeds
lard
up to 1 tablespoon tomato concentrate
½ kg (1 lb) tomatoes, peeled, chopped
½ kg (1 lb) okra, trimmed
salt, pepper
cayenne, or 1 dried chilli, chopped, with seeds

Stew the ham, onion, celery and sweet pepper in a little lard. When they are beginning to soften, raise the heat so that they colour lightly. Put in 1 teaspoon of tomato concentrate and the tomatoes. Stir for a moment or two, until the tomatoes begin to subside into a stew, then put in the okra and seasonings. Cover and simmer slowly until cooked, checking from time to time in case a little water is required. The okra should not be left high and dry, neither should they be completely covered with liquid. Taste and adjust the seasonings, adding extra tomato concentrate if it seems a good idea.

CHICKEN GUMBO

250 g (8 oz) gammon rasher, cubed
1 farm chicken, jointed
125 g (4 oz) chopped onion
1 clove garlic, chopped
1 red pepper, chopped, minus seeds, or 1 dried chilli chopped, with seeds
375 g (12 oz) okra, trimmed, sliced
lard and any fat from the chicken above
1 tablespoon flour
250 g (8 oz) chopped tomato
1 tablespoon tomato concentrate
chicken stock or water
bouquet garni
salt, pepper, cayenne or Tabasco sauce
1–2 dozen oysters (optional)
parsley

Brown the gammon and chicken, onion, garlic, red pepper or chilli and okra in the lard and chicken fat. You will have to do this in batches, transferring each item as it colours to a large pot and adding more lard as necessary; start with the meat and colour it over a sharpish heat, then lower the temperature for the vegetables, so that they soften and do not become too brown. When the last batch is ready, stir in the flour, cook for a couple of minutes, then add the tomato, concentrate and enough stock or water to make a slightly thickened sauce. Tip this over the contents in the large pot, adding extra liquid if need be, barely to cover the meat and vegetables. Put in the bouquet and seasoning (if you use chilli rather than red peppers, go lightly with the cayenne or Tabasco). Simmer, with the pot covered, until the chicken is tender – about one hour or longer. Meanwhile open the oysters, being careful to save all their juice. Ten minutes before serving the gumbo, mix in the oysters and their liquor to heat through. Taste and adjust the seasoning. Remove the bouquet, and add a good chopping of parsley. Serve with boiled rice.

If you cannot afford oysters, use mussels instead. This is better than having no shellfish flavour at all.

ONION, SHALLOT and SPRING ONION

Somewhere, aeons ago, perhaps in the north Asiatic homeland of the Indoeuropeans, men started to grow the onion – started to grow and diversify and improve some unidentified wild species – and turn it into the essential vegetable which could be nibbled green and white as spring onions, or eaten on its own, boiled or baked and stuck with cloves or added to dishes by the score as a combination of lubricant and flavouring. Of course one great merit of onions was that they could be dried off and hung up in strings to keep through winter and spring.

Since then no civilization, or should I say no developed cuisine, has been able to do without onions. I opened a book of Senegalese cooking, from Dakar. There was hardly a recipe in it without onions, '2 gros oignons, 1 très gros oignon, 250 grammes d'oignons' – in fish and meat dishes with such names as Gniry-bouna or Caldou or Lakh-lalo; and I remembered, from half a thousand years ago, Chaucer's court officer, his summoner, pink-faced, lecherous as a sparrow, who so loved 'garleek, oynons and eek lekes', with wine which was strong and 'reed as blood'. No doubt officers of the Old Bailey, tipstaffs, tuck into bread and cheese and pickled onions in their local (cheese with onions was already a Greek and Roman habit).

HOW TO BUY AND PREPARE ONIONS

Rather than buy onions every week, lay in a couple of strings to see you through the winter. Hang them up in as dry a place as possible and they will last through until February or March. When you want a specially sweet onion for baking or slicing into salads, or small onions for glazing, you will have to make a special purchase of Spanish or pickling onions, but at least basic supplies are assured.

The tear-inducing properties of onion vary and can be mitigated either by skinning the onion under a cold tap, or by sitting down well back from the onion, so that your face isn't hanging over it.

The easiest way to chop an onion, after pulling the skin downwards and cutting off the root, is to halve it down the middle. Place the halves, cut side down on the board. Make several cuts from top to tail of each half, quite close together and vertical to the board. Then slice horizontally across twice or three times, starting from the top end; finally cut downwards at right angles to the first cuts you made, and there you are.

Remember when you are putting whole onions into stock, that their brown skin adds a beautiful golden colour so leave it on.

Two frugal points – when you extract the onion from the bread sauce, rinse it and keep it complete with cloves for adding to the carcase stock. And when your onions sprout towards the end of the winter, use the green shoots as if they were chives.

WHITE SOUBISE

This soothing onion soup comes from a booklet of 1912, sold in aid of the Marlow Cottage Hospital. Its ingredients are simple; the unusual part is the bread thickening, which gives a slightly grainy texture. It's almost a bread sauce soup.

> ½ kg (1 lb) onions, chopped
> 60 g (2 oz) butter
> 300 ml (½ pt) light stock or water
> 300 ml (½ pt) milk
> 60 g (2 oz) white breadcrumbs
> salt, pepper

Stew the onion in butter in a covered pan, without browning it, for an hour. Keep the heat really low, or use an asbestos mat. The idea is that the onion should gently dissolve into a purée. Mix in the stock or water, the milk and the breadcrumbs. Raise the heat and bring just to boiling point. Sieve through the medium plate of the *mouli-légumes* (a blender would make the purée too smooth). Season. Return to the heat to simmer for a further hour; again keep the mixture barely at simmering point. Check from time to time and give the soup a stir up. Add more water if it becomes too thick. Serve with cubes of bread fried in butter.

This is an ideal soup to make if you have a solid fuel or oil-fired stove that is on all the time. The pan can be moved to the edge for the long slow cooking, then put on the hotter side of the stove that has been protected by the second lid.

SAUCE AND PURÉE SOUBISE

Soubise, as you will have gathered, means onions. The Hôtel Soubise in the Marais district of Paris, that once belonged to the great Soubise family, now houses the French archives. It is a building of elegance, beautiful statues and colonnade but not an onion dome in sight.

The sauce has a particularly delicate flavour. Chop a large onion. Put it into a pan and cover it with boiling water. Cook for three minutes, then drain and return it to the pan with a tablespoon of butter. Cover the pan and stew the onion, without colouring it, until soft. Pour in 300 ml (½ pt) béchamel sauce and simmer for 15 minutes. Sieve or blend to make a smooth sauce. Stir in four or five tablespoons of cream – the better the cream, the better the flavour – and heat through. Add salt. Serve with veal, lamb, chicken or fish.

The purée is more oniony, with a knobbly smoothness that comes from the rice. Blanch ¾ kg (1½ lb) chopped onions as above, drain and put with a generous tablespoon of butter into the pan. Add 150 g (5 oz) Italian long grain rice and stew gently for a few minutes, stirring. Add ¾ litre (1¼ pt) chicken or veal stock, salt, pepper and a pinch of sugar. Simmer uncovered until the rice is very tender – this is why Italian rice is best – and then sieve. Reheat with two tablespoons of butter and six tablespoons of double cream. Correct the seasoning. The purée should be very thick. Serve with roast or grilled meat – it is especially good with lamb chops and fried mushrooms.

You can make a good vegetable dish by mixing cooked chestnuts – stew them in stock – into either sauce or purée and surrounding the whole thing with Brussels sprouts and mushrooms. Or you could put the chestnuts and purée into baked tartlet cases.

GLAZED OR CARAMELIZED ONIONS

This is one of the most delicious ways of cooking onions when they are to be eaten with beef, lamb, venison or pigeon. Glazed onions with glazed carrots form part of several classic garnishes – *bourguignonne* means with glazed onions, small fried mushrooms and bacon dice, *nivernaise* glazed onions, carrots and turnips, braised lettuce and boiled potatoes, *bourgeoise* glazed onions, carrots and bacon dice. The *bourguignonne* garnish is an excellent way of cheering up a stew, especially if you add small pieces of fried bread as well. The *nivernaise* and *bourgeoise* garnishes are for larger pieces of meat. They are arranged round it on the serving dish and you need no other vegetables, though a green salad is usually served afterwards.

Glazed onions are never served in large quantity on account of their rich sweetness – allow three or four per person according to their size and the rest of the dish. Pickling onions do very nicely when cooked this way, though you can quite well use small onions so long as they are of an even size.

Peel enough onions to make a single layer in the bottom of a large pan. Cover them with water. Add a pinch of salt, a heaped tablespoon of butter and a level tablespoon of sugar. Put the lid on the pan and bring to a vigorous boil. Remove the lid and allow the liquid to reduce until it ls no more than a rich brown syrup in the base of the pan. As this stage comes near, turn and shake the onions so that they brown as evenly as possible. Naturally you must keep an eye on them, to make sure they do not burn. Sprinkle on a little more salt and plenty of pepper.

ROASTED ONIONS

'An honest laborious country-man,' wrote Evelyn, 'with good bread, salt, and a little parsley, will make a contented meal with a roasted onion.'

I imagine that Evelyn's country-man speared his onion and turned it in front of the fire until the outside was dark and crackling, the inside tender. Nowadays we roast onions differently. The best thing is to peel and blanch them for a good five minutes in boiling salted water, then to set them in a pan of lard at the top of the oven until they are brown. Remember to turn them in the fat from time to time. They can be cooked with potatoes and parsnips, above the joint of roasting meat, or on their own.

Another way, when meat is to be served, is to arrange the blanched onions around the joint so that they finish cooking in the juices. Turn them from time to time so that they cook and colour evenly. Allow 30–45 minutes, according to the size of the onions and the temperature of the oven.

STUFFED ONIONS

The best way of cooking Spanish onions, a way that shows off their huge size and mild flavour. Choose six of more or less equal weight, about 350 g

(¾ lb) each, and examine the base of each one to make sure it is undamaged and able to hold the onion together.

> 6 Spanish onions
> butter or beef dripping
> breadcrumbs
> beef or veal stock

STUFFING
> 350 g (¾ lb) minced lean chuck steak or rump
> 100 g (3–4 oz) white breadcrumbs
> 100 g (3–4 oz) fat streaky bacon
> 2 teaspoons tomato concentrate
> 1 scant level teaspoon each dried thyme and oregano
> 6 tablespoons béchamel sauce, or mornay, mushroom, velouté
>   sauce
> 1 tablespoon grated Parmesan
> salt, pepper, cayenne

Cut odd roots from the onions, then remove the brown skin carefully so as not to pull away the base. Cook the onions, base down, in a large pan with plenty of boiling, salted water, for 50–60 minutes. Test them by pushing a skewer or larding needle into the centre of one of the onions. It should not be completely tender, but nearly so. Remove them carefully from the water, trying not to dislodge the brilliant yellow outer skins, and run them under the cold tap. Drain upside down. Now take a sharp pointed knife and cut in a circle round the stem end. Remove the circle and with a pointed spoon empty out the centre of each onion; the trick is to remove the centre oval bulb, then the outer skins come away easily. The difficult part is to remove as much as you can, without the onion collapsing. Do not despair, though, if your first one is not perfect. It can be re-formed round the stuffing: just make sure you know which one it is, so that you can take extra care when serving it (use two tablespoons, or slide a palette knife carefully underneath).

Chop the inner part of the onions, draining all surplus liquor away. Put half in a basin, season it, then add the stuffing ingredients, seasoning to taste. The flour-based sauce binds and softens the mixture; use a little more if you like. Fill the onions, doming the stuffing up like a top-knot. Range them in an ovenproof baking dish, in which you have melted four or five tablespoons of butter or dripping. Cover the top-knots of stuffing with the remaining chopped onion, scatter a few breadcrumbs on top of that, and put into the oven at 200°C/400°F/Gas Mark 6 for 45–60 minutes. Baste the onions with the juices from time to time. If they seem to be drying up too much, pour in a few spoonfuls of beef or veal stock.

Stuffed onions are sometimes served with a sauce – tomato, for instance – but this is not necessary.

VARIATIONS Substitute one or another of the stuffings from pp. 562–567.

FRIED ONION RINGS

The first time I ate crisp onion rings was at the Frying Pan restaurant opposite the BBC in Bristol. After an evening broadcast – this was 25 years ago, when programmes went out live – everyone would troop across to relax and restore themselves with grilled steak or chops and these splendid onion rings. They were a speciality of the Frying Pan in those days. Now they are a popular dish everywhere.

You do not have to eat them only with meat. They go well with vegetable dishes, too – the tomato and oatmeal tart on p. 516 for instance, or with gratins and creamed root vegetables. Nor do you have to make them in huge quantity – though whatever quantity you care to provide will always vanish – as they are more of a relish than a vegetable. Their crispness livens more solid fare. Two huge Spanish onions should be enough for six people.

Peel the onions. Slice a centimetre ($\frac{1}{2}$ in) from each end and set aside for another occasion. Slice the central part of the onions about 3 mm ($\frac{1}{8}$ in) thick, then push out the rings. Remove the smallest core part and put it with the ends. Now soak the rings in salted milk for about 20 minutes. Drain a few of them and shake them in a bag of seasoned flour. Deep-fry them until they are crisp and brown, then spread them out on absorbent kitchen paper and keep them warm while you fry the next batches.

Serve piled up on a warm plate, or as a garnish for meat or vegetables.

Note The onion rings can also be dipped in batter before they are fried (no need to soak them in milk first). This makes them heavier, more filling altogether, but they are good all the same.

PISSALADIÈRE

This is the onion and tomato tart of Nice, which has a family relationship to Italian *pizza*. The similar names are a coincidence – *pizza* means pie and *pissaladière* comes from *pissalat*, a Niçois form of anchovy sauce.

First you need to make up half a litre (good ¾ pt) of tomato sauce – according to the season and the tomatoes available, follow the fresh tomato sauce on page 512, or the Italian tomato sauce on page 511.

Next stew ¾–1 kg (1½–2 lb) sliced onions in a little salted water until tender. Drain and mix with the sauce

Line a Swiss roll tin with shortcrust pastry. Spread the onion and tomato mixture over it evenly. Lattice the top with split anchovy fillets (you will need two tins) and dot with stoned black olives. Brush over with olive oil. Bake at 220°C/425°F/Gas Mark 7 for 25–35 minutes, until the pastry is crisp. Serve hot, tepid or cold. Wonderful picnic food.

Note Instead of shortcrust pastry, you can use thinly rolled bread dough. Allow the dough to rise in the normal way until it doubles in bulk. Then roll it out and spread with the onion mixture. Prove for 20 minutes, then bake at 230°C/450°F/Gas Mark 8 for about 25 minutes. But keep an eye on it.

SMALL ONION TARTS
TARTES À L'OIGNON

These small tarts have a rich flavour from the cheese and the beef sauce that is used to coat the onion lightly. They are quite unlike the traditional type of onion tart that is set solid with egg custard.

> *6 tartlet cases baked blind, about 10 cm (4 in) in diameter*
>
> FILLING
> *8 medium onions, sliced*
> *125 g (4 oz) butter*
> *salt, pepper*
> *1 heaped tablespoon flour*
> *½ litre (¾ pt) beef stock*
> *3 heaped teaspoons grated Cheddar*
> *1–2 heaped tablespoons grated Parmesan*

Keep the tartlet cases warm, or reheat them, while you make the filling.

Cook the onions very slowly in half the butter until they soften, then raise the heat a little so that they turn an even, rich brown. Season. Make the sauce by melting the remaining butter and stirring in the flour when the butter turns a golden-brown *noisette* colour; gradually add the stock and simmer steadily until you have 300 ml (½ pt) rich brown sauce. Mix in the onions, check the seasoning. This filling should be a tumble of moist-looking onion held together lightly with sauce – if there seems to be a surplus of sauce, spoon it into a little pot for later use with another dish. Divide between the pastry cases. Mix the cheeses together and sprinkle over the filling. Place in a very hot oven or under the grill so that the cheese melts and colours slightly.

ALSATIAN ONION TART
TARTE AUX OIGNONS À L'ALSACIENNE

A rich and filling tart, best eaten at midday if there is a tender stomach in the family. Another way out is to blanch the onions in water for seven to ten minutes until they are soft, rather than cook them in lard. This gives a paler taste.

> 1 kg (2 lb) onions, chopped
> 125 g (4 oz) lard
> 250 ml (8 fl oz) single cream
> 3 eggs
> salt, pepper, grated nutmeg
> 60 g (2 oz) smoked bacon cut into thin strips
> 23–28 cm (9–11 in) tart tin lined with shortcrust pastry

Cook the onions slowly in the lard for 15 minutes, until they are golden but not browned at all. Stir occasionally. Remove from the heat and drain the onions of their cooking juices (keep them for another dish). Mix the cream with the eggs, seasoning and bacon. Add the onions and put into the pastry case.

Bake at 220°C/425°F/Gas Mark 7 for 15 minutes, then lower the heat to 180°C/350°F/Gas Mark 4 until the filling is set. Be careful not to overcook; check after 15 minutes at the lower temperature – if the filling seems liquid under the crust in the centre, put it back for another five or ten minutes.

TIAN OF ONIONS

A *tian* is a provençal gratin dish, a shallow round affair, wider at the top than the base so that you get the maximum amount of brown crust on top of whatever is cooked in it, usually vegetables. The food takes the name of the dish, rather as a meat stew is called a 'casserole'.

> 6 huge, or 12 medium onions
> 1½ tablespoons olive oil
> ½ clove garlic, uncrushed
> 2½ cm (1 in) piece of carrot, chopped
> ½ bay leaf
> sprig thyme
> 1½ tablespoons flour
> 300 ml (½ pt) milk
> 150 ml (¼ pt) cream, double or single
> 1 egg, beaten
> salt, pepper
> breadcrumbs
> extra olive oil

Peel, slice and blanch the onions in boiling, salted water until they are almost cooked, but still a little crisp. Drain well in a colander. Meanwhile make the sauce – put oil, garlic, carrot and herbs in a pan. Set over a low heat to simmer for a moment or two, then stir in the flour to take up the fat. Moisten with the milk; let the sauce boil down a little, then add the cream. Aim for a thick but not gluey sauce.

Off the heat, whisk in the egg rapidly. Season. Strain just under half the sauce into an oiled, gratin dish. Spread the onions on top, then cover them with the rest of the sauce, straining it too. Sprinkle with an even layer of crumbs and dribble a little olive oil over them. Bake at 160–180°C/325–350°F/Gas Mark 3–4 for half an hour or a little longer. If the top isn't as golden-brown as you would like, put it under the grill. The olive oil gives a delicate and unusual flavour.

ORIENTAL ONION AND CARROT SALAD
OIGNONS ET CAROTTES À L'ORIENTALE

This rich and savoury adaptation of Escoffier's *oignons à l'orientale* can be served on its own as a first course, or as part of a mixed hors d'oeuvre or cold table for a party. The sweet-sour flavours are unusual and good. Choose small onions or large pickling onions – tiny pearl onions will disintegrate in the cooking. Peel them carefully so as to leave intact the firm base that holds the onion together. Carrots should be in the middle range, with plenty of flavour, not woody and not so tender that they will dissolve into the sauce.

350 g (¾ lb) carrots, diced
olive oil
¾ kg (1½ lb) small whole onions, peeled
600 ml (1 pt) water
150 ml (¼ pt) white wine vinegar
150 ml (¼ pt) dry white wine
150 g (5 oz) sultanas
150 g (5 oz) sugar
6 tablespoons chopped tomato
1 tablespoon tomato concentrate
pinch cayenne
1 level teaspoon salt
freshly ground black pepper
bouquet garni
2–3 bay leaves
chopped parsley

Put the carrots into a large heavy pan with just enough olive oil to coat the base. Stew slowly, with the lid on the pan, until the carrots begin to soften. Then remove the lid, so that they turn a light golden colour. Add the remaining ingredients. Cook, uncovered, until onions are done, and the sauce reduced by two-thirds to a dark, thick consistency. This can take 45 minutes or more – do not hurry the cooking. Correct the seasoning to your taste – more pepper, a drop more vinegar and so on. Remove the bouquet and chill. Serve in a shallow dish, with two or three small fresh bay leaves on top, a sprinkling of parsley and a tablespoon of olive oil.

SHALLOTS

I used to be worried a bit by the way Tennyson called his fairy-woman the Lady of Shalott, when shallots were undoubtedly growing in neat rows in the kitchen-garden of the Tennysons at Somersby Rectory. Did anyone ever tease him and say that no one visited her because (see below) she had been eating too many spring onions? Tennyson's shalott came from an Italian story about a Damosel of Scalot who died at Camelot for love of Lancelot. Kitchen shallots, which are an ancient clustering form of the common onion, derive their name if not their existence from Ascalon in Palestine, which is still a very good place for growing onions. Greeks and Romans both called the shallot the Ascalonian onion. We have been eating shallots since the Middle Ages – perhaps they came with the wine from Bordeaux – and the point of them is that they provide an essentially onion flavour, stronger than the bland sweet flavour of the large onions, but without quite such a powerful smell. And since shallots are small, it is easy to apportion the flavour to steaks, liver, stews and so on.

One thing to realize is that the flavour of shallots can vary from region to region and according to the variety. There are three main kinds – the smallish grey-brown skinned shallot, the kind the French call the *échalote grise*, then the pinkish red ones, and the bronze-skinned yellowish variety which keep well – these the French call *cuisse de poulet* because they look like a chicken drumstick cooked to a shiny golden-brown. The great centre for shallots is Bordeaux, where you have the fine culinary partnership of shallots and claret in sauces to go with steak. The idea of a reduction of shallots and wine, or even wine vinegar, to make a flavouring purée for a sauce, worked its way northwards to form the basis of the *beurre blanc* of Nantes, the Bercy and *marchand de vin* sauces of Paris, and sauce béarnaise. You might object that the Béarn is on the south-west border of France, well to the south of Bordeaux. True, but the sauce was not invented there. It was first made in the north of France at Saint-Germaine near Paris, with the very northern ingredients of butter and shallots, in the kitchen of the Pavillon Henri-Quatre, which is still an hôtel. Henri IV was from the Béarn, from the royal family of Navarre, and the sauce was named in his honour.

The most outspoken expression of Bordeaux, claret and shallots, is something of a surprise to English people. The first time we came across it, we were having lunch with a friend in Touraine, who had a tiny farm and vineyard. In our honour he had bought a bottle of claret and a large sirloin steak (normally most of his meals were supplied from his own fields, animals and cellar). While the steak grilled, he chopped up handfuls of shallots, and put some evenly over a hot metal serving dish. The steak was seasoned and placed on top. The rest of the shallots were scattered over the steak. Weren't they going to be cooked? We looked at each other. No they were not. '*Entrecôte bordelaise*, as it should be,' said Maurice, as he put the dish on the table and divided the steak and shallots between us. After the first cautious mouthful, we found the mixture of crunchy shallots, rare steak and the meat juices delicious, especially with the claret. I can recommend this very simple dish. Nonetheless the shallot sauces are good, too, and not well enough known here which is odd in view of our fondness for beef, and – in the past at least – our fondness for shallots.

Incidentally, never brown shallots if you want to use them instead of onions. Frying makes them bitter. Stew them in the fat slowly, or add them in the form of a purée boiled down with wine or wine vinegar, or whole.

SAUCE BORDELAISE

A sauce to serve above all with grilled steak, an entrecôte for instance or sirloin steak.

> *100 g (3–4 oz) butter*
> *50 g (scant 2 oz) chopped carrot*
> *50 g (scant 2 oz) chopped onion*
> *1 rounded tablespoon flour*
> *600 ml (1 pt) light veal or beef stock*
> *bouquet garni*
> *60 g (2 oz) chopped shallots*
> *300 ml (½ pt) red wine*
> *75 g (2½ oz) beef marrow*

Melt a third of the butter. Put in the carrot and onion and leave to brown. Stir in the flour and cook for a minute or two, to make a brown roux. Add stock slowly, put in the bouquet and simmer steadily, uncovered, for 30–40 minutes. Strain into a clean pan, bring to simmering point and skim.

In another pan, boil the shallots and wine until they are reduced to about a teacupful of moist purée. Tip it into the skimmed brown sauce, and simmer 15 more minutes, skimming. You should end up with 400 ml (a scant ¾ pt).

While the sauce is finishing, slice or cube the beef marrow and poach for a few minutes in salted water. Slices can be put on the steak, or the dice can be mixed into the sauce.

Last of all, whisk the remaining butter into the sauce and serve.

SAUCE MARCHAND DE VIN

The wine-merchant's sauce made with a good red wine, preferably a claret. Again, it is for beef.

Reduce four finely chopped shallots to a moist purée with a large glass of red wine, until the liquid has gone down by two-thirds. Keep it warm while you grill the meat and put it on a serving dish.

Reheat and season the shallot purée, stir in a heaped teaspoon of chopped parsley, a squeeze of lemon and four generous tablespoons of butter. The creamy sauce should be poured over the steaks immediately. Overheating or waiting around will liquify it.

SAUCE BERCY

Make the *marchand de vin* sauce with a dry white wine, preferably from the Loire. In spite of the red wine with beef convention, this is a favourite sauce for grilled beef and you drink whichever white wine you used in the sauce – a Sancerre if you are lucky, Quincy if you are luckier still.

SHALLOT SAUCE
SAUCE À L'ÉCHALOTE

Reduce six finely chopped shallots to a purée with seven tablespoons white wine vinegar, as in the sauces above. Add 300 ml (½ pt) good beef stock and simmer for an hour; add a little more stock if it reduces much below its original level. Thicken with *beurre manié* (page 552). Finish with chopped parsley, salt and pepper.

SHALLOT BUTTER
BEURRE À L'ÉCHALOTE

Peel and weigh four shallots. Blanch them for 20 seconds in boiling water, then cool and chop them as finely as you can manage. Mash into twice their original weight of butter, until you have a homogeneous paste. Add salt and pepper. Form into a roll and chill briefly. Cut into slices and serve on top of grilled steaks or lamb chops.

RED MULLET IN THE STYLE OF NANTES
ROUGET-BARBETS À LA NANTAISE

This recipe gives a version of *beurre blanc* that cannot fail. The cream binds with the butter into a rich smoothness, that prevents it melting to oil. When you buy the mullet, be sure they still contain the livers, a great delicacy.

> 6 red mullet, scaled, cleaned
> salt, pepper, oil
> 4 chopped shallots
> 150 ml (¼ pt) Muscadet or dry white wine

3 tablespoons double cream
125 g (4 oz) soft lightly salted butter
squeeze of lemon juice

Season the mullet inside and check that the livers are in place. Arrange them on an oiled grill rack. Brush them with oil and season the outside. Switch the grill on to heat up.

Reduce the shallots and wine until there is no visible sign of liquid and the shallots are soft and moist.

Grill the fish on both sides. Remove the livers and chop them. Keep the fish warm on a serving plate.

Add the cream to the shallots and bring to a vigorous boil. Add the butter bit by bit until you have a rich amalgamation of shallots, cream and butter. If it shows signs of oiling, stir in a tablespoon of very cold water off the heat and mix vigorously. Season, adding a squeeze of lemon to bring out the flavour, and mix in the chopped livers. Reheat, stirring all the time, pour over the fish and serve with bread, and Muscadet to drink.

Note Shallots are essential to a number of fish stews and sauces of northern France – *moules marinière, mouclades, migourée* (for recipes see my *Fish Cookery*) and Boulogne mussel and potato salad (p. 407).

HOW TO IMPROVE A CHEAP BEEF STEW

Make up a shin of beef stew in the usual way, browning an onion and carrot in bacon fat or lard with the meat, stirring in flour and stock and so on. Also put in several small rolls of pork skin (the butcher will often give you the skin free); they much improve the texture of the sauce. Belly of pork has the same effect, but costs more.

An hour before the end of cooking time, add two tablespoons wine or sherry vinegar, two teaspoons of sugar and 12–18 peeled, whole shallots. Vinegar and sugar compensate for a lack of wine, the shallots add a melting and delicious savouriness.

At the very end, the liquid can be strained off and reduced to a rich sauce and then poured over the meat, pork skin rolls and shallots. Surround with boiled potatoes, or triangles of fried bread, scatter with parsley and you will have a dish that anyone would enjoy.

SPRING ONIONS

For salad, 'green onions' are generally raised from seeds of the common onion (especially of the kind known as White Lisbon) or as side-shoots of the shallots, blanched before pulling. I like the White Lisbons of the English greengrocery or market stall, leaf shoot and immature bulb together, which you can eat from end to end, without waste.

Spring onions are not much used in European cookery, although the Irish put them into nettle broth and champ (p. 333 & 405). You might try them in this delicious tart from the west of France. They work best when the bulb is growing large into a decidedly onion roundness. Line a 25 cm (10 in) tart tin with shortcrust pastry and cover the base with a layer of chopped large spring onions sweated in butter for about five minutes. Mix 250 g (8 oz) Jockey *fromage frais*, or half curd cheese and half yoghurt, with four egg yolks, three tablespoons of double cream, salt and pepper. Whisk the four egg whites until they are stiff, fold them into the cheese mixture and spread over the spring onions. Bake at 200°C/400°F/Gas Mark 6 for 25–30 minutes in a hot oven.

If you find spring onions mentioned in Chinese or Japanese recipes, the onion you should use – if you can get it or if you happen to grow it in your garden – is of the different species we call the Welsh onion (*Allium fistulosum*). Germans and others call it the winter onion, the French *ciboule*. This is the onion most liked and cultivated in the Far East, where it was developed from another wild species now unknown.

It is also the onion of the best onion poem I know, to set against *The Lady of Shalott*:

From Juan Fang the Hermit on an Autumn Day, Thirty Bundles of Winter Onions

Potherbs in the autumn garden round the house
Of my friend the hermit behind his rough-cut
Timber gate. I never wrote and asked him for them
But he's sent this basket full of Winter Onions, still
Damp with dew. Delicately grass-green bundles,
White jade small bulbs.
Chill threatens an old man's innards,
These will warm and comfort me.

By Tu Fu, written in 759 A.D.

The French chop these onions and cook them very lightly in butter, then add them to a clear chicken soup with vermicelli. Warm and comforting indeed on a cold day, more so than chopping them into a salad like spring onions or chives.

For the Chinese, the rule is onions with beef, garlic with pork; which more or less fits our own practice – if you take onions to include shallots and spring and Welsh onions.

ORACHE
or MOUNTAIN
SPINACH

Orache, an ancient pot-herb of Asia and temperate Europe, was introduced into Western Europe in the Middle Ages. Its green, yellow or red leaves, according to variety, can be cooked like spinach, though you are more likely to find them colouring a flower border than a kitchen garden or allotment. Seedsmen class orache with the flowers, HA hardy annual, under its botanical name *Atriplex hortensis*.

Orache is one of the vegetables Rabelais was concerned with in the 1530s when he was sending his friend and patron, Geoffroy d'Estissac, abbot of Maillezais, seeds for the splendid vegetables he came across in Italy. He decided not to include orache or cress in the bundle. The Neapolitan kinds were hot and harsh, not nearly so sweet or digestible as the kinds already being grown near Maillezais, at the abbey of Ligugé in Poitou.

A century earlier, round about 1430, orache was on the English court menu, one of several leafy vegetables and herbs in a dish called *joutes* (see page 500). It continued in our vegetable gardens for three centuries longer – Evelyn grew it among his pot-herbs at Sayes Court – but it faded out. As Philip Miller said in his *Gardener's Dictionary* of 1741, 'there are very few in England that are fond of it'. He gives a warning against eating anything but the young leaves, 'for when it is run up to seed, it is very strong'.

PARSNIPS

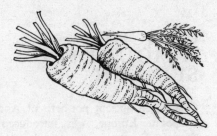

On the prose of parsnips – if you
think they are prose – a little
poetry is sprinkled in the name
of one of the great poets of our
time. The Russian for parsnip is
pasternak. Perhaps Boris Pasternak, Boris Parsnip, reflected that parsnips have a flavour entirely of their own and are more delicious than their reputation allows – as long as that flavour is properly brought out and combined.

I used to think that the parsnip as a vegetable had been developed in England. But now I feel sure it is one of the more ancient of European root vegetables, given up elsewhere, but retaining a popularity here in the north where it grows so well. That is not to our discredit. French dictionaries give a phrase '*Des panais*!' – 'Parsnips!' 'Nothing doing!', 'Damn that!' But if the French don't care much for parsnips, except as a flavouring root in *pot-au-feu*, I suspect they fancied them much more in the past, in the Middle Ages. 'Parsnips' comes to us from the Old French, anyway.

Nonetheless by the 16th century they had been demoted to animal fodder; nothing was thought to be 'more delicate .. nothing more odoriferous than the flesh of young pigs fed on parsnips and roasted with a stuffing of fine herbs,' (pigs, in Italy, destined for *prosciutto crudo* are fed still on parsnips, and what is more delicate and odoriferous than fine Parma ham?). In France now the carrot has entirely superseded the parsnip; in England we have kept them both – parsnips, having a certain precedence of treatment, being grown in market gardens rather than in fields, like carrots. Americans cook parsnips with brown sugar and fruit juice, so that they glaze beautifully. They serve them with turkey and so on, in the manner of sweet potatoes.

Medieval doctors credited parsnips with several virtues. They kept off adders, they cured toothache, they gave men an appetite for women, they reduced swollen testicles, they were eaten for stomach-ache and milk they were boiled in was given against dysentery. A vivid image of parsnips comes from the letters of Edmund Burt, Chief Surveyor to General Wade during the making of roads in the Highlands in the 1720s. 'I have seen women by the river's side washing turnips, parsnips and herbs, in tubs with their feet.' In wintertime, remember.

HOW TO CHOOSE AND PREPARE PARSNIPS

Choosing parsnips is not always easy. Like most roots, they are sold by weight. They need to be chosen by eye. A kilo (a good 2 lb) of parsnips, prepared and packaged on the supermarket shelf, will be enough for six people. At the greengrocer's, where whiskery roots, protuberances, bits of earth are included, you will need at least 1½ kg (3 lb). Why go to the greengrocer's? Because parsnips with earth and all their faults displayed are likely to have kept more of their flavour than the over-washed chastities sweating away in cling film.

The ideal is to be able to pick out unwrapped parsnips, all of an even medium size, parsnips without too many bits cut off, parsnips without a hard centre, without a deep top ring of earth to be gouged away. In other words what the French might call, if they ever ate them, *panais portions*, on the analogy of trout, one parsnip per person, weighing between 200 and 250 g, at least 6 oz but nearer 8 oz, to allow for waste.

If you see tiny parsnips about as we did sometimes after the dry summer of 1976, do not despise them. In their skins, they cook to a gentle deliciousness of flavour.

If you have to make do with huge parsnips, cut them up no more than is necessary to fit them into the pan. Remove the tough core after the preliminary blanching, when you are cutting the parsnips up for the final stages of cooking.

On the whole avoid peeling the parsnips before cooking them. Top and tail them, yes, but no more unless there are blemishes to be removed. A good way to make sure they are evenly cooked is to stand them head down in boiling salted water to cover the fattest part. Then the narrowing part can steam away at a slower rate, so that it does not turn to a flopping collapse before the thicker part is ready. Once the parsnips are just cooked, or cooked to the stage demanded by the recipe, drain them and run them under the cold tap until you can peel away the skin. Now the parsnips are ready for cutting up and finishing.

Never forget the proverb about kind words buttering no parsnips. Butter, or a good enriching element – cream, olive oil, beef dripping – is essential. Parsley, chives, tarragon help to make parsnips even more delicious; they are obvious additions. Less obvious are spices, nutmeg, cinnamon, even curry powder if you are making parsnip soup.

Parsnips may have to be boiled as a preliminary to some finishing process, but never serve them straight from the water, any more than you would appear at the dinner table dripping from a bath. Only asparagus and Aphrodite can get away with it.

BUTTERED PARSNIPS

Scrub, boil and cool the parsnips under the cold tap. Peel off the skin and slice or quarter them as appropriate to their size. Cut away any hard core. Melt a good generous knob of butter in a frying pan and turn the pieces over and over in it, adding plenty of pepper. They should not brown or fry in any proper sense of the word. They just need to absorb butteriness. Add a little extra butter and plenty of chopped parsley, or a mixture of parsley, chives and tarragon, and serve with meat, with cod, or on their own.

CREAMED PARSNIPS

Follow the recipe for buttered parsnips, but pour on some whipping or double cream before you add the herbs. Not a great deal, just enough to make a coating sauce, with the pan juices.

JOHN EVELYN'S BUTTERED PARSNIPS, 1699

Scrub, top and tail the parsnips – one medium-sized parsnip per person – and cook them in boiling salted water until just tender. Run them under the cold tap, and strip off the skins. Cut across into slices, about a centimetre ($\frac{1}{2}$ in) thick if you like them soft in the middle, thinner if you like them crisp all through, then roll the slices in seasoned flour and cook them in butter until they are golden-brown. If you can regulate your burners to a steady moderate temperature, you can finish the parsnips in butter straight from the packet; if you feel uncertain, clarify the butter first. You should end up with nice clean fat in the pan and the parsnips a golden-brown. No charring. The flour will form a light delicious coating. Use two pans, so that the cooked parsnips do not have to wait around a long time; this is also convenient if you are making both thick and thin slices. If they do have to be kept warm in the oven, line the dish with kitchen paper to absorb surplus butter.

Have ready two bowls, one with melted butter in it, the other with 1 heaped tablespoon of sugar mixed with 1 heaped teaspoon of cinnamon. Serve these with the parsnips for those who like the extra flavouring.

I find that a watercress salad, or watercress sandwiches, goes well after this dish. Pepper-tasting greenery refreshes your appetite, after the sweet spice and buttery softness. If you intend to serve the parsnips with beef or ham, make a watercress and chicory salad, with a few raw sliced mushrooms, to eat afterwards.

Note You can also sprinkle the sugar and cinnamon over the parsnips towards the end of cooking time.

FRIED AND BAKED PARSNIPS

Parsnips need richness, as I have said; this need not come only from butter and cream, but from dripping or from the neutral agency of a deep-frying oil.

1) SARATOGA CHIPS Cut the parsnips into wedges downwards. Remove any woody core. Parboil them until they are almost tender. Drain them, then deep-fry them until golden-brown.

2) Peel, and slice the parsnips. Cut out the core when necessary, to turn the slices into rings. Parboil them until almost tender and drain. Dip them in egg, roll them in breadcrumbs and fry them in clarified butter until golden.

3) Boil, peel and sieve 1 kg (2 lb) parsnips. Mix with two large eggs, a tablespoon of flour, 100 g (3½ oz) melted butter, 125 g (4 oz) chopped walnuts and enough milk to make a soft but coherent mixture. Deep-fry tablespoons of the mixture. Serve with fish, or on their own.

4) Peel and cut parsnips into pieces a good centimetre (½ in) thick. Boil for five minutes, drain well. Bake them in a tin of dripping – along with potatoes, if you like – on a top shelf in the oven when you are roasting beef. Or put them round the beef to cook more slowly in the juices.

CURRIED PARSNIP SOUP

1 heaped tablespoon coriander seeds
1 level teaspoon cumin seed
1 dried red chilli, or ½ teaspoon chilli flakes
1 rounded teaspoon ground turmeric
¼ teaspoon ground fenugreek
1 medium onion, chopped
1 large clove garlic, split
1 large parsnip, peeled, cut up
2 heaped tablespoons butter
1 tablespoon flour
1 litre (2 pt) beef stock
150 ml (¼ pt) cream
chopped chives or parsley

Whizz the first five ingredients in an electric coffee mill, or pound them in a mortar; mix the ground with the unground spices in the mill or mortar, so that they have a chance to blend well together. Put the mixture into a small jar – you will not need it all for this recipe, but can use it up with lentils or spinach.

Cook the onion, garlic and parsnip gently in the butter, lid on the pan, for ten minutes. Stir in the flour, and a tablespoon of the spice mixture. Cook for two minutes, giving the whole thing a turn round from time to time. Pour in the stock gradually. Leave to cook. When the parsnip is really tender, purée in the blender and dilute to taste with water. Correct the seasoning. Reheat, add the cream and serve scattered with chives or parsley. Cubes of bread fried in butter can be served as well.

PARSNIPS AU GRATIN

I came to this dish one winter's evening as a way of stretching an inadequate supply of sausages round the family. It turned out to be so delicious that we have it regularly now for its own sake. This quantity is enough for six.

> 6 medium parsnips
> ½ kg (1 lb) sausages
> pepper, thyme
> scant ½ litre (¾ pt) mornay sauce
> 3 tablespoons each grated Gruyère and breadcrumbs
> 2 tablespoons melted butter

Boil, peel and slice the parsnips until they are just tender. Place half of them in a buttered gratin dish in as even a layer as possible. Remove the skins from the sausages and crumble them into a basin, adding plenty of pepper, and a little thyme if they are short on flavourings. Spread this sausagemeat over the parsnips and place the remaining parsnips on top. Heat the sauce to something under boiling point and pour it over the dish. Mix cheese and breadcrumbs and scatter them over the top; pour on the melted butter. Bake at 180–190°C/310–375°F/Gas Mark 4–5 for 45–60 minutes, or at a lower or higher temperature if more convenient. Complete the browning under the grill if necessary. Serve bubbling hot.

PARSNIP AND MUSHROOM GRATIN

Good on its own, or with white fish – cod, hake, halibut and so on.

1 kg (2 lb) parsnips, peeled
125 g (4 oz) butter
250 g (8 oz) mushrooms, sliced
150–175 ml (5–6 fl oz) whipping cream, or double and single
  cream mixed
salt, pepper, grated nutmeg
2 tablespoons breadcrumbs
2 tablespoons grated Parmesan cheese

Cut the parsnips into chunks just over a centimetre across (about ½ in), and boil in salted water until almost cooked. Drain and return to the pan with half the butter. Cover tightly and leave over a low heat to finish cooking; shake or stir the pan from time to time. Cook the mushrooms in half the remaining butter, slowly at first, then more briskly to evaporate the juice to a slight moisture. Stir in the cream(s) and season well. Put the parsnips into a gratin dish, pour the mushrooms and cream over them. Mix the crumbs and cheese and scatter over the top. Melt the last of the butter and pour it over the crumbs. Brown under the grill, not too fast, so that the whole dish is golden and bubbling.

OTHER SUGGESTIONS

Parsnip soufflé See the recipe for *Jerusalem artichoke soufflé*, on p. 277. Instead of the hazelnuts, use either chopped walnuts or 100–125 g (3–4 oz) chopped mushrooms cooked in butter.

PEAS

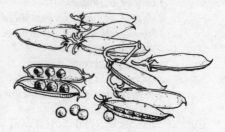

Peas and other pulses are some of man's oldest vegetables. They have been helping to keep our stomach's quiet for thousands of years. Because they can be dried for storing until next season, longer if necessary, they make a sensible food, if a dull one. It was this dried pea that everyone ate and that Roman writers mentioned when sighing for the days of antique virtue. You might think that occasionally someone, somewhere, for a week or two in the spring, must have gobbled a few young green peas. If they did, we do not hear about it until the 16th century, when Italian gardeners developed tender varieties for eating fresh and small. Garden peas at last, not field peas only.

Sadly I have to concede that the great days of young peas are over in this country at least, after only three centuries. We started growing them in the 17th century. Now in the 1970s – *finis*. Freezer companies have put an end to the enjoyment of young peas, to the expectation of their arrival. The frozen pea caricatures the real thing, but so closely that it spoils it. Processing exaggerates sweetness, turns flavour to uniformity. The frozen pea has come to occupy something of the position of grey dried peas of earlier times. One sees it everywhere, all year round and takes it for granted, avoiding it if possible, eating it with contempt (except in soup when its characteristics are modified by other ingredients and dilution). Now it seems as if greengrocers can hardly be bothered with the real thing. Invariably it is picked too large, so that its toughness throws people back to the freezer. It is only fair to add that the fault does not lie with the freezing itself. If you grow a good variety of peas in your garden and freeze what you cannot eat immediately, you will find them a lot better than commercial brands. And you can organize a long gap, so that your own first peas come along as a special pleasure of early summer.

I should like to go back in time to the great days of the garden pea, and invite myself to dinner with Thomas Jefferson. It would have to be when he was no longer the President of the United States, and was free to enjoy the garden he had made at Monticello in Virginia. More specifically I would choose late spring for my visit, when everyone was waiting to see who would be the first to put peas on the dinner table. I have a particular occasion in mind.

Growing the first peas was an annual competition in Jefferson's neighbourhood among his friends. It is mentioned in his Garden Book and letters, often, as peas were his favourite vegetable and he grew several varieties. The winner, almost always Mr George Divers, had the privilege of inviting the others to a dinner at which the first peas were the star item. Little notes were rushed round to the circle of friends, 'Come to-night – the peas are ready'. One year Jefferson's peas were ready before he received Mr Divers' invitation. His family reminded him that he had the right to give the special dinner and urged him on to the preparations. 'No, say nothing about it,' replied Jefferson, 'it will be more agreeable to our friend to think that he never fails.' Jefferson was a lovable man, unusually so for a President of the United States. For him money was not the name of the game, or vanity. I should like to have shared that secret Jefferson family meal when they ate their first peas, keeping the news from George Divers, so as not to upset his pride.

CHOOSING PEAS

Everybody knows that peas should be on the small side, bright green and plump. You may well have to drop this standard slightly when out shopping, but at least avoid peas with a wizened or brown or greyish look.

How to cook freshly picked peas

When peas have just been brought in from the garden, or have the right juicy look, the best way of cooking them is to boil them in their pods. This means that the peas steam to tenderness inside and the outer part of the pod is not wasted.

Serve the peas with melted butter in a separate bowl. Everyone picks up their peas by the stalk end, dips them into some butter and sucks and nibbles out the peas together with the delicious outside of the pod.

If you feel that the occasion is too grand for such uninhibited eating, save the pods when you shell the peas and keep them to make purée or soup. Turn to the broad bean purée recipe on page 105 for the method.

For the butter, you could substitute an hollandaise sauce or any cream and egg sauce of the kind you serve with asparagus or seakale, but I think the simplicity of melted butter with the peas takes some beating.

How to cook young, bought peas

There is no beating the English style. Shell the peas. Put 1 cm ($\frac{1}{2}$ in) of water into a saucepan with a very little salt, pepper, a bouquet of mint and sugar if the peas seem to need it. When this boils fast, put in the peas and cover tightly. Leave over a moderately fast heat for about five minutes, or until tender.

Extract the bouquet and strain off any liquor that may be left. Taste for seasoning. Stir in a knob of butter and some chopped mint. Or a knob of butter and enough cream to make a coating sauce, with chopped parsley rather than mint.

If the peas are on the short side, add a few small young carrots that have been separately cooked.

How to cook older peas in the French style

I think that the French style gives older peas a better chance, as it provides for a gentle stewing. It can be used for young, tiny peas too,

> 1 small lettuce, or 8 outer leaves of a larger lettuce, shredded
> 6 spring onions, chopped
> 1 tablespoon chopped parsley
> ¼ teaspoon salt
> 3 tablespoons butter
> 4 tablespoons water
> ½ kg (1 lb) shelled peas
> 2 very young carrots, chopped (optional)
> sugar

Make a bed or nest of the lettuce in the bottom of a pan. Put the onions, parsley, seasoning, butter, water, peas and carrots, if used, on top. Cover tightly and stew for about 20 minutes. There should be hardly any juices left in the pan. Taste and season with extra salt, and a pinch of sugar if necessary. Serve the peas with lettuce and juices from the pan.

DRIED PEAS

There are two main kinds of dried peas – the round wizened greyish-green ones, and the split kind that may be yellowish or a beautiful deep green. The former need soaking. The latter very often do not, though this can vary with different brands. Should there be instructions on the packet, follow them as they will be the best way of dealing with the peas inside. If there are no instructions, soaking will not hurt.

For a general discussion on the ways of soaking dried vegetables, turn to page 83. There are a few dodges for speeding up the process.

Some of the lentil recipes can be adapted to dried peas successfully: I can recommend the dhall curry in particular. So, too, can some of the bean dishes.

Carlin peas

In the north-east of England, you will find another kind of dried pea, the carlin. It is small and round, a blackish-brown colour, quite unlike the normal split dried peas. It takes its name from the Sunday before Palm Sunday, Carling or Care Sunday, because it was much eaten then – and still is. In the playgrounds you will often hear children singing and naming the Sundays of Lent

> Tid, Mid, Misere,
> Carlin, Palm, Pase-Egg Day.

On Carlin or Carling or Care Sunday, now known as Passion Sunday, the church anticipates in its ritual the care, i.e. the pain, of Good Friday.

Whether the mass of people in the long past ate carlins as a graceful religious gesture, or as a holiday dish, I do not know. I suspect they ate them because they were thankful to eat anything. Carlins are really a pea grown for feeding to birds and animals. Towards the end of winter, when stores ran low and famine came closer in, people shared the fodder put aside for animals.

Should you be in Thornaby or Stockton-on-Tees in Lent, look out for packets of these peas put up by Amos Hinton and Sons, complete with cooking instructions. They have been selling them for over 100 years, and are now the only source, I believe. Annual sales for Carlin Sunday run into several tons, so north-easterners keep up some memory of the care of their ancestors, even if it is soothed into the prettiness of folk custom.

JUNE PEA SOUP

A young pea soup has a fresh sweet flavour for midsummer meals. It is essential to use the best peas you can find, not over-large ones. In winter, frozen peas do quite well; although something of the pure directness is lost, the soup tastes refreshing by contrast with the heftier food of the season.

1½ kg (3 lb) young peas
175 g (6 oz) spring onions, or young green-stemmed onions,
    chopped
1 clove garlic, chopped
175 g (6 oz) butter
250 ml (8 fl oz) whipping cream
chopped mint, or parsley and chives
salt, pepper

Shell the peas on to the weighing scale pan, stopping when you have 600 g
(1¼ lb), more or less. Stew the onions and garlic in half the butter in a
heavy, covered pan, without browning them. Add the peas together with
three or four of the brightest and best pods and 1½ litres (2¾ pt) water.
Simmer until the peas are just tender – do not overcook. Purée in the
blender, then sieve into the rinsed out pan (the sieving is necessary to
catch any tough skins or fibre from the pods; it doesn't take a minute as
the peas are already puréed). Reheat to just under boiling point. Stir in
the remaining butter, then the cream, herbs and seasoning to taste.

If you wish to serve the soup chilled, leave out the second half of the
butter and instead of 250 ml (8 fl oz) whipping cream, pour in 150 ml (¼
pt) each whipping and single cream. Add the herbs just before serving.

For a very different, but equally delicious soup of a more rustic kind,
turn to the recipe for Purslane and pea soup on p. 431.

ONE-EYED BOUILLABAISSE WITH PEAS
BOUILLABAISSE BORGNE AUX PETITS POIS

The idea of the title seems to be that when you are too poor or low to
afford the fish deemed necessary for bouillabaisse, you make do with
vegetables and eggs. You will notice that this humble version of the
famous provençal soup needs the same fast boiling to bring oil and water
together. It makes an appetising lunchtime dish. As leeks are difficult to
buy in this country in the summer, use a small bunch of spring onions
instead with an ordinary onion.

6 tablespoons olive oil
2 leeks, or spring onions and an onion, or a leek and an onion,
    chopped
2 large tomatoes, skinned, chopped
4 cloves garlic, crushed
1 sprig garden fennel

> *finger-length strip of dried orange peel*
> *pinch of saffron filaments*
> *6 new potatoes, scraped and sliced*
> *½ kg (1 lb) shelled peas*
> *salt, pepper*
> *1 egg per person*
> *1 slice stale French bread per person*
> *chopped parsley*

Heat the oil in a large saucepan and brown the leeks or spring onions and onion very lightly. Put in the tomato, aromatics and potatoes, stirring them well together. Add 2 litres (3½ pt) boiling water. Set over the fiercest, reddest heat you can contrive. When the liquid bubbles hard, put in the peas – choose middle-sized peas of good quality for this recipe, rather than tiny young ones. Add seasoning. Boil hard until the potatoes are almost cooked, then lower the heat and poach the eggs in the soup. Check the seasoning.

Put the bread into individual soup bowls. Remove the eggs and vegetables with a pierced ladle to a serving dish and sprinkle them with parsley. Pour the soup into the bowls. Serve everything together if you like, in the French style, or keep vegetables and eggs warm while you drink the soup.

DUCK STEWED WITH GREEN PEAS

Surely duck and green peas is an unassailably English combination? Hannah Glasse and Parson Woodford would never have criticized it as a 'foreign kickshaw'. But that is what it was. It is first described in *The Accomplisht Cook*, by Robert May, in 1660, as a dish in the French style.

Robert May knew what he was talking about. If he said that duck with green peas was a French idea, he had more reason to know than any other cook of the time, so well was he grounded in the cookery and eating habits of both countries.

Eighty years after the early editions of *The Accomplisht Cook*, duck and green peas were an established pairing, with no 'French' in the title. Two centuries after, when it comes to Mrs Beeton, naturalization is completed.

This 18th-century recipe is convenient in the summer, as it can be cooked on top of the stove over a very low heat. If you put some new potatoes on top of the peas to steam, the whole main course of the meal is cooked in one pot with no trouble. If you have to put the oven on for other things, you can follow a slightly different version of the recipe and half-roast the duck, then transfer it to a deep pot and add stock, lettuce and peas. Complete the cooking either in a cooler oven, or on top of the stove.

> 300 ml (½ pt) duck giblet stock
> 1 large duck
> bouquet of parsley, bay, thyme and rosemary
> oil
> 1 large crisp lettuce, shredded
> ½ kg (1 lb) shelled peas
> salt, pepper
> 2 egg yolks
> 5 tablespoons double cream
> lemon juice

Do not include the liver when making the giblet stock. Tuck the bouquet into the bird and prick the skin over with a fork. Fit the duck into a large flameproof pot and brown it all over in a little oil. Pour off the fat. Turn the duck on to its breast and add stock. Cover and simmer for an hour and a quarter. Turn the duck over, push the lettuce down the sides and pour the peas in on top. Season, replace the lid and cook for a further three-quarters of an hour or until the duck is tender and the peas are soft and impregnated with the delicious juices.

Carve the bird, drain the vegetables and arrange them on a dish. Skim the fat from the juices in the pot. Taste them and boil down if need be, to concentrate the flavour. Beat the egg and cream together and stir into the juices – keep the heat low so that the sauce thickens without curdling. Season and add lemon juice to taste. Pour some over the duck and serve the rest separately.

ARTICHOKES CLAMART

Prepare and cook six artichoke bottoms (see page 19). Add to peas cooked in the French style (page 368) to heat through at the end of their cooking time. Arrange on a dish with a knob of parsley butter in each artichoke heart. Or put round a roast joint of lamb or a chicken.

Clamart in French cookery language means peas, because that is where the best peas in Paris were grown in the old days.

WINTER PEA SOUP
POTAGE SAINT-GERMAIN

The name of Saint-Germain goes back to the time when Paris was a smaller city, surrounded by market gardens that supplied Les Halles, the 'belly of Paris'. Every day before dawn, carts trundled in with produce from the different districts. Clamart and Saint-Germain were famous for their peas; when you see these names in a cookery book or on a menu, it means that the dish has peas as a main ingredient.

I find that the French *pois cassés* (split peas) have by far the best colour and flavour. They need no soaking and are cooked within the hour. In the following recipe I use frozen peas in small quantity to give an extra freshness to the taste. This does not spoil the delicious mealiness of the texture at all.

> 250 g (8 oz) split green peas
> 2 small onions, halved
> about 100 g (3–4 oz) frozen peas
> 150 ml ($\frac{1}{4}$ pt) milk
> salt, pepper
> 6 tablespoons cream
> 1 good tablespoon butter

Put the split peas and onions into a pan with 1½ litres (2½ pt) water. No salt. Bring to the boil, cover and simmer until cooked. Meanwhile cook the frozen peas in a little salted water and set the whole panful aside. When the dried peas are tender, put them with their liquor into the blender, with the frozen pea liquor and half the frozen peas. Whizz at top speed until very smooth. Return to the pan, add the milk and seasoning with the remaining cooked peas. Reheat gently to just below boiling point; take the pan from the heat and skim away the white foamy scum. Mix in cream and butter. Serve with small croûtons of bread fried in butter.

VARIATION Add a level tablespoon of dried mint to the soup when you reheat it, stirring it well in. Taste and add a little more if necessary.

DRIED PEA SOUP WITH EGG
POTAGE DE POIS CASSÉS AUX OEUFS DURS

A splendid winter soup, with an attractive appearance and flavour in spite of the unpromising tapioca. Non-devotees of tapioca pudding, those who refuse to have tapioca in the house after childhood experiences, should think again. As a finish to soup, the texture of tapioca is most agreeable.

> 175 g (6 oz) split green peas
> 2 litres (3½ pt) water
> 100 g (good 3 oz) tapioca
> 60 g (2 oz) butter
> salt, pepper
> 3 hard-boiled eggs, chopped
> chopped parsley

Simmer the peas in half the water until cooked. Sieve finely, or blend, adding the rest of the water to achieve the consistency you prefer, keeping the soup on the thin side. Return it to the pan, bring to the boil and stir in the tapioca. Cook for 15–20 minutes, stirring from time to time so that the tapioca does not huddle together in a sticky mass at the bottom of the pan. When it is at last cooked to small transparent pearls – the French call tapioca, *perles de Japon*, which does improve its reputation – stir in the butter and season to taste. Pour into a hot tureen or bowls, scatter the egg and parsley on top and serve immediately.

ENGLISH DRIED PEA SOUP

In this country we have nearly always added some kind of ham or bacon flavouring to dried pea soup. Follow the recipe for the Saint-Germain soup on page 373, but cook the peas in ham liquor, or in water with a ham bone or a piece of bacon. Omit the cream and the freshener of peas at the end.

Any pieces of meat on the bone can be cut up and added to the soup just before serving. Or you can cook some fatty bacon until crisp, then reduce it to crumbs.

DRIED PEA PURÉE WITH GREEN PEPPERCORNS

Hot spicing and plenty of butter or meat juices from the roasting pan, turn a heavy purée of dried peas into pleasant eating. In this case, I recommend tiny whole green peppercorns. These are the berries straight from the pepper vine, which have been canned or brined instead of dried into the familiar black peppercorns of everyday cooking. The flavour is different, peppery of course, but more juicy and green-tasting. They are put up in various ways. I find them best when they have been put into wine vinegar (the Wiltshire Tracklements Company do them). They last well, and keep their crispness. Canned green peppercorns have been softened by the processing and have to be used up fairly rapidly once the can is opened.

> ½ kg (1 lb) split green peas
> 1 medium onion, chopped
> 1 medium carrot, sliced
> bouquet garni
> large knob of butter, or meat juices with some of the meat fat
> 1–2 tablespoons green peppercorns
> salt, sugar

Soak the peas if necessary. Drain them and put with plenty of water into a pan. Bring to the boil, skim and add the vegetables and bouquet. Cover and simmer until tender. Drain, discard the bouquet, and put through the *mouli-légumes*. Reheat with butter or meat juices, adding the peppercorns gradually to taste, along with some salt. A little sugar might also be a good idea.

Serve with duck or lamb, gammon, bacon or sausages.

PEASE PUDDING AND PORK

I was brought up on pease pudding made with yellow split peas, in the north-eastern manner. Butchers had great pots of it in their windows, and would slice off as much as you wanted for reheating at home. Good housewives, though, made their own, and boiled it in a cloth along with the pork. If you were not so given to pease pudding, you would make a smaller one in a pudding basin. Less picturesque, but a lot easier to manage.

It would perhaps upset my grandmother if she saw me using green split peas to make the pudding. I think their flavour is better than the yellow peas' and the colour looks more attractive with the salt pork or bacon. You can also make the pudding quite on its own, and serve it with separately cooked black pudding, white pudding, sausages or bacon rashers. Extra butter is a good idea, too.

> 1–1½ kg (2–3 lb) salt pork, bacon or gammon joint, or a meaty
>   ham hock
> ½ kg (1 lb) dried split peas, soaked if necessary

60 g (2 oz) butter
1 large egg, or 2 small ones
salt, pepper

Put the meat and peas into a pan with plenty of water. Bring to the boil and simmer until the peas are tender. This may be anything from 40 minutes for green split peas, to an hour for yellow peas that are not in their first youth. When they are tender, lift out the pork or other meat into another pan and strain the cooking liquor over it so that it can continue to cook (allow 40 minutes per ½ kg or 1 lb). Put the peas through the coarse plate of the *mouli-légumes*, or mash them down with a fork. Mix with butter, eggs and seasoning and put into a buttered basin. Cover with foil, tie it down, and boil or steam the pudding for an hour.

Turn it on to a hot serving dish, and surround it with slices of the cooked meat and a little of the broth. If any pease pudding is left over, it can be sliced and fried in butter next day, and served with bacon rashers.

Whatever the old song says, do not be tempted to keep it for nine days in the pot.

OTHER SUGGESTIONS

Green peas with boiled ham, gammon, bacon, etc., see Boiled beans and ham with parsley sauce, p. 107.

Risi e bisi A Venetian dish made by adding young peas cooked as on page 367 to a risotto. Stir in two tablespoons grated Parmesan and serve extra Parmesan in a separate bowl for people to help themselves.

Egg and peas Jewish style Cook young peas as on page 367 in a frying or sauté pan. When they are almost cooked, make four or six depressions in them with the back of a spoon and slip an egg into each one. Cover the pan again. Beat up another egg with six tablespoons cream. When the eggs in the pan are set, pour over the cream and put under a hot grill for a moment to set the top. Season and serve.

PEPPERS and CHILLIS – THE CAPSICUMS

The variety of capsicums is enormous if you live in Mexico and Latin America; four or five different kinds may be required for a dish, and they will be graded on a heat scale. We see little of this in Europe, even if we live in Spain or Hungary where they make an industry of peppers and of paprika, which is the dried and ground powder of peppers. (Cayenne is the dried and ground powder of chillis; chilli powder – confusing, this – is a blend of chillis and other spices, specially devised for Texas dishes with a Mexican tinge such as *chilli con carne*.)

India and Hungary, Italy and Spain had to await Columbus to develop what are now their most typical dishes. Even after the discovery of America it took time – the red dishes of Hungarian cookery, *paprikás* and *gulyás*, date from the 17th and 18th centuries only. In this country we have waited longer still. Peppers seem only to have been on sale here for about 20 years, first as an expensive exotic, more recently as a commonplace. Greengrocers love them for their agreeably long 'shelf life', to the extent that one can now buy them all the year round in the smallest towns, along with the cabbage. One cannot say as much of our older vegetables such as spinach or watercress, let alone seakale.

Peppers have added a new note to our meals, a sweet vigour that even the blandest Dutch greenhouse produce is able to bestow. Chillis are getting about rather more the last few years: for this we should be grateful to West Indian, Indian and Pakistani immigrants.

With peppers, as with courgettes, I look back to Elizabeth David's first book, *Mediterranean Food*. It came out in 1950, when, as she later remarked, every item of good cookery was either rationed or unobtainable. Many of her readers would not have known a pepper if they had seen one, let alone a courgette or an aubergine. 'I hope,' she wrote, 'to give some idea of the lovely cookery of those regions to people who do not already know them, and to stir the memories of those who have eaten this food on its native shore, and who would like sometimes to bring the flavour of those blessed lands of sun and sea and olive trees into their English kitchens.' These things she did. We should thank her for it.

Mrs David used the word pimento, equivalent to the Spanish *pimiento*. Nowadays I would say that her word has been superseded by pepper, with or without the adjective sweet. European names do not help the confusion. French cookery writers use *piment* usually for chillis, but occasionally for peppers as a short form of *piment doux*, sweet pimento (the more common word is *poivrons*, large peppers). This uncertainty is carried through sometimes into translations into English. You have to look at the quantities carefully, think about the final effect, and work out which is the more likely. It is usually fairly obvious if you have some idea of French food, but to a novice it can mean problems. In American cookery books, bell or bullnose peppers mean large sweet peppers.

If you watch vitamins, you may be happy to know that peppers are rich in vitamin C.

HOW TO CHOOSE AND PREPARE PEPPERS AND FRESH CHILLIS

When choosing peppers and fresh chillis, avoid any that are wrinkled or patched with brown. They should always be smooth and sleek, with a glossy brightness.

In this country peppers are always mild (apart from their seeds, which must always be removed). You will find the ripe red peppers sweeter and more mellow than the green ones. Yellow peppers are closer in flavour to the red kind. Sometimes in a tray of green peppers, you will notice that some of them are beginning to turn yellow and red; they are ripening and will eventually turn red, though as they are off the plant they may wrinkle as they do so.

Chillis are always hot; the smaller the variety, the hotter they seem to be. Never be tempted to try even the tiniest bit of chilli to see how fiery it is. Its heat can be taken for granted. Again, the red ones are riper than the green, so that they do have a slightly different flavour, but as their contribution is heat, choose red or green to blend in with the colours of the dish you intend to make.

Beware of the capsicin in chillis when you prepare them (I have never suffered from peppers, though I can imagine people with extra sensitive skins could feel some discomfort). It has an onion effect, with heat rather than stinging tears. It can make your skin tingle, your nose and throat burn. Should you rub your eyes while preparing chillis, they will soon be very painful. Always rinse your hands well afterwards and keep the chillis under water (in a bowl or from the tap) as you prepare them. The seeds are usually removed, unlike the tiny dried chillis that one buys at the spice counter. The first time I prepared them, there was a bowl of stewed pears not far away. At the end of supper, I was still suffering from capsicin, and my husband said 'What a good idea to pepper the pears: brings out their sweetness.' The volatile oils had carried the heat to them as well as into me.

For many dishes it is worth skinning peppers. You can do this in several ways. Put them into a really hot oven, if you happen to have it on for something else, until they blacken and blister. Or spear them with a wood-handled fork and turn them over a gas flame. Or turn them under a very hot grill. Or put them on to an electric spit for about 20 minutes.

Having reduced the skins to the blistered state by one or other of these means, you should be able to peel off the thin outer skin. Do this under the cold tap or in a bowl of water, so that the tiny charred bits are rinsed away. Then cut away the stalk, slit open the pepper and wash away the seeds. This charring of the outside reduces the flesh of the pepper to the most delicious rich softness – obviously you lose the crispness. When you are making a salad, weigh up which texture you prefer.

The natural companions of peppers and chillis are tomato, anchovies, olive oil – above all olive oil.

Canned and frozen peppers are useful storage items, with a good resemblance to the fresh kind. Of course you must take into account that the canned peppers are already cooked; they are the equivalent to peppers that you have grilled and skinned. Frozen peppers have been lightly blanched, so are only half cooked, if that.

SPANISH PEPPER SOUP

An unusual, and spicy soup. One of the best vegetable soups I know, light but full of interest and flavour.

> 250 g (8 oz) tomatoes, chopped
> 1 dessertspoon wine or sherry vinegar
> 1 rounded teaspoon dark brown soft sugar
> 2 litres (3½ pt) beef stock
> 2 large peppers, grilled, skinned, diced
> 100–125 g (3–4 oz) green beans, cut in 1 cm (½ in) lengths
> 3 tablespoons long grain rice
> good pinch saffron
> small pinch cayenne
> 1 heaped teaspoon Spanish or other mild paprika
> salt, pepper
> chopped chervil or parsley

Reduce the tomatoes, vinegar and sugar to a small amount of concentrated purée. Sieve it into a large pan. Add the stock and bring to the boil. Put in the peppers, beans, rice, saffron and seasonings. When the rice is cooked, the soup will be ready. Correct the seasoning and serve with chervil or parsley scattered on top.

Canned peppers can be used – though the flavour will not be quite as good; a small can, with four peppers, will be needed.

GRILLED PEPPER SALAD

Slice four large peeled tomatoes and put them into a shallow dish. On top arrange four grilled skinned peppers, cut into strips. Pour over some olive oil vinaigrette. On top arrange a lattice of anchovy fillets, split down lengthways (one tin is enough), and put a stoned black olive in the centre of each gap. Round the edge arrange quartered hard-boiled eggs – two or three according to the number of people and the rest of the meal. Scatter with fresh chopped basil or parsley and serve well chilled.

FONTINA AND PEPPER SALAD
ADA BONI'S INSALATA DI FONTINA

Try to get hold of yellow peppers. If this is not possible, the red ones are next best. Grill and skin them, then cut into strips. To every three peppers, allow 125 g diced *fontina* cheese (Gruyère is a substitute, but not a very good one, as *fontina* is much creamier), and six or eight chopped, stoned green olives. Make an olive oil vinaigrette, lightly flavoured with Dijon mustard and add to it three tablespoons of double cream. Pour it over the salad and chill thoroughly before serving.

The thing that makes this salad is the three contrasting tendernesses – pepper, *fontina* and olives. It looks beautiful, too.

MARINADED PEPPER AND OLIVE SALAD

This is one of the very best of the pepper salads. Prepare it 24 hours in advance: if you leave it two days in the refrigerator, it won't come to any harm. Only the eggs are added at the last moment, if you decide to use them.

Green peppers are not mellow enough for this recipe, but you could use a mixture of red and yellow peppers. Olive oil is essential.

3 large red peppers
juice of a lemon
salt, pepper
3 cloves garlic
small bunch parsley
18 black olives or stuffed green olives
150 ml (¼ pt) olive oil
3 hard-boiled eggs, sliced (optional)

Grill and skin the peppers. Cut them into strips, discarding seeds and stalk. Lay the strips in a shallow dish. Pour over the lemon juice and add seasoning.

Chop the garlic and parsley leaves together and scatter evenly over the peppers. Halve and stone the olives if necessary and arrange them on top, not too formally. Pour over the oil. Cover the dish over with cling film and leave in the cool until next day. Arrange halved slices of egg round the dish just before serving.

EASTERN SALAD

A typical chopped salad of the Near East and North Africa that appears in a number of variations. The ingredients are always much the same, but their quantities can be varied – sometimes you may be startled by the amount of chilli, at other times it will be only a hint of pepperiness. This kind of mixture is particularly good when stuffed into a pocket of pitta bread with *felafel* or *hummus*, as a relish: then you can afford to increase the quantity of chilli. The thing to be careful about is the quality of the ingredients. The tomatoes must be the huge craggy kind, Marmande or Eshkol are sometimes to be bought in the shops. The olive oil should taste strongly of its source, the Greek Minerva brand or a Cypriot olive oil is best. The lemon should be large and juicy.

1 large green pepper, seeded
1 chilli pepper, seeded
2 huge tomatoes, skinned
half a bunch of spring onions or 1 medium onion
2 large cloves garlic, crushed with salt
2 tablespoons chopped parsley
1 tablespoon chopped mint
5 tablespoons olive oil

juice of a large lemon
salt, pepper
1 English cucumber (or 2 small cucumbers)

Chop the first five ingredients together to a speckled mash – the pieces should be about 3 mm (⅛ in). Mix in the parsley, mint, olive oil and lemon with salt and pepper to taste. The salad can now be left for a while. Chop the cucumber only half an hour before the meal, salt it and add it to the salad; if you include it in the original chopping, the salad will be too wet, should it have to hang about at all. Taste and correct the seasoning.

Note I assume in this recipe that you are using firm beefy tomatoes. If yours are on the wet side, tip the juice into a basin and use it for something else, so that the salad is not awash.

PEPERONATA

For *peperonata* the proportion of the two main ingredients can be varied. The peppers can be red, yellow or green – or all three. If you have to use tomatoes of the underprivileged kind, allow five to every two peppers, as so much will disappear. If the tomatoes are your own growing, large, sweet, firm all through, allow one to each pepper. But however you start, the final result should be moistly juicy, even a little dry, never sloppy or wet. You can eat *peperonata* hot with lamb, chicken, pork, veal or cold as a salad, scattered with parsley and black olives. If any is left over, it can be reheated, divided between pastry cases and topped with eggs, poached or mollets.

Cut the peppers into strips, discarding seeds and stalks. Lightly colour a large sliced onion and one or two cloves of chopped garlic in olive oil; they should not really turn brown. Put in the pepper strips, cover and simmer for 10–15 minutes. Add the peeled and chopped tomatoes. Boil steadily, without a lid, for 30–40 minutes, until the pepper lies in a rich vivid tomato stew. At this point you will need some vigorous seasoning – salt, pepper, even sugar and a dash of vinegar. If you like, a teaspoon of crushed coriander can be added towards the end, just give it long enough to release its orange fragrance.

PIPÉRADE BASQUAISE

A version, from the Atlantic corner where France and Spain meet, of the favourite Mediterranean stew of tomatoes and peppers. It differs from Italian *peperonata* in the final inclusion of eggs, and from Tunisian *chakchouka* in the pork embellishments, which would not appeal to an Arab.

> 3 peppers, preferably red
> 1 huge onion, chopped
> 3 cloves garlic, chopped
> lard or bacon fat
> ¾ kg (1½ lb) tomatoes, peeled, chopped
> good pinch dried marjoram or thyme
> salt, pepper
> 6 slices Bayonne ham, or paper-thin gammon
> butter
> 6 eggs

Either grill, skin and slice the peppers, or cut ungrilled peppers into strips, discarding the seeds and stalks. The first is more trouble, but it enhances the smooth, spiced richness of the *pipérade*: the second gives a lighter, more rustic feel to the dish. Soften the onion and garlic in the fat. Add the pepper pieces, tomatoes, marjoram or thyme and a little seasoning. As the vegetable juices start to flow, raise the heat and cook vigorously to get rid of wateriness. Towards the end, heat through the Bayonne ham or gammon in a little butter and arrange round a serving dish. Quickly stir the eggs into the vegetables, allowing them to scramble creamily. Turn on to the dish and serve.

VARIATION Put the vegetable stew into a dish. Heat the bacon or ham in butter and lay them on top, then fry the eggs in the juices, adding extra butter if need be, and put an egg on each piece of ham.

BASQUE PEPPER AND EGG TARTS

Make the tomato and pepper stew as above, but with two-thirds of the ingredients. Have six shortcrust tartlet cases, measuring 10–12 cm (4–5 in) across, baked blind.

Flavour the vegetable stew with six tablespoons Madeira or port and divide between the pastry cases. Rapidly cook four to six eggs into an omelette (or two) roll it up and cut across into strips. Arrange on top of the tarts and sprinkle with parsley before serving.

TUNISIAN CHAKCHOUKA

Make a tomato and pepper stew, as for *pipérade*. When it is nicely reduced and seasoned, make six depressions in it with the back of a tablespoon. Slide an egg carefully into each one. Cover the pan and leave to simmer until the egg whites are opaque. Season and serve.

PEPPER OMELETTES

Peppers cut in strips and reduced to a soft melting mass in lard, goose fat or olive oil, can make a delicious flavouring for omelettes.

Choose three large ones, and cook them in a covered pan for ten minutes or longer. Then season them with salt. Quickly mix this *fondue* into six or eight large eggs, beaten together in a basin with salt and finely chopped garlic, then return the whole thing to the warm pan.

If you are eating the omelette straightaway, cook it lightly in the French style – it should be *baveuse*, i.e. liquid and bubbly in the centre. If you want to take the omelette cold on a picnic, or to work, cook it gently to firmness like a Spanish *tortilla* or Persian *kuku* (pp. 406 and 471).

Or you could use the pepper mixture to fill a straightforward *omelette nature*.

STUFFED PEPPERS
PEPERONI RIPIENI

As a general rule, allow one pepper for each person if the dish is a major part of the meal. Try to choose them of an even size, and with flat bases, otherwise you will be in practical difficulties. Flat is perhaps not quite the word for the base of a pepper. What I mean is, make sure that the peppers will stand upright, without a list to one side: this is often difficult to achieve, but is essential to the serene presentation of the dish.

Method 1) Slice the stalk end from the pepper. Clean out the seeds in the usual way. Fill the pepper with stuffing to the top; if the mixture is at all dry, put a dab of butter on top. Replace the sliced off stalk end. Arrange the pepper with its mates in a shallow ovenproof dish. Sprinkle them with olive oil, pour a very little water round them, and bake at 190°C/375°F/Gas Mark 5 for an hour. 20 minutes before the end, tomato sauce can be poured round the base of the peppers.
Method 2) Method 1 can be shortened by blanching the peppers in a pan of boiling salted water for five minutes, before stuffing them. Drain them, run them under the cold tap and slice off the tops etc, as above.

Method 3) Put the peppers under a hot grill, or thread them on to an electric spit, and leave them until the thin outer skin is charred enough to be peeled off. Either slice the top off or cut the peppers down into two even halves, before removing the seeds and stuffing them. Bake at 180°C/350°F/Gas Mark 4 for 45 minutes. This is a particularly good method if you want to eat the peppers cold. It means more trouble at the beginning, but the resulting dish is more succulent; stuffing and peppers melt together.

CHEESE AND RICE STUFFING

Il Talismano della felicità, by Ada Boni, first published in the twenties, is the modern Italian counterpart to our Mrs Beeton's *Book of Household Management*. There never seems to have been a complete English translation of the *Talismano*, but a few years ago we had *Italian Regional Cooking* from Nelson, a collection of her recipes from an Italian women's magazine. I often make this and the following stuffings from the two books, adapting them slightly. If you cannot buy or afford Italian Provola or Mozzarella cheese, substitute a crumbling piece of Cheshire or Lancashire cheese. Parmesan is essential for its sweet pungency, but a little goes a long way so it does not turn out as expensive as it seems.

> *1 medium onion, chopped*
> *2 tablespoons each butter and olive oil*
> *150–175 g (5–6 oz) Italian long grain rice*
> *up to ¼ litre (½ pt) stock or water*
> *250 g (½ lb) Provola or Mozzarella cheese, diced*
> *3 tablespoons grated Parmesan*
> *1 teaspoon dried oregano*
> *2 hard-boiled eggs, crumbled*

Cook the onion in the fats until soft and yellow. Add the rice and stir it about for two minutes, then pour in half the stock or water. Cover and simmer until the rice is cooked, adding more stock or water if needed to prevent sticking. Remove the pan from the heat, mix in the remaining ingredients and season to taste.

NEAPOLITAN STUFFING

A light and piquant mixture that makes a good stuffing for skinned peppers (method 3, above). Pour a little tomato sauce round the peppers.

> 150 g (5 oz) dry white breadcrumbs
> 75 g (2–3 oz) raisins
> 18 stoned black olives, chopped
> ½ tin anchovies, chopped
> 2 tablespoons capers
> 2 tablespoons chopped parsley
> 1 tablespoon chopped fresh basil
> 6 tablespoons olive oil
> salt, pepper

Mix all the ingredients together, adding a very little salt.

SICILIAN STUFFING

This mixture is normally used to stuff sardines, but it is most delicious with skinned peppers.

> 100 g (3–4 oz) chopped onion
> 1 large clove garlic, chopped
> olive oil
> 100 g (3–4 oz) pine kernels
> 200 g (7 oz) white breadcrumbs
> 100 g (3–4 oz) sultanas
> 2 heaped tablespoons chopped parsley
> 1 tin anchovies, chopped
> 1 teaspoon sugar
> lemon juice, salt, pepper
> juice of 2 oranges

Cook the onion and garlic in enough oil to cover the base of the pan. Go slowly at first, then raise the heat to colour them slightly. Put in the pine kernels and then the breadcrumbs, stirring them about until they begin to turn a nice brown. Remove from the heat and add the remaining ingredients except for the orange juice, putting in the seasonings little by little to taste. Divide between the peppers. Bake in a well oiled dish. Just before serving remove the caps of the peppers and pour the orange juice over the stuffing. Replace the caps and serve.

ROUILLE, ROULHO

The famous red pepper sauce of Provence that is always served with *bouillabaisse* and *aigo-sau* and other fish soups. It goes well, too, with a mild sweetcorn soup (p. 485) or with cream of tomato soup. I sometimes add a fresh red chilli pepper – a small one – which is not at all correct, but works quite well.

The usual recipe – which can be made in a blender – is two cloves of garlic crushed with the flesh of two sweet red peppers. Squeeze out a thick crustless slice of bread 60 g (2 oz) in a little fish stock or water (depending on the dish it is to accompany) and add that to the peppers, then stir in gradually about four tablespoons of olive oil, and some extra fish stock if you like.

If you want a lighter, richer sauce, substitute a large egg yolk for the bread. Whizz it with the garlic and peppers, then add a larger quantity of olive oil until you have a red mayonnaise.

ROMESCO SAUCE
SALSA ROMESCO

One of the famous sauces of Spanish cookery, from the province of Tarragona where the small hot Romesco peppers are grown. In this country I would recommend you buy fresh red chillis and hang them up to dry in your kitchen. The other essential ingredient – except of course for tomatoes – is hazelnuts, which add texture and their oil to the sauce.

> *2 huge tomatoes, Marmande or Eshkol type*
> *3 fine fat cloves garlic*
> *24 hazelnuts, blanched (or 12 each almonds and hazelnuts)*
> *2 large red home-dried chillis, seeded*
> *salt, pepper*
> *¼ litre (8 fl oz) olive oil*
> *2 tablespoons sherry vinegar, or 1 each wine vinegar and dry sherry*
> *chopped parsley*

Bake the whole tomatoes and whole, peeled garlic in a moderate oven for about 15 minutes. After ten minutes add the nuts and chillis. Transfer to the blender, scraping the tomato pulp free of its skin, and blend to a purée, adding the seasoning and the olive oil – gradually – to make a smooth sauce. Finally stir in the vinegar or vinegar and sherry, and correct the seasoning.

Salsa romesco makes a delicious sauce with lightly cooked cauliflower, potatoes, cardoons or celery, although it is usually served with fish and chicken. Try it stirred into soups as well, much as you would *rouille* (previous recipe).

EAST-WEST SAUCE

A Lebanese sauce invented at the Yildizlar restaurant in Beirut. ·Use it for marinading a jointed chicken overnight, then grill the chicken with the sauce still on it – preferably over charcoal. Serve with a green salad.

> ½ kg (1 lb) sweet red peppers
> 125 g (4 oz) fresh red chillis
> 6–8 tablespoons olive oil
> juice of a large lemon
> salt

Slice and seed the peppers and chillis. Cook them in boiling salted water until tender, then put through the medium plate of the *mouli-légumes*, or purée in a blender. Mix the olive oil and lemon into this hot purée and add extra salt if necessary. This freezes well.

COLORADO SAUCE

A delicious sauce to use when making *chilli con carne*, rather than the chilli powder sold in small bottles. It can also be used as a marinading mixture like the East-West sauce above.

> 6–7 small dried red chillis, or 4–5 large fresh ones
> 1 large red pepper
> 1 large onion, chopped
> 1 large clove garlic
> salt

If the chillis are dried, soak them in a little water for an hour, then slit them and wash out the seeds. Discard the stalks. Do the same with the large pepper. Purée with the other ingredients, using the soaking water if necessary to moisten the vegetables. If you use fresh chillis, you might need a tablespoon or two of cold water. Season with salt. You can keep this sauce in a covered container in the fridge for two days, or you can freeze it.

CHILLI CON CARNE

Buy stewing meat for this dish, beef, veal, mutton or pork, rather than the finest cuts. Underdone left-overs can be used, as well. Avoid minced beef. You can use tinned red kidney beans, they are delicious, but it is far cheaper to buy them loose and uncooked at a good grocery or delicatessen. Another alternative is to omit the kidney beans from the stew and serve them separately in a salad, or as part of three-bean salad (see page 206).

> 500–725 g (1–1½ lb) meat, cubed
> olive oil
> 1 large onion, chopped
> 2–3 cloves garlic, crushed
> 1 small green pepper, seeded, sliced
> Colorado sauce (see page 388)
> 1 tablespoon tomato concentrate (optional)
> 1 teaspoon ground cumin
> 125–250 g (4–8 oz) red kidney beans, cooked
> salt, brown sugar

Trim the meat where necessary and brown it in olive oil. Transfer to a casserole. Brown the onion and garlic lightly in the same oil, and scrape on to the meat. Add the pepper, sauce and just enough water to cover the ingredients. Cover tightly and leave to stew until cooked, keeping the heat low. Check the liquid level occasionally. By the end of the cooking time it should have reduced to a brownish red thick sauce. If it reduces too soon because the lid of the pan is not a tight fit, or you had the heat too high, top it up with water.

Last of all add the tomato if used, the cumin, the kidney beans if you are not serving them separately as a salad, with salt and brown sugar to taste. Simmer a further 15 minutes, correct the seasoning and serve with rice and a green salad.

I believe that Americans drink coffee with this on occasion. Beer or a vigorous red wine or plain iced water seem preferable to me.

GULYÁS

Gulyás, or goulash if you prefer an anglicized spelling, is flavoured with paprika – i.e. peppers ground to a powder – but here is a version that includes peppers as well. You will find it worth making a visit to a good delicatessen for Hungarian paprika to flavour this and other dishes in the book: it has a richer, darker flavour than the Spanish kind often sold in supermarkets, though it is not necessarily any hotter.

> 1½ kg (3 lb) chuck or rump steak, cubed
> 4–5 rashers streaky bacon, cut into strips
> 4 large onions, chopped
> 3 tablespoons Hungarian paprika
> 2 green peppers, seeded, roughly chopped
> salt

Sear the beef in its own fat in a heavy iron frying pan. Put it into a casserole. Deglaze the pan with 300 ml (½ pt) water, and pour it over the beef when it is bubbling hot. Fry the bacon in the pan, adding the onions when the fat begins to run. Brown everything lightly, then stir in the paprika. Cook for a further two or three minutes, then tip it over the beef. Add the peppers. Cover closely and stew until the beef is tender. Add salt, stew a further five minutes, then correct the seasoning if necessary. Serve with separately cooked ribbon noodles, and plenty of green salad.

PAPRIKA CHICKEN HUNGARIAN STYLE
PAPRIKÁSCSIRKE

Paprikás has a family resemblance to *gulyás*, but chicken and veal are the preferred meats, and the sauce is finished with soured cream.

> 1 chicken jointed, seasoned
> lard
> 2 medium onions, chopped
> 1 tablespoon Hungarian paprika
> salt
> chicken stock
> 2 green peppers, seeded, coarsely chopped
> 250 ml (8 fl oz) sour cream
> 1 level tablespoon flour

Brown the chicken to a beautiful golden colour in the lard, then remove it from the pan while you brown the onions in the same fat. Pour off any surplus fat, sprinkle on the paprika and a little salt and stir over the heat for two minutes. Pour in a little chicken stock to deglaze the pan and make

½ cm (¼ in) depth. Replace the chicken, add the peppers and cover tightly. Leave on a low heat until the chicken is tender. Turn the pieces occasionally. You should not need to add any more stock: *paprikás* and *gulyás* are not sloppy stews at all – by the time they are ready the sauce should coat the meat richly rather than float it. Pour off any surplus fat carefully.

Whisk the cream gradually into the flour to make a smooth mixture. Add this to the pan of chicken and simmer for ten minutes, turning the meat so that it has even benefit of the sauce. Finally correct the seasoning.

Serve with boiled ribbon noodles (*tagliatelle*), preferably the soft home-made kind.

PIKE WITH SAFFRON AND PEPPERS

Pike steaks flecked with green and red pepper strips in a pool of saffron yellow sauce, a Zurbarán still life, and as delicious as it looks. The recipe was given me by Mr Tom Hearne of the Hole in the Wall, at Bath.

If you do not know a pike fisherman, substitute bass or cod steaks, or fillets of a firm white fish such as halibut, turbot, brill or sole. The sauce can be made in advance and reheated. Freeze it if you like, but keep the final addition of cream, Madeira and lemon juice until you have reheated it for serving.

> SAUCE
> 1 kg (2 lb) fish bones, heads, trimmings
> 1 medium onion and carrot, sliced
> 5 cm (2 in) piece celery
> bouquet garni
> 10 peppercorns
> 1 dessertspoon wine vinegar
> 150 ml (¼ pt) dry white wine
> generous pinch saffron
> salt
> 125 g (4 oz) butter
> 60 g (2 oz) flour
> 1 red and 1 green pepper, sliced
> 1 clove garlic, finely chopped
> 150 ml (¼ pt) cream
> 2 tablespoons Madeira
> lemon juice

FISH
6 steaks, or fillets for six people
butter
2 shallots, chopped
1 bay leaf
salt, pepper

For the sauce, start by making a fish stock. Put the first seven ingredients from the sauce list, down to and including the wine, into a large pan. Cover generously with water. Boil not too fast for 30–40 minutes. Strain into a clean shallow pan and reduce to 750 ml (1¼ pt). Set aside 150 ml (¼ pt) for cooking the fish. Add saffron and salt to the rest. Make a roux of half the butter and the flour, add the saffron stock and simmer to the consistency of double cream. Meanwhile cook the pepper strips and garlic gently in the remaining butter. When they are soft, add them with their juices to the sauce and simmer for a further ten minutes. Last of all pour in the cream, Madeira and lemon juice to taste, with any extra seasoning required. The sauce should have a good pouring consistency, not too thick.

To cook the fish, butter a shallow baking dish generously. Put in it the steaks or fillets, with the shallots and bay leaf. Season, and pour round them the fish stock you set aside when making the sauce. Bake in the oven until cooked – 15–25 minutes according to the thickness of the fish – at 190°C/375°F/Gas Mark 5, or simmer on top of the stove until cooked. Place the fish on a large shallow serving dish and keep it warm. Boil down the cooking juices to a couple of tablespoons or so and pour them into the sauce through a sieve. Rinse out the sieve and strain the sauce carefully round the fish. Divide the pepper strips evenly between the pieces and spoon a little sauce over them. Serve with bread to mop up the remains – or put out spoons, as they do in some French restaurants.

OTHER SUGGESTIONS

Ratatouille, see page 52.

POTATOES

When you think of the place potatoes now occupy in our European diet, when you think of the Irish dying and emigrating because of the potato famine, or of Van Gogh's black-lodged peasant family munching potatoes, it seems incredible that they should have taken so long to become accepted. Not quite as long as the tomato – also from the New World – but much longer than beans or Jerusalem artichokes.

The way they came into different countries in Europe is amusing. In Germany, or to be precise in Prussia, after the famine in 1774, Frederick the Great sent a free load of potatoes to the starving peasants of Kolberg. They refused to have anything to do with them – until Frederick sent along an armed soldier to communicate his enthusiasm. The psychology was right. Three-quarters of a century later, potatoes were so established and appreciated in Germany that the citizens of Offenburg raised a statue to Sir Francis Drake, potato in hand, whom they credited with introducing potatoes to Europe. Alas, the Nazis removed it. In France Antoine Parmentier tried a different tack. He started at the top, with courtly flattery. His experience of potatoes had come when he was a prisoner of the Prussians in the Seven Years War. He would have died without them, and saw how they would be a benefit to the French. On his return he presented a bouquet of potato flowers to Louis XIV. Marie Antoinette thought they were so pretty, that she tucked one into her hair. Now potato dishes in France bear the name of Parmentier.

As far as we were concerned, they seem to have sidled into the country in a thoroughly inconspicuous manner. Perhaps their early success in Ireland put us off. There is a fine blunt comment by Stephen Switzer, from 1733, to this effect, 'That which was heretofore reckon'd a food fit only for Irishmen, and clowns, is now become the diet of the most luxuriously polite'. Switzer was writing from his experience as garden designer and seedsman to the aristocracy. He lived in London. In 1778 – forty years later – Gilbert White declared that potatoes had only become common around Selborne in Hampshire during the last 20 years. Premiums had helped persuade the poor to grow them – a better idea for the practical English than guns or bouquets – and they were esteemed by people 'who would scarce have ventured to taste them in the last reign'. Obviously their acceptance was patchy. In Scotland they were not much grown before 1770, though they made a rapid conquest after that.

This country has produced many fine varieties of potato, all suited to different methods of cooking. We have the climate and skill for raising potatoes. We like potatoes. Yet greengrocers and farmers have managed to reduce this treasure of the Incas – did you know that they measured their units of time by the time it takes a potato to cook? – to three or four varieties whose only virtue is yield. So if you want a good floury potato, a fresh tasting new potato, a new potato that doesn't disintegrate when you slice it for salad, a good baking potato or a good potato for chips, you will have to grow it yourself. And if you want advice and a real choice of potatoes, write to Donald MacLean at Dornock Farm, Crieff, Perthshire. He can supply two hundred varieties in small quantity, from one tuber (his Museum Collection) to several kilos. His stock includes lemon-fleshed salad potatoes such as Aura, the kidney-shaped Kipfler, as well as Pink Fir Apple and Purple Congo which is a strange dark bluish colour all through and especially delicious. While you wait for your potato harvest, look out for farmers and greengrocers who sell Golden Wonder and Desirée, or the new Croft variety.

Potato varieties and their uses

Although a variety may taste and behave differently according to the soil it's grown in and the person who cooks it, discriminating potato-eaters agree on the qualities of some of the best known kinds.

Maincrop all-rounders	Croft	Golden Wonder
	Desirée	King Edward
Floury	Arran Chief	Maris Piper
	Kerrs Pink	Ulster Prince
	King Edward	(when mature)
Mashed	Desirée	Pentland Ivory
	Golden Wonder	Redskin
	Pentland Hawk	plus floury varieties
Potato cakes, etc.	Arran Victory	Kerrs Pink

Baked	Arran Comrade	King Edward
	Epicure	Maris Piper
	Golden Wonder	Redskin
Boiled	Craigs Royal	Pentland Javelin
	Epicure	(the whitest potato)
	Maris Peer	Sharpes Express
		Ulster Prince
		plus salad varieties
Irish stew	Bishop	
Salad	Arran Pilot	Golden Wonder
	Aura or Lemon-	Jaune de Holland
	fleshed Salad Potato	Kipfler
	Bintje	Purple Congo
	Desirée	Red Cardinal
	Di Vernon	Sirtema
	Eigenheimer	
Frying	Croft	Maris Piper
	Majestic (tasteless,	plus salad varieties
	chip-shop favourite	
	as it does not absorb oil)	

HOW TO CHOOSE AND PREPARE POTATOES

Choosing potatoes is not easy. So many are sold in printed plastic bags, that successfully conceal size and condition. These potatoes will often have been washed, too, and this does not help them. Of course you do not want to buy half a field of earth with your potatoes, but if flavour is what you are after then you will do better to find a greengrocer who sells his potatoes loose and unwashed. Best of all, if you have storage space, is to make an expedition into the country to find a farmer who sells a good variety by the sack.

With new potatoes, you will do better to buy them in small quantities at frequent intervals. If they are freshly dug, the skins should rub away easily in your fingers. I find sometimes that a greengrocer will let you have the small ones a little cheaper (apparently many people are too lazy to scrape them and prefer mammoth sizes). They are well worth the extra trouble, as they look and taste better from being cooked whole. New potatoes should either be steamed or put into boiling water with sprigs of mint, unless of course you intend to stove them.

Never peel old potatoes, unless you wish to deep-fry them. Scrub them well, cut away any bad patches, and boil or steam them as they are. Not only do they keep their shape better, but you lose less of them as the skin – and the skin alone – can be stripped off easily at the end. A bay leaf helps flavour.

If you are making a salad, potatoes are best steamed. Even so, they will fuzz and crumble at the edges, but less. There is no proper salad new potato on sale in this country. Of the old potatoes, Desirée does quite well.

Instructions for cutting chips, gaufrette potatoes and so on, are given on page 401.

POTATO SOUP

I should think that potatoes are more used than any other vegetable in northern Europe for soups. Apart from their availability, they are so easily varied as a soup flavouring by adding leeks and onions and so on, or by cheese, spices and herbs. In this recipe the additions are lard and garlic, which both blend wonderfully with potato. Instead of the croûtons, or even with them, you could provide a small bowl of crisp bacon pieces.

> *375 g (12 oz) peeled, diced potato*
> *100 g (3–4 oz) chopped onion*
> *2–3 large cloves garlic, finely chopped*
> *2 big tablespoons lard or pork fat*
> *1 litre (1¾ pt) light beef or veal stock, or water*
> *salt, pepper*
> *chopped parsley*
> *cubes of bread fried in lard, with garlic*

Cook the potatoes, onion and garlic in the lard over a gentle heat for about ten minutes, turning them over occasionally. This process should not be hurried, or the vegetables will brown and the special flavour of garlic and lard will be spoiled. Add the stock or water and simmer until the potato is tender. Sieve or purée in the blender. Reheat and check the seasoning. Stir in the parsley, and serve with the bread in a separate bowl.

DUCHESSE, CROQUETTE AND DAUPHINE POTATOES

These three types of potato are all made with the same basic mixture:

½ kg (1 lb) potatoes, scrubbed
2 small egg yolks
25 g (scant 1 oz) butter
salt, pepper

Potatoes should be freshly cooked for these dishes; they are much easier to work and better in flavour than leftovers from the day before. As the sieved purée should be as dry as possible, it is a good idea to bake the potatoes in the oven, before peeling and putting them through a *mouli-légume*. If you boil them first, return the purée to the pan to dry it out if necessary. Beat the two egg yolks into the hot potato, together with the butter. Season.

DUCHESSE POTATOES Pipe the potato mixture into swirled pyramid shapes on a buttered baking sheet. Brush them over with beaten egg, then brown them in a hot oven at 220°C/425°F/Gas Mark 7. The mixture can also be piped as a border round dishes or scallop shells, or it can be made into nest shapes to take creamed peas or a broad bean purée.

CROQUETTE POTATOES The mixture can be flavoured in all sorts of ways, but soften it very slightly with a little hot milk. Mix in chopped hazelnuts, or parsley and chives, or grated Parmesan, or shrimps if you like. Leave the whole thing to get cold, then form it into croquette shapes, or cutlets; dip them in egg, roll them in breadcrumbs and fry until golden-brown all over in butter. You can also roll them rather firmly in slivered almonds and then fry them.

POMMES DAUPHINE are delicious and simple enough: you mix the potatoes, after adding egg and so on, with a choux paste. Then you deep-fry spoonfuls of the mixture, which puff up lightly.
   For the choux paste you will need the following ingredients:

125 ml (4 fl oz) water
¼ teaspoon sugar
30 g (1 oz) butter
60 g (2 oz) strong plain flour
2 large eggs
¼ teaspoon salt

Bring the first three ingredients to the boil, stirring them together so that the butter melts. Tip in the flour off the heat, mix it in vigorously, then return to the heat for a minute or two until the paste leaves the sides of the pan. Remove and allow to cool for several minutes, then beat in the eggs one by one and the salt. Add to the hot potato mixture, stirring them together to make an even paste. Deep-fry and serve with grilled and roast meat. They are lighter by far than chips. Their texture and flavour please everyone.

POTATO CAKES

It is important to use freshly cooked potatoes if you want potato cakes to taste as they should. I was delighted and surprised to come across a version of our familiar potato scones and cakes, in a pastrycook's shop in Orléans, and even more delighted to find the recipe, which is very close to ours.

Whatever kind of potato cake you want to end up with, you start with a basic mixture:

> 250 g (8 oz) potatoes, scrubbed
> 30–60 g (1–2 oz) butter
> salt, pepper
> about 60 g (2 oz) plain flour

Boil, peel and sieve the potatoes. Mix in the butter while they are still warm, then season well. Gradually add the flour until you have a firm dough that you can roll out – the best way of doing this is to mix everything together with your hands.

THIN POTATO CAKES Roll out the dough into a circle about ½ cm (¼ in) thick or a little more. Cut it across into triangles and prick them with a fork. Cook them on a greased griddle or frying pan or hot plate, turning them so that they brown on both sides. Eat with butter, while they are still hot.

THICK POTATO CAKES Shape the dough into a roll about 4 cm (1½ in) in diameter. Cut it into slices a good 1 cm (½ in) thick. Cook as above, only rather more slowly. Good with bacon and egg.

POTATO GNOCCHI
GNOCCHI DI PATATE

Italians poach their potato cake mixture and serve it as a first course with butter and Parmesan cheese, or *pesto*, or a sauce.

> 1 kg (2 lb) potatoes, scrubbed, boiled
> 250 g (8 oz) plain flour
> 1 teaspoon baking powder
> 2 eggs
> salt, pepper
> butter
> Parmesan cheese
> tomato or meat sauce, or pesto

Peel the hot potatoes and sieve them to make a smooth dry purée. Sieve the flour and baking powder together, mix into the potatoes and add the eggs, and seasoning if necessary. Spread out on a dish and chill.

When you want to cook the gnocchi, form the dough into long finger-thick pieces on a floured board, then cut them into short lengths of about 2–3 cm (1 in). Press each one gently round your finger to curve it.

Bring a large pan half full of salted water to a steady but not too rumbustious boil. Put in a few gnocchi. In about four minutes they will bob to the surface. Remove them with a slotted spoon to a buttered serving dish and keep them warm. Repeat until the dough is used up. When adding extra layers to the serving dish, pour a little melted butter and some grated Parmesan cheese between each one.

Serve with more butter and cheese, or with tomato or meat sauce. Genoese *pesto* is made by pounding together 60 g (2 oz) fresh basil leaves, two cloves of garlic, 30 g (1 oz) pine kernels and three tablespoons each grated Parmesan and pecorino cheese, to a mass. Then gradually add up to a teacupful of olive oil. Die-hard perfectionists use a pestle and mortar: an electric blender produces a very good result and that is what I use most of the time.

ROAST POTATOES

Scrub and boil the potatoes in well salted water for six minutes. Remove them, allow them to cool and strip off the skins. Cut the potatoes into convenient pieces and rough up the surfaces slightly with a fork; this helps to make the golden-brown coating at the end even crisper.

Heat lard or dripping in a shallow roasting pan either in the oven or on the stove. Put in enough to give you 1 cm ($\frac{1}{2}$ in) depth if possible. Turn the potatoes in the fat, add any other root vegetables that you want to cook in the same way – parsnips, for instance, which should also have been blanched for six minutes and then peeled and cut up. Put on to a top shelf above the roasting meat 40–50 minutes before the end of cooking time.

You will get the best results if the oven is at 190–200°C/315–400°F/Gas Mark 5–6. Keep an eye on the potatoes, turning them over from time to time, so that they acquire an even coating.

PAN-FRIED OR SAUTÉ POTATOES

A warning – do not believe recipes that instruct you to fry the potatoes over a high or fairly high heat. You need a moderate heat, so that the frying takes 15–20 minutes to produce a golden-brown crust. If you are using electric rings, you will soon find out which number on which ring gives the steady heat required. With gas rings, use an asbestos mat to give an even distribution of heat beneath the pan.

First of all wash, peel and cut up enough potatoes to form a single layer in a large heavy frying pan. They should be a little smaller than the normal size for roast potatoes. Blanch them for seven minutes in boiling salted water, then drain them. (If you prefer sliced potatoes, blanch them for five minutes.)

Heat up enough olive oil, beef dripping or lard to make a half centimetre ($\frac{1}{4}$ in) depth in the pan. Put in the potatoes, and leave them to cook for 15–20 minutes, turning them occasionally. They should be crisp outside and tender in the middle. I find that olive oil gives by far the best result of all, a superb flavour and a light crisp brownness. This is expensive, but if you eat a few potatotes and value their flavour, you may well think it worth while. If you want to use butter, clarify it first, or it will burn.

Potatoes fried in this way really need no more flavouring than salt sprinkled over them at the end. But for a change, add chopped fennel leaves or tarragon for the last five minutes of cooking time. A good addition is separately fried onions – 125 g (4 oz) for $\frac{1}{2}$ kg (1 lb) potatoes – which should be mixed into the drained cooked potatoes just before serving; this is *pommes de terre lyonnaise*.

DEEP-FRIED POTATOES

I have to declare my interest in the matter of deep-fried potatoes. They are most unsuitable for making in the ordinary family kitchen. The telephone rings, a child crawls round your feet, the cat demands his supper with a paw, and your attention is diverted. I once set a pan of oil on fire in this sort of way. The horror was appalling, out of all proportion to the black ceiling. At least there were no children about. I know we should not over-protect them, but I draw the line at boiling oil.

Having said this, I admit the book would be incomplete without instructions for making these delicious kinds of potato.

Check with the *Cutting up Vegetables* section on page 548 and prepare the potatoes accordingly. Dry them well. Put a first batch into the deep-frying basket, keeping the rest covered with a cloth; never cook more than 200–250 g (6–8 oz) at once.

CHIPS Heat up the pan, half full of oil, to 196°C/385°F. Lower the frying basket into the fat, stir the potatoes with a wooden spoon once or twice and cook them until they are tender but pale. Drain and repeat with the remaining potatoes. Raise the heat until the oil is at 200°C/395°F, and lower the first batch of potatoes into the pan again. Remove when they are golden-brown, and keep them hot on a baking sheet lined with absorbent kitchen paper, while you brown the rest. Sprinkle them with salt and serve. When the oil is cold, pour it through a paper coffee-filter, set in a funnel, back into a clean bottle for further use.

GAUFRETTE AND MATCHSTICK POTATOES These are cut so thin that they only need frying once, at the higher temperature, until golden brown.

GAME CHIPS OR CRISPS Slice peeled potatoes on the cucumber blade of the grater, dry them well, and cook like *gaufrettes*.

SOUFFLÉ POTATOES do not work unless you make them with waxy potatoes. Peel them, cut them into slices 3 mm (⅛ in) thick, and soak them in water for an hour. Dry them carefully. Heat the oil in the pan to about 160°C/325°F and lower in the first batch of potato slices. When they rise to the top of the fat, remove them and spread them out to dry and cool on absorbent paper. Just before the meal, heat the oil to 200°C/395°F, and plunge the potatoes in. At this second cooking they should puff out into golden-brown balloons.

DANISH SUGAR-BROWNED POTATOES
BRUNEDE KARTOFLER

These sound odd, but taste good with ham, poultry, veal, lamb, pork – meats that are often served with a sugared sauce or fruit. Remember that when potatoes first came in they were as often used for puddings as for a vegetable. For this recipe have new potatoes that are evenly sized. Naturally you could use large potatoes cut up, but it will be more difficult to get an even glaze without burning the sugar. Rather than weighing the potatoes exactly, put them into your frying pan to make sure that you are not using too many for comfort. They must fit in a single layer, but it should not be so tightly packed that you cannot turn the potatoes over and keep them moving.

> about ½ kg (1 lb) new potatoes
> 50 g (1½ oz) butter
> 50 g (1½ oz) granulated sugar
> 1 teaspoon salt

Wash, boil, then peel the potatoes, being careful not to overcook them. Melt the butter in a heavy frying pan and stir in the sugar. Cook to a light golden-brown caramel – keep stirring, or you risk a black caramel. Lower the heat and put in the potatoes, turning them about until they are coated and brown with the sugary sauce. Turn them into a dish and sprinkle them with the salt.

SCOTS STOVED POTATOES
STOVED TATTIES, STOVIES

All her life Lady Clark of Tillypronie, on Deeside, collected recipes. From friends in Europe as in England, from chefs she employed when her husband was in the diplomatic corps in Paris, Brussels and Turin, from hotels she visited, from her Scotch neighbours, she took notes, ideas, housekeeping hints. In 1901, after her death, her husband handed the vast jumble to a friend to be put in order and edited for publication. Reading between the lines of the polite and deferential introduction, this was a nightmare task (Bagshot clear soup came with five versions and 13 variations). Perhaps it helped that Lady Clark was no shrinking Dora Copperfield: her recipes show that she could cope as enthusiastically with salting a pig or kippering a salmon as with a soufflé. No doubt others did the work, but she supervised with gusto.

As she and her husband built Tillypronie, and spent much time there, she had a chance to collect a number of Scots recipes. This one for potatoes is excellent. I often use it for new potatoes, giving them about half an hour on the plate of an oil-fired cooker. 'Stoved' from the French *étuvé*, meaning stewed, in this case potatoes stewed in their own juices, with only a tiny amount of water and butter to prevent them sticking in the first stages. With gas or electric cookers, an asbestos mat helps to keep the temperature evenly low. Keep the lid of the pan, which should be a heavy one, jammed on tightly with foil.

They must be good potatotes to begin with. Peel them, and put them in a pan with about 2 tablespoonfuls of water. Sprinkle them with a little salt and add a tiny bit of butter here and there. Cover close and simmer till soft and melted.

FRENCH STOVED POTATOES
POMMES DE TERRE ÉTUVÉES À L'AIL

The combination of lard and garlic with potato is one of the most savoury I know. The whole house fills with the appetizing smell. The quantity of garlic should be varied according to your own taste and the quantity of potatoes. A winter recipe for full-grown potatoes of fine flavour.

Follow the Scotch recipe above, putting at least three peeled and halved cloves of garlic in with the water. Instead of butter, dot the potatoes with lard, using it fairly generously. They should be cut to even-sized pieces of 2–3 cm (1 in) after peeling.

If you are cooking a large quantity, shake the pan from time to time and turn the potatoes about, say every 20 minutes. If cooked very slowly, the potatoes can take over an hour. At no point should you risk browning them.

BAKED POTATOES

After scrubbing the potatoes and cutting the worst blemishes out neatly, push a skewer through the centre of each one. This will conduct heat to the middle and speed up cooking time. Brush the potatoes with melted butter or olive oil or lard, sprinkle with salt – though this does little more than flavour the skin slightly – and lay on foil on a baking sheet. For potatoes weighing on average 250 g (8 oz) allow 1–1¼ hours at 200°C/400°F/Gas Mark 6, turning them over occasionally to prevent the skin forming a hard patch on one side. If you like the skin crustier, leave the potatoes longer.

When you want only half a dozen potatoes for the family, domestic economy dictates that you use the oven for other things as well. When a stew is planned, allow extra time for the potatoes and put them at the top of the oven – 2 to 2½ hours will be all right. When pastry is to go in to a very hot oven, put the potatoes at the bottom, on the floor of the oven, and give them less time, 45 minutes or a fraction longer.

When all is said and done, you are not timing a soufflé. Baked potatoes are good-tempered. If they are ready early, they will not spoil. You will know when they are done by the slightly crackling noise that the skin makes as you squeeze the potatoes gently, and by the yielding softness inside.

TO SERVE BAKED POTATOES To me there is only one thing better than salty farm butter with potatoes, and that is pressed caviare. It sounds expensive, and it is, but it makes a fine start to a meal. I had the idea from the head of the caviare importing firm, T. G. White's in Soho. Every Saturday, he told me, he and his wife bake potatoes the size of duck's eggs, split them and eat them with pressed caviare out of a tube (this is the way it often comes, like toothpaste, though colour and flavour are not at all the same). Next best is lumpfish caviare or smoked cod's roe mashed up with butter. Many people like soured cream and chives; I do, too, but prefer butter any day.

Endless elaborations are possible. Some restaurants devote themselves to baked potatoes stuffed in many ways, just as a number of French restaurants these days serve pancakes with 50 and more different fillings. Many people, and not just children, will enjoy a baked potato lunch party with lots of different items for them to make up their own mixtures – sausages, eggs, crumbled bacon rashers, mustards, cream cheeses such as *boursin à l'ail et aux fines herbes* or a pepper-covered cream cheese, a variety of smoked and salted fish. One good easy flavouring is Angostura bitters: scoop out the pulp from the halved potato, mix it with plenty of butter, some salt and pepper, then add the bitters drop by drop to taste. Pile the potato back into the skins, return to the oven to heat through and serve with more butter.

MASHED POTATO

Scrub, boil and then peel 1 kg (2 lb) potatoes. Put them through the coarse plate of the *mouli-légumes*, or through a potato ricer, or mash them with a potato masher. Any of these methods are better than over-cooking the potatoes and completing their collapse with a fork. *They should always be hot*.

I have tried sieving potatoes electrically, but find that however low a speed, however large the plate, the potatoes turn to an unpleasant starchy glue.

Season them, especially with pepper, and gradually beat in about 150 ml (¼ pt) of hot cream, milk or stock. Do this over the fire so that the whole purée warms through. The quantity of liquid needed can vary with the variety of potato, and also with the texture you prefer. Finally add a good knob of butter.

VARIATIONS

Use half each of potatoes and onions, cooking them together.
Use half each of turnip or parsnip and potatoes, cooking them separately, then mashing them together with cream, milk or stock and butter.
Use half each of celeriac and potatoes, cooked separately, then mashed together.

CHAMP

Like many other peasant dishes, Ireland's champ depends for success on using good ingredients. The potatoes should be well flavoured. There must be plenty of salty butter, no niggling. The spring onions, or other greenery, should be lively. If you buy vegetables sweating in plastic from the supermarket, you will probably be at a loss to understand why anyone eats it above subsistence level.

> *6 large fat spring onions*
> *300 ml (½ pt) milk*
> *¾ kg (1½ lb) potatoes*
> *salt, pepper*
> *butter, buttermilk or creamy milk to drink*

Chop the spring onions, including the green leaves. Put them into a pan with the milk and simmer for about 20 minutes, while the potatoes are cooking. Rice or mill the potatoes, or mash them with a fork. As the lumps disappear, mix in the milk and onions, adding extra milk if you need to. The mashed potatoes should be firm enough to hold their shape, but creamy. Season. Divide the mixture between individual heated plates, shaping it into mounds. Make a well in the middle of each mound and put in a huge lump of butter. It melts rapidly. The idea is to fork up some of the champ from around the edge and dip it into the central pool of butter before eating it. The final mouthfuls are glorious, when you get to the buttery centre. With champ, drink buttermilk, which can be bought from health food stores and some groceries, otherwise Channel Island milk.

VARIATIONS The method is always the same, but sometimes a large handful of chopped cabbage or peas will be used instead of spring onions, or a teacupful of chopped chives/parsley/young nettles. It is good, if unorthodox, with leeks.

SPANISH TORTILLA

A *tortilla* when made in Spain – it is something quite different in Latin America – is a solid vegetable omelette, the most popular vegetable being the potato in combination with onions and with other cooked vegetables. Ingredients and methods are the same as a Persian *kuku* (see pages 51, 471), though Spaniards do sometimes use bacon, smoked ham, chorizo as well as or instead of vegetables, which would not do for the Muslims. Presumably the dish was introduced by the Arab conquerors of Spain.

It is not surprising that the potato should be so popular. It came to Europe from the New World in the Spanish ships, and was being grown in Spain at least as early as 1570, probably much earlier. The Hospital de la Sangre of Seville, according to Salaman was buying potatoes regularly by 1576, but only by the pound – in other words as a luxury. In 1584, the purchase was being made by the *arroba*, a unit of 25 lbs.

The detailed method for a *tortilla* is given on page 471, the recipe for *kuku*. For six people, beat up six to nine eggs and season them. Prepare a filling based on two large diced potatoes and two large chopped onions; cook them gently at first in olive oil, then more rapidly so that they brown nicely. Add any extra cooked vegetables – peas, peppers, spinach, green beans – or some chopped bacon or ham, and fry for a minute or two longer. Transfer with a slotted spoon, so that the oil falls back into the pan, to the beaten eggs. Mix well. Reheat the oil in the pan and put in the egg mixture, smoothing it down into a thick cake. When the underneath is golden-brown and the *tortilla* just firm, turn it over with the help of a large plate and brown the other side; or put the frying pan under a hot grill to brown the top. Serve hot, warm or cold, on its own, or with piquant dishes of kidneys, sweetbreads, bacon and so on.

POTATO SALADS

The important thing, the difficult thing with all potato salads, is finding a good variety of potato that doesn't crack up when you slice it. Nearly as important is to leave the potatoes to cool down in a well flavoured liquid – you can use dry white wine, dry cider or vinaigrette – and to finish them with onions of one kind or another. One of the first potato salads I ever made was for an Anglo-French friend of ours who loves food. He looked at the genteel scatter of chives across the top and said 'Where do you keep the onions?' I told him, wondering what he intended. He returned with a huge Spanish onion and a sharp knife. The onion was chopped, not too small, and tipped into the bowl and mixed in. 'Taste that,' he said. After a few second's huff, I was grateful to him. He had taught me that a potato salad is not a ladylike dish: it should have a direct appeal, from the delicate earthiness that characterizes good potatoes and the sweet fire of a good onion.

WITH VINAIGRETTE

> *1 kg (2 lb) new potatoes*
> *150 ml (¼ pt) dry white wine (optional)*
> *olive oil vinaigrette*
> *plenty of chopped raw onion/spring onion/chives*
> *chopped parsley*

Wash and boil or steam the potatoes, then peel them while still hot. Cut them up into the wine, if used, or into the same quantity of vinaigrette. Turn the whole thing carefully from time to time, as the potatoes cool.

If you used white wine, drain off any surplus liquid and dress the potatoes with just enough vinaigrette to coat them nicely.* Mix in the onion to your inclination. Serve chilled with parsley on top.

WITH MAYONNAISE

Follow the recipe above to the *, making sure that the potatoes are evenly coated but not paddling in a sea of vinaigrette. Now mix in the mayonnaise, starting with 150 ml (¼ pt). This will be easy to do on account of the oil-coated potatoes; it also means that you need to use less mayonnaise, which makes for a lighter salad. Now add the onion flavouring(s), and serve chilled with parsley.

These salads can be served on their own, or with cold meats and a green salad. In France and Germany, in particular, they are eaten with hot smoked sausages and cured sausages of all kinds, a good partnership.

POTATO AND MUSSEL SALAD
SALADE À LA BOULONNAISE

Open 2–3 kg (4–6 lb) mussels with 200 ml (7 fl oz) dry white wine, in the usual way (page 188). Discard the shells. Put the mussels aside and strain the liquor into a small pan. Cook, peel and cut up 1 kg (2 lb) potatoes into a basin. Bring the mussel liquor to the boil, pour it over the still hot potatoes and leave to cool. To serve, arrange the drained potatoes on a dish, with mussels scattered on top. Pour over a few tablespoons of vinaigrette or mayonnaise, and cover generously with chopped parsley, shallot and chives. Serve chilled. One of the best potato salads. A winter delight.

POTATO PANCAKES
KARTOFFELPUFFER

German potato pancakes are a delicious indulgence, best eaten on their own, an ideal family dish for the weekends. The friend who supervised my first efforts assured me that the odd bluish tinge of the raw potato batter is all part of the charm and that the pancakes should be soft in the centre to contrast with the crisp brown edges.

> ¾ kg (1½ lb) peeled potatoes
> 100 g (3½ oz) grated onion
> 2 eggs
> 175 g (6 oz) plain flour
> good pinch salt
> lard

Shred the potatoes on the coarse side of the grater into a bowl. Mix in the onion, eggs, flour and salt, one by one, and stir smoothly and thoroughly into the potato shreds. Heat enough lard in a heavy frying pan to make ½ cm (¼ in) depth. Put in four tablespoons of the mixture in a heap, then flatten it down with the back of the spoon to make a pancake about 12 cm (5 in) or so across. If you use a large enough pan, two or three *kartoffel-puffer* can be cooked at once. When the whiskery edges are golden-brown and crisp, turn the pancakes and cook the other side. The centre will have a smooth, paler look, more like pancakes. Drain well and serve with apple or cranberry sauce, or with slices of Cox's Orange Pippins cooked in butter, and scattered with sugar.

POTATOES AND PEARS FROM LUCERNE

A nice thing about the great Montaigne was his love of food and drink. This is not, I know, an essential for literary excellence, but it makes for sympathy. On his journey from France into Italy in 1580, he enjoyed the

Swiss habit of serving pears with meat. It still persists there – as in Germany – in delicate dishes like this one.

> *1 large onion, sliced*
> *2 tablespoons butter*
> *175 g (6 oz) dried pears, soaked, cut up*
> *½ kg (1 lb) small new potatoes, scraped*
> *honey, pepper, salt*

Cook the onion slowly in butter until soft, in a covered pan. Add the pears, their soaking water, the potatoes, a tablespoon of honey, pepper and a little salt. Pour in more water so that the potatoes are barely covered. Simmer with the lid on the pan until the mixture is tender and slightly glazed. Drain, correct the seasonings to taste and serve hot with ham, pork, salt pork or beef.

POTATO CAKE FROM LIMOUSIN
GÂTEAU DE POMMES DE TERRE

Judging by the acres of moss-covered tree trunks, the district of France around Limoges has more rain than we do. Food needs to be comforting. This is the dish for a wet summer day. Similar pies are made in the Bourbonnais and the Berry.

> *puff pastry made with ¼ kg (½ lb) flour*
> *¾ kg (1½ lb) new potatoes, scraped*
> *150 g (5 oz) chopped onion*
> *4 cloves garlic, chopped*
> *salt, pepper, grated nutmeg*
> *60 g (2 oz) slightly salted butter*
> *250 ml (8 fl oz) cream*
> *1 egg, beaten*
> *parsley, chives, chervil, chopped*

For this recipe you need ½ kg (1 lb) total weight of puff pastry, i.e. pastry made with ¼ kg (½ lb) flour, a scant ¼ kg (½ lb) butter and so on, or one large packet of frozen puff pastry which works well.

Roll out the pastry and use just over half to line a greased 25 cm (10 in) flan or shallow cake tin, preferably with a removable base. Slice the potatoes on a mandolin slicer, or the cucumber blade of the grater, then blanch for two minutes in boiling, salted water. Drain and arrange on the pastry in layers with onion, garlic and seasonings. Dot each layer with butter. Pour in half the cream. Cover with the remaining pastry, pinching the edges firmly together and knocking them up. Make a central hole and decorate with the trimmings. Beat the remaining cream into the egg and brush over the pastry. Bake at 230°C/450°F/Gas Mark 8 for 30–40 minutes; cover the pastry with butter paper if it is browning too quickly, you can also lower the heat a little after about 20 minutes, but the potatoes need a good heat to cook them. Meanwhile stir the herbs into the remaining egg and cream. When the pie is cooked, pour in the egg and cream through the central hole, using a funnel – go slowly. Put it back into the oven for five minutes, then serve.

POTATO AND CHEESE DISHES

Everybody who grows potatoes has seen the point of mixing them with cheese, to make solid, appetizing dishes for family meals. Differences come with different local cheeses, the quantity used, the method of cooking. They are very much peasant dishes, or dishes of the poor who cannot afford meat. When one comes to grander cookery, the quantity of cheese is reduced to a seasoning, or gratin topping, so that the potatoes can be served with meat.

PAN HAGGERTY from Northumbria – a favourite dish of my childhood. Slice up 1 kg (2 lb) peeled potatoes and ½ kg (1lb) onions. Grate 125 g (4 oz) Cheddar or Cheshire cheese. In a heavy frying pan melt a tablespoon or two of beef dripping. Put in layers of potatoes, onion and cheese, seasoning each layer, and beginning and ending with potatoes. Cover and fry gently until the underneath is brown and the vegetables are tender. You can brown the top, either by turning the pan haggerty over with the aid of a plate, or by putting the frying pan under the grill.

POTATO CAKE WITH CHEESE, from Touraine and Anjou. Melt a large lump of butter in a non-stick metal pan – a sauté pan or cake tin will do. Fill it with slices of potato, layered with thin slivers of Gruyère cheese, seasoning the potato layers as you go. Cook for an hour or a little longer until the cake is cooked. Turn it on to a dish, scatter with parsley and serve. Use about 175 g (6 oz) cheese to 1 kg (2 lb) of potatoes.

POTATO GRATINS
GRATINS DE POMME DE TERRE

The most famous gratins of all, gratins which are now part of the classical tradition of French cookery, come from the Savoie and Dauphiné, neighbouring Alpine provinces on the Italian border. The Dauphiné, which runs down to the Rhône, was added to France in 1349 and became an appanage of the king's eldest son, who from then on was known as the Dauphin. There is some variation in the recipes for *gratin dauphinois*, and much argument, but its home is said to have been the Vercors, a forested mountainous district scarred by river gorges (and which offered so strong a natural shelter to the Maquis resistance fighters in the last war that the Germans were driven to a dreadful and methodical devastation of revenge). One may imagine that the damp rigours of the winter climate are slightly alleviated by the regular appearance of *gratin dauphinois*.

The deliciousness of any gratin lies in the contrast of soft but well flavoured vegetables with the crusty, brown surface. To achieve this harmony one needs a certain depth and moisture, but not enough to deprive people of a good share of crispness. No wonder special dishes have been evolved to get the proportions right, without anxiety, and without too much reliance on weighing. In this country we are most familiar with the oval, brown and yellow dishes from Berry, which are sometimes called 'sabots'. But there are round ones, too, from other parts of France, some with straight sides, some with sides sloping in to the base, and some only glazed inside and on the rim. Like our pudding basins, they are made in a complete range of sizes, from $2\frac{1}{2}$–7 cm (1–$2\frac{1}{2}$ in) deep and from 16–40 cm ($6\frac{1}{2}$–16 in) at their widest point. Although gratin dishes are usually round or oval, and made in an ovenproof earthenware, there is no reason why you should not use a shallow square or oblong dish in Pyrex or stoneware.

As gratins are all made by the same method, I shall give that first, and follow it up with lists of suggested ingredients. Remember that they are suggestions, and not Holy Writ.

GRATIN METHOD

Butter the dish generously before you start. Then prepare and slice the potatoes and other vegetables, and layer them into the dish with their seasonings and flavourings, ending always with a smooth layer of potato which can be arranged decoratively in circles of overlapping slices. Pour in the liquid, and cover the top with grated cheese and breadcrumbs. Dribble a little melted butter over the whole thing as evenly as possible. Bake at any temperature convenient for you from 220°C/425°F/Gas Mark 7 to 160°C/325°F/Gas Mark 3, bearing in mind that the time the gratin will take will be partly affected by the temperature. I say 'partly' because it is a most accommodating sort of thing to cook.

The method can be fiddled with in various ways. It is a good idea, for instance, to slice the potatoes into a bowl of water on the wide cucumber blade of the standard grater. This means that the slices come thin and even. They can then be blanched for two minutes in boiling salted water, which has two advantages – it saves a little time and gets the potatoes started, and it means that the gratin dish can be properly packed without gaps and without too much final subsiding. Onions can also be blanched, or stewed in butter, if they are a major item in the dish. If they are being used as a flavouring, they should be added raw.

If I am uncertain when people are likely to turn up for supper, I make a gratin with a béchamel-based sauce as liquid. Then the dish goes under the grill for the top to turn a pleasant golden colour. It is laid to rest in the oven at 150°C/300°F/Gas Mark 2 and I know that if it has to stay there an hour or an hour and a half, nothing dreadful will happen to it. But if the family turns up in half an hour, it will be enjoyably eatable then, too. In the following lists, the ingredients are given in the order in which they are put into the gratin dish, not necessarily in the order in which they were prepared.

GRATIN DAUPHINOIS

This is the version given by Edouard de Pomiane, said to be the 'correct' thing as it contains neither eggs nor cheese. He adds a teaspoon of flour to the cream to prevent it separating: other chefs have added a little water instead. The top browns on its own, without the assistance of bread-crumbs or grated cheese.

> ¾ kg (1½ lb) potatoes, peeled, sliced thinly
> salt, pepper
> 4 cloves garlic, finely chopped
> scant ½ litre (¾ pt) boiling milk
> 200 ml (7 fl oz) double cream mixed with 1 teaspoon flour

Cook for 50 minutes at 200°C/400°F/Gas Mark 6 then raise the heat to complete the browning.

GRATIN AUX CÈPES

The best gratin of all, because ceps, even more than other mushrooms, have a perfect affinity with potatoes. If your woods do not run to ceps, the plump rounded *Boletus edulis* with spongy gills, you could use field or even cultivated mushrooms.

A better substitute would be a couple of packets of dried Italian, Polish or German ceps. They should be soaked for 20–30 minutes in hot water first, and the liquid should be added to the cream and water. This quantity, much smaller than the ¾ kg (1½ lb) of fresh ceps in the ingredients, is enough to make a good flavour. If you like, you could include 250 g (½ lb) cultivated mushrooms, to give an extra presence and texture. Remember that fresh ceps give up a lot of moisture, so be prepared when using packaged ceps on their own, or with cultivated mushrooms, to add extra liquid if the gratin looks at all dry as it cooks.

> *1 kg (2 lb) waxy new potatoes, or Desirée potatoes, peeled and*
>   *sliced thinly*
> *about ¾ kg (1½ lb) ceps, sliced, stalks chopped*
> *salt, pepper*
> *4 tablespoons finely chopped onion*
> *4 tablespoons finely chopped parsley*
> *6 tablespoons grated Gruyère*
> *2 cloves garlic, finely chopped*
> *300 ml (½ pt) double cream*
> *60 ml (2 fl oz) water*
> *extra cheese*
> *melted butter*

When making up the gratin, mix all the flavouring items (onion, parsley etc) together evenly and sprinkle them over the layers of potato and ceps.

GRATIN SAVOYARD

The liquid in the Savoyard method is usually beef stock, but water, or water and white wine can be used instead.

> *1 kg (2 lb) potatoes, peeled, sliced thinly*
> *125 g (4 oz) grated Gruyère or Emmental*
> *1 clove garlic, finely chopped*
> *scant ½ litre (¾ pt) stock or water*
> *salt, pepper*
> *extra grated cheese*
> *melted butter*

JANSSON'S TEMPTATION, JANSSON'S FRESTELSE

A deservedly famous Swedish dish, one of the best gratins of all. The anchovies form a central criss-cross layer, with onions on either side, potatoes top and bottom.

> *1 kg (2 lb) potatoes*
> *2 really large onions*
> *2–3 tins anchovy fillets, or 15 Swedish anchovies, filleted*
> *150 ml (¼ pt) single cream*
> *salt, pepper, butter*
> *150 ml (¼ pt) whipping or double cream*

Peel the potatoes and shred them into matchstick strips on a mandolin grater, or with a sharp knife. Slice the onions. Make up the gratin in the usual way, going easy with the salt, and dotting the top with butter. Pour on any oil from the anchovies and the single cream, and bake for an hour at 220°C/425°F/Gas Mark 7. Taste the juices and adjust the seasoning. Pour over the whipping or double cream and serve. Do not worry if the cooking juices have a curdled appearance – the final addition of thicker cream pulls the sauce together.

BEST END OF NECK IN THE STYLE OF THE BAKER'S WIFE
CARRÉ D'AGNEAU À LA BOULANGÈRE

Lamb *boulangère* is a reminder of the very different lives of ordinary French people, until recent prosperity. Even now they do not use ovens as much as we do for cooking meat, and they rarely make cakes and baked puddings. Certainly in the summer meat is nearly always pot-roasted on top of the stove, even fillet of beef which will be turned over in a deep pot over a sharp flame – no lid – until the outside and a little of the inside is cooked but the middle part rare.

Certainly this is economical with firing or electricity, but it is also the result of long habit. In the old days housewives, who often lacked an oven and even a kitchen, in the modern sense, would send prepared dishes of food along to the bakery so that they could be put into the ovens after the first bread of the day had been taken out. It would then be ready by midday, along with the second lot of bread which would be put in when the meat was taken out. Some people would even say exactly where in the oven they wanted their dishes put.

Nowadays nearly everyone has an up-to-date cooker, should they wish to use it, but in some country districts the old habit has not been entirely lost. When we first went to work in France in 1961, we had a primitive cave house with a couple of Calor gas burners. One of the children would take the Sunday meat sitting on its vegetables down to the bakery, and the *boulangère* would put it into the ovens which were in a cave directly under ours (the smell of bread hovered twice a day over our tiny terrace, which incorporated the bakery chimney in its wall: the top of the chimney opened three feet above our heads). At midday someone would go down for our lunch, to bring it up the steps carefully covered with a white cloth to protect it from dust (and picking fingers). Now we have moved round the cliff to another cave, which is just as primitive, but further from the bakery. I pot-roast the meat on Sundays like my neighbours, but it does not taste nearly as good as it used to.

According to the number of people, buy half a best-end of neck of lamb, or a whole one. Ask the butcher to trim the bones and chine them, for easy carving.

Brown the meat lightly in butter, and set it in a large gratin dish in a bed of thinly sliced, blanched potatoes, interlayered with thinly sliced blanched onions and seasoning. Bake in a moderate oven, 180–190°C/350–375°F/Gas Mark 4–5. The juices from the meat will blend deliciously into the vegetables, and they can be augmented by a few tablespoons of stock made from lamb bones and trimmings.

Some recipes suggest that the onions should be stewed gently in butter, rather than blanched. This can make the dish a little too fatty, depending on the lamb, but it does give it a good flavour. You can get round the problem by being careful to remove the outer layer of fat from the joint, but do not denude it altogether or the lean meat will be tough.

OTHER SUGGESTIONS

Potato moussaka. Substitute 1–1½ kg (2½–3 lb) of good quality potatoes for the aubergines in the recipe on p. 57. Peel them and slice them very thinly into a bowl of cold water, using the cucumber blade of a grater or mandolin slicer. Swish the slices about, then drain them and dry them in a cloth. Fry them in oil, just like the aubergines, until they are nicely browned. It doesn't matter if some of the slices become slightly crusty at the edges.

For a change, assemble the *moussaka* in six or even eight ovenproof pots, rather than one large one. To do this, you will need 1½ kg (3 lb) of potatoes, weighed after any blemished parts have been cut away.

Leek and potato soup, page 302.

PUMPKIN

One of the best sights in autumn, in countries south of England, is a pile of pumpkins in a farmyard, tumbled in giant profusion against the round strawstacks. Colours go from greenish white and green, through yellow and ochre, to the most intense orange. Sometimes the colour is mottled or spotted or streaked. The shapes are as varied as the colours, but they all taste the same.

We grow pumpkins, too, but on a smaller scale, in gardens. Every year in Swindon, and in other towns as well, there is a prize for the largest one. People inject them with milk, water them with kind words, protect them against blemishes, and finally stagger along to the competition with their immense burden.

Pumpkins have been cultivated since ancient times in Europe, the Middle East and the Americas. The word 'pumpkin', though, goes back only to the 17th century, deriving ultimately from the Greek for melon, *pepon* (the earlier word we used was pompion). This means 'cooked by the sun', ripe in other words. The French *potiron*, originally meaning 'huge mushroom' from the Arabic word for morel mushrooms, settled down to become pumpkin only in the 17th century, the century of Perrault and his tale of Cinderella and her coach. The older French word, the word much used by farmers in our part of France still, is *citrouille*, referring to the citrus yellow colour. All derivations lead to the sun. And all recipes come from climates hotter than ours.

A surprising thing is that, as with sweet corn, there seems to be more recipes for using pumpkin from the south west of France than from America, which has the reputation for squashes, pumpkins, the edible gourds. It goes into soup, certainly, and pies, but also into bread and to eke out apples, to emphasize the colour of sweet corn and cornmeal dishes with its blazing orange. In spite of its watery appearance, and its ability to cook in its own liquid, pumpkin has a lower percentage of water, weight for weight, than either cabbage or tomatoes. Nonetheless bad cooking can make it seem wet and pointless as a form of nourishment; the careless cook can quite destroy its fine, delicate flavour.

See chapters on *courgettes*, *marrows* and *squashes*, for other recipes that can be adapted to pumpkin.

HOW TO CHOOSE, STORE AND PREPARE PUMPKIN

Although pumpkins grow to enormous sizes – you can buy seeds for a variety called Hundredweight – greengrocers prefer those that weigh 3–5 kg (7–12 lb) which they can sell whole, or in halves and quarters. Once a pumpkin is sliced up, flavour and nourishment rapidly diminish and mould can appear in a few days. It is worth buying a whole one to make a magnificent Hallowe'en lamp, or a *carbonada criolla* for a bonfire party. Surplus flesh from the hollowed out pumpkin can be cooked in a tiny amount of water, just enough to set it going without burning any, drained and stored in the freezer.

Whole pumpkins can be stored for months, like marrows and squashes. Choose perfect ones that have ripened evenly, and be careful as you move them not to knock them about. In *Gourds*, John Organ recommends placing pumpkins and squashes in a relative humidity of 80% and a temperature of 27–30°C (80–85°F) for ten days. Then put them in a regular temperature of 10–15°C (50–60°F), and they will keep up to six months. Our pumpkins have to put up with less cosseting. They stay in the back kitchen where temperature and humidity vary, but we find they last for three or four months. I check them from time to time, and if they show any soft patches, use them immediately, freezing the left-over flesh.

Most pumpkin recipes start with a prepared purée. Quarter the pumpkin, scrape out the seeds and loosest fibres; the seeds can be dried and their kernels eaten, if you like nibbling. The Greeks call the various edible seeds *pasatempo*, literally a pastime. Spread pumpkin or other large squash seeds out on a baking sheet, sprinkle them lightly with salt, and bake for 20 minutes at 190°C/375°F/Gas Mark 5. In some Mediterranean countries, seed and nut vendors put up their stalls close to a bus stop. The route between will be tracked on the pavement with a scatter of outer husks, as only the kernels are eaten. They pass the time as you wait.

Having got rid of the seeds, cut the flesh into chunks, removing them from the skin. Put them into a heavy pan, with the thinnest layer of water possible, cover it, and place it on a low heat. As the juices come from the pumpkin, raise the heat to cook it as quickly as possible. Strain off the liquid, keeping it for soup. Now the purée is ready for use in a variety of recipes. Or it can be frozen.

Another way of cooking the flesh, is to cut the pumpkin in two and bake it until the inside is soft enough to be scooped out. If you mash it up with plenty of butter and seasoning, it makes a good vegetable with grilled or roast meat. The flavour is better, too.

As you will see from *carbonada criolla*, pumpkin will thicken meat stews. No need for flour or cornflour. Add 250 g (8 oz) of pumpkin chunks to the usual ingredients, rather more if you are cooking large quantities.

To cook pumpkin flowers, turn to the courgette flower recipe on p. 228.

PUMPKIN SOUPS

Simplest of all are the farmer's soups in the Berry and Vendômois districts of France, around the Loire and the Loir. Ideal for a winter's night, all the ingredients close at hand. This kind of homely dish often carries the older name of *'soupe'* rather than the later and more elegant *potage*.

For *soupe à la citrouille*, allow a kilo (2 lb) slice of pumpkin for six people. Remove the peel, seeds and woolliest fibres; cut the rest into chunks. Put them in a pan with just enough water to cover them. Add the salt. When the pumpkin is tender, drain it well, then sieve it through a *mouli-légumes*. Dilute it to the consistency you prefer with milk, or milk and water, or with milk and single cream. Season with a little grated nutmeg.

Divide the quantity according to children and adults. Sugar the children's soup. Add salt and pepper to the adults', and serve theirs with a bottle of white wine. Toasted rye bread goes well with the soup; it should be a bread of the lighter kind, not one of the dark sticky ryes. Small golden croûtons of white bread fried in butter look beautiful with the orange soup, too.

For a more elaborate soup, follow this recipe for *potage au potiron* from the *Orléannais*. Cook slowly in butter the white parts of two or three leeks, a large onion, chopped, and a couple of small turnips, diced. Add the cubed flesh of a 1 kg (2 lb) slice of pumpkin and cook for ten minutes longer. Moisten with 2 litres (3½ pt) stock or water and season. Simmer for 45 minutes, then sieve through the *mouli-légumes*. Correct the seasoning, add a pinch of sugar and a ladleful of cream.

A variation – omit the extra vegetables, and add ¾ kg (1½ lb) peeled and sliced tomatoes, after the pumpkin has been softened in butter.

EAST CRAFTSBURY PUMPKIN OR SQUASH SOUP

A recipe from Evan Jones's magnificent and informative book, *American Food*.

> *1 pumpkin or other winter squash, about 20 cm (8 in) in diameter*
> *   and 18–20 cm (7–8 in) high*
> *2 tablespoons very soft butter*
> *sea salt, pepper*
> *1 medium onion, sliced thinly*
> *60 g (2 oz) long grain rice*
> *generous ¾ litre (1½ pt) chicken stock*
> *freshly grated nutmeg*
>
> **GARNISH**
> either *4 bacon rashers, crisply fried, crumbled*
> *2 tablespoons grated Mozzarella cheese*
> or *125 ml (4 fl oz) double cream*
> *1 heaped teaspoon chopped chives*

Cut a lid from the stalk end of the pumpkin or squash, then remove the seeds and cottony fibres (it is easier to do this with your hand rather than a spoon). Rub inside the walls with the butter, then sprinkle them with salt and pepper. Put in the onion and rice. Set the pumpkin in a pan or ovenproof dish. Bring the stock to boiling point, pour it into the pumpkin and replace the lid. Bake for two hours at 190°C/375°F/Gas Mark 5. Remove the pumpkin – or squash – from the oven, and with a pointed spoon scrape some of the pumpkin flesh from the walls into the soup. Taste and correct the seasoning. Add the nutmeg to your liking – a good pinch will probably be about right. Either scatter the bacon and cheese on top, or heat the cream and pour that in with the chives. Replace the lid and bring to the table. As you serve out the soup, be careful to scrape in more of the pumpkin so that everyone has a good share.

Like the baked pumpkin and *carbonada criolla* later in the section, this makes a most attractive dish if you are careful to pick out a perfectly-formed and umblemished squash or pumpkin. If you want to prepare this soup for a large party, it is prudent to make it with several pumpkins rather than one huge one, which will certainly crack with the weight of the stock before it is ready.

ROAST PUMPKIN

Like parsnips, carrots or potatoes, pumpkin can be roasted in the oven with a joint of beef or veal or pork.

Cut a slice of pumpkin into convenient size pieces, leaving the skin in place to hold the pumpkin together. Cut away the seeds and woolly fibres. Salt and pepper the pieces, and either arrange them in the fat round the joint, or cook them in lard or dripping in a separate pan above the joint. Pumpkin needs about an hour at 190°C/375°F/Gas Mark 5, the normal roasting temperature. With beef, which needs a faster oven, three-quarters of an hour should be enough. Turn the pieces over occasionally.

Root vegetables are often blanched for a few minutes before they are roasted in the oven. Don't do this with pumpkin, or it will turn to a mush.

FRIED PUMPKIN

Peeled and seeded slices of pumpkin can be lightly coated with seasoned flour and fried in butter, as a vegetable. If this is done carefully, so that the orange slices are tinged with golden-brown, they will taste delicate, not watery.

Pumpkins make splendid fritters. Dip peeled and seeded slices, about 1 cm (½ in) thick, in light batter, p. 568, and deep-fry, or shallow-fry in a good depth of fat.

AMERICAN BAKED PUMPKIN

A simple, good dish from early settler days, when whole pumpkins were wrapped in cabbage leaves to protect them from the worst of the ashes as they baked in the fire.

For six people, pick out a nicely shaped pumpkin measuring about 25 cm (10 in) in diameter. Remove a lid from the stalk end and pull out the seeds and fibre. Smear two or three tablespoons of soft butter round inside the walls.

The usual thing is to season it next with coarse or sea salt and pepper ground in from the mill. But try it as a change sprinkled with sugar. Replace the lid. Then stand it in an ovenproof dish, or a pan.

Bake it for upwards of an hour at 190°C/375°F/Gas Mark 5, until the flesh inside is soft and fragrant. Remove the lid and pour in a generous quantity – about 300 ml (½ pt) – of nearly boiling double cream, either salted and peppered, or sugared; whipping cream could be used instead, or an equal quantity of single and double mixed together.

TOULOUSE-LAUTREC'S GRATIN OF PUMPKIN
GRATIN DE POTIRON

In the collection of Toulouse-Lautrec's paintings, in his native, rose-brick town of Albi, one tiny canvas takes the eye with a brilliant patch of deep yellow, topped by an even more brilliant patch of red. It is not by Lautrec, but a portrait of him done by his friend Vuillard in 1898, who caught him as he was attending to something in the oven. The sad, bearded, familiar face turns towards us looking over his brilliant red-shirted shoulder. The baggy yellow trousers conceal his short legs.

Lautrec loved food, he loved meals with his friends, he loved making dishes of his own and was a skilful cook who could hack up a live lobster and cook it *à l'américaine* in a friend's elegant drawing room without making a scrap of mess or spattering the furniture. His tastes were, as you would expect, for good straightforward food, simple, beautiful, carefully prepared and simmered gently for hours, coaxed into perfection. This gratin of pumpkin is an example of his taste and it has the added pleasure of colours like the ones in that Vuillard painting. A close friend of his, Maurice Joyant, made a collection of his recipes, traditional French ones that he enjoyed making, and also dishes that he had invented himself, under the title *L'Art de la Cuisine*. An annotated and illustrated translation was published in 1966 by Michael Joseph, as *The Art of Cuisine*.

> *1 kg (2 lb) piece pumpkin*
> *seasoned flour*
> *oil*
> *½ kg (1 lb) onions, sliced*
> *250 g (8 oz) tomatoes, peeled (or use canned ones)*
> *salt, pepper, sugar*
> *breadcrumbs*
> *butter*

Peel the pumpkin thinly, remove the cottony inside and pips, and slice the wedge into pieces about ½ cm (¼ in) thick 'and as wide as half of your palm'. Turn the pieces in the flour and fry them in the oil until they are golden but not brown. Do this in batches, so that you have no more than one layer in the pan at a time. Drain them well.

In another pan cook the onions gently in some oil until they are soft, but not coloured. Add the tomatoes, and raise the heat as their juices begin to run. You should end up with a moist mixture of onions, bathed lightly in tomato. Season it with salt, pepper and a little sugar.

In a gratin dish, layer slices of pumpkin and the onion mixture, adding a little extra seasoning. Finish with pumpkin, then scatter it evenly with breadcrumbs. Melt a knob of butter and dribble it over the crumbs. Bake at 180°C/350°F/Gas Mark 4 until the gratin is bubbling at the sides, and the flavours are well mixed – about 45 minutes. Finish browning the top under the grill, if necessary. Serve immediately.

BAKED PUMPKIN
GRATIN DE POTIRON

Like vegetable marrow, pumpkin cooks fairly rapidly to a mush and needs a strengthening element if it's to be eaten as a vegetable. In this recipe, potato adds body; in the German recipe following, apple does the same job, though with a very different result as far as flavour goes.

> ¾ kg (1½ lb) piece pumpkin, peeled, seeded
> 250 g (½ lb) potatoes, peeled, cubed
> 2 eggs
> 100 g (3 oz) butter
> 100–125 g (3–4 oz) grated Gruyère
> 1 tablespoon grated Parmesan
> salt, pepper

Boil pumpkin and potatoes together in salted water. Drain them conscientiously and mash them well to a coherent purée. Beat the eggs and stir them into the vegetables with 60 g (2 oz) of the butter. Mix the cheeses together and add about half, or a little more, to the vegetables. Season and taste to see if a fraction more cheese should be added. Be generous with the black pepper.

Grease a shallow gratin dish with a butter paper and spread the mixture over it. Sprinkle with the remaining cheese. Melt the last of the butter and dribble it over the top as evenly as possible. Bake in a moderate to fairly hot oven, until the top is nicely browned and the edges bubbling.

PUMPKIN AND APPLE
KÜRBISBREI MIT ÄPFELN

A most savoury German recipe for cooking pumpkin with apple. It goes
well with sausages, or it can be eaten on its own with croûtons of fried
bread. The important thing is to see that the apples retain a certain
solidarity, and appear as small golden yellow lumps in the orange pumpkin, a little larger than the pieces of fat bacon and onion. For this reason it
is better to use a good eating apple like Cox's Orange Pippin, rather than
a sour cooking apple which is likely to collapse and lose identity with the
pumpkin, which inevitably cooks to a soft mash.

> *1 kg (2 lb) slice pumpkin*
> *60 g (2 oz) streaky bacon, e.g.* geräuchter bauchspeck *from a*
> *German delicatessen*
> *1 heaped tablespoon butter*
> *250 g (8 oz) chopped onion*
> *350 g (¾ lb) eating apples of good flavour*
> *lemon juice and rind*
> *salt, sugar*

Peel, seed and cube the pumpkin. Cut the bacon into strips and cook them
gently in a large pan with the butter. As the fat begins to run from the
bacon, add the onion, and allow both to brown lightly to a golden colour.
Add the pumpkin and about ½ cm (¼ in) of water. Cover closely and
simmer for about 10–15 minutes, until the pumpkin begins to collapse.
Add the apples, peeled, cored and cut in chunks, a good squeeze of lemon
juice and a teaspoon of chopped or grated rind. Season with a little salt
and some sugar – be careful not to oversalt the dish, just add enough to
bring out the flavour. Continue to cook gently with the lid off the pan, so
that the apple pieces disintegrate to small yellow lumps and the juices
evaporate. Pour off any surplus liquid and correct the seasoning with
more lemon juice and sugar, and salt if necessary; the seasoning is always
the tricky part with sweet and savoury dishes.

CARBONADA CRIOLLA

There's a rich cartoon by Daumier of a husband and wife in their garden.
Under his waistcoat he swells like one pumpkin, under her blouse she
swells like two, and they look down at even more enormous pumpkins at

their feet. No caption required. But I grow pumpkins and I sympathize. Autumn soup or gratin dishes or pies would have been the destiny of those French pumpkins, inevitably, in recipes of the kind given at the beginning of the chapter. I use mine for November firework parties or the children's Hallowe'en junketings, providing either pumpkin pie from America, or from South America (the Argentine) this glorious communal stew, *carbonada*, served inside the hollow of an orange pumpkin.

If you have to buy one, don't choose a Daumier pumpkin too enormous to go into your oven. And avoid the kinds that only just fit in, they have a tendency to collapse. The most suitable weigh about 5 kg (10–11 lb), no more. This should fit easily, but it is wise to measure the oven and take the measure with you to the greengrocer's. It doesn't matter from the point of view of flavour what shape or colour the pumpkin is – they all taste much the same. For *carbonada* at a party, though, choose a spectacular orange pumpkin, the kind that gathers in curved ridges to the small stem.

TO PREPARE THE PUMPKIN Wash it, then cut out a lid from the top – the stalk makes a useful handle, so don't remove it. Nick both lid and base, so that they can be fitted together precisely when the dish is to be served.

Using your hands, a spoon and a small sharp knife, pull out and discard the centre fibrous part with the seeds. Now go carefully as you come to the more solid, seed-free part, so that you do not tear away the intermediate layer which can quite well be cooked.

Cut and scoop away the solid part of the flesh, leaving a good wall to the pumpkin. Be careful not to pierce through the skin. Measure out the weight of pumpkin required for the recipe and cut it into chunks (any left over can be used for soups etc.). Brush the inside of the pumpkin with melted butter, then sprinkle it lightly with sugar. Replace the lid and put the whole thing on to a baking sheet or ovenproof dish. Set aside while you start the stew.

If you have bought a smallish pumpkin, do not worry if there is a lower weight of pumpkin chunks than listed in the ingredients. The variations in *carbonada* recipes are many. The role of the pumpkin, apart from acting as container, is to thicken the juices of the stew without heaviness; it dissolves into a grainy texture that is most agreeable to eat. The pronounced sweetness of the dish makes it a great success with children, yet it is clear and fresh-tasting. The colours, ranging from creamy white to brown to brilliant yellow and orange, make an autumn glory appropriate to Hallowe'en or Bonfire night.

THE MEAT STEW

If your party consists mainly of children, the following list of ingredients should make enough *carbonada* for 20, or even more. It is a most filling dish, plenty for about 15 adults. With a small pumpkin, you will not get all the stew into it at once. Keep the remainder warm and use to refill. The meat can be either beef or lamb. On balance, I think beef is better. You could use shin rather than chuck steak, allowing more time for it to cook. It is a help if you do most of the cooking in advance the previous day, up to the point of adding sweetcorn and peaches, then you can allow for a little more time and not feel rushed.

> 2 large onions, chopped
> 4 large cloves garlic, chopped
> olive oil
> 1½ kg (3 lb) chuck steak, cubed
> ½ kg (1 lb) tomatoes, peeled, chopped
> 1 tablespoon tomato concentrate
> 2 litres (3½ pt) beef stock
> bouquet garni
> 1 heaped teaspoon dried oregano
> salt, pepper
> 1 kg (2 lb) sweet potato, peeled, cubed
> 1 kg (2 lb) potatoes, peeled, cubed
> 1 kg (2 lb) pumpkin, cut in chunks
> 2 tins (200 g, 7 oz) sweetcorn, drained, or ½ kg (1 lb) frozen
>   sweetcorn kernels
> 2–4 heads sweetcorn, cut in 2 cm (¾ in) slices (optional)
> 12 canned white peach halves, sliced (or yellow peaches)
> syrup from canned peaches

Cook the onion and garlic in a little oil until soft but not browned. Transfer to a large saucepan. Brown the beef in the oil, adding more oil if necessary, then put it with the onion in the saucepan. Stir into the meat the tomatoes, tomato concentrate, half the stock, herbs, a little salt and plenty of pepper. Cover and simmer until the meat is almost cooked – an hour or so. Add the sweet potato, potato and pumpkin, with more stock so that the contents of the pan are covered. Return to the boil and simmer covered for 20–30 minutes, until the meat is tender, the potatoes are cooked and the liquid thickened with dissolved pumpkin. Stir in the sweetcorn and peaches and simmer for a further 15 minutes. Taste and correct the seasoning, adding a little of the peach syrup to bring out the sweetness. Don't overdo this. Remove the bouquet, if you can find it.

When you have added the potatoes and pumpkin to the stew, put the prepared pumpkin shell, with its lid, into the oven at 190°C/375°F/Gas Mark 5. Leave it for half an hour, or longer if the walls are thick. Keep checking to make sure that it is not on the point of collapse, bearing in mind that it has to spend ten minutes in the oven at the end, with the completed stew inside it. If you have a large casserole into which the pumpkin can be settled, it is prudent to use it as a kind of support to the base and lower walls.

When the stew is completed, ladle it into the pumpkin, and put into the oven for 10–15 minutes before serving.

Note Some recipes give parsnips as an alternative to sweet potato, but it doesn't do. It is too sweet, too individual in flavour, and dominates the other ingredients. Best to look out for sweet potato in good time. Like pumpkin it keeps well, and it is essential. Doubling up the potato is not the same.

AMERICAN PUMPKIN PIE

Like the Shaker tomato custard on p. 521, this American dish of Thanksgiving feasts is an old English recipe, adapted to pumpkin. Where we might have used quince or apple, they took pumpkin. To make the pumpkin purée, either cook it with hardly any water until soft, or bake it in the oven for at least an hour, at 160°C/325°F/Gas Mark 3, then scrape the flesh from the skin and put it through a *mouli-légumes*. This version is from *The Joy of Cooking* by Irma S. Rombauer and Marion Rombauer Becker.

22 cm (9 in) baked pie crust

FILLING
375 ml (12 fl oz) cooked pumpkin
375 ml (12 fl oz) double cream
6 level tablespoons brown sugar
2 level tablespoons white sugar
½ teaspoon salt
1 teaspoon ground cinnamon
½ teaspoon ground ginger
⅛ teaspoon ground cloves

> 125 ml (4 fl oz) dark corn syrup or light molasses (or golden syrup)
> 3 slightly beaten eggs
> 1 teaspoon vanilla extract or 2 tablespoons brandy or rum
> 175 ml (6 fl oz) shelled walnut (measured in a measuring jug)

To make the filling, put all the ingredients except the last two into the top half of a double boiler, or in a basin over a pan of simmering water. Stir together until the mixture thickens. Allow it to cool slightly, then add the last two ingredients. Pour into the pie crust, and serve with whipped cream.

You may find it easier to mix the filling ingredients and pour them into a lightly cooked pie crust, then bake for an hour at 160–180°C/325–350°F/Gas Mark 3–4. Serve cool. The walnuts and alcohol can be omitted: some people like chopped ginger mixed with the cream.

The snag about American recipes for pumpkin pie is their insistence on brown sugar. However light fawn a sugar I use, it seems to boss the flavour. Undoubtedly the Americans have an even sweeter tooth than we do, judging by their cake and biscuit recipes generally. Perhaps pumpkin from the States has a stronger flavour (though I have never noticed that pumpkins from our hot corner of France taste any more forceful than the ones we grow at Broad Town). I suggest you use white sugar instead, or try the delicious recipe below, also from America.

PUMPKIN MOUSSE

This is a typical fruit mousse recipe, using pumpkin rather than strawberries or raspberries. The only other difference is the strong dose of spices.

> 1 packet of gelatine 15 g (½ oz)
> 4 tablespoons hot water
> 3 large egg yolks
> 2 tablespoons lemon juice
> 100 g (3½ oz) caster sugar
> 300 ml (½ pt) pumpkin purée
> 1 level teaspoon each ground cinnamon, allspice
> ½ level teaspoon grated nutmeg
> 150 ml (¼ pt) each double and whipping cream
> 3 large egg whites

Dissolve the gelatine in the water. Whisk the yolks, lemon juice and sugar electrically, or over a pan of simmering water, until thick and straw-coloured. Add the gelatine, pumpkin and spices. Whip the creams and fold them into the mixture. Then whip the egg whites stiff and fold them in gently. Pour into individual glasses – there should be plenty for eight flûtes or tulip glasses – and when the mixture is set, run a layer of cream over the top and scatter with chopped walnuts. Serve with Barmouth biscuits or with sponge fingers, though a crisp biscuit is better.

TURKISH CANDIED PUMPKIN

A friend who lived in Ankara for several years gave me this excellent recipe. The tricky part is to cook the pumpkin slowly enough. The sugar in the recipe has a hardening effect that prevents the pumpkin pieces dissolving into a purée.

> *1 kg (2 lb) slice pumpkin*
> *250 g (8 oz) sugar*
> *150 ml ($\frac{1}{4}$ pt) water*
> *about 125 g (4 oz) walnuts, coarsley chopped*
> *whipped cream*

Peel and seed the pumpkin, then cut it into chunky slices, about 3 cm (1$\frac{1}{4}$ in) long. Bring the sugar and water to the boil, stirring, in a wide shallow pan. Put in the pumpkin and cook very gently for at least an hour. Cover the pan at first, then remove the lid towards the end so that the juices evaporate to a syrup. Or you can remove the candied pumpkin to a dish and boil down the juices to make them syrupy. Serve well chilled with walnuts and cream on top.

This is a cheap dish that looks exotic and luxurious, and tastes delicious. Serve in small quantities as it is on the sweet side.

OTHER SUGGESTIONS

Beans with corn and pumpkin, p. 88

PURSLANE

Although purslane was introduced into this country in the
Middle Ages and was popular as
a pickle and salad, it is now to be
found in seedsmen's catalogues
rather than greengroceries. It is worth growing for its young leaves;
one authority remarks that in Malawi the name means 'buttocks of the
wife of a chief' which helps in remembering their shape and fleshy
texture. Purslane is lightly cooked in boiling salted water in the Far
East, and reckoned to be more delicate than spinach, Swiss chard or
beet greens. In the Middle East it is often included in the Syrian peasant
salad *fattoush*. The French use *pourpier* as a salad with a vinaigrette
dressing.

FATTOUSH

Based on a recipe of Claudia Roden's from her *Middle Eastern
Food* (Penguin). The ingredients and their proportions can be varied
to taste.

> *1 large cucumber or 2 small ones*
> *salt*
> *½ pitta (Arab bread) or 4 thin slices white bread*
> *juice of 2 lemons*
> *4 firm tomatoes, peeled, chopped*
> *1 medium, mild onion, chopped, or 1 bunch spring onions*
> *3 tablespoons each chopped coriander and purslane*
> *2 tablespoons freshly chopped mint, or 1 tablespoon dried*
> *2 cloves garlic, chopped finely*
> *6–8 tablespoons olive oil*
> *black pepper, sugar*

Slice the cucumber. Put it into a colander with a light sprinkling of salt, to
drain for an hour. Dry and chop. Toast the bread, break it into small
pieces and put into a salad bowl; moisten it with lemon juice. Mix in the
cucumber and remaining ingredients. Taste and adjust the flavours to
your liking.

Purslane in France

A neighbour of ours at Trôo sent me along some purslane one day. We were instructed to make it into a salad with beetroot, like the lamb's lettuce salad on p. 284.

A book on Périgord cookery describes purslane as having fleshy leaves with a fine flavour of hazelnuts. In those parts they leave them 24 hours with a sprinkling of salt. Drain them well and mix in a salad bowl with olive oil, wine vinegar, fresh milled black pepper and some slices of sweet red pepper (remove the skins first by grilling them, p. 379). Hard-boil two eggs, shell them, and chop the yolks and whites separately: scatter the two decoratively over the salad.

PURSLANE AND GREEN PEA SOUP

A splendid recipe from *La Cuisine de Madame Saint-Ange* (Larousse 1927, reprinted 1958). Madame Saint-Ange points out that purslane gives the soup 'une onctuosité toute speciale': she also insists that the peas should be especially tender skinned. The soup is not sieved or puréed so it must not be marred by using peas that have developed something of a celluloid coating.

> 30 g (1 oz) purslane leaves
> 80 g (scant 3 oz) sorrel
> 10 g ($\frac{1}{3}$ oz) chervil leaves
> 40 g (generous 1 oz) chopped onion
> 80 g (scant 3 oz) butter
> 1$\frac{1}{2}$ litres (2$\frac{3}{4}$ pt) water
> 2 teaspoons salt
> $\frac{1}{2}$ litre ($\frac{3}{4}$ pt) shelled peas

First chop the purslane leaves on their own, as finely as you can. Then remove the stalks from the sorrel, and chop that; then chop the chervil. Put with the onion into a heavy pan, with half the butter. Put over a low heat, then cover and leave to stew gently to a purée for three-quarters of an hour. Stir with a wooden spoon from time to time, to make sure nothing sticks or browns.

Add the water, which should be hot, and the salt. When it comes to the boil, put in the peas and leave to simmer gently for a further 45 minutes. Take the pan from the fire, stir in the rest of the butter and pour into a tureen.

RADISH

It insults radishes, the most ancient of appetizers, to chop them up and bury them in a salad. Radishes, John Evelyn wrote with his usual sense of authority, 'are eaten alone, with Salt only, as conveying their Peper in them'. Or eat them with salt and butter or with and bread and butter.

In England we have been eating radishes for more than a 1000 years. They ate them in Rome, they ate them in Greece and in Egypt (where labourers working on the Great Pyramid were given radishes and garlic in their rations). About the Greek liking for radishes there is a right and proper story in Pliny's *Natural History*, though Pliny rather turns it against the Greeks. *Ex voto* models of radishes, beetroots and turnips were dedicated to Apollo, in his temple at Delphi; the turnips were made of lead, the beets of silver, the radishes of gold.

Radishes, the milder or more peppery kinds, clear the taste and prepare for drink or food. That has been their historical, traditional function. You find Horace on his farm writing about

> . . . . . lettuces and radishes such as excite
> The languid stomach,

or Ben Jonson in one of his plays sending his characters off to a tavern to have 'a bunch of radish and salt to taste our wine'. In Westmorland the owners of the great Elizabethan mansion Levens Hall outside Kendal, used to stage a Radish Feast every spring on May 11th, the old May day after the change in the calendar, serving radishes with brown bread and butter – to act as a relish and mitigator of the strong ale that was drunk.

The worst to be said of radishes is that Romans and English have spoken of them with a tinge of moral superiority. In his retreat at Olney, Cowper's idea of a winter meal – the last meal of the day, after his books were put by was a 'Roman meal' – 'Spare feast! a radish and an egg' – radishes followed by a boiled egg, I suppose.

My preference among the radishes are the mild kinds, scarlet with white tips – a dish of them each with enough green leafage to hold them by as we nibble or bite (some people claim that this tiny tuft of stalk should be eaten too, as it makes the radish more digestible). The look of them on the dish is as appetizing as the taste.

The winter radishes or Chinese radishes, are favourites in the East, in India and in Africa, as well as China and Japan. Black, red, or violet outside and white inside, each weighing up to a pound or several pounds, are sliced and eaten raw, or grated, or boiled or made into soup. Seeds are in regular supply, they are not difficult to grow, but I find them a rather coarse food, far from matching up to a nibbler's plate of French Breakfast or Cherry Bell or White Icicle – with salt, bread and butter.

If the radishes you buy are not as crisp as they could be, put them into a bowl of water with ice cubes, and leave them for a couple of hours. This improves them enormously. Then arrange them in a bowl or on a plate, with a small bowl of coarse sea salt, and a basket of bread. The idea is to butter the radish rather than the bread, then dip it into the salt.

The one radish salad I wholeheartedly enjoy is a Moroccan dish quoted by Claudia Roden in her book on Middle Eastern food. This is a plate of sliced radishes and sliced oranges, cut into smallish pieces and the whole thing sprinkled with salt and lemon juice. It can be made either with winter radish or the more familiar summer kind.

Both summer and winter radishes can be cooked – follow turnip recipes as a general guide – but I find they are rarely eaten with enthusiasm.

RED CABBAGE

If you lived in the Lake District before the early 19th century, your garden contained no more than onions and red cabbages (potatoes, cabbage, turnips, beans were grown in the fields). Not an invigorating prospect for the cook. Even less for her victims, as the red cabbage was invariably pickled. No one thought of stewing it.

This extraordinary national taste, which in the north amounted to a passion, darkened many high teas of my childhood. In three centuries of cookery books, few writers have given any hint that red cabbage might have a destiny beyond salt and the vinegar barrel. Odd, especially in the 19th century when the German influence was strong, and the other ingredients needed to turn red cabbage into a good dish were all to hand – apples, brown sugar, cured pork, onion.

Red cabbage cooked in the German style and eaten with hare or venison, or with a mixture of smoked pork and sausages, is really delicious. The longer it stews, the more nutty and succulent it becomes, without losing its character: in fact it seems to taste better with reheating. It is one of the best dishes to come home to, after a long walk on a winter's day; if you are a bit late, no harm can come to it.

Indeed red cabbage goes well with all the rich dark meats of wintertime – venison, hare, partridge, goose, duck and beef – and benefits from smoked pork or poultry being cooked in with it. They add a subtle flavour and depth to the dish that set off perfectly the spiced, sweet-sour cabbage, and the softened but biteable texture of the stalk and ribs. Such dishes are splendid, too, for the magnificent colour of richest Burgundy red. The various textures and shapes of the meaty ingredients, piled on to the cabbage, perhaps with cream-coloured dumplings or potatoes, give a superb feeling of northern winter comfort.

HOW TO PREPARE RED CABBAGE

Trim off any damaged outer leaves. Quarter it, then cut away and slice the hard stalk first, thinly, then the rest of the cabbage into 1 cm (½ in) thick shreds. Rinse the cabbage and do not be dismayed by the blue and purplish juices that stain the water (they come from a red pigment called anthocyanin).

Even if you have a small family to feed, it is worth buying a large red cabbage and cooking the lot. Any left over can be kept in the refrigerator for a couple of days and then reheated. Alternatively put it into the deep-freeze – it freezes well.

RED CABBAGE STEWED IN THE ENGLISH MANNER

Both the following recipes are taken from Maria Rundell's *Domestic Cookery*. It first came out in 1806, and laid the fortunes of John Murray, the publisher (later consolidated by their sales of Byron), going into 70 editions over the next 90 years. Maria Rundell was good at vegetables (she gave an early recipe for tomato sauce), rather against the contemporary European opinion of English cooking.

1) Slice very thinly and then rinse a red cabbage. Put it into a pan with salt, pepper, a large knob of butter and 'no water but what hangs about it'. Cover the pan tightly and stew until tender (keep the heat low, especially at first: in fact you would do best to put the whole thing into the oven at about 150°C/300°F/Gas Mark 2, if you have a solid fuel or oil-fired cooker). Add two or three tablespoons of vinegar, and give it a boil through. Check the seasoning. 'Serve for cold meat or with sausages on it.'

Eliza Acton, writing in 1845, adds that the stalk should be taken out first, then it should be quartered and sliced separately. If you omit the sausages, season the cabbage with lemon juice, cayenne and a half-cupful of 'good gravy', i.e. a rich stock. She gives the cooking time as three or four hours.

2) Into a pan or casserole, put a tablespoon of butter, a slice of ham, 250–300 ml (about half a pint, or a little less) beef stock and 125–150 ml (4–5 fl oz) vinegar. Add the cabbage. Cover and stew for three hours. When tender, add a little more broth, salt and pepper, and a tablespoon of sugar. Mix well and boil without a lid until 'the liquor is wasted'. Put it into a hot serving dish and lay fried sausages on it.

Smoked bacon and sausages or more ham can be put into the pot with the cabbage, in the style of the recipe for red cabbage with smoked meat, on page 437.

CARÊME'S RED CABBAGE WITH APPLES
CHOUX ROUGES À LA VALENCIENNE

A variation of the German style, from Antonin Carême's *L'art de la cuisine française au XIXe siècle*, published in three volumes from 1833 to 1835. The main feature in the recipe is the large quantity of lard and bacon used; the apples are not the usual cookers, but the crisp sweet Reinette, a favourite French apple, not unlike our Cox's Orange Pippin, which can be used instead. Carême used a *daubière* or similar kind of pot with a lid specially shaped to take red hot coals, heat above as well as below. Cooking the dish in the oven comes closest to this method.

> ½ kg (1 lb) smoked streaky bacon in a piece
> 250 g (8 oz) lard
> 1 huge red cabbage, sliced
> salt, pepper, grated nutmeg
> bouquet garni
> 250 ml (8 fl oz) beef stock
> 2 measures eau de vie or brandy
> 6 large Reinettes, cored and sliced

Cut the bacon into dice, and brown them lightly in the lard. Pack the cabbage and aromatics on top, then pour in the stock and brandy, and 'throw on top of the cabbage' the six sliced apples. Cover with a round of butter paper. Cover and simmer for two and a half hours. Check from time to time that the liquid hasn't evaporated, but do not stir the pot. The bacon should remain below, the apple on top.

When the cabbage is tender, skim the fat from the top, remove the bouquet and stir the whole thing up. Put it on to a serving dish, if the cooking pot is not reasonably respectable. 'Sometimes small grilled chipolata sausages are placed on the cabbage.'

THE DUCHESS OF ORLÉANS'S RED CABBAGE

'Cook a medium sized, sliced red cabbage in four pints of bouillon with two slices of cooking apple, and an onion stuck with clove, and add two glasses of good red wine. Sprinkle generously with spices and let it simmer for several hours.'*

* Quoted by Theodora Fitzgibbon in her *Taste of Paris*, Dent 1974.

Four pints may seem a lot of liquid for one medium-sized red cabbage, even if it does bubble away for several hours. But remember that the old wine-measure pint, which is still the American pint, is intended, rather than our 20 fl oz Imperial pint of 1826. Charlotte Elizabeth's pint would therefore have been a fraction under half a litre, about 16 fl oz. The metric system was introduced by Napoleon, over three-quarters of a century later, but the old measurements lingered on years after that. Copies of this recipe are said to have been handed out after the funeral of the German-born Duchess of Orléans, in 1722, by a nurseryman with a glut of red cabbages to get rid of. A valedictory note, purporting to be by the Duchess, said that this was her favourite recipe for red cabbage.

RED CABBAGE WITH CHESTNUTS

Follow the recipe above, but put into the pot a tablespoon of brown sugar, and 200–250 g (6–8 oz) peeled chestnuts, layering them in with the cabbage. Dried chestnuts are ideal – weigh out 125 g (4 oz) or a little more, cover them with water just off the boil, and leave overnight before you cook the cabbage.

RED CABBAGE WITH SMOKED MEAT

A complete one-pot main course, excellent midday food in wintertime. Although the origin and general method is German, the idea of including smoked poultry came from a friend in France. There *poulet fumé* is sold in most supermarkets (and rather despised). In this country it is rare; for us it turns the dish into something specially enjoyable. If you like frank-furters, they can be added as well. Indeed any of the smoked sausages are suitable, or even fried uncured sausages of quality. But whatever smoked meats you choose, the essential is a piece of smoked belly bacon with a strong, genuine flavour. There is nothing to beat the German *geräuchter bauchspeck* sold in good delicatessens. It permeates the cabbage in an irresistible way, whereas most smoked bacon on general sale, the kind in plastic packets, is far too genteel to make a proper contribution.

> 1 red cabbage, sliced
> ½ kg (1 lb) piece smoked belly bacon, sliced (include rind and bones)
> ½ kg (1 lb) sour apples, peeled, cored, sliced
> 375 g (12 oz) sliced onion
> 60 g (2 oz) dark brown sugar

salt, pepper
1 teaspoon ground cloves
4 tablespoons malt vinegar
150 ml (¼ pt) water
175 ml (6 fl oz) red wine

THE MEATS
smoked pork boiling sausage or rings
slices of garlic sausage, thickly cut
frankfurters
fried fresh sausages
smoked chicken or other poultry
smoked ham

Layer the cabbage, bacon, apple and onion into a commodious pot, sprinkling each layer with sugar, seasonings and cloves. Pour on vinegar and water. Cover and cook for upwards of three hours in the oven at 150°C/300°F/Gas Mark 2, or all day at a lower temperature if this suits you better and you have a solid fuel or oil-fired stove. The cabbage can also be cooked on top of the stove as in the previous recipe. If the cabbage is very liquid by this time, remove the lid and boil off the surplus. At this point, you can set the whole thing aside for later reheating with the meats.

40 minutes before the meal, have the cabbage boiling and add the wine and smoked pork boiling sausage or rings. 15 minutes before serving, arrange garlic sausage, frankfurters, fresh sausage, poultry, ham – some or all of them – on top, to heat through properly.

Finally check the seasonings and adjust them to taste.

Serve in the cooking pot with potatoes boiled in their jackets.

PICKLED RED CABBAGE

Raw red cabbage in a salad and pickled red cabbage are two things I detest. For those who don't share my feelings, here are two English recipes that they will find interesting:

1) from the *Art of Cookery*, by Hannah Glasse, first published in 1747. She gives a simple recipe for pickling the sliced cabbage in vinegar and salt with allspice, then goes on to remark, 'It is a pickle of little use but for garnishing of dishes, salads and pickles, though some people are fond of it'.

But she also gives, though one can still feel a faint sniff, a more elaborate recipe for pickling 'the fine purple cabbage, so much admired at the great tables'. This is what you do:

'Take two cauliflowers, two red cabbages, half a peck of kidney beans, six sticks, with six cloves of garlic on each stick; wash all well, give them one boil up, then drain them on a sieve, and lay them leaf by leaf upon a large table and salt them with bay salt; then lay them a-drying in the sun, or in a slow oven, until as dry as cork.'

2) the modern style of pickling red cabbage is not much different from the simple recipe given by Hannah Glasse. Instructions are more specific, that is all.

Weigh out 1 kg (2 lb) sliced red cabbage: use the firm hearts, not the stalks or outer leaves. Put it in layers into a pot, sprinkling sea salt between the layers; the usual quantity for dry-salting vegetables is a heaped tablespoon for 1 kg (2 lb). Leave for 24 hours, then pour off the salty liquor and drain the cabbage well in a colander. Pack into jam or bottling jars. Pour over cold spiced vinegar, enough to cover well. Stretch cling film wrap over the jars, if they do not have their own lids, and store for ten days before eating. Do not leave for more than eight or ten weeks, as the cabbage will lose its crispness.

ROCKET

A few years ago now, in Kyrenia before the partition of Cyprus, we started casually into the green salad provided with our grilled lamb chops and *talattouri* (cucumber and yoghurt salad, p. 239), and stopped short at the first mouthful. It had a delicious, slightly hot, peppery flavour, which made the humdrum lettuce greenery of the rest of the salad taste specially good. 'Rokka,' said the waiter, and explained that it was also used with spring onions to enliven tomato salads. Rocket is the English name. It grows in long leafy stems, a plant much to be recommended for a garden corner.

Rocket as an unremarkable item of salads is now eaten only in Mediterranean countries. John Evelyn grew it once in his kitchen garden, along with corn salad, clary, purslane, and all the other greens we have sadly allowed to disappear from our salad bowls.

SALAD BURNET

Three letters survive from Rabelais's correspondence in the mid-1530s with his friend and patron, Geoffroy d'Estissac, Bishop of Maillezais in Poitou. He was writing from Rome to send all the information he could on Papal politics, but had also been charged as a pleasant diversion with sending seeds to France. Italy was in those times the best country for food in Europe. The general attitudes of the Renaissance, combined with the stimulus of plants from the New World, had had their effect on gardening. The sunny patch of garden in the monastery at Maillezais, kept moist through the hot summer by the drainage canals of the reclaimed Marais, must have been a good ground for the seeds from Naples that Rabelais sent back – 'the best seeds from Naples, the same as the Holy Father has had sown in his secret garden of the Belvedere'. Rabelais doesn't detail the salads he sent, but apologizes for not being able to get hold of any *pimpinelles*, i.e. *pimprenelles* or salad burnet. This slightly bitter, cucumber flavoured plant was once, as Evelyn remarked, 'a very common and ordinary sallet furniture'. It is still sold in Italy in bunches of mixed salad greens.

In France, as in England, its use declined in the 19th century. Nowadays one has to go to a specialized nurseryman for the seeds, and I imagine that one of its last distinguished appearances was in Napoleon's daily salad on St Helena. This was a salad of haricot beans: the recipe was given by Carême. He had it from Monsieur Chandelier, chef to the Princess Borghese, whom she sent out to cheer her brother's exile as best he could, with the poor provisions available and the lack of equipment.

NAPOLEON'S DAILY SALAD

Soak and simmer 250 g (8 oz) dried haricot beans in unsalted water with peppercorns, a bouquet garni and an onion. Drain them when they are cooked and mix them with:

> *a good handful of green herbs, chopped –*
> *salad burnet, chervil, tarragon, chives, parsley*
> *6 tablespoons olive oil*

1 tablespoon tarragon vinegar
1 heaped teaspoon moutarde de maille
salt, freshly ground pepper, pinch of sugar

Leave for a few hours, or overnight, for all the flavours to blend together.

WHITE WINE CUP WITH SALAD BURNET

In *Herbs for All Seasons*, Rosemary Hemphill remarks that the first part
of the botanical name *Poterium sanguisorba* comes from the Greek for a
drinking cup, *poterion*, and gives this recipe:

2 bottles dry white wine
1 wineglass brandy
1 wineglass blackcurrant juice
90 ml (3 fl oz) soda water
1 dozen strawberries, hulled
a bunch of salad burnet

Mix together in a large bowl. Chill and add some ice cubes before serving.
   If you have a bottle of *Crème de cassis* use this instead of blackcurrant
juice, but go slowly, stirring it in to taste, as it is more concentrated in
flavour.

OTHER SUGGESTIONS
Beurre Montpellier, p. 540.

SALSIFY and SCORZONERA

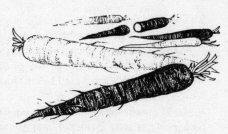

Salsify and scorzonera have much in common. They both provide us with long carrot-shaped roots of a succulence unusual in root vegetables. They have the same delicious flavour, though people in a position to know say that scorzonera has a slight edge over salsify. They are both natives of southern Europe, and belong to the same family. The difference between them is that salsify's skin has a whitish, parsnip look, while scorzonera's is a brownish black.

Italian gardeners developed the wild scorzonera root into its present state of tender firmness, although it early became popular in Spain and so acquired its botanical name of *Scorzonera hispanica*.

Salsify came in to this country about 1700, rather later than scorzonera, probably via France, though it was originally developed in Italy too. Often it was called vegetable oyster, a name one still finds in seedsmen's catalogues, or oyster plant. One authority says that this must have been because it had a slightly oysterish flavour, and could be used in meat pies instead of oysters which were often added for piquancy in the days when they were cheap (see the recipe for *tourtière*, on p. 447). If the flavour was once there, modern varieties have it no longer. Salsify tastes of nothing but itself – and scorzonera, of course. Other people have compared it to the parsnip, but this won't do either. Parsnip has a softer texture than the clean waxy bite of salsify, and is much sweeter.

The odd thing is that neither vegetable has ever really caught on, at least with the general public. Intelligent gardeners, from John Evelyn onwards, have always grown either salsify or scorzonera. People who write books on gardening have been pushing them from the 17th to the 20th century, but outside a few specialized and resourceful greengrocer's shops one can rarely buy them.

HOW TO CHOOSE AND PREPARE SALSIFY AND SCORZONERA

Choose roots that are as smooth as possible, and firm: avoid flabby ones.

SALSIFY

Top and tail the roots, then run them under the cold tap and scrub away any soil with a small scrubbing brush.

At this point the French scrape away the dark brown skin, putting the pieces into acidulated water as they go to keep them white. Then they cut them up into pieces 5–8 cm (2–3 in) long, and cook them for at least an hour in a *blanc à légumes*, p. 549.

I find it just as satisfactory, and a good deal quicker, to cook the roots in their skins in salted water. Leave them whole if possible, only cutting them in half if absolutely necessary to fit them into the pan. All the salsify I have cooked has been tender in half an hour: to have left it longer would have reduced it to a stringy mush. Perhaps French salsify is thicker and tougher; the kind I have bought has usually been about 2 cm (¾ in) thick, or a little less. After cooking, run the roots under the cold tap until you can peel away the skin. Then cut the creamy roots into whatever lengths the final preparation may require. By this method you cannot use the cooking liquor in a sauce, as it turns dark brown. It follows that for soup it is better to follow the French method of scraping, as you want the liquor to draw out the flavour.

SCORZONERA

Top, tail, scrub and simmer the roots as above. When tender, peel and cut them up.

If you grow your own salsify and scorzonera, you may like to follow the French habit of using the young green leaves in salad. I have never had the chance to try this, but one French cookery book describes the flavour as 'exquisite'. Salsify and scorzonera can also be grown to produce blanched shoots in the manner of chicory. For instructions, turn to John Organ's splendid book, *Rare Vegetables for Garden and Table*, published by Faber and Faber in 1960.

All the recipes following are suitable for both scorzonera and salsify. A number of turnip, parsnip and carrot recipes can be used too, but bear in mind that salsify and scorzonera are more delicate in flavour, so choose the finer recipes – for instance, the carrot and mushroom baps on p. 166 – and subdue any strong ingredients that might overwhelm them. They make a good gratin, too.

If you are buying rather than growing salsify, you will probably conclude that 1 kg (2 lb) has to be enough for six people at the price.

Gardeners will be able to take a more liberal view and dig up 1½ kg (3 lb).

CREAM OF SALSIFY

> ½ kg (1 lb) salsify
> 1 onion, chopped
> 1 clove garlic, chopped
> 3 tablespoons butter
> 1 heaped tablespoon flour
> 1 litre (1¾ pt) veal or chicken stock
> 150 ml (¼ pt) single or whipping cream
> salt, pepper, chopped parsley

Prepare and scrape the salsify in the French manner and put the pieces into acidulated water so that they do not discolour. Soften the onion and garlic, without browning them, in the butter. Stir in the flour, then the stock. Add the salsify, cover and simmer until it is very tender. Blend the soup to a purée, or put through the *mouli-légumes* twice (first the coarse plate, then the medium or fine one) to get the texture really smooth. Reheat to just under boiling point, adding more stock or water if you find the soup on the thick side. Mix in the cream and correct the seasoning. Serve scattered with parsley.

Fingers of toasted cheese go well with this and the following soup. Toast slices of bread on one side only and cut off the crusts. Spread the untoasted side with grated Parmesan or Cathedral City Cheddar, mixed to a paste with a little butter. Place under a hot grill until the cheese is melted and lightly coloured. Cut the toast into fingers.

CREAM OF SCORZONERA

Cook the scorzonera in the usual way, until barely tender. Peel off the skin, cut into lengths, then complete the soup as above. Naturally the cooking time will be much briefer – about 20 minutes.

This method can be applied to salsify if you are in a hurry. It also provides a useful method of turning canned or cooked, left-over salsify into soup.

SALSIFY SALAD

Cook, peel and cut up the salsify while it is still warm: in other words, when it is drained just run it under the cold tap briefly, so that you can remove the skin without burning yourself. Cut it into 3 cm (1 in) pieces and put them into a good olive oil vinaigrette, with plenty of chopped parsley and chives or spring onion. Leave the salsify to cool before serving it.

SALADE MONTFERMEIL

Cut up and mix in a bowl 375–400 g (13–14 oz) cooked salsify, 200 g (7 oz) cooked potato and four large or six smaller artichoke bottoms. Dress with vinaigrette. Scatter two crumbled hard-boiled eggs over the top and some chopped parsley.

SALSIFY WITH FINES HERBES

If the salsify is to go with grilled and fried meat, chops, escallops and so on, this is the way to finish it.

Fry the blanched and cut up salsify in butter until golden-brown, scattering over it at the start the leaves of a good sprig of thyme, or a small branch of rosemary. When it is ready, put it into a hot dish and scatter generously with chopped parsley, tarragon and chives.

Instead of this typically French mixture, you can use the Italian *gremolata*, which is easier from the point of view of ingredients, as autumn turns to winter. Mix two good tablespoons of chopped parsley with the grated rind of half a large lemon and a small clove of garlic, finely chopped.

SALSIFY WITH SAUCE POULETTE

Serve the cooked and cut up salsify with this version of *sauce poulette*. Make a velouté sauce with butter, flour and about 400 ml (¾ pt) of the cooking liquor. Boil it down to a good consistency. Whisk together five tablespoons dry white wine, or two good tablespoons of dry white vermouth, and two egg yolks. Pour in some of the sauce, then return the whole thing to the pan. Stir steadily until the sauce thickens a little more. It must not boil, or the egg will curdle, though it will do so less easily than in many egg sauces as the flour keeps the smooth consistency. Taste and correct the seasoning. Sharpen if you like with a little lemon juice and pour over the salsify. Scatter with parsley.

SALSIFY PARMESAN

Cook the roots as above, then fry them gently in butter (preferably clarified) until they are nicely golden-brown. In another pan, fry baps or other soft rolls until crisp; slice them in half and remove some of the crumb, being careful not to break the outsides. Sprinkle the cooked salsify, off the heat, with plenty of grated Parmesan cheese – about three tablespoons – and a good pinch of cayenne. Put it into the rolls and serve it immediately. Allow ¾ kg (1½ lb) for six, or a little less than the normal amount.

CREAMED SALSIFY
SALSIFIS À LA CRÈME

Boil and cut up the salsify. Complete the cooking in butter, but do not allow the salsify pieces to colour. Pepper them well. Stir in six tablespoons of double cream and a little lemon juice for every 1 kg (2 lb). Keep stirring until the salsify is lightly bathed in the sauce. Serve sprinkled with parsley. Or cover with grated Parmesan cheese and brown lightly under the grill – especially good.

Instead of double cream, use clotted cream for an even better flavour.

Although salsify cooked this way does go well with chicken and light meat, it tastes even better on its own.

SALSIFY FRITTERS

A most delicious way of serving salsify, and an economical one, too. ¾ kg (1½ lb) will provide enough fritters for six people as a first course. If you intend to serve them with grilled or roast veal, lamb or chicken, ½ kg (1 lb) will be all right. They are also good with fish, soft roes in particular.

Cook and peel the salsify, then cut it into slightly longer pieces than usual – 7–9 cm (3–4 in). Dip them into one of the fritter batters on p. 568 and fry in deep oil until golden brown. Serve with lemon quarters.

Escoffier recommends marinading the cooked and cut up salsify in one part lemon juice, two parts olive oil, with plenty of parsley and pepper. If you do this, peel the salsify while it is still warm so that it has a better chance of absorbing the flavours. Leave for several hours, then drain off the marinade, dip in batter and fry. Serve with large sprigs of parsley, deep-fried to crispness with the last of the fritters.

BOURBONNAIS CHICKEN AND SALSIFY PIE
TOURTIÈRE DE POULET DE GRAINS

The *tourtière*, the utensil that has given its name to this pie and the one following, is an old style of copper pie dish. It belongs to the days before ovens were common, when much cooking was done *au foyer*, on the hearth. With the pot cranes and hooks, grill racks and huge iron pot, cooking *au foyer* was quite a difficult technique. It is still in use on some small farms where the old people are hanging on. Occasionally you will meet an elderly Parisian with country roots, who can still produce a splendid meal all from the fireplace, just as her mother and grandmother had done every day. This kind of thing is still much closer to the French than it is to us: their equivalent phrase to our 'house-warming' is *'pendre la crémaillère'*, to hang up the pot-hook. The *tourtière* was used at the side of the fire, where it would stand on a bed of hot cinders raised up on its three legs. More hot cinders would be pulled on to the rimmed lid. This is what is meant in old French cookery books, when they stipulate 'heat below and above', an instruction that also applies to the stew pot known as a *daubière*, which has no legs, and a far more deeply concave lid.

This recipe comes from Roger Lallemand's *La Vraie Cuisine de Bourbonnais*. The ingredients were listed without quantities, so I have worked out the ones that seem best, but there is a certain amount of give and take. The recipe following is very slightly adapted from *La Bonne Cuisine du Périgord*, by La Mazille, which was first published in 1929 by Flammarion.

> shortcrust pastry made with ½ kg (1 lb) flour
> 2–2½ kg (4–5 lb) farm chicken, jointed
> butter
> ½ kg (1 lb) salsify, cooked
> 250 g (8 oz) mushrooms, sliced
> 18 small onions, blanched 5 minutes
> 18 black olives, stoned
> salt, pepper, ground nutmeg, ground cloves
> chopped parsley, tarragon
> beaten egg to glaze

Line a metal pie dish with two-thirds of the pastry, keeping the rest back for the lid. Brown the chicken in butter. Put into the dish in a good jumble with the remaining ingredients. Cover with the pastry lid, and decorate with a few pastry leaves. Brush over with egg glaze. Bake at 200°C/400°F/Gas Mark 6 for 20 minutes, then at 180°C/350°F/Gas Mark 4 for half an hour. Push a larding needle or skewer carefully through the central hole in the lid and see if the chicken is tender: it should be by this time. If not, put the pie back in the oven for a further ten minutes, and protect the lid with paper if necessary.

You will notice that no liquid is added to this pie. The juices of bird, mushrooms and onions provide enough moisture. The mixture of flavours is delicious, even if you have to resort to canned salsify.

PÉRIGORD CHICKEN AND SALSIFY PIE
TOURTIÈRE AUX SALSIFIS

The problem about cooking a pie in a *tourtière*, was turning it out afterwards. Turn is perhaps not quite the word, as it was more of a sliding, tipping, coaxing movement. To make things easier the *tourtière* was well buttered and the pastry had a couple of eggs added to the usual quantity of ½ kg (1 lb) flour and 250 g (8 oz) lard, to give it extra firmness. I use a large metal flan tin with a removable base, and a simple flaky pastry for a crisp lightness. A shortcrust pastry of the usual kind works perfectly well, too, as in the previous recipe. If you do not feel like making a pie, the filling can be eaten as a delicious meal on its own; give it a longer cooking time in the pan, and don't bone the meat.

> 2–2½ kg (4–5 lb) farm chicken, jointed
> goose, duck or chicken fat
> 12 small onions
> 2 tablespoons flour
> 150 ml (¼ pt) dry white wine
> 200 ml (6–7 fl oz) water
> ½ kg (1 lb) salsify, peeled, cut up
> 200 g (6–7 oz) skinned tomatoes, chopped
> salt, pepper
> flaky or shortcrust pastry made with ½ kg (1 lb) flour
> beaten egg to glaze

Brown the chicken in fat with the onions, allowing about 15 minutes. Sprinkle with the flour. Pour in the wine and water mixed, then tuck in the salsify and add the tomatoes, with seasoning. Cover and leave to simmer steadily for 30 minutes. Remove the chicken, leaving the vegetables to stew on a little if the salsify is still on the hard side. Cut away and discard the bones and divide the chicken into convenient pieces. The sauce in the pan should be thick and on the short side; if it is becoming gluey, add a little chicken stock but not much. Leave the chicken and vegetables to cool down while you make the pastry.

Line the flan tin, then lay the chicken on top and pour over the salsify and the sauce. Cover the pie, decorate and brush over with beaten egg to glaze. Slip a baking sheet into the oven, heated to 220°C/425°F/Gas Mark 7, and leave it for five minutes, before putting the pie into the oven. This brings immediate heat to the underneath crust and prevents it getting soggy from the sauce. After 20 minutes' cooking, check that the pastry is not becoming too brown. If it is, protect it with a butter paper. Leave the pie for another 15 minutes.

SAMPHIRE, see MARSH SAMPHIRE

SCAROLE, see BATAVIA and ENDIVE

SCORZONERA, see SALSIFY

SEAKALE

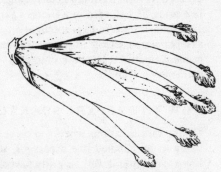

It is a pity that seakale, our one
English contribution to the basic
treasury of the best vegetables,
should not be more eaten. It is
not often in the shops, so you
have to grow it yourself, buying
plants from such firms as
Thompson and Morgan of Ips-
wich, or Paske's of Kentford, in Suffolk, or else find a neighbour to
give you a bundle or two. To have called it a kale, a cabbage, is not
quite fair. Seakale is no *Brassica*, rank like the swede or the turnip or
all the cabbages which descend from the wild cabbage of the sea-cliffs.
A related crucifer it may be, but it is an aristocrat of the northern
coasts, its shoots wonderfully delicate in flavour once they have been
blanched.

Seakale was being grown in English gardens by the early decades of the
18th century, transferred from its natural habitat, which Gerard in his
Herball had evocatively described as 'the bayches and brimmes of the sea,
where is no earth to be seen, but sande and rolling pebble stones'. By
1699 John Evelyn, in his *Acetaria*, had come to speak of 'our sea-keele' as
'very delicate', meaning in flavour. On the beaches along the drift lines
where seakale grew, from seeds which floated around with the tide,
country people bleached it by heaping sand round the shoots, which they
cut and carried to market. The practice continued – particularly along the
Hampshire and Sussex sands – long after seakale was domesticated in the
gardens.

The French took to it cautiously, not having thought of it for themselves. But the great Carême appreciated it, calling it 'Sickell' – not a
word which has established itself in French – mentioning it in a breath
with celery and asparagus and describing several ways of serving it. He
first came across it in England, in 1816, when he was cooking for the
Prince Regent at his Pavilion in Brighton (where you may still see the
kitchen much as it was in his time).

Thomas Jefferson was one of those who grew seakale regularly in
America. It first appeared in his *Garden Book* in 1809. In 1821 he was
anxious to get hold of earthenware pots to cover the young shoots in the
garden. I suppose they must have been earthed up in previous years. He

ordered 50 from a pottery near Richmond; they were to be made exactly like some he had seen in a friend's garden.

In spite of Jefferson's preoccupation with seakale, it has never taken off as a popular vegetable in America any more than it has here. Commerce remains unmoved by the enthusiasm – the long enthusiasm – of keen gardeners and cooks.

HOW TO PREPARE SEAKALE

On the rare occasions when you come across seakale in the shops, it is likely to be bundled like asparagus. Remove the raffia when you get home, and rinse it well, as gritty particles can remain obstinately in the grooves. Trim off any earthy stalk. The young tender leaves can either be left on, or cut off and used in a salad.

Re-tie the seakale into one or several bundles for cooking. Put them into the boiling water on their side, and cover the pan. When the stalks are tender, drain the bundles well. Victorian cooks served them on folded napkins or pieces of toast to catch the last moisture, but this is unnecessary.

Sauces that go with asparagus go with seakale, from plain melted butter flavoured with lemon juice, to a béchamel sauce with plenty of cream, or an hollandaise.

Here are two more suggestions:

SEAKALE WITH SAUCE AURORE
CHOU MARIN À L'AURORE

Drain the seakale carefully and place it in a long dish so that it lies one way. Make the *sauce aurore* on p. 553, pour some of it over the heads and serve the rest in a sauceboat. If you like you can scatter a little grated Parmesan over the sauce in the dish and put it under the grill for a little while to colour slightly – be careful not to let the seakale itself catch the heat, just the sauce with the cheese.

SEAKALE SAUCE

From *Food for the Greedy, a collection of receipts*, by Nancy Shaw, published in 1936.

> 4 shallots, finely chopped
> 50 ml each (¼ pt in all) white wine, white wine vinegar and tarragon
>   vinegar

300 ml (½ pt) béchamel sauce
salt, pepper, cayenne
3 egg yolks
3 good tablespoons butter

Boil the shallots and wine and vinegars until you have no more than one
good tablespoon of purée. Gradually add the béchamel sauce, mixing it in
well. Taste and adjust the seasonings, with a good pinch of cayenne.
Keeping the sauce under boiling point, whisk in the yolks one by one and
stir until it is thick and rich. Off the heat, add the butter bit by bit, so that it
mixes in without turning the sauce oily. Strain over the cooked seakale
and serve immediately with everything very hot.

A good sauce, too, for Swiss chard stalks, boiled celery or asparagus.

SORREL

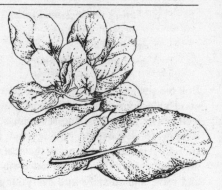

I learnt the advantages of sorrel in France, where every garden however small – or at least every garden in Touraine and Anjou – has a patch of it as close to the kitchen door as possible. We often visit one friend in the early evening, after she has finished working away at her 'ruin', an old farm cottage, or rather two cottages, barn and wine cave, that she has hauled back into life from a rapacity of brambles and nettles. We talk round the kitchen table, door open, or under the lime tree if it is hot. We smile at each other across pots of wild flowers, a scatter of crayons, pebbles, shells, matchstick models. Suddenly reviving, she remembers food and rushes to the nearest corner of the vegetable patch to grab a couple of handfuls of sorrel. In a quarter of an hour we are eating a lively soup, fresh and agreeably sharp, that she has made with the minimum of trouble.

In France everyone, it seems, knows about sorrel, and the homely standby of quick sorrel soup. In Wiltshire, at home again, I find it useful for unexpected visitors because it is unfamiliar. Sorrel soup, sorrel omelette, sorrel sauce with fish or eggs mollet, turn a frugal meal into something special. People wonder if it's spinach with lemon juice? No that's not quite right. Only if they were brought up in the country do they have a chance of recognizing the taste from nibbling wild sorrel leaves in the fields. This wild sorrel is *Rumex acetosa*. Garden or French sorrel, *Rumex scutatus*, is far more succulent, the acidity muted to a pleasanter level.

French sorrel is easily grown. It needs no gardening skill to speak of, and demands no more than a patch of modestly reasonable soil. Year after year it returns in March with its bright red stalks and rolled-up young leaves that soon uncurl to an arrow-sharpness. Keep it well picked, so that you are never left with only the older stronger leaves, and you will find it a splendid standby until the hard frosts of November.

HOW TO USE AND PREPARE SORREL

Sorrel comes half-way between vegetables and flavouring. A few chopped leaves can be added to a green salad for zest. A larger quantity can be cooked down to fill omelettes, like spinach, or to make soups and sauces. The best-known of French sorrel soups is made purely of sorrel and water, with an enriching finale of egg yolks and cream. In other soups it provides an accent to sharpen several milder vegetables, or blends in with a collection of leaves from parsley to lettuce.

Do not be irritated by the 'handful' measurement. Sorrel is rarely, if ever, on sale: one always has to go into the garden or yard to pick it for oneself. So 'handful' is more useful than a weight. However as a general guide I find my handful varies between 125 and 150 g (4–5 oz), and that for most occasions a couple of handfuls is enough. You will find that in most sorrel recipes there is plenty of give-and-take as far as quantity is concerned. The final taste is all that matters, and that can be adjusted easily enough. Another point to remember is that young bright leaves, the ones that are barely unfurled, taste freshly sharp. When the leaves are big and dark green, they become very strong indeed and a little coarse in flavour. Then they must be used with discretion. As most handfuls involve a mixture of leaves at various stages, the whole thing balances out.

When you bring in the sorrel, wash it well and tear away blemished bits of leaf. Cut off the stalks and the larger ribs. Stack the leaves in one or two piles, then roll them up and cut across with scissors. The strips will melt rapidly into a purée, without stringiness or the need for sieving.

Sorrel is usually cooked in butter, but many people prefer it cooked in lard or bacon fat when they are making soup. The great Carême always recommended this.

MARGARET COSTA'S GREEN SOUP

One of the best springtime soups has a sharp acid flavour, which dispels the stodgy feelings left by winter food. Most recipes for sorrel soup cook the leaves in with potato and onion, but Margaret Costa had the excellent idea of using raw sorrel leaves and a blender, so that the fresh flavour and colour remain unspoilt. If the soup is on the thick side, dilute it with water or more stock. Always serve it with small cubes of bread fried golden-brown in butter.

45 g (1½ oz) butter
½ onion, finely chopped
2 medium potatoes, peeled, diced (about 250 g, 8 oz)
1 litre (1¾ pt) good chicken stock, not from a cube
salt, pepper, grated nutmeg, pinch of sugar

2 good handfuls of sorrel
4 tablespoons double cream
chopped chives

Melt the butter and cook the onion in it until soft but not brown. Stir in the potatoes and stock. Add seasoning. Simmer until the potatoes are cooked. Meanwhile cut away any thick stems from the sorrel and wash it well.

Purée the soup in the blender with the sorrel leaves, until smooth and bright green. You will have to do this in batches. Return it to the pan, check the seasoning and consistency, adding more stock or water if necessary, and reheat without boiling (if you boil the soup, the sorrel will become dark green and lose the full vigour of its flavour). Add the cream and chives. Pour into a tureen and serve with the cubes of bread.

This soup can also be served chilled, but it will need a little more cream and quite a lot more liquid.

SORREL SOUPS FROM FRANCE
POTAGES D'OSEILLE

The following soups are all simple to make, though none can quite compare with Margaret Costa's green soup, of the previous recipe, for freshness of taste. Light home-made stock can be used instead of water, or the liquor from cooking vegetables, but keep these flavours unobtrusive – particularly the meaty ones. Serve the soups with small cubes of bread fried in butter, or cubes of toast. If you can buy a French loaf, cut some of it into thin diagonal slices, fry them in butter, and put them into the tureen just before pouring on the soup. You will notice that most sorrel soups are thickened with egg yolk.

1) Wash 250 g (8 oz) sorrel. Remove the stalks and cut into strips. Cook to a purée in 60–75 g (generous 2 oz) butter. Add 1½ litres (2½ pt) water, which should be boiling hot. Season and simmer for 15 minutes. Beat together two egg yolks and four tablespoons of cream. Pour in a little soup, then return it to the pan and stir over a lowish heat for a minute or two. Make sure the soup stays below boiling point, well below, or the egg will curdle.

2) POTAGE GERMINY – a richer version of the first soup, not a homely soup but part of the classic repertoire. Cook 250 g (8 oz) prepared and sliced sorrel leaves in 30 g (1 oz) butter. Add to a litre (1¾ pt) chicken

consommé, heated but not boiling. Whisk together six egg yolks and 150 ml (¼ pt) cream. Thicken the soup in the usual way and, just before serving, stir in 60 g (2 oz) butter. Serve thin slices of baked bread with this soup. Fried croûtons would be too much.

CARÊME'S SORREL SOUP
POTAGE À L'OSEILLE CLAIRE

'Wash and chop a good handful of sorrel, a lettuce and some chervil. Then melt them in a little scraped and sieved bacon fat, or in butter. Tip them into some consommé, prepared in the usual manner. Add a pinch of sugar and skim the soup free of fat. After simmering for half an hour, pour the soup on to some little croûtons of bread dried hard in the oven. Serve.

'Sorrel cooked in bacon fat has a much more savoury taste: I owe this hint to Monsieur Riquette.'

The only point that is in the least tricky, is the business of scraping and sieving the fat from a piece of bacon. Nowadays, thanks to electricity, we need only chop it roughly, then pulverize it in a chopper or blender. As an alternative you could use lard, or butter, as Carême suggests.

SORREL SAUCES FOR FISH, VEAL AND EGGS

Sorrel sauces can be made in three ways. In every case you start off with a purée of sorrel. Wash the leaves, cut off the stems, and leave them to drain. When more or less dry, cook them for a few minutes in a good tablespoon of butter, stirring until they turn to a dark green mass. You can chop the leaves beforehand if you like, but it is hardly necessary.

1) To make the simplest sauce of all, bring 250 ml (8 fl oz) whipping cream to the boil, stir in sorrel to taste off the heat and add a few spoonfuls of juice from the fish or meat cooking pan. Season to taste. If the fish was grilled, add a little water to the juices and boil them up vigorously to give you a spoonful or two of flavoured liquor. If the sauce is to go with eggs, just stir in a knob of butter.

2) Add sorrel purée to a small amount of creamy béchamel sauce. Flavour again with the fish or meat juices, or a knob of butter.

3) Add the purée to a hollandaise sauce (see page 556). A superb sauce for salmon.

SORREL OMELETTES

Sorrel is one of the best flavourings for an omelette. Its sharpness goes beautifully with eggs. There are two ways of tackling the recipe:

1) The usual French style. Cook two good handfuls of sorrel in a tablespoon of butter, until they flop to a purée. Season. Make omelettes and put a tablespoon of the purée into the middle of each one before flipping it over and on to the plate. Avoid putting too much sorrel in; think of it as a flavouring rather than a filling.

2) Boulestin gave this recipe from the south of France. 'No omelette done with cooked sorrel can compare with it.'

First prepare the sorrel. Pick two or three handfuls, choosing young leaves. Wash them and cut away the thick stalks and ribs. Dry the leaves, then put them one on top of the other and roll them up. With scissors snip half-way down the roll, then across in thin slices. Snip down the rest of the roll and slice again. You should end up with small pieces that are not too juicy. I find scissors better for this than a knife or parsley mill. Beat 8 to 12 eggs in a bowl with salt and pepper. Add the sorrel, with a finely chopped clove of garlic and a little chopped chervil. Make one large omelette or several small ones; keep the mixture on the thick side so that the surface sorrel barely cooks. This gives the omelette a 'peculiar acid taste, extremely pleasant and fresh'.

SOYA BEANS, MUNG BEANS and BEAN SPROUTS

If you want to grow bean sprouts at home – an easy business, in jam jars, which fill up and look as if mice had been shredding some strange substance or other inside them – you need seeds either of the soya bean or the mung bean. The sprouts can be eaten on their own as a salad, or in a sandwich like mustard and cress (they are rather good that way), or as a vegetable simmered for a few minutes only in salted water. Lightly cooked bean sprouts are often included in the stir-fried mixtures of meat and vegetables which are such a part of Chinese eating. Elegant Chinese speak of bean sprouts as 'coolie food', but the soya bean has been grown in China at least since the Western Chou dynasty (1027–770 B.C.).

Everyone knows that soya beans give more protein, weight for weight, than most other living things, animal or vegetable. But the modern manufacturers of T.V.P., 'textured vegetable protein', masquerading as steak, don't tell us that the Buddhist vegetarians of China in their temple kitchens (in the T'ang empire) were the first to turn soya bean-curd into imitation meats, imitation poultry and imitation fish, which they prepared with great tastiness. The practice continues in China after more than 1000 years.

Soya beans can be grown – with a little care and trouble – in northern gardens. Cook them shelled, like other beans, but don't be tempted to try them raw, in which state they do not agree with most stomachs. Dried soya beans and dried mung beans (the mung bean originated in India) can be treated like haricots and lentils.

Soya beans, fresh or dried, taste milder, less beany than haricots or kidney beans. Finished with cream, butter and parsley or a good tomato sauce, they taste finer, less rustic than similar preparations of other beans. Add them to meat stews, especially when you are a little short on the meat, and use them for soups.

BEAN SPROUTS can be bought these days in country supermarkets as well as in speciality shops in cities. Often they are sold in $\frac{1}{2}$ kg (1 lb) bags. This, I find, provides a vegetable for six people, to serve with grilled or roast meats, barbecued sweet-sour pork spare ribs or lamb. Or you can use one bag for two dishes, part as an ingredient of chop suey, the remainder as a salad (the Chinese dressing is soya sauce, or soya and wine vinegar plus sesame oil, sometimes in equal proportion sometimes as little as a quarter, seasoned with salt and sugar). A handful added to soup-stews instead of shredded leek (see page 293), changes the flavour and gives an even better crunchiness – the liquid can be seasoned with some soya sauce.

A warning – bean sprouts must be used the day they are bought, or the next day (though by that time they are diminished in freshness). Thereafter they degenerate to a rotting mush. They should always be rinsed and drained well.

STIR-FRIED BEAN SPROUTS

Unless you have a Chinese *guo* (*wok*), you will need to cook the bean sprouts in two batches if you are using a whole bag at once.

> *1 bag bean sprouts*
> *oil, preferably groundnut oil*
> *1 clove garlic, crushed*
> *250 ml (8 fl oz) chicken stock*
> *pepper, salt, sugar*
> *1 dessertspoon soya sauce*
> *1 heaped teaspoon cornflour*
> *2–3 spring onions, cut in 2 cm (¾ in) lengths*

Drain the rinsed sprouts well. Cover the base of the pan with a thin layer of oil, add the garlic and heat up. Stir in the sprouts and keep them moving about, over a high heat, for one minute. Add the stock and cook hard for a further minute, still stirring. Season to taste and add the soya sauce, plus the cornflour slaked in three tablespoons water, and the spring onions. Cook for one minute more, there should not be much liquid, just a binding creamy juice to set off the cooked, but still crisp, sprouts. Eat with rice or as a pancake filling.

This recipe can be varied in many ways. Reduce the stock by half, and put in 250 g (8 oz) skinned, chopped tomato with it. The whole thing can be turned into a light main course by frying 200-250 g (7–8 oz) thinly sliced lean pork or chicken, with or without an equal weight of peeled shrimps (avoid frozen ones, they are too watery), for one minute before putting in the sprouts.

CHOP SUEY

There are many recipes for chop suey, a dish apparently invented by the Chinese to keep undiscriminating Westerners happy. The cultivated Pekin taste would despise it heartily, I suspect.

Assemble, in roughly equal quantities, onion, Chinese leaf, celery and bean sprouts, sliced as appropriate; blanch the Chinese leaf for two minutes in boiling salted water. Slice 125 g (4 oz) fresh mushrooms or five soaked Chinese mushrooms, and about 60 g (2 oz) each bamboo shoots and water chestnuts.

Chill and slice paper thin, or cut into strips, lean pork tenderloin or boned chicken – the weight should be a quarter of the total weight of vegetables.

Heat up a pan with oil and garlic, as for the stir-fried bean sprouts. Stir in a large skinned, chopped tomato, and when it reduces to a liquid state, put in the onion. Cook for one minute, then add the Chinese leaf. Cook for one minute then add the remaining vegetables. When they are just about done, add the meat and continue to stir for another minute. Finish with the soya cornflour mixture described above. Serve with rice.

Note Stir-fried recipes of this kind should be juicy, but never wet. Keeping the heat high once the juices start to run copes with this problem. Do not keep the cooked food lingering in the oven, either, or it will lose its fresh flavour and gradually become sodden.

SPINACH

King's and emperors are celebrated for odd things. In the past
it was glory and conquest, maximum praise for maximum mis-
ery inflicted Nowadays it is
ordinariness and the ability to ride a bicycle or to shake a thousand hands
with a smile. Let me offer you something better. The great T'ang emperor
T'ai Tsung, no mean soldier by anyone's standards, had an idea we can all
appreciate though it is not an idea to inspire chroniclers and historians.
He asked tributary rulers to send the best plants their countries grew.
And in 647 – this at least was chronicled, though briefly – the King of
Nepal sent him spinach. Now spinach is not a native plant there. It had
come from Persia fairly recently, and must have been a rapid success or it
would not have been thought worthy of T'ai Tsung. Imagine this powerful
man forgetting his campaigns, forgetting his beloved horses, the magnificent Bayards that he celebrated in verse, as he tasted the first dish of
spinach served in China.

This is also the first record of spinach anywhere, though we know the
Persians were the people who developed it, sometime in the 6th century.
The name the Chinese still use, *poh ts'ai*, Persian vegetable, reflects its
origin. So do our European names, though in a different way: they are all
derived from the Persian word *aspanākh*, via Arabic. The Arabs, who
were great gardeners like the Persians, made their way through North
Africa to Spain, and by the end of the 11th century *espinaca* was being
cultivated around Seville.

In England its success seems to have been rapid. William Turner in his
New Herball of 1568 described it as 'an herbe lately found and not long in
use'. Yet by the end of the century it had become so familiar that no one
thought of it as new any more.

As spinach was unknown to the early Greek physicians, to Galen and
Dioscorides, whose theories dominated medicine and vegetable-eating
until the Enlightenment, it arrived without any baggage of 'medicinal
virtues'. Disconcerting for 16th-century herbalists. I hope it added zest to
the hot and cold spinach salads of the Tudors, and to the sweet spinach
tarts they often made. We have dropped the use of spinach to make tarts:
it seems to have gone out in the 18th century. In Provence, though, on
Christmas Eve when everyone sits down to the prolonged *gros souper*

before Midnight Mass, a spinach tart is often put on to the table. In the days when storing fruit was tricky and one had to rely on dried fruit for winter sweetness, spinach tart must have been a refreshing and juicy alternative. One can see how such a dish could survive in Provence, which has only come to prosperity gradually in the last 100 years.

Modern doctors have done what Galen and Dioscorides were unable to do. They have found medicinal virtue in spinach, vitamins and iron. In my youth it was shovelled into children as if their survival depended on it. Now it is forced on children no longer. The oxalic acid it contains is reckoned to be undesirable. Though who, I wonder, eats enough spinach often enough to be affected by it?

The thought of spinach is pleasure. French cooks, Chinese cooks, Italian cooks, Indian cooks would all rate spinach the best of leaf vegetables. In writing about food, the word spinach becomes a term of praise, a standard of vegetable aspiration.

HOW TO PREPARE SPINACH

When buying spinach, assess its liveliness. It should have a bouncing, bright appearance. As you stuff it into your basket or string bag, it should crunch and squeak. Although frozen leaf spinach is useful, it bears no comparison with the fresh for flavour and texture. Frozen chopped spinach seems to me to be useless, except possibly for soup. 1 kg (2 lb) fresh spinach will give you $\frac{1}{2}$ kg (1 lb) when cooked: enough for four people as a vegetable. A 1 kg (2 lb) pack of frozen leaf spinach is the equivalent to $1\frac{1}{2}$ kg (3 lb) of fresh, because it has already been blanched and so lost some though not all of its moisture; it will give you $\frac{3}{4}$ kg ($1\frac{1}{2}$ lb) when cooked, enough for six.

Pick over fresh spinach, discarding withered or alien leaves and tough stalks. Wash it in several changes of water, as it can be dirty after a period of rain. Stuff the spinach into a large pan, with no extra water (enough clings to the leaves to prevent it burning at first, then the leaves give out their juice). Put on the lid with a weight if necessary to keep it down. Stand it over a low to moderate heat, so that the bottom leaves do not catch. After five minutes, give the whole thing a stir about. Judge the amount of liquid in the pan and raise the heat so that the spinach cooks more rapidly. Turn the panful into a colander, cut it with a fish slice or the edge of a plate, and leave to finish draining.

As the meal comes near, reheat the spinach with a really large knob of butter. Spinach absorbs masses of butter. In one famous French recipe, spinach was cooked and reheated over five days. Each day butter was added, so that by the end half a kilo (1 lb) had absorbed about 300 g (10 oz) of butter. It really is delicious, but when I published a version of the recipe in the *Observer* colour magazine I was chided by a reader not for the amount of butter and cholesterol levels, but for the toxic effect of reheated spinach, particularly on babies. All I can say is that the mixture is obviously unsuitable for young children in any case, and that we have never suffered from it. The final purée is so rich and so full of flavour that one eats it in tiny quantity, more as a sauce with grilled and roast meat than a vegetable. When the editor of the *Observer* magazine in those days came to lunch so see if I could cook, he ate it with pleasure and still lives. And I got the job.

This is the basic English way of cooking spinach, leaving it to stew in its own juice, the method we have used since the 16th century when spinach first arrived. In France and Italy it is often blanched in a quantity of boiling salted water until just tender, then it is drained well. For some recipes it is the best method, but as a rule I think that our way conserves flavour and virtues better. When the following recipes demand cooked spinach, I mean spinach cooked in the English style.

Wateriness is the enemy of spinach, as I have implied. Its natural allies are butter, olive oil, cream, cheeses hard and soft, yoghurt, ham and bacon, anchovies, nutmeg, pepper. A little sugar, too, will bring out its flavour, particularly in soup.

Although you need not feel bound by the cheeses named in the recipes, they do give the best results. Salty, crumbly Lancashire goes best of our own cheeses with spinach, but for essential piquancy of flavour there is no substitute for Parmesan. In the small quantities required, it is not expensive. Ricotta has the best milky flavour of the soft cheeses, though curds, Petits Suisses and Gervais squares do well. Full cream cheese dissolves when heated, so is best avoided.

CREAM OF SPINACH SOUP

> 3 tablespoons chopped onion
> 2 cloves garlic, finely chopped
> 4 tablespoons butter
> 1 heaped tablespoon flour
> ½ litre (scant 1 pt) veal or chicken stock or water
> ½ kg (1 lb) spinach
> ½ litre (scant 1 pt) milk
> 125 ml (4 fl oz) whipping cream
> salt, pepper, grated nutmeg, sugar
> chives (see recipe)

Cook the onion and garlic gently in the butter, with a lid on the pan. When the mixture looks golden, stir in the flour, then the stock or water. Add the spinach, cover and simmer until the spinach is cooked. Sieve through a *mouli-légumes*, or purée in a blender for a really smooth soup. Reheat, adding milk or milk and water to achieve the desired thickness, then the cream. Check the seasoning – if you used water rather than stock for instance, you will need a good bit of salt. Use sugar to bring out the flavour, but don't overdo it. The traditional extra flavouring is nutmeg, but in early summer I think chives taste better on account of their freshness.

Serve with a bowl of croûtons, if the soup is part of a fairly large meal. If the soup is the meal, serve it with toast covered with cheese and set under a hot grill for a few moments; it makes a good supper.

SPINACH AND YOGHURT SALAD

This dish from Persia and the Middle East tastes better when home-made yoghurt is used.

> 3 tablespoons chopped onion
> 60 g (2 oz) butter
> ½ kg (1 lb) spinach, cooked, roughly chopped
> 150 ml (¼ pt) natural yoghurt
> 1–2 cloves garlic, finely chopped
> 1 level teaspoon ground cinnamon
> salt, pepper, sugar

Cook the onion in butter until yellowish but not soft. Add the spinach, stir it about well and give it a few moments longer to heat through properly and absorb the oniony flavour. Mix the yoghurt, garlic and cinnamon in a basin. Add the spinach and onion mixture while still hot, stirring everything well together. Season to taste. Serve chilled.

CREAMED SPINACH

A favourite and easily varied way of serving spinach. It is a good way of rescuing frozen leaf spinach, disguising its lack of crisp freshness and emphasizing its virtues. I find it too sloppy a treatment for chopped frozen spinach, which is really only suitable for soufflés or soups.

The basic principle is to chop the cooked and drained spinach not too finely, just enough to make it easy to eat, then to mix it with enough creaminess to make a smooth, rich mixture. The creaminess can be a thick béchamel or cream sauce – say 150 ml ($\frac{1}{4}$ pt) to $1\frac{1}{2}$ kg (3 lb) of fresh spinach or a 1 kg (2 lb) packet of frozen spinach. With high quality spinach that crunches under your hand, a few spoonfuls of cream – double, whipping or soured – and a good knob of butter make a lighter finish.

The final seasoning, apart from pepper and salt, can be nutmeg or mace, with grated Parmesan or a little lemon juice. One seasoning I like does sound peculiar, but it works: at the end of cooking time, mix the spinach with béchamel or cream sauce as usual and heat through. Have ready a medium clove of garlic, finely chopped then crushed with a little salt, and a teaspoon of Marmite. Quickly stir these into the spinach and serve straightaway. Instead of Marmite, you could use strong beef jelly from under the dripping, or quarter of a stock cube dissolved in a couple of teaspoons of water. In a family household, though, Marmite is the easiest to come by, and this is how the recipe was given to me. Use just enough to add an unidentifiable, savoury accent. The garlic, too, should be the lightest of breaths over the spinach, barely recognizable: when it is reduced by chopping and crushing with salt, it mixes in with subtlety.

In the 18th century, Seville orange was often used as a flavouring rather than lemon. The creamed spinach was then turned into its dish and surrounded with a fence of fried bread triangles. Quarters of Seville orange made a central boss. This is a fine dish for the dark days of January and February when these bitter oranges are in season. As a change, flavour the spinach with Seville juice, but in the centre put wedges of sweet orange to blend in with the astringent sharpness and make one of the most beautiful of all vegetable dishes. To it you could add rolled fillets of sole poached in a white wine stock, or small turbot steaks that have been lightly fried. Fish, spinach and orange flavour combine well.

Small quantities of creamed spinach can be used in elegant ways – for instance, as a filling for omelettes – but add a little extra milk or cream to soften the mixture and some Parmesan to bring the flavour through the blandness of the batter covering. Or you can use it as a base, on fried bread, for eggs poached or mollets, or – this is especially delicious – poached and lightly fried brains topped with a spoonful of cheese sauce. Another good idea is to serve the hot spinach in tartlet cases baked blind; top with strips of ham or a few curls of grated cheese and arrange round a roast leg or loin of lamb, or between grilled lamb chops, or fried noisettes of lamb.

To me, these are the best ways of eating spinach; a small quantity, attractively presented. The piquant seasoning and crisp stylishness make one's mouth water.

MADAME RAFAEL'S SPINACH PANCAKES

The people of Israel eat more vegetables per head than anyone else. If you visit their markets, say at Tel Aviv, you can see why. Mile after mile of vigorous and brilliant vegetables line the streets, including spinach which is a favourite throughout the Middle East. This way of using it was invented by the wife of the Israeli ambassador in London. When she makes the dish in London, she uses Cheddar cheese for the flavouring.

4 eggs
2 tablespoons oil or melted butter
125 g (4 oz) flour
150 ml (generous ¼ pt) milk
salt, pepper
1½ kg (3 lb) spinach
2 good tablespoons butter
2 tablespoons flour
extra milk
250 g (8 oz) grated Cheddar
150 ml (¼ pt) soured cream

Make a pancake batter with the first four ingredients and season it. Cook, drain and roughly chop the spinach. Make a thick sauce with the butter, flour and milk. Add it to the spinach and sieve the whole thing, or purée it in the blender. Make 20–24 pancakes in a 20 cm (8 in) omelette pan, diluting the batter with more milk if the first pancake is too thick and clumsy – the pancakes should be a little firmer than usual on account of the extra eggs.

Grease a deep dish, about 22 cm (9 in) in diameter, with a butter paper. Layer in the pancakes and spinach and sprinkle the spinach layers with grated cheese. Keep the last of the grated cheese to sprinkle over the last layer, which should be pancake, and pour over the cream. Bake at 180–190°C/350–375°F/Gas Mark 4–5 until bubbling. Complete the browning under the grill if necessary.

SPINACH AND MUSHROOM PANCAKES

For this dish use an ordinary pancake batter, rather than the four-egg mixture of the previous recipe. The anchovies add an unusual but appetizing edge to the mushroom sauce.

18 pancakes, p. 567
¾ kg (1½ lb) spinach, creamed
1 small onion, chopped
1 clove garlic, chopped
4 tablespoons butter
75 g (2–3 oz) mushrooms, chopped
1 tablespoon flour
250 ml (8 fl oz) milk
125–150 ml (4–5 fl oz) cream
½ tin anchovies
salt, pepper, cayenne, ground mace
breadcrumbs browned in butter

Fill the pancakes with creamed spinach. Roll them up and place closely together in a buttered ovenproof dish. Make a sauce by cooking the onion and garlic gently in butter, then add the mushrooms; when they soften a little, stir in the flour and moisten with milk and cream. Simmer for ten minutes. Chop the anchovies and add them gradually, to taste. Season. Pour over the pancakes and bake for 30 minutes or so in a moderate oven, 180–190°C/350–375°F/Gas Mark 4–5, until bubbling. Scatter with breadcrumbs and serve.

If you have no anchovies, use 3 tablespoons of grated Parmesan.

SPINACH SOUFFLÉ

One of the best vegetable soufflés. The flavour can be varied by mixing in about 150 g (5 oz) diced smoked ham with the thick sauce at the beginning, or a tin of chopped anchovies and 60 g (2 oz) pine kernels or slivered almonds. The ideal with a soufflé is to have a crisp outside and a creamy centre. If by some ill chance you have to keep a cooked soufflé waiting, leave it in the oven with the door barely open for a further five minutes.

½ kg (1 lb) spinach, cooked, chopped
3 tablespoons butter
2 tablespoons flour
150 ml (generous ¼ pt) hot milk
salt, pepper, grated nutmeg
4–5 egg yolks
3 tablespoons grated Parmesan
4–5 egg whites, beaten stiffly
3 tablespoons breadcrumbs

Reheat the spinach with one tablespoon butter. Make a thick sauce with the remaining butter, flour and milk. Stir in the spinach and season well, remembering that the eggs will soften the flavour. Remove the pan from the heat and whisk in the egg yolks one by one, then about half the cheese to taste. Fold in the egg whites carefully – the best way to do this is to beat in a tablespoon of egg white fairly vigorously to slacken the mixture, then the rest of the egg white can be folded in gently with a metal spoon. A few small blobs of egg white won't matter, better to leave them than turn the mixture about too much. Butter a 1½ litre (2½ pt) soufflé dish and sprinkle it with breadcrumbs, tipping out the surplus. Pour in the soufflé mixture. Scatter the remaining crumbs and cheese on top. Have the oven heated to 200°C/400°F/Gas Mark 6, with a metal baking sheet inside. Place the soufflé dish on the hot sheet and close the door. Turn the heat down immediately to 190°C/375°F/Gas Mark 5 and leave for 30 minutes.

SPINACH ROULADE WITH TOMATO SAUCE

A roulade is a soufflé baked flat, then rolled round a filling, Swiss roll principle. Usually it is served hot with a sauce and if the three items are well contrasted in texture and flavour, the final result looks and tastes superb. Spinach roulade, with a soft cheese and onion filling, I owe to a reader in Norfolk who has worked on the original recipe from Poland in a way that adds to its success. This is what cooking is all about; often a small change – in this case the substitution of onion for mushrooms – can alter a dish to great effect. Her use, too, of English cottage cheese, very different in texture from similar continental cheeses, gives the filling a bulky deliciousness that will surprise those who share my normal dislike of cottage cheese as produced by our factory dairies.

Like soufflés, meringues and éclairs, roulades belong to the class of chef's miracle dishes. Miracles they must have seemed in the 18th century, when most of them were invented. This was the time when controllable ovens first began to appear, though they were rough and ready by comparison with our modern thermostat-controlled ovens. I remember the tense atmosphere at home, in the early 1930s before we had a gas stove, when the wind blew from the wrong direction and the kitchen range was playing up and a dinner party was to start with cheese soufflés. Such dishes trail a legacy of awe even now, although a child could make them with a modern gas or electric cooker.

If you are making this dish entirely from scratch, do the tomato sauce first, so that it can bubble down to a good consistency while the roulade and onion filling are being prepared. The recipe is on p. 511. If you keep a supply of tomato sauce in the deep-freeze – a wise thing to do – you will only need to reheat it at the end.

> 30 g (1 oz) each butter and flour
> milk
> 125 g (4 oz) finely chopped onion
> 175 g (6 oz) drained cottage cheese
> salt, pepper, grated nutmeg
> 175–200 g (6–7 oz) cooked spinach, chopped
> 4 eggs, separated
> 1 level tablespoon grated Gruyère or Cheddar
> 1 level tablespoon grated Parmesan
> ½ litre (about ¾ pt) home-made tomato sauce

First attend to the roulade filling. Melt the butter in a heavy pan, stir in the flour and then enough milk to make a thick but not gluey sauce. Remove a tablespoon of this sauce and set it aside for the roulade. Add the chopped onion to the pan and cook for three minutes, stirring all the time. Add the cottage cheese and cook for a further three minutes, again stirring. Season. Put the pan aside and attend to the roulade.

Sieve or blend the spinach with the spoonful of sauce and the egg yolks. Season and fold in the cheeses gradually, to taste. Whisk the egg whites until stiff and fold in as for a soufflé. Line a 33 × 23 cm (13 × 9 in) Swiss roll tin with foil. Spread the spinach mixture over it as evenly as possible. Bake for 15 minutes at 190°C/375°F/Gas Mark 5. Remove from the oven, put a clean cloth over the whole thing and invert it on to a table. Leave for five minutes, then lift off the tin and remove the foil. Trim the edges of the roulade; this makes it easier to roll and prevents the rolled edges cracking in an unsightly way. Quickly reheat the onion filling and spread it over the roulade, leaving the edges free. It will spread sideways with the pressure from rolling, and you do not want it to overflow.

Now ease the roulade gently into a roll, using the cloth to turn it. Finally hold a hot serving plate close to the roll, at an angle of 45°, and give a final flip so that the roulade lands on the dish with the join down. This is fun to do, not nearly as difficult as it sounds. I am not good at this kind of thing, and hold my breath with anxiety, but it always works beautifully. So take courage.

Reheat the tomato sauce until it bubbles thoroughly and pour it round the completed roulade. If any blemishes need disguising, pour a little tomato sauce decoratively over them. Serve on very hot plates. If need be, the whole thing can be left in the oven for a few minutes to keep warm:

it won't hurt if the temperature is not too high; 150°C/300°F/Gas Mark 2, is about right, or a little lower.

The three separate items of this dish can be frozen successfully. They can then be reheated and put together just before the meal.

PERSIAN SPINACH KUKU
KUKUYE ESFANAJ

A *kuku* is a solid kind of omelette, containing one or more vegetables, sometimes meat as well. It is more like a Spanish *tortilla* than a French omelette, which should be lightly cooked and almost liquid in the centre. This is not surprising, as the *tortilla* is a close relation of the *kuku*, brought to Spain by the Arabs in their conquering progress westwards. The great temptation with a *kuku* is to use it as a dust-bin for left-overs. This may work out, if the left-overs are of high quality and if you are skilful at balancing flavours. Be cautious, though, as you could end up with a stale tasting disaster.

Persians serve *kuku* as a side dish, or cut into small squares to go with drinks. They take it on picnics, too, as it is one of those convenient dishes that can be eaten hot, warm or cold. To get the best of the flavour I think one should eat *kuku* between the two extremes, not hot from the oven, not chilled from the refrigerator, but somewhere in between.

There is a *kuku* recipe on p. 51 that is made rather differently. The mixture contains flour and is baked in the oven. This is because the rich softness of fried onion and aubergine needs a different treatment from the thinner, lighter flavour of a green vegetable. Of course if it is more convenient, you can bake this spinach *kuku* in the oven, but I think it is better when fried on top of the stove with plenty of clarified butter or olive oil.

> *1 large potato*
> *clarified butter or olive oil (not corn oil)*
> *1 medium onion, chopped*
> *½ kg (1 lb) spinach, cooked, chopped*
> *salt, pepper, pinch sugar, lemon juice*
> *5 eggs, beaten*

Peel the potato and cut it into small dice. They should be fairly regular, slightly larger than a coarse chop, but a good deal smaller than the normal size of dice. Fry in a little clarified butter or olive oil until they begin to brown. Add the onion and continue to cook until the potato dice are golden-brown. Put into a bowl with the spinach and season well – plenty of pepper – adding lemon juice to taste. I feel that the lemon should be fairly strong to lift the solidity of the dish. Beat in the eggs.

Take a frying pan, preferably non-stick, of about 20 cm (8 in) diameter at the base. Pour in six tablespoons clarified butter or oil and heat it up. Tip in the mixture, smoothing it down to a thick, fairly even layer. Cover with the lid of the pan or foil. Cook for 15 minutes. By this time, the centre should be just firm. Slide the whole thing under the grill to brown the top. Alternatively slide the *kuku* on to a plate, then turn it upside down back into the pan and cook a few moments longer.

To bake the *kuku*, turn the mixture into a buttered or oiled gratin dish. It should be 1½–2 cm (½–¾ in) thick, or even a little more. Give it 30–45 minutes at 180°C/310°F/Gas Mark 4, until it is just firm, with a slight crust on top.

If you want to eat the *kuku* as a main dish, or a fairly substantial first course, serve it with tomato salad and yoghurt for sauce.

SPINACH GNOCCHI
GNOCCHI VERDI

These small, dark green dumplings, served sizzling hot with Parmesan cheese, are one of the best ways of eating spinach. In some parts of Tuscany, where the dish comes from, they are called *ravioli verdi*, although they are quite unlike what we have come to think of as ravioli. They are usually eaten as a first course, but they make a good main course if served with a tomato salad and those potato crisps that are cut like a lattice.

> ¾ kg (1½ lb) spinach, cooked, chopped
> 125 g (4 oz) butter
> 375 g (¾ lb) ricotta or curd cheese
> salt, pepper, grated nutmeg
> 2 large eggs
> 3 tablespoons flour
> up to 100 g (3 oz) grated Parmesan cheese
> extra flour, extra Parmesan

Mash the spinach in a pan with one-third of the butter and all the ricotta or curd cheese (do not use a full fat cream cheese, as it will melt completely). Heat this mixture for four minutes, stirring all the time. Season to taste. Off the heat, whisk in the eggs, then the flour and one-third of the grated cheese. Spread on a dish and chill until firm.

Flour your hands and form tablespoons of the mixture into croquettes. Bring a large pan, half-full of salted water, to a steady simmer. Lower four or five croquettes into the pan. Poach for four to five minutes, until they bob up to the surface of the water and look softly puffy. Do not let the water boil hard, or the croquettes cook too long; in either case they will tend to disintegrate. There is bound to be a certain loss into the water, but it should not become really soup-like. Drain the cooked *gnocchi* on kitchen paper.

Butter a shallow ovenproof dish with half the remaining butter. Arrange the *gnocchi* on top, in a single layer. Scatter them with the last of the cheese and dot with the last of the butter; or, and this is better still, melt the butter and dribble it as evenly as possible over the whole thing. Place under a hot grill for a few moments, then serve with a small bowl of extra grated Parmesan cheese, if you like.

This dish sounds fiddly, but once you get the idea it is quite straightforward. The spinach mixture can be made well in advance. Or one can freeze the uncooked croquettes. Naturally they will need a longer cooking time, and the water should first be at a good boil so that it does not cool off too much as the croquettes are slipped in.

EGGS, HAM OR SOLE FLORENTINE
OEUFS, JAMBON OU SOLE À LA FLORENTINE

The name of this thoroughly French gratin is a gesture to the superiority of Florentine gardeners in the matter of spinach. It is a most useful recipe to know, as the proportions of the three main items can be adjusted to suit supplies.

First of all put a bed of cooked, drained and chopped spinach into a well buttered gratin dish. It should be a centimetre ($\frac{1}{2}$ in) thick. On top of it, arrange one of the following:

1) six eggs mollet, i.e. eggs boiled for precisely six minutes, so that the whites are well set and the yolk still runny. Shell them carefully, so as not to crack the whites.

2) six poached eggs – this is the classic style, but I think that eggs mollet taste better.

3) six slices of a good quality cooked ham, rolled up loosely.

4) six fillets of sole (or plaice) rolled up and poached in equal quantities of white wine or cider and water, plus a good knob of butter and seasoning. Keep the liquid, boil it down to a concentrated essence and add it to the final item, which is:

A mornay sauce, about half a litre (¾ pt), made to the recipe on p. 553. Pour it over the whole thing. Scatter the top with grated Parmesan and bake in a hot oven until bubbling and browned, or brown under the grill if the various elements were combined while still hot.

Note If you are making this dish for four people only, you will need four eggs, four slices of ham or four fillets of fish.

SPINACH PYRAMIDE
EPINARDS À LA MODE DE CHEZ NOUS

The Pyramide is a fine restaurant at Vienne, on the Rhône.

> *1 kg (2 lb) young fresh spinach*
> *175 g (6 oz) butter*
> *salt, pepper, sugar, grated nutmeg*
> *6 slices good ham*
> *6 slices white bread*
> *300 ml (½ pt) béchamel sauce*
> *6 tablespoons double cream*
> *3 egg yolks*
> *lemon juice*

Wash, cook, drain and refresh the spinach under the cold tap. Then press it with the edge of a small plate to expel all moisture. Reheat it in a clean pan with 60 g (2 oz) butter. Season to taste with salt, pepper, a hint of sugar and grated nutmeg.

Meanwhile trim the ham to the shape of the bread and heat it through gently in 30 g (1 oz) butter. Fry the bread in the remaining butter. Place the ham on the bread, arrange them on a hot serving dish and keep them warm. Bring the sauce to boiling point with the cream; off the heat, beat in the yolks one by one, then cook gently to thicken the sauce, without boiling it. Season to taste with salt, pepper, nutmeg and a squeeze of lemon.

Divide the spinach between the slices of bread and ham, pour some sauce over each one and serve immediately.

MIDDLE EASTERN PASTRIES WITH SPINACH

All kinds of dough are used for making tiny pastries in this part of the world, but the most delicious are puff, flaky and fila pastry.

Commercial frozen puff pastry works well; it should be rolled out very thinly and cut into round or square shapes about 8 cm (3 in) across. Teaspoonsful of the filling are placed in the centre of each piece of pastry, which is then folded òver to make either a half-moon, envelope or triangular shape. Brush a baking sheet with water and put the pastries on top, brushing them over with beaten egg. Bake them at 260°C/500°F/Gas Mark 10 for five minutes, then lower the heat to 190°C/375°F/Gas Mark 5 and leave until they are well risen and golden.

Even more delicious, deep fry the pastries in hot oil.

Whichever method you choose, these pastries or *börek* are best eaten immediately. If you want to keep them, freeze them uncooked.

Fila pastry can be made at home, but it can be a problem as it has to be stretched to transparent thinness. It is better to buy it in ½ kg (1 lb) packets from Greek and German shops (it is the same thing as strudel pastry). They can be stored in the freezer. Any of the fillings below will require about 12 sheets of prepared fila, which is about half the packet.

If the pastry is frozen, thaw it completely or it will be too brittle to use. Cut the pile of sheets into long strips about 8 cm (3 in) wide, using scissors or a sharp knife, then separate the strips carefully.

Brush the first strip of pastry with melted butter. Place a teaspoon of filling (see below) about 2 cm (¾ in) in from one end. Fold the top corner down over the filling, to meet the base, in a triangular flap. Now fold the triangle enclosing the filling over on to the strip, so that the short edge is straightened again. Repeat until you end up with a plump triangular shape. Tuck the end under the last fold, then repeat with the remaining strips of pastry, which should be kept under a damp cloth.

Deep-fry at not too high a temperature until golden-brown for the best results. Or arrange on an oiled baking sheet, brush over with melted butter and bake at 200°C/400°F/Gas Mark 6 for 35–40 minutes. Eat the pastries straightaway with drinks, or as a first course, or as a supper dish.

FILLINGS All these fillings can be used in larger pastries, or double crust pies using puff pastry or fila pastry.

SPINACH AND SOFT CHEESE

> ½ kg (1 lb) spinach, cooked, drained
> 2 tablespoons clarified butter
> 125 g (4 oz) curd, cottage or ricotta cheese – not cream cheese
> salt, pepper, grated nutmeg

Reheat the spinach with the butter. Sieve the cheese meanwhile, then remove the spinach from the heat and stir in the cheese. Season well.

SPINACH AND CHEESE

> ½ kg (1 lb) spinach, cooked, drained
> 2 tablespoons clarified butter
> 125 g (4 oz) grated feta, Gruyère or Lancashire cheese
> 1 beaten egg, salt, pepper, grated nutmeg

Reheat the spinach with the butter. Mix in the cheese, off the heat, then egg and seasoning.

SPINACH, HAM AND CHEESE

To either of the fillings above, add 75 g (generous 2 oz), chopped cooked ham.

SPINACH AND BRAIN

In this filling, poached brains are used instead of cheese as the softening, binding part of the mixture. It is delicious, too, as a filling for an omelette or pancakes.

> 1 medium onion, chopped
> 2–3 tablespoons clarified butter
> ½ kg (1 lb) spinach, washed
> 250 g (½ lb) poached brains
> salt, pepper

Cook the onion until soft and golden in the butter, then add the spinach, cover the pan and leave until the juices run. Remove the lid, raise the heat and complete the cooking; see that the liquid has boiled away leaving a juicy mass. Mash the brains and mix them into the spinach with seasoning. If there is any surplus liquid, pour it off.

SPINACH, NUT AND RAISIN FILLING

> 1 medium onion, chopped
> 2–3 tablespoons clarified butter
> ½ kg (1 lb) spinach, washed

>   *3 level tablespoons chopped walnuts or almonds, or whole pine*
>   *kernels*
>   *2 tablespoons seedless raisins*
>   *salt, pepper*

Cook the onion and spinach as above. Mix in the remaining ingredients, adjusting the quantities to your taste. The mixture is delicious.

MUSSEL AND SPINACH GRATIN

A delectable and unusual recipe from Evan Jones's *World of Cheese*, published by Knopf in 1976. Many cheese cookery books are disappointing, but not this one – perhaps because it makes clear the special delight and relationship between vegetables and cheese, cheese of different and specified kinds. For this dish you may not be able to get Italian *fontina*, the creamy cheese from Piedmont. It will not be quite the same with Gruyère or Emmental, but they are the nearest thing to it on general sale. See the recipe on p. 380, for a pepper and *fontina* salad.

>   *3 kg (6–7 lb) mussels*
>   *1½ kg (3 lb) fresh spinach, or 1 kg (2 lb) pack frozen spinach*
>   *2 tablespoons chopped shallot or spring onion*
>   *125 ml (4 fl oz) dry white wine*
>   *3 tablespoons butter*
>   *2 tablespoons flour*
>   *125 ml (4 fl oz) whipping or double cream*
>   *generous pinch saffron*
>   *salt, pepper*
>   *125–150 g (4–5 oz) grated* fontina *cheese*

Wash and scrape the mussels, discarding any that are broken, or that remain open when tapped sharply. Wash and cut the spinach into shreds with scissors, or divide the frozen block, slightly thawed, into smaller pieces. Put the shallot or onion into a large pan with the wine. Place the spinach on top, then the mussels. Cover tightly, set over a very high heat, and boil for five minutes. Remove the cooked mussels and discard the shells. Take the pan off the stove if the spinach is also cooked. Drain the spinach, keeping the liquor carefully.

Make a sauce with the butter, flour, liquor from the spinach and mussels, and cream. Stir in saffron and seasoning to taste. Boil for a few minutes to release the saffron yellow. Spread out the spinach in a buttered casserole, put the mussels on top and then the sauce. Finish with grated cheese. Bake in the oven at 200°C/400°F/Gas Mark 6 for ten minutes, then complete the browning under the grill. The dish should be heated through properly, without the mussels being overcooked to rubber.

SWEET SPINACH TART
TARTE D'ÉPINARDS AU SUCRE
TARTE AUX ÉPINARDS PROVENÇALE

On Christmas Eve, before Midnight Mass, French families in Provence settle down to the first meal of Christmas. The children have set up the crib and decorated the house. Preparations for next day's food are properly advanced. For a meal before mass, meat is proscribed, so the main course will be fish. First hunger satisfied, everyone sits back to talk, drink and nibble at the *Treize Desserts*, the enormously extended pudding course and the star of the feast, until it is time to go to church.

The basis of the *Treize Desserts* – the number is magic, not obligatory – includes almonds, hazels, walnuts, raisins, dates, pears, grapes, tangerines, nougat, sweets and petits fours, things we all eat at Christmas in Europe. But from Carpentras to Apt, one is likely to find this sweet spinach tart on the table as well.

Do not blench at the idea. Take courage from the thought that sweet spinach tart was a thoroughly English delicacy in the days before modern fruit storage – as one 18th-century writer remarked, 'This is good among tarts in the winter for variety'. Tudor recipes might include rosewater as a flavouring, but later we inclined more to the candied orange and lemon peel of this modern recipe from France. It is best to use fresh spinach, but the tart can be made in advance up to baking point, and frozen.

> 250 g (generous 8 oz) spinach
> 125 ml (4 fl oz) each milk and single cream, or 250 ml (8 fl oz) milk
> 60 g (2 oz) caster sugar
> half vanilla pod
> 2 small egg yolks
> 30 g (1 oz) flour
> shortcrust pastry
> candied orange and lemon peel

Cook, drain and chop the spinach, being careful to reduce or remove larger pieces of stalk that have not dissolved into the mass. Make a pastry cream by bringing the milk, cream, sugar and vanilla pod slowly to the boil. Whisk the yolks together in a bowl meantime and remove a dessert-spoonful to be used as a final glaze. Stir the flour into the yolks gradually and smoothly. Tip on the boiling milk, then return the whole thing to the pan and stir until the custard cream thickens and is almost boiling. Cool and remove the vanilla pod. Mix into the cold spinach.

Line an 18 cm (7 in) tart tin, with a removable base, with the pastry. Gather up the trimmings, roll them out and cut long strips. Put the spinach mixture into the pastry case and make a lattice over the top with the pastry strips. Brush over the pastry with the spoonful of egg. Cut the peel into neat triangles (or use ready cut peel) and put some into each green gap. Have a metal baking sheet in the centre of the oven when you switch it on to 190°C/375°F/Gas Mark 5, so that you can slip the tart straight on to it. This extra heat helps the pastry to bake crisply underneath the filling. Eat hot or warm, with cream.

OTHER SUGGESTIONS

Lentils with spinach, p. 309

SPRING and WINTER GREENS

A number of leafy vegetables are referred to loosely as 'greens'. 'Eat up your greens. Finish your greens,' they said to us when we were small. Greens were never asparagus or tiny French beans or sugar peas, but the nastier aspects of the cabbage clan.

Kale I have always hated though curly kale will pass. I have a slight affection for sprout tops, but this is because I can buy them in Marlborough market in tip-top freshness and because one of the Italian members of the family has shown me a fine way of finishing them (see p. 115, the recipe for Red hot broccoli). In the early months of the year 'spring greens' come along, cabbages that have failed to develop a heart. Heartlessness is never a desirable quality, but they will pass too. Turnip tops have their devotees among gardeners, Romans and Virginians (recipe on p. 531).

It goes without saying that there is no point in buying any of these varieties of herbage unless it is beautifully fresh, with the dew on it. Or to be truthful the drops of morning rain and fog. Once the leaves flag, such pleasure as they might have given is no longer to be hoped for.

As to their treatment, you can look at it two different ways. If you dislike them and only give them to the family to do them good (or to work off your sadistic impulses), you may well resent lavishing trouble and buttery attentions on them. On the other hand you may take the view that anything you put on the table should be treated with the best means and skill you can bring to the job. Seeing that few of us are desperate for vitamins and do not need stoking up with greenery like sheep in a kale field, I feel that there is something to be said for the second attitude.

The first thing is to buy in generous quantity. Greens are cheap enough. This means you can be ruthless in cutting out tough and stringy stalks, keeping only the best leaves.

When you have washed and picked them over, range them on a chopping board with all the heads together at one end. Cut the greens across if necessary to get them into the pan. Put on some salted water – about 1 cm (½ in) or a little more – and when it's at a rolling boil, put in the stalk ends of the greenery, with the more delicate leaves on top so that they steam. Jam on the lid and leave to boil vigorously. After five minutes test one of the thicker pieces and take the pan off the heat when you judge that the stalks are just cooked. Drain in a colander. In a wide frying pan melt a good knob of butter, with a little chopped onion and garlic. Cook over a low heat to soften the onion, but do not let it boil. Put in the well drained greens and stir them about, raising the heat slightly. Put them into a hot serving dish and stir in a knob of fresh butter.

Kale soups are made with oatmeal or barley, and the water or meat broth the kale has cooked in; the greenery must be chopped. Finish with cream and serve with oatcakes. A Scotch speciality. See nettle soups on p. 333, for the general idea.

SWEETCORN,
or MAIZE

I have the impression that eating corn-on-the-cob was not much known in this country, or in Europe either, until a few years after the last war. 'Indian corn', this from Edward Bunyard, writing in 1937, 'unusual in England but most usual in America . . . The principle of the lathe is adopted in eating them.' The war, then the hard years after it, held such things up. In any case one could not take to sweetcorn in the years of butter rationing. Butter it must have, plenty of it, to bathe the yellow grains and dribble down one's chin, as one chews away. Special dishes for holding the cob, small pronged handles to push in at each end and make the lathe-turning easier, belong to the fifties and the first enthusiasm for a new food.

Although corn-on-the-cob has come late to our tables, maize has been a delight of childhood since the time of Cobbett. Its orange-gold, plump grains, clean, rustling, dry were kept on a meal-bin to be thrown twice a day to hens. The hard grains skittered and bounced over the yard. The hens rushed after them. Fattened with such food, they really were worth eating. With Seamus Heaney, the poet and a farmer's son, one might see an image of love in the brilliance of the corn, the shine of the scoop:

> And here is love
> like a tinsmith's scoop
> sunk past its gleam
> in the meal-bin.

From the opposite end of Europe a more exuberant poet, the Romanian Lucian Blaga, turns corn into an image of summer and light. But then to a Romanian, corn means corn-bread, *mamaliga*, corn fritters, corn-cakes, and not just hen-food:

> . . . As a child I used to love to jump
> naked into the maize barrel,
> drowning up to my neck in golden grain.
> On my shoulders I felt like a load like a river,
> and even now, in my late days, sometimes
> seeing mounds of maize grain on the threshing-floor,

I have to make a great effort to restrain myself
from caressing them with my cheek.
I suppress the desire to fondle them
only through fear of solar gods
with their firm, reasonable dreams.

Glory to all seeds, past, present and forever!
The thoughts of summer, a great heaven of light
is hidden in all of them as they sleep.
Throbbing in the dreams of seeds
there are fields sighing and gardens at noon,
aeons of woods
nations of leaves
and the murmur of a people of singers.

In northern France, maize is still for hens, even more for cooking oil, even more still for the plastic factories. Farmers round us laugh when we pick the very young greenish cobs to eat – the full-grown yellow ones are far too tough. To them maize is for business, for making money, not for gastronomy. Hen-food on the table? Leave that to the English.

Maize came to both east and western Europe in the 16th century from Central America and was soon valued for the yellow meal it provided, as well as for the cobs. It flourished in the heat of Spain, France, Italy, the Balkans, Portugal, which is not surprising as maize is a hot climate cereal. It seems it was first grown as a crop in Mexico about 7,000 B.C. The tiny wild heads, not as long as an ear of wheat, were developed by Maya and Inca farmers almost to the size of the cobs we know to-day. In North as well as in Latin America more varieties, colours and shapes of corn are grown than Europeans can guess at. In Peru there is even a purple corn, that adds colour as well as a particularly delicious flavour to sweet dishes and drinks.

So for southern Europe maize meant a new cereal, rather than animal food, or a vegetable. The grainy, yellow stiff mass of corn porridge, known as *polenta*, or *milhas*, or *mamaliga*, according to language, has become as essential to meat dishes in the Po valley, or Aquitaine, or Romania, as potatoes have in this country (but then potatoes were developed by mountain-dwellers in the Andes, in a cooler, wetter climate altogether). Corn-bread is familiar, too. Corn-on-the-cob depends on growing a more tender variety than field corn. Jefferson in Paris in the 1780s could not buy the right seeds for his garden in France or Italy, and thanked a friend warmly for sending him some from America. He would have had the same problem in the 1950s. The habit of eating corn-on-the-cob seems to belong most of all to this country, as far as Europe is concerned, no doubt because we have felt the strongest influence of the United States on our diet and food trade. I would say that canned corn is in every grocery. Frozen corn, packets of frozen rice mixed with corn and peppers, are becoming a commonplace.

I feel that sweetcorn is a vegetable to be eaten occasionally, preferably from the cob, whether you chew it off, or scrape it off with a knife before putting it into the pan. I am always surprised to find that I like it. The sweetness, the texture, do not appeal to me when I think about them. Then the occasional cob, or a creamy corn pudding, or corn soup, come as a delight. The corn grains cooked fresh have too much character for a daily food; to be successful as a staple item, the grains must be reduced to cornmeal or treated with lime and turned into *tortilla* flour.

HOW TO CHOOSE AND PREPARE SWEETCORN

An American friend told me that corn should be cooked the moment it is picked, as the sugar soon converts to starch. The juicy sweetness is lost. In an ideal world, you should put a pan of water on the stove to heat up, then go out and cut the cobs.

This knocks out fresh corn cobs on sale at the greengrocer's. At the most optimistic assessment, they will be 24 hours from the plant before they encounter boiling water. Certainly I have always found deep-frozen corn-on-the-cob has more flavour: presumably it has been rushed from field to processing factory, like the peas we hear about in television advertising. One attributes this difference to our lack of sun, or the variety grown, but it seems more likely to be the time between grower and cook. One has to conclude that only the gardener knows exactly how sweetcorn tastes at its best.

If you need corn as one ingredient among others, you will do better to scrape grains from a cob rather than open a tin. Allow roughly one cob for every 125 g (4 oz) of corn required for one recipe; it is sensible to weigh the amount you have from the first cob, to be able to judge how many more you will need, because there can be quite a variation of size. To scrape the corn, you must first remove the husk and silky threads. Then stand the cob upright on a board, holding it at a slight angle, and scrape down as close to the hard inner core as you can. The ideal is to use a special scraper, but I have never seen one on sale in this country. Most of us have to do the best we can with a knife. Once the grains are off the cob, scrape it down again, rather more vigorously, to get the last of the bits and pieces, and the juice. This may sound fiddly and time-consuming, but it can be done quite rapidly. The grains are far less resistant than one might think.

The broad-leaved husks covering the cob can be used to make the wrapping for Mexican *tamales*. These are little packages of food enclosed in a *tortilla* dough, then in the husks of the corn. They are steamed until the mixtures inside are cooked firm enough to be picked up and eaten. The problem about *tamales* in this country, is buying the right flour, which is made from maize treated with lime and then ground. Corn-meal sold in delicatessens and good groceries is not at all the same thing. Another problem is the poor range of peppers and chillis to flavour the filling. With hefty food, piquancy is essential. Americans in the southern states take the way out with a *tamale* pie. This is a spicy meat mixture, topped, and sometimes enclosed as well, in corn-meal dough. Good, but not so delectable as the husk-wrapped *tamales*.

Cans of corn and bags of deep-frozen corn kernels are both good items for storing. I do not much care for the canned corn with peppers added, preferring to make such additions myself, and I have a positive dislike of creamed corn. The sweetness and texture seem unpleasantly curdled, unless you intend to reduce the corn to soup in a blender or crisp it up in fritters.

SWEETCORN SOUP

The addition of sweet pepper and cayenne spices the mild delicacy of sweetcorn soup; without such assistance, it can – to some people – taste a little too sweet, even sickly. In summer, this recipe makes a delicious chilled soup: keep the hot seasoning for either a *rouille* sauce to go with it, or to blend with cream.

> 250–300 g (8–10 oz) sweetcorn kernels, preferably fresh or
>   frozen, rather than canned
> ½ litre (generous ¾ pt) veal, beef or chicken stock
> 1 small onion, finely chopped
> 1 tablespoon butter
> 1 level tablespoon flour
> ½ litre (generous ¾ pt) milk
> water
> 4 tablespoons whipping or double cream
> salt, pepper, cayenne
> 1 red pepper, grilled, skinned, seeded
> chopped parsley

Cook the sweetcorn in the stock, then purée it. The mixture will not be completely smooth, but this improves the texture so do not worry about it. If you have no blender, put the sweetcorn and stock through the coarse, then the medium, blade of the *mouli-légumes* to get it as fine as possible.

Meanwhile make a béchamel sauce by softening the onion in the butter, then stirring in the flour and finally the milk. Simmer for five or ten minutes gently, then add the sweetcorn and stock purée. Dilute to the consistency you like with water; if the soup is to be chilled, it should not be thick. Stir in the cream and reheat if desired. Season with salt, pepper and cayenne. Dice the pepper fairly small and add to the hot soup just before serving it, with a good sprinkling of parsley. For cold soup, mash the red pepper with some more cream and add a good seasoning of cayenne to make it vigorously hot. Alternatively make the *rouille* on p. 386.

CORN-ON-THE-COB

This is the corn you turn like a lathe to eat. Provide plenty of butter, sea salt and a pepper mill – and cloth napkins. Paper ones disintegrate into revolting tatters under the strain of such direct eating. Those little prongs that can be stuck in at each end of the cob are frowned on as an unnecessary refinement by some people. And they can let you down badly, if they are not well pushed in. You are supposed to get messy when eating corn-on-the-cob, and enjoy it.

There are several ways of cooking the corn. First, and most useful is:

1) BOILING Strip the husks and silky thread from the cobs and allow one decent-sized cob per person. Put on a large pan of unsalted water, or half water and half milk. When it comes to a rolling boil, slide in the cobs one by one, so as not to cool the water. Put a lid on the pan, and leave for five to ten minutes according to tenderness. Drain and serve immediately with butter, salt and pepper: do not salt the boiling water, as it prevents the corn from softening as it should.

2) CODDLING The second way, for very young, tender corn, is only suitable for your own growing. Bring unsalted water to a rolling boil in a pan with a tight-fitting lid. Slip the prepared cobs into the water, so that it does not go off the boil. Jam on the lid (use foil to make a tight fit). Remove the pan from the heat and leave for five to ten minutes. Drain and serve with butter, salt and pepper.

3) GRILLING A good method if you are cooking out of doors on char-

coal. Do not remove the husks. Just bend them back so that you can cut away the silky threads, then pull them forward again to their original position. Swish the cobs about in cold water, making sure that plenty gets in between the husks and grains. Twist the husk ends tightly together. Put on the grill over the hot coals, and cook for 25–45 minutes, turning the cobs four times to cook all sides. Use tongs. Of course the time will depend on the size of the cobs. Remove the husks before serving them; this may seem to diminish the picturesque cookery, but it's a job requiring sharp knives and oven gloves. It would be unkind to foist it on visitors who may be balancing a plate on their knees. For children it would be dangerous.

If you have bought cobs without husks, or are using frozen ones, wrap them in pieces of buttered foil, then cook them in the same way. Let the cobs thaw out first.

Serve with butter, salt and pepper.

4) ROASTING Or rather, baking. If the cobs are in their husks, prepare them as for grilling. Lay them on a rack in a pan, and give them 25–45 minutes at 200°C/400°F/Gas Mark 6. Remove the husks before serving.

If the cobs are without husks, or frozen (thaw them first), wrap them in pieces of buttered, peppered foil, making loose but tightly sealed packages, and bake at 220°C/425°F/Gas Mark 7 for 30 minutes, or until tender.

SWEETCORN CREAMED

If you do not fancy the usual style of eating corn-on-the-cob, scrape off the grains into a pan, then scrape down the cobs firmly to extract all the juices and bits possible. Add a knob of butter. Cover tightly and set over a moderate heat for a few minutes. It will not take long for the corn to become tender. Salt it, add some pepper, and a few tablespoons of double cream. Serve it in small ramekins or bowls, with toast.

This method can be adapted to frozen sweetcorn kernels. Cook them in boiling water to cover for a few minutes, until they are just tender. Pour off the water, add butter and cream. If you like a spicy hot flavour, devil the corn with Worcester sauce and a few drops of Tabasco. (I would not recommend this with scraped corn, as it is too vigorous for the gentle creamy flavour.) Simmer for a few moments until the corn is quite soft, then taste and add salt.

CREAMED CORN PUDDING

The most delicious of all fresh corn recipes, even better than corn-on-the-cob. It provides just the right amount of moisture and richness and buttery flavour. Try to use corn scraped from the cob. If you have to use the whole canned or frozen grains, break them down slightly in the blender or electric chopper. Do not overdo this: they should not be turned into a purée, but have a thick, clotted look when mixed with the cream.

> *approximately 4 corn cobs, or ½ kg (1 lb) corn grains*
> *150 ml (¼ pt) each whipping and double cream, or 300 ml (½ pt)*
> *   whipping cream*
> *1 level dessertspoon sugar*
> *salt, pepper*
> *60–90 g (2–3 oz) butter*

Mix the corn with the cream, sugar and enough salt to bring out the flavour. Add a little pepper. Rub a gratin dish, or six small pots, with half the butter. Put in the corn cream and dot the remaining butter over the top. Bake in a warm oven, at 160°C/325°F/Gas Mark 3 for 30–45 minutes, until the sides and top are nicely crusted with golden-brown. Serve as a first course.

SWEETCORN PURÉE
HUMITAS

Humitas is an Indian dish from the Argentine much eaten with roast meat and grilled steaks. It goes well, too, with sweetbread, chicken and veal, acting as a vegetable-cum-sauce. Sometimes the *humitas*, or spoonfuls of it, are parcelled up in corn husks and steamed, like *tamales*. Then they call it *humitas en chala*, *humitas* in a shawl. The proportions of the ingredients can be varied according to taste.

> *¾ kg (1½ lb) kernels cut from fresh cobs of corn, or defrosted frozen*
> *   sweetcorn*
> *5 tablespoons milk*
> *2 large eggs*
> *2 teaspoons paprika*
> *¼ teaspoon cayenne*

salt, pepper
75 g (generous 2 oz) butter
75 g (generous 2 oz) roughly chopped green pepper
75 g (generous 2 oz) roughly chopped spring onion
1 large ripe tomato, skinned, chopped
2 heaped tablespoons grated Parmesan

Blend the sweetcorn with the milk in a blender, and as it breaks down to a purée, add the eggs and seasonings. If you do not have a blender, put the sweetcorn through the *mouli-légumes*, then add milk, egg, seasonings. You should end up with a thick and slightly bobbly looking mass, as the sweetcorn skins will not entirely disintegrate. Melt the butter and in it cook the pepper and onion slowly until they soften, then add the tomato. Raise the heat so that the tomato cooks down to a pulp. Stir in the sweetcorn and lower the heat again, so that the mixture doesn't thicken too quickly and curdle the eggs by boiling. Keep stirring it for five to ten minutes, then add the cheese; mix it in well, so that it melts, but do not let it cook. Remove the pan from the heat, adjust the seasonings if necessary, and serve.

EGGS AND SWEETCORN HENRI IV
LES OEUFS VERT GALANT

The Vert Galant, the eternal lover, was Henri IV of Navarre, grandfather to our Charles II and France's Louis XIV. The recipe is an invention of another Béarnais, Guy Mouilleron, who runs the fine restaurant, Ma Cuisine, in Walton Street, London. It is an elegant but simply made dish, combining eggs, sweetcorn and *foie gras*, which are all found at their best in south-western France. I suggest you substitute a fine chicken or duck liver pâté for impossible *foie gras*.

The method is really an assembly job; prepare or thaw the items from the deep-freeze, and put them together at the last moment. Even the sauce can be made in advance and reheated.

6 baked pastry cases, 10 cm (5 in) across
375 g (¾ lb) cooked sweetcorn
6 poached eggs or eggs mollets
6 good teaspoons foie gras *or liver pâté*
sauce béarnaise, p. 557

Put the reheated pastry cases on a serving dish. Divide the sweetcorn between them and make a depression in the centre to take the egg. Mix the *foie gras* or pâté into the hot sauce; spoon it over the top of each egg. Place under a fierce grill for a few seconds to glaze and colour slightly. Serve at once – an elegant first course for a meal with a main course that can look after itself.

FRIED CHICKEN MARYLAND WITH CORN FRITTERS

This dish has lost its charm by over-exposure in cheap restaurants. At home, it tastes quite different and is a delicious way of serving chicken. The classic accompaniments are fried bacon and sweetcorn fritters. Sometimes I have had fried banana with it, too, which is not at all correct but popular with the children. Fried chicken Maryland is an ideal family dish.

> *1 large or 2 small farm chickens, jointed*
> *milk*
> *seasoned flour*
> *bacon fat or lard*
> *2 tablespoons plain flour*
> *½ litre (¾ pt) milk*
> *4 tablespoons double cream*
> *2 egg yolks*
> *salt, pepper*
> *1–2 streaky bacon rashers per person*
>
> **CORN FRITTERS**
> *60 g (2 oz) plain flour*
> *2 tablespoons milk*
> *good pinch paprika*
> *good pinch cayenne*
> *1 large egg yolk*
> *¼ teaspoon salt*
> *175 g (6 oz) drained canned or cooked corn*
> *1 large egg white, stiffly whipped*

Dip the chicken pieces in milk, then turn them in seasoned flour. Brown them all over in bacon fat or lard. Transfer them from the pan to an oven dish. Cover it tightly and put into the oven at 180–190°C/350–375°F/Gas Mark 4–5 for half an hour, or until cooked. Add no liquid. With the pan juices make a sauce: stir the plain flour into them, cook for a moment or two, then pour in the milk and cream. Simmer gently for 15 minutes. Just before serving, when everything is ready, beat in the egg yolks and thicken without boiling over a low heat. Season. Grill the bacon rashers.

While the chicken is in the oven and the sauce simmering, make the corn fritters. Mix the ingredients in the order given, then drop table-spoons of the thick batter into hot butter and fry until golden-brown both sides.

Arrange the cooked chicken, bacon rashers and corn fritters on a large hot serving dish. Put the sauce into a jug.

If you like the idea of bananas as well, peel, halve and quarter three large ones, and fry the pieces gently in butter. The snag about this is that you have too many things to take care of at the last minute, unless you can call on some help for frying the bananas.

MAMALIGA, POLENTA OR MILHAS

At home in the kitchen, prudent cooks use a double boiler, so that they can turn their backs on the *mamaliga* without fear of it catching.

This is a simple recipe – bring 1 litre (1¾ pt) water to the boil in the upper part of the double boiler. Stir in a ½ litre (generous ¾ pt) cornmeal gradually to avoid lumps. Season it with salt and pepper, or with just a hint of salt if you want to eat it with jam or honey. Now place the upper part of the boiler over the lower part, which should be half full of boiling water. Leave for 20 minutes or longer, stirring occasionally. If you want a softer mush, add extra water. If you want to make a sweet dish, add sugar and some liqueur when the consistency is right.

Turn the mixture on to a wooden board if you want to serve it imme-diately. Otherwise pat into a shallow buttered dish to cool, for later use as cornmeal fritters.

CORNMEAL FRITTERS
MAMALIGA FRIPTE, CRUCHADES, ESCOTONS

If you are not brought up to it, *mamaliga* can seem dull. I confess to preferring it made over in fritters, the crispest one could possibly eat and altogether delicious.

Slice the cornmeal porridge when cold into finger lengths. Fry them in very hot corn oil until crisp and lightly browned. Serve with the garlic sauce on p. 562, or a home-made tomato sauce. These fritters can also be eaten with rich meat stews.

CORN-CAKE

A mellow, buttery cake that keeps well.

>   90 g (3 oz) butter
>   175 g (6 oz) sugar
>   2 large eggs
>   175 g (6 oz) cornmeal
>   rind and juice of a lemon

Cream the butter and sugar, beat in the eggs, then the cornmeal and rind.
Add lemon juice slowly to taste. Put into an 18 cm (7 in) cake tin, lined
with Bakewell parchment, and cook for 45 minutes at 180°C/350°F/Gas
Mark 4. Test in the usual way. This cake will only rise slightly, owing to
the cornmeal. The top should be lightly brown and crisp.

OTHER SUGGESTIONS

Beans with corn and pumpkin, p. 88
Carbonada criolla, p. 424, in which sweet corn is an essential ingredient.

SWEET POTATOES

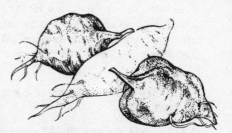

People look at sweet potatoes, at their grooved irregular shapes, at their purplish red skin, and wonder what to do with them. Three, even four hundred years ago, they would have known because sweet potatoes were then the 'common' potato.

Batatas, as the Haitians called them, were brought to Europe by Columbus in 1493 after the discovery of America. They took happily to the warm soil of Spain, being a tropical plant, and were soon known as Spanish potatoes.

Their chestnut flavour and mealiness make them deliciously versatile. In Shakespeare's day, for instance, they were sold in crystallized slices, a kind of Tudor *marrons glacés*, as an aphrodisiac – hopefully – along with candied sea holly. In the *Merry Wives of Windsor*, at his meeting in the park with Mistress Ford, Falstaff exclaims, 'Let the sky rain potatoes; let it thunder to the tune of "Green Sleeves", hail kissing-comfits, and snow eringoes' – eringoes being sea holly.

Their last great burst of popularity came when Napoleon's Josephine, a Creole, introduced them into the gardens of Malmaison. Soon the beauties of Paris were serving sweet potatoes at intimate suppers, to stimulate the ardour of their lovers. I should not hope for too much in this direction. Better to rely on the heart-warming qualities of a well cooked meal, than on any particular item in it.

I have seen it said that sweet potatoes may have been introduced into the United States from Europe, rather than directly from Central America. This may be so, but it is easier to imagine sweet potato plants creeping up into what are now the southern states, along with cultivated maize perhaps, and being grown by the Indians there. However they arrived, they are now very much a part of American food. Everyone serves glazed sweet potatoes with the Thanksgiving turkey, for instance. In the South they are an integral part of the soul food that has become so fashionable in recent years; in the North they are a point of nostalgia for black people who have had to leave the South to find a living.

Incidentally there is a point of confusion – sweet potatoes are often called yams, though I do not think that yams are ever called sweet potatoes. When you use an American cookery book, especially if it deals mainly with the food of the Southern States, look out for this. Yams are not as sweet as sweet potatoes, yet the sweetest variety of sweet potato is called Louisiana yam (the best substitute is an ordinary sweet potato with extra sugar: you might for instance boil it first in slices in a syrup, rather than plain water, as most recipes start with preliminary cooking).

HOW TO CHOOSE AND PREPARE SWEET POTATOES

Pick out uncut and unscratched sweet potatoes, matching them as far as possible for size. Wash them gently, remembering that their skins are much more delicate than the skins of ordinary potatoes. Boil them whole and unpeeled.

If you cannot avoid buying cut sweet potatoes, wrap them tightly in foil before committing them to the pan or oven, to keep in as much flavour as possible.

BAKED SWEET POTATOES

'The Spanish Potatoes' – this is John Parkinson, the great botanist, gardener and royal apothecary, writing in 1629 – 'are roasted under the embers, and being pared or peeled or sliced, are put into sacke with a little sugar, or without, and is delicate to be eaten.' One must agree with him that this – *batatada* apart – is the best way of eating sweet potatoes, even if the oven has to be used instead of embers. Try to buy four or six small potatoes, or two or three larger ones, so that sharing out does not become difficult.

> *about 1½ kg (3 lb) sweet potatoes*
> *salt, pepper*
> *100 g (3–4 oz) butter*
> *4 tablespoons cream*
> *glass sweet sherry, or orange juice and grated rind, or teaspoon*
> *Angostura bitters*

Push a meat skewer into each potato through the fattest part before wrapping it in lightly buttered foil (this takes the heat rapidly into the centre, and speeds up the cooking). Bake like ordinary potatoes. Temperature – between 150°C/325°F/Gas Mark 3 and 200°C/400°F/Gas Mark 6 – will depend on whatever else is cooking in the oven at the same time, and the time taken will depend on the size of the potatoes. Err on the side of generosity, as the final reheating can always be delayed.

When the potatoes are ready, cut them in halves, and remove the pulp leaving a good shell. Season with salt and pepper, then beat in the other ingredients to taste. Divide between the shells, and put into a hot oven 220°C/425°F/Gas Mark 7 to reheat.

Note Angostura bitters give a delicious spicy flavour.

CANDIED SWEET POTATOES

We have all heard about Pilgrim Fathers feasting at their first harvest home on wild turkey (a tastier creature, I am sure, than domesticated birds of modern times). We know about cranberry sauce and often serve it ourselves with the turkey at Christmas. Another Thanksgiving dish, candied or glazed sweet potato, is less well known than turkey or cranberry sauce. A pity, because it can be delicious. Try it also with boiled ham, pork, venison and so on, but go carefully. Candied potatoes are best thought of as a relish, like chestnuts, rather than as a vegetable. Too much can be cloying.

> 1 kg (2 lb) sweet potatoes
> 100 g (3–4 oz) butter
> 125 g (4 oz) brown sugar, or a teacup of maple syrup, or 2
>   tablespoons honey
> juice of half a lemon or orange
> 6 tablespoons brandy or rum

Wash and boil the potatoes in their skins. Peel, then slice them up. Arrange in a single layer of overlapping slices in a large shallow dish, that has been greased with a buttery paper. If one dish is not large enough, use two; you want to expose as much of the surface of the slices as possible.

Dab the butter over the top, and the sugar – or pour over the maple syrup or the melted honey. Cook in the oven to brown the top of the potatoes, basting with the juices. As they begin to colour, sprinkle on the juice of the lemon or orange.

Temperature is not important, as long as it is not too low, but the lower the oven the longer it will take to glaze the slices. The best way is to cook them on a shelf above the poultry or meat, say at 190°C/375°F/Gas Mark 5 – they will take 45 minutes. As a last resort, finish the browning under the grill.

Note The sugar, syrup or honey can be increased or decreased according to taste. Crushed pineapple is sometimes added. If you do not care for so much sweetness with your bird, eat the candied potatoes as a pudding, with cream.

SWEET POTATO BAKED WITH APPLE

A variation of the candied recipe, which goes particularly well with pork instead of apple sauce.

> ½ kg (1 lb) boiled sweet potatoes
> 375 g (12 oz) cored, sliced apples
> 125 g (4 oz) brown sugar
> 125 g (4 oz) butter

Peel the potatoes and put half of them into a buttered casserole. Put half the apples on top (they can be either sweet or cooking apples) and sprinkle with half the sugar, or a little less if eating apples are being used. Dab half the butter over the top. Repeat with the remaining ingredients.

Cover the dish, and put it into the oven above the joint. After about 30 or 45 minutes, remove the lid or foil and leave until the top is glazed and the apples thoroughly cooked.

MASHED SWEET POTATO AND HAM CASSEROLE

Sweet potatoes are mashed in precisely the same way as ordinary potatoes. In case this has become one of the lost domestic arts under the avalanche of dehydrated potato, like goffering frills or shelling peas, I offer some 17th-century advice: 'both sorts of them ... being very mealy when they are drest ... require *a great deal of Butter*'. My italics.

Having mashed about ½ kg (1 lb) of sweet potatoes with butter and a little milk, you can go on to embellish them, American style, with one, or at the most two of the following, before you set up the casserole: apple purée, crushed pineapple, walnuts, pecans, sherry, orange juice, Angostura bitters (I hesitate to try marshmallows, but pass the information on in case your inclinations lean in that direction).

Arrange two slices of good ham or salt pork per person in a lightly buttered ovenproof dish. Pour over about 250 ml (½ pt) béchamel sauce and top with a layer of the mashed sweet potato. Scatter the top with some chopped walnuts or pecans, if you like. Bake at

180–190°C/300–375°F/Gas Mark 4–5, for about 45 minutes, until well heated through. A much better Monday dish than shepherd's pie.

Note You can also do this with chicken or turkey, or half ham and half chicken.

BATATADA

Here is an exquisite dish, delicate and delicious, the perfect answer to those who think that sweet potatoes have too sturdy an air. It comes from Portugal.

> *175 g (6 oz) boiled, peeled sweet potato*
> *150 g (5 oz) sugar*
> *9 large egg yolks*
> *ground cinnamon*
> *125 g (4 fl oz) double or whipping cream*

Sieve the sweet potatoes. Cook the sugar with six tablespoons of water, to make a fairly thick syrup, and stir in the sweet potatoes. Continue to cook over a low heat, stirring all the time, until the mixture becomes transparent and gluey. The colour will vary according to the variety of sweet potato used: some will turn the mixture to a subtle greenish-yellow tone.

Remove the pan from the heat. When it has cooled to tepid, beat in the yolks gradually. Put back over a very low heat indeed, stirring until you have an opaque cream. It is ready when the spoon parts it, like Moses with the Red Sea, leaving a clear path on the base of the pan. The moment this happens, dip the pan into cold water to arrest the cooking and lower the temperature as fast as possible. Taste and season lightly with cinnamon. Divide between six small pots or custard glasses. Chill and top before serving with a swirl of whipped cream, slightly sweetened, and a dusting of cinnamon. Serve with sponge finger biscuits or *langues de chat*.

SWEET POTATO PIE

This good pudding has a resemblance to pumpkin pie. I think it is best eaten cold, with plenty of whipped cream.

> *sweet shortcrust pastry made with 250 g (8 oz) flour*
> *250 g (8 oz) boiled, peeled sweet potato*
> *100 g (3–4 oz) melted butter*

> *100 g (3–4 oz) dark brown sugar*
> *175 g (6 fl oz) single cream*
> *2 large eggs*
> *½ teaspoon grated nutmeg*
> *1 teaspoon ground cinnamon*
> *3 tablespoon brandy*
> *good pinch salt*
> *100–150 g (3–5 oz) shelled walnuts*

Line a 23–25 cm (9–10 in) tart tin with a removable base with the shortcrust pastry.

Sieve the sweet potato into a basin and beat in the remaining ingredients, in the order given, except for the walnuts. Chop one-third of the walnuts coarsely and add them to the sweet potato mixture. Pour this into the pastry case and arrange the rest of the walnuts round the edge of the filling. They may sink in slightly, but don't worry about this.

Bake at 200°C/400°F/Gas Mark 6 until the filling has risen and set. Test it with a warm metal knife or skewer just as you would a baked custard. When it comes out clean, remove the pie from the oven. Allow 40–45 minutes. The pie can be baked at a lower temperature if this is more convenient.

SWISS CHARD and OTHER LEAF-BEETS

Several forms of beet (or rather of *Beta vulgaris* var. *cicla*) are grown not for their roots but for the enlarged midribs of the leaves, which are boiled, and have something of the delicacy of seakale. Usually these broad stems that run up into the green are a creamy white, although seedsmen do sell red-stalked varieties. In the seakale beet, the stalks are even larger and more pronounced, as its name suggests, than in Swiss chard. The green leaf provides a second, though coarser tasting vegetable. Two for the price of one. An exception to this is the spinach beet; its stalk is much smaller and is cooked together with the leaf as if it were true spinach.

We may follow early Dutch practice in calling the commonest chard Swiss, but the great country for cooking it is undoubtedly France. There the stalks are prepared in many ways, with delicious sauces, in the manner of cardoons and celery. The coarse green leaves are cleverly incorporated into soups, black puddings, stuffed cabbage dishes, tarts, pâtés and pancakes, in which their resilience shows to advantage: spinach is often given as a substitute in such recipes, but its delicacy and melting qualities are at a disadvantage, and chard leaves are to be preferred.

Chard, like the French *cardon* and *carde*, comes from the Latin for thistle, referring to the prickly head, the teazle head which might be used for *card*ing wool. It has come to stand for the white edible rib of chard. In French cookery books you will find *cardes de bette*, or *de blette*, means the ribs of beet, i.e. of chard, as opposed to the leaves; sometimes the word *côtes* is used instead, meaning ribs more literally.

HOW TO PREPARE CHARD, ETC

Cut off the green part of the leaves; sometimes scissors are better than a knife for doing this. Wash it well. It can be cooked according to spinach recipes, with a few sorrel leaves to liven the flavour, but is better used for the dishes described later in the section.

Wash the stems. Cut them into 10 cm (4 in) lengths, stripping off the fine skin and stringy part as you go. These pieces can then be cooked in salted water until tender, or in a *blanc* (see page 549). They can also be cut smaller still and used with the greenery in the recipe for *farci*. These stems are the best part, and are usually served with a sauce on their own, once they have been cooked. For instance, with a cream or mornay sauce, a sauce aurore, or a velouté seasoned with curry powder; some of the cooking water is used in making these sauces, along with milk. Another way is to spread the cooked chard in a shallow buttered dish, cover it with a flour-based sauce and turn it into a gratin.

Lightly cooked chard stems are sometimes included in an *ailloli garni*, instead of cardoons. *Charcutiers* in western France may also chop up the whole stalk, stem and leafy green, and add it to pig's blood when they are making black puddings. There are recipes for this in my *Charcuterie and French Pork Cookery*. In Périgord, chard stalks are included in soup, along with cabbage, and the green leaves accompany pork, veal or chicken as a vegetable.

LE FARCI

In most parts of France *farci* is short for *chou farci* (see the cabbage section). In the Charentes and Poitou, *farci* implies cabbage of course, but it also includes a large quantity of chard.

In those parts, chard is often called *joutes* or *jottes*, rather than *bettes*, or even *joutes à farci*. This old word, a Gallic word from the popular Latin *iutta*, originally meaning wild mustard, was used in Anglo-French, too. At first it meant pot-herb generally, then it gradually became attached to chard, as it still is in the west of France. At the beginning of the 12th century the Cambridge Psalter, which was written in Anglo-Norman, had the wicked flourishing like the *joute verte*, the green chard, rather than the green bay-tree. And in a cookery manuscript of the 1430s, there is a recipe entitled *Joutes*. Borage, violet leaves, mallows, parsley, young vegetables, herb Bennet or wood avens, bugloss, orache and beet, were all boiled together, chopped small and reheated in broth with bread-crumbs. This thick soupy mixture was strained. Marrowbones were put in and saffron to colour it, perhaps some more beef broth. Finally it was put into a dish and taken to the table to eat with bacon, 'as men serveth frumenty with venison'. *Joutes* would, I think, recognize a Charentais or Poitevin *farci* as its descendant.

The usual vegetables are cabbage, chard or spinach beet, sorrel, some-times spinach, with a good mixture of green herbs and chives. The bacon is now included in the vegetable mixture with the breadcrumbs, and the whole thing is boiled in beef broth or water. I confess to preferring a baked version, that we bought at the *charcuterie* of Maillezais, near the abbey where Rabelais was a monk for a while. But first the traditional boiled *farci*:

> *1 round tight cabbage, about 1 kg (2 lb)*
> *15 large chard or spinach beet leaves*
> *good handful of sorrel*
> *250 g (8 oz) piece fat streaky bacon, cubed*
> *2 tablespoons lard*
> *4 medium onions, chopped*
> *2–3 cloves garlic, crushed with salt*
> *4 eggs*
> *50 g (scant 2 oz) breadcrumbs*
> *chopped parsley, chives, tarragon, savory etc.*
> *salt, pepper*
> *quatre-épices, or ground nutmeg, cloves, cinnamon*

Cut away the outer leaves of the cabbage, blanch them in boiling salted water for five minutes, then drain them and spread them out on a double muslin cloth, overlapping each other to make a large circle (see the recipe for *sou fassum*, page 138).

Chop the inside of the cabbage with chard and sorrel.

Cook the bacon gently in the lard, allowing it to colour slightly once the fat begins to run. Put in the onions and garlic, cover the pan until the onions are soft, then raise the heat and remove the lid so that they can cook to a golden colour.

Put the chopped greens into the frying pan, bit by bit. Stir them round and cook until they wilt down to a reasonably solid mass. Boil off excess liquid if necessary, leaving a moist rather than a wet mixture. Put it into a bowl. Add the eggs one by one, then the crumbs, herbs, spices and seasoning to taste. Put the mixture on to the cabbage leaves, tie up the cloth, after pressing the leaves round the stuffing, and boil for two hours in beef stock or water. Eat hot or cold.

BAKED FARCI Keep the outer cabbage leaves for another dish. Make up the stuffing as above, but roll it in a piece of softened caul fat and fit it into a baking dish. Cook for one to one and a half hours, according to the depth of the *farci*, at 180°C/350°F/Gas Mark 4. Finish off the browning under a hot grill if necessary. Eat hot or cold.

PROVENÇAL FAGGOTS
GAYETTES (CAILLETTES) DE PROVENCE

A rustic dish from Provence and the neighbouring provinces, with many variations of detail. The ideal greens for the mixture are the leaves of Swiss chard or spinach beet, though spinach is often used instead, mixed with some sorrel. The slightly tougher chard or beet leaves do undoubtedly contribute more to the recipe. One cookery book I have takes the derivation of *gayette* or *caillette* from *gaio*, which is provençal for pork sweetbread, another from *gaille*, which is the flead or flair of the pig, that inner sheet of fat, enveloping the kidneys and covering the inner side of the loin. Perhaps both were once included in the mixture, just as our own faggots are often a medley of the bits and scraps of the carcase. The essential ingredient is caul fat. This you may have some difficulty in finding. Go to a small family butcher of long establishment – he will at least know what you are after. In a supermarket your request is likely to be met with a cool stare of total ignorance. *Gayettes* are very much a lunchtime family dish for a cold day. They are sometimes served with root vegetables such as carrots and potatoes, rather than the peas of our faggot tradition.

> 250 g (½ lb) high quality sausage meat, or spare-rib pork, minced
> 60 g (2 oz) very fat bacon, minced
> 200–250 g (6–8 oz) pork liver, minced
> 1 medium onion, finely chopped
> 1 clove garlic, finely chopped
> 2–3 tablespoons butter
> 250 g (½ lb) chard or beet leaves, or spinach
> 30 g (1 oz) chopped parsley
> 60 g (2 oz) black olives, or sorrel leaves
> salt, pepper, grated nutmeg
> caul fat, lard

Mix the meats together well. Sweat the onion and garlic in butter until soft. Tip into the meat. Blanch the green leaves for three or four minutes in boiling salted water, drain well, chop and add to the meat, along with the parsley and olives – stone and chop them – or sorrel. Season to taste.

Divide the mixture into knobs the size of a golf ball, or a little larger. Soften the caul fat in tepid water, stretch it out carefully and cut as many squares as you have knobs of meat: they should just cover the palm of your hand and lower part of the fingers. Take up a square. Lay a knob of meat in the middle and wrap the caul fat round it. Repeat until all the knobs are swathed in the white fatty veiling. Grease a gratin dish with lard and place the knobs in it, close together, with the smooth sides up and the gathered edges down. Bake at 160–180°C/325–350°F/Gas Mark 3–4 for 40–60 minutes, until the tops are slightly browned.

Gayettes can be eaten cold as a kind of homely pâté, with plenty of bread. If you find the liver flavouring too strong, use less next time, and a little more pork. Don't be afraid to juggle the quantities – some recipes for instance give 200 g (6 oz) of lean pork, 100 g (3 oz) of salted belly of pork or streaky bacon, and only 100 g (3 oz) pork liver, plus 'two or three handfuls of chard or beet leaves'. The idea of including black olives I took from Escudier's *True Provençal and Niçoise Cooking*, but they can quite well be left out. I like their small sharpness with the rather solid mixture.

CHARD AND PORK PIE
TOURTE AUX FEUILLES DE BETTES

If you find the rustic mixtures of the *gayettes* or *farci* daunting, start by making this pie, or the pâté of the recipe following. They have few ingredients and are simple to make.

> **FILLING**
> *1 kg (2 lb) chard or spinach beet leaves*
> *2 tablespoons butter*
> *250 g (8 oz) Cumberland sausage, or other high quality sausage*
> *2 cloves garlic, crushed*
> *salt, pepper*
> *plenty of parsley*
>
> **PASTRY**
> *300 g (10 oz) plain flour*
> *pinch salt*
> *150 g (5 oz) butter*
> *1 egg*
> *iced water*

Wash, cook and chop the chard leaves. Melt the butter and cook the skinned sausage in it, breaking up the meat with a fork. When it begins to brown, put in the garlic and chard. Cook for five minutes. Taste for seasoning. Stir in plenty of chopped parsley, and remove from the heat.

Sift the flour and salt into a bowl. Rub in the butter in the usual way. Beat the egg with a tablespoon of water. Set a tablespoon aside for glazing the pie and use the rest with extra water to make a soft dough. Roll it out and line a tart tin or pie plate about 2½ cm (1 in) deep, and about 20–23 cm across (8–9 in). Put in the filling. Cover with pastry. Make a central hole and brush over with the egg and water you set aside. Score a diamond pattern on the pastry.

Put into the oven at 190°C/375°F/Gas Mark 5 for half an hour. Serve straightaway.

CHARD AND PORK PÂTÉ

The quantities given below make a delicious small pâté that many people find more agreeable than the all-meat recipes. If you have no chard or spinach beet, spinach may be used instead, as it can in the pie recipe above.

> ½ kg (1 lb) chard or spinach beet leaves, or spinach
> 300 g (10 oz) boned pork belly
> 60 g (2 oz) lean unsmoked bacon
> quatre-épices, or ground nutmeg, cinnamon, cloves
> salt, pepper
> piece of caul fat, or fat pork (not bacon) strips

Wash, cook and chop the greenery. Mince the pork and bacon. Mix them all together and season with ¼ teaspoon quatre-épices, or a pinch of each of the spices, salt and plenty of pepper. Wrap the pâté in a piece of softened caul fat and fit it into a small ovenproof pot. If you cannot get caul fat from your butcher, put some fat pork strips over the top of the pâté to keep it basted. Stand it in a pan of hot water.

Bake for 45 minutes at 180°C/350°F/Gas Mark 4. Brown under a hot grill afterwards if necessary. Keep for 24 hours in the refrigerator before eating.

TOMATOES

Many foods we think of as Mediterranean, the ones we buy to remind us nostalgically of holidays in the sun, are comparative newcomers to Europe. As one historian has remarked, Herodotus – or Helen of Troy, for that matter – would not recognize their native haunts if they returned there to-day. Orange and lemon groves, peach orchards, eucalyptus forests, cypresses pointing the silver slopes of olive trees, would cast an unfamiliar shade. The colour and taste of the food would seem alien. Even Lorenzo the Magnificent never set eyes on the magnificent tomatoes, sweet peppers and beans of present day markets in Florence. In his day basil, the royal herb, was for keeping bedrooms free of flies; it had not yet encountered its soul-mate, the tomato.

Can one imagine Mediterranean food without tomatoes? Even in the north of Europe it has become almost a staple food. Every week most of us will be making use of tomatoes in one form or another several times. This is a late conversion in England. For centuries tomatoes were feared as being chill to the stomach, a possible cause of gout and cancer, or excessive sexual appetite, in every way unsuitable to the national stomach. Their great advance began after the First World War, I would say, and has reached a peak in the last 15 years.

The word tomato now embraces the best and the worst of the vegetable kingdom. It means the huge red tomatoes of the Mediterranean, that burst with sun and flavour into great curves that are firm to the centre as you cut into them. It also means the pale, underprivileged rotundity of the northern shops, the dreaded Moneymaker and similar varieties, whose only virtues are regular size and vast yield. You will find a number of growers who have a couple of rows of Marmande or other well flavoured variety for their own consumption, while their main crop is Moneymaker.

The sad thing is that we need not suffer in this way at all. Countries with superb climates, Spain, the Canary Islands, Israel, are growing specially tasteless tomatoes for the London market. They would not dream of foisting them on their own people, and their other clients in Germany and France. I was asked to join a party of journalists on a visit to Israel in the

February of 1976. We were driven a long way into the reclaimed Negev desert, and proudly shown a large acreage of glasshouses, growing tomatoes especially for our readers. In their usual efficient way, the Israelis had conducted a thorough investigation of the demand by discussing the project with all the leading supermarkets. 'This is the British housewife's choice!' we were told. We looked at each other in dismay, and as politely as possible put our point of view – which was that we are given no choice, despite years of battering at obdurate supermarkets.

None the less we stand a better chance of success than the American housewife, who will soon be subjected to a cubic tomato which can be dropped six feet without splitting, and which has been gassed to turn it red. You might conclude that Americans have odd tastes. Do they build them up into pyramids on the kitchen table, then send them flying, as a preliminary to making a tomato salad? Is this some strange ethnic rite that the whole-food cranks have resuscitated? The answers of course are, 'They do not', and 'No!'. It's the big growers and important supermarket people who have worked out that such a tomato would suit them very well, never mind the customers. To this end the delicious open-air tomatoes, ripened on the plant, that were imported from Mexico, or grown on small American farms, are being squeezed out.

The red and golden apple or the love apple as it was first called in Europe – our name is a version of the Mexican *tomatl* – needs someone to save it from the dragon of commerce. Though I doubt if 'someone' could do the job. The days of Perseus and St George and swords are over. This is the age of the common man. We all have to shove.

You might also consider whether you couldn't possibly grow some tomatoes of your own. A local nursery will be selling tomato plants, but if you want to try the particularly good Marmande tomato, you will have to get the seeds from Sutton's, of Torquay, or on holiday in France.

HOW TO STORE AND PREPARE TOMATOES

Tomatoes should be stored at around 9°C/48°F, stalk side down, if you want to enjoy their best flavour. This means small, frequent purchases, as the refrigerator box will be too cold. If you have bought tomatoes in a plastic wrap, remove it as soon as you get home. Small tomatoes can be put into an egg box, perched in the compartments.

Tomatoes should always be peeled for a salad: pour boiling water over them, count ten if they are ripe, 15 or 20 if they are harder and greenish, then run them under the cold tap, and strip off the skin. Another way, which I find useful when water is in short supply, is to rub the tomatoes all over with the back of a knife-blade.

Many French and Italian recipes are meticulous about removing the seeds from tomatoes before using them. This is a reasonable proposition with large beefy tomatoes. If you removed the seeds and the surrounding juiciness from the average tomato on sale here, there would be almost nothing left. A feature of many French recipes, *tomates concassées*, are tomatoes peeled and pipped and roughly chopped. This is a good way to prepare tomatoes for freezing: cook them hard and briefly to drive off the water, or else drain off the water if there is too much to evaporate quickly. Season them lightly, cool and pack them in useful quantities for making soup and sauces, or adding to stews.

When tomatoes become expensive, you will find it cheaper and better to use canned tomatoes. They need salt, sugar and plenty of milled pepper.

CREAM OF TOMATO SOUP

A proper tomato soup – keep it for summer dinner parties and serve it hot or chilled.

> $\frac{1}{2}$ kg (1 lb) ripe tomatoes of good flavour
> 100 g (3 oz) chopped onion
> 125 g ($\frac{1}{4}$ lb) chopped young carrot
> bouquet garni
> 1 litre (2 pt) chicken or light beef stock
> salt, pepper
> 300 ml ($\frac{1}{2}$ pt) single cream
> chopped chives, or basil

Simmer the vegetables and bouquet with 900 ml (1$\frac{1}{2}$ pt) stock until they are tender. Remove the bouquet and purée the soup in a blender, or sieve until it is very smooth. Taste, add seasoning and remaining stock. If it is to be served cold, stir in the cream and chill. If to be served hot, bring cream to the boil in a clean pan and add the soup gradually. In both cases, serve scattered with chives, better still with basil.

If you like a spicy flavour for a change, add nutmeg to the soup as it cooks, and sprinkle a little on top when it is served.

TOMATO AND ORANGE SOUP

As a good variation of the recipe above, add to the soup as it simmers, the finely grated rind of two oranges and some nutmeg. When it has been sieved or puréed, pour in the juice of the oranges. Use a little sugar to bring up the flavour if necessary. Serve with three or four thin slices of orange floating on top and a sprinkling of chives and parsley.

TOMATO AND MUSSEL SOUP

A most delicious and simple soup, a meal in itself. Wash 2 kg (4 lb) mussels, scrape them and discard any that remain open when tapped hard with a knife. Put 250 ml (8 fl oz) water, a chopped onion and a chopped clove of garlic into a large pan. Add the mussels, cover and set over a high heat for five minutes. Remove the mussels as they open and throw away the shells. Strain the liquor and make it up to ½ litre (¾ pt) with fish, chicken or veal stock or water. Reheat with 250 ml (8 fl oz) tomato sauce. Put the soup through the medium plate of a *mouli-légumes*, correct the seasoning and reheat. Add the mussels, scatter with parsley and serve with triangles of bread fried in olive oil or butter.

TOMATO AND HARICOT BEAN OR LENTIL SOUP

This more solid tomato soup is full of flavour and interest and particularly restoring in summer when the heat dies down and everyone feels tired. The ideal is to use a richly flavoured variety of tomatoes from the garden, and to serve the soup with a thick scatter of chopped basil leaves. If you have to substitute parsley and chives, for instance in wintertime, use butter rather than olive oil as the cooking fat.

>125 g (4 oz) haricot beans or lentils
>100 g (3 oz) smoked streaky bacon or geräuchter bauchspeck
>125 g (4 oz) chopped onion
>1 small stem celery, chopped, or celery leaves
>1 large clove garlic, chopped
>3–4 tablespoons olive oil (or butter)
>½ kg (1 lb) peeled, chopped tomatoes (or 400 g/14 oz can)
>2 tablespoons tomato concentrate (or ketchup)
>1 litre (2 pt) beef stock
>salt, pepper, brown sugar
>chopped basil (or parsley and chives)
>bread cubes fried in olive oil (or butter)

Put the dried vegetables in a basin and pour over enough boiling water to cover them by a good 2½ cm (1 in). Leave for an hour or two until they begin to soften. Cook the bacon, onion, celery and garlic gently in the oil or butter, in a covered pan so that they do not brown. Add the beans or lentils and their water, tomatoes, concentrate or ketchup and stock. Simmer until the dried vegetables are cooked. Now add salt, plenty of pepper and sugar if necessary. Simmer for three minutes. Purée half the soup in a blender, then pour it back into the pan – this makes a slightly thickened but not too heavy soup; if you need to, dilute it further with more stock or water. Reheat and correct the seasoning. Scatter with herbs and serve with the bread cubes.

ROBIN McDOUALL'S TOMATO ICE CREAM

An unusual first course, from *Cookery Book for the Greedy*, now unfortunately out of print. It can be served on its own with brown bread and butter, or embellished with prawns, shrimps, lobster or crab. No mayonnaise is required. If the tomatoes are not quite the best, use a pinch of sugar to help their flavour.

> *2 huge, or 3 large tomatoes*
> *juice of ½ lemon*
> *tomato ketchup to taste*
> *125 ml (4 fl oz) double cream, whipped*
> *salt, pepper*

Put the tomatoes through a *mouli-légumes* or blender, then sieve to eliminate any last trace of the skin. Add the remaining ingredients in the given order. Freeze at the lowest possible temperature, until firm.

RUTH LOWINSKY'S TOMATO ICE AND TOMATO CONSOMMÉ

Two simple recipes from a beautifully produced cookery book of the thirties, *Lovely Food*, published by the Nonesuch Press in 1931. They are for home-grown tomatoes only, unless you have brought some home from a Mediterranean holiday. The slight freezing turns the first mixture to a delicious mush: serve it in glasses with wholemeal bread and butter.

'Make a purée of raw tomatoes and strain it through a sieve. Flavour with a little salt and pepper and freeze slightly.'

CONSOMMÉ FRAPPÉ AUX TOMATES is equally simple. Cut up ½ kg (1 lb) fresh tomatoes and simmer them with consommé for an hour. (Use a good veal or beef stock and clarify it, p. 550: start with half a litre and dilute to taste at the end.) Serve well chilled with fried bread croûtons spread with whipped cream. Try it with Cornish cream, too.

GAZPACHO

An odd word, *gazpacho*. Perhaps the meaning is bread or food eaten from a wooden bowl, a *kaz* in Spanish Arabic. Some have derived it from a Spanish verb meaning to be stuffed, crammed, full up and suffering indigestion (it is the easiest dish in the world to eat too much of). A *gazpachiero* was someone who took *gazpacho* round to men at work, particularly in the heat of Andalusia where most recipes come from.

> ¾ *kg (1½ lb) ripe tomatoes, peeled, chopped*
> *1 cucumber, peeled, chopped*
> *2 green peppers, seeded, chopped*
> *3 cloves garlic, chopped*
> *250 g (8 oz) wholemeal breadcrumbs*
> *5 tablespoons red or white wine vinegar*
> *5 tablespoons olive oil*
> *2 heaped teaspoons salt*
> *1 tablespoon tomato concentrate*
> *½–¾ litre (1–1½ pt) water*
>
> GARNISHES
> *small croûtons, fried in olive oil*
> *hard-boiled eggs, crumbled*
> *spring onion, chopped*
> *cucumber, peeled, chopped*

Mix the first nine ingredients together in a bowl. Put them in the blender and purée them gradually, transferring each smooth batch to another bowl. Use the water to help the ingredients reduce easily to a purée. Correct the seasoning. Chill for an hour or two. Ice cubes and a little more water are added just before serving. Some or all of the garnishes should be put into little bowls and arranged on a tray. People help themselves to whatever they fancy.

Gazpacho is really a liquid salad, a salad-soup, and is best eaten as a main lunch dish on a hot day. Drink white wine with it, Spanish white wine. When you become used to making *gazpacho*, experiment with the proportions to see if you come up with a *gazpacho* you like even better than this one.

BASIC TOMATO SAUCE

A splendid sauce, one of the great freezer standbys if you have a glut of home-grown tomatoes of a good variety, e.g. Marmande, which are firm all through and full of flavour. Don't bother to use the Dutch and Guernsey tomatoes normally on sale: a large can will give a much better taste.

> *3 large cloves garlic, chopped*
> *1 large onion, chopped*
> *125 g (4 oz) streaky bacon or Italian* coppa, *chopped (optional)*
> *2 tablespoons butter, lard or bacon fat, or 3 tablespoons olive oil*
> *1 large carrot, diced*
> *1 kg (2 lb) tomatoes, skinned, chopped, or 1 large can (800 g, 1 lb 12 oz)*
> *150 ml (5 fl oz) dry white or red wine, or 5 tablespoons Marsala or brown sherry*
> *salt, pepper, sugar*
> *dried oregano, chopped fresh basil*

Soften the garlic, onion and bacon, if used, in the fat. Add the carrot, tomatoes and wine, breaking down canned tomatoes with a wooden spoon. Raise the heat and cook hard, uncovered, for 15 minutes, or more slowly for 45 minutes if the tomatoes are canned. Add seasoning and oregano after ten minutes. The chopped basil should be put in just before serving and a little more scattered over the top.

As a general principle aim to finish with a chunky purée that is moist rather then watery. If you are freezing the sauce, you will find this the most useful texture. If you want to smooth out the sauce a little, without spoiling its vigorous character, put it through the coarse plate of a *mouli-légumes*. If you are using some of the purée to flavour other sauces, e.g. sauce aurore, put it through the medium plate. If you want to make cream of tomato soup, p. 507, a blender is the thing.

SALSA ALL'AMATRICIANA (sauce in the style of Amatrice, a small town in the Sabine hills near Rome). Soften a chopped onion in lard. Add 150 g (5 oz) diced lean bacon, raise the heat and brown lightly. Pour in 150 ml (¼ pt) dry white wine, then ½ kg (1 lb) good fresh tomatoes, or a medium (400

g, 14 oz) can. Season, and boil hard for 15 minutes. Serve with spaghetti
or rice, or with a variety of vegetable dishes.

SALSA ALLE VONGOLE (tomato sauce with mussels). Open 1½ kg (3 lb)
mussels, see page 508, and add, with their reduced liquor, to either of the
tomato sauces above. If you use the basic recipe, only half the quantity
will be required. Serve with pasta. Although the sauce can be made with
frozen tomato sauce, do not refreeze it after the mussels have been added
as their texture will deteriorate in the cold.

UNCOOKED TOMATO SAUCE

There are few nicer ways of eating tomatoes than this uncooked sauce, so
long as you make it with tomatoes of flavour and firmness.

Skin and chop 1 kg (2 lb) tomatoes. Chop a small sweet onion or several
large spring onions with a clove of garlic as finely as possible. Mix with
tomatoes and season well. Chill for three hours. Sprinkle with olive oil
and chopped fresh basil or parsley. Serve chilled, with very hot spaghetti
or egg noodles.

FRESH TOMATO SAUCE

A good and quickly made sauce for serving with other vegetables, or for
making *pissaladière* (see page 349), or for bottling or freezing for the
winter. Again, you need first-class tomatoes.

In a covered frying pan, sweat a chopped onion and a clove of garlic in
60 g (2 oz) butter; when they are soft, put in 1 kg (2 lb) skinned, chopped
tomatoes. As the juices begin to run, raise the heat to maximum and cook
hard for five to ten minutes to make a chunky purée. Pour off the juice
rather than prolong the cooking. Season with salt, pepper, a hint of sugar
and plenty of fresh basil.

BASIC TOMATO SALAD

The best seasoning for tomatoes is basil, fresh green basil, the king of the
herbs. Alas, it is difficult to grow, and if you do succeed the flavour is paler
than it would be in Italy where the hot sun gives it strength. If basil is
out of the question, use parsley and chives.

Skin the tomatoes, slice them across and arrange on a dish. Pour over
an olive oil vinaigrette, seasoned with black pepper and a little sugar if the
tomatoes were not as sweet as they could be. On top put thinly sliced

onions or chopped spring onion and finally a good chopping of basil or parsley and chives. Serve straight away, with bread and butter, as a first course.

When you have good tomatoes, sweet onions, good olive oil and plenty of basil, there is no nicer salad. It is my favourite dish of August and September.

TOMATO AND EGG SALAD

One of the simplest and most delicious things it is possible to make in the salad line. Like everything simple, it depends on using ingredients of quality for success – large Marmande tomatoes grown out of doors, free range eggs, a good vinaigrette with plenty of chopped parsley and chives.

I first realized how good this mixture can be when we were invited to have lunch with various neighbours in France. The usual start is a salad like this, perhaps with a beetroot dish as well (see p. 96). An attractive plate is almost covered with a layer of skinned, sliced tomatoes, then comes a slightly smaller layer of hard-boiled egg slices, each layer seasoned. This is repeated according to the number of people, finishing with a central rosette of egg. Over the top a good scatter of chives and coarsely chopped flat-leaf parsley, with a mild vinaigrette.

From the *charcutier*'s window, town dwellers may choose to buy an elaborated version. Large fine tomatoes are turned over and sliced down almost but not quite through. Egg slices are slipped into the cuts, and the whole thing is topped with a swirl of mayonnaise.

TOMATO, EGG AND TUNNYFISH SALAD

Another popular dish at French summer lunch parties. Make a tomato and egg salad as above, but leave a clear circle in the centre. Season the tomato and egg and dress with a little vinaigrette – not enough to swamp them. Drain a tin of tunnyfish, break it up with a fork and mix it lightly with mayonnaise and plenty of chopped parsley and chives. The mayonnaise should preferably be made with olive oil. Put the fish in the centre of the salad.

If you want to serve this as a main course, make a rice salad to go with it: boiled rice, mixed with vinaigrette and a few sharp bits and pieces – chopped anchovies or chopped olives, some strips of red pepper, capers and chopped spring onion.

STUFFED TOMATOES

Two things are essential if stuffed tomatoes are to be worth eating (very often they are dull and watery). Firstly the tomatoes must be firm and of a good variety. Secondly the seasoning must be thoroughly savoury and positive.

> 6 large, craggy tomatoes, or 12 small ones
> salt, pepper, sugar
> stuffing from pp. 562–7
> butter or olive oil
> 6 or 12 rounds of white bread

Slice a lid from each tomato at the stalk end. Scoop out the pulp with a pointed spoon, being careful not to pierce the outer flesh and skin. Chop the pulp roughly, and leave it to drain in a sieve. Season the tomatoes inside with salt, pepper and a little sugar, then turn them upside down in a colander or on a perforated plate to drain.

Make up the stuffing, incorporating the tomato pulp as indicated.

Fill the tomatoes about two-thirds full of stuffing. Put a dab of butter on top, then the tomato lids. Butter a baking dish and arrange the tomatoes in it. Bake at 180°C/350°F/Gas Mark 4 for 20–30 minutes, until the tomatoes feel soft (do not cook them to a collapsing state). If you use a cheese stuffing, bake the tomatoes at 200°C/400°F/Gas Mark 6 so that the cheese melts but does not overcook – 10–15 minutes should be long enough, but be guided by the feel and appearance. Different varieties of tomato need different cooking times.

Meanwhile fry the bread rounds to a nice golden colour in butter or olive oil. Put the stuffed tomatoes on top and serve either on their own or round a joint of meat or a piece of baked fish.

GRATIN OF SLICED TOMATOES

A simple, but most successful dish from Evan Jones's *The World of Cheese* (Knopf, 1976). The first time I tried it, we had been cherishing the last leaves on two basil plants and it seemed a good way to enjoy them. I have no enthusiasm for dried basil, though it could be used instead.

> ¾ kg (1½ lb) firm tomatoes, peeled, sliced
> 6 teaspoons dry sherry
> fresh basil, chopped

salt, pepper
12 tablespoons double cream
6 heaped tablespoons grated Cheddar
parsley, chopped

Divide the tomatoes between six buttered gratin dishes, about 10 cm (4 in) in diameter. Dribble a teaspoon of sherry over each. Scatter on chopped basil. Bake for 30 minutes at 150°C/300°F/Gas Mark 2. Remove from the oven and pour two tablespoons of cream over each dish of tomatoes and sprinkle the cheese on top. Bake for 15 minutes longer, at the top of the oven, and serve with a little parsley over the melted cheese. Plenty of good bread is required to mop up the juices.

TOMATO TART (1)

Line a 23–26 cm (9–10½ in) tart tin, with a removable base, with short-crust pastry. Bake blind until it is set firm and very lightly coloured. For the filling you will need:

1 onion, chopped
60 g (2 oz) butter
1 teaspoon chopped parsley
½ teaspoon thyme
½ kg (1 lb) tomatoes, skinned, sliced
salt, pepper, sugar
4 eggs
150 ml (¼ pt) each milk and cream
2 tablespoons each grated Parmesan and Cheddar
2 tablespoons chopped basil (see note)

Cook the onion in the butter until soft and golden. Stir in the parsley and thyme. Drain off any surplus moisture and spread the onion over the pastry case. Arrange the tomatoes on top and season them. Beat the eggs, milk and cream together, with salt and pepper. Pour over the tomatoes. Bake in the oven at 200°C/400°F/Gas Mark 6 for 20 minutes. Mix the cheeses with half the basil, and scatter over the top. Return the tart to the oven for about ten minutes, until the cheese is bubbling and golden-brown. Sprinkle with the remaining basil just before serving hot, or warm.

Note If fresh basil is out of the question, add a level teaspoon of dried basil to the cooked onion, and the same to the grated cheeses.

TOMATO TART (2)

This recipe is based on a Piedmontese dish, hence the use of cornmeal, one of the main foods of northern Italy. It gives a delicious crusty texture to the tomatoes, not crisp exactly but richly sandy. The other essential is firm tomatoes of the Marmande type.

> 1 kg (2 lb) tomatoes
> salt, pepper, sugar
> puff pastry made with 250 g (8 oz) flour
> cornmeal
> butter
> 2 tablespoons grated Parmesan cheese
> chopped fresh basil (optional)

Cut the tomatoes, without peeling them, into slices about ½ cm (¼ in) thick; set aside the end slices to use up in another dish. Season the tomatoes with salt, pepper and a very little sugar. Leave them to absorb the flavours while you roll out the pastry not too thinly and line a 25–28 cm (10–11 in) flan tin with a removable base.

Now turn the tomato slices in cornmeal so that they are nicely coated, then fry them slowly in butter so that the meal turns to a golden-brown crust. When all are done, pile them into the pastry case in closely fitting layers. Sprinkle the top with melted butter, cheese and basil if possible. Dried basil can be used, but it has lost its soul. Bake at the top of the oven at 230°C/450°F/Gas Mark 8 for 15 minutes, then lower the heat to 190°C/375°F/Gas Mark 5 for a further 20 minutes. A little longer will not hurt – the important thing is to serve the tart straight from the oven so that the juices do not have time to make the pastry soggy.

TOMATO AND OATMEAL TART

Savoury tarts, so-called 'quiches' that would make a chef from Lorraine pallid with outrage, have become a cliché of institutional catering. This recipe makes no claims at all to being a quiche. I was trying out oatmeal pastry and wanted to work out a good filling that would be complemented by its rough crispness. The secret is to make a good thick reduction of the tomatoes and onion, with a fiery seasoning to give the whole thing lightness.

PASTRY
125 g (4 oz) plain flour
125 g (4 oz) rolled oats
good pinch salt
125 g (4 oz) butter or lard, or both mixed
1 large egg, beaten

FILLING
1 medium to large onion, chopped
1 large clove garlic, chopped
60 g (2 oz) butter
1 medium tin (400 g/14 oz) tomatoes
1 large egg
single or whipping cream
1 heaped tablespoon grated Parmesan
1 teaspoon harissa, or chilli sauce, or 1 small seeded chilli, or cayenne
60 g (2 oz) Cheddar, grated

Make the pastry in the usual way and use it to line a 20–30 cm (8–9 in) tart tin with a removable base. Put it in the larder or fridge to rest while you make the filling.

Cook the onion and garlic until soft in the butter, without browning them. Tip in the tomatoes with their juice (and the small chilli, if used: chop it first). Boil hard until fairly thick and not at all watery. Meanwhile break the egg into a measuring jug and bring it up to 150 ml (¼ pt) with the cream, mixing them thoroughly together. At this point switch on the oven to 190°C/375°F/Gas Mark 5 and put a baking sheet on to the centre shelf to heat up at the same time.

When the tomato mixture is nicely thick, remove it from the heat. Stir in the cheeses, then the egg and cream. If you have not used a chilli, stir in the harissa (which I think is best) or chilli sauce or cayenne; do this gradually, to taste. Add more if you think the mixture could be hotter, and a little salt.

When the oven is at the right heat, turn the tomato mixture into the pastry case and use the trimmings of pastry to make a simple lattice. Or sprinkle the top with some grated Cheddar and breadcrumbs. Put the tart on to the heated baking sheet and leave for 30 minutes, or a little longer. Serve straight from the oven, or warm, with a green salad.

TOMATO MOUSSE

A rose-coloured mousse of fine flavour, a centre-piece for lunch in the garden. Set it in the middle of a large old-fashioned meat plate, preferably a pink one, and surround it with slices of avocado dressed with vinaigrette, hard-boiled eggs, prawns, or chicken and ham. The recipe can also be used to make a pudding if you omit the Worcester sauce and pepper and increase the sugar. Whichever version you choose, serve dark brown bread with it, rye, wholemeal or a good granary, and lightly salted butter.

> *250 ml (8 fl oz) sieved tomatoes*
> *175 ml (6 fl oz) tomato juice*
> *1 packet gelatine (to set 600 ml, 1 pt)*
> *4 tablespoons very hot water*
> *good pinch salt*
> *Worcester sauce*
> *freshly ground black pepper*
> *sugar*
> *250 ml (8 fl oz) whipping cream, or two-thirds double and*
> *one-third single cream*

Mix the tomatoes and tomato juice with the gelatine dissolved in the hot water. If the gelatine goes knobbly, add a little of the tomato juice to the mixture and stand the bowl in or over a pan of barely simmering water. Stir until smooth. Mix in the rest of the juice and sieved tomatoes and tip rapidly into a cold basin before the heat has a chance to alter the fresh flavour. Season with salt, then about two teaspoons of Worcester sauce, a half teaspoon of pepper and a dessertspoon of sugar. The flavour should be on the strong side, as the cream will mute the sharpness. Whip the cream(s) until thick and add to the tomato jelly when it is the consistency of thick egg white. Check the seasoning. Pour into an oiled ring or decorative mould. Chill until set.

GAME WITH TOMATO AND CHOCOLATE SAUCE

I have based this recipe on the Spanish *perdices a la bilbaina*, partridges Bilbao style, adapting it to a casserole method for hare, pigeon and older game birds. The unusual interest of the sauce depends on two items that the Spaniards brought back from the New World, tomatoes and chocolate. Tomatoes have become a cliché of everyday cookery but chocolate as a flavouring for meat dishes has barely survived. There is the Italian hare with sweet-sour sauce and an odd dish of pasta eaten with butter, pine kernels and grated chocolate on All Souls' Day, a dark dish of earthy flavour known too appropriately as *vermicelli atterati*, that sustains recently bereaved families in Naples. This sauce is more cheerful altogether: the flavour of unsweetened chocolate and the essences of game give splendour to the tomato. One could even adapt it to beef, though one really needs game for the fullest richness.

> *3 partridges, or casserole grouse, or 6 pigeons, or 1 hare, jointed*
> *3 rashers of fat bacon, chopped*
> *lard or bacon fat or oil*
> *2 large onions, sliced*
> *2 carrots, diced*
> *2 huge tomatoes, skinned, chopped, or 1 medium can (400 g/14 oz), or 375 g/13 oz smaller tomatoes*
> *2 cloves garlic, crushed*
> *beef or game stock*
> *1 tablespoon chopped parsley*
> *salt, pepper, 2 cloves, grated nutmeg*
> *1 tablespoon sherry vinegar (or wine vinegar)*
> *¼ square Baker's chocolate, or 2–3 teaspoons grated bitter chocolate*
> *1 glass dry sherry*

Brown the game and bacon quickly in a little fat. Put them into a casserole, breast-down if you are cooking birds. Add the vegetables, garlic and enough stock barely to cover. Stir in the parsley and seasoning, going carefully with the nutmeg, and the vinegar. Bring to the boil, cover and simmer until the game is tender. Transfer it to a dish, carving the birds into two pieces each (keep the carcases for stock). Keep them warm. Put the sauce through the coarse plate of the *mouli-légumes* and reheat. Add the grated chocolate slowly to taste, and the sherry. Simmer for five to ten minutes, then taste again and make any adjustments to the seasoning you feel necessary. Pour over the birds. Scatter with a little extra parsley and serve with buttered noodles.

BEEF AND TOMATO STEW WITH OLIVES

This recipe is loosely based on the provençal *daubes*, those mellow stews of beef that used to be cooked for hours in the ashes of a fire, with red hot coals in the saucer-shaped lid. One can sometimes buy enamelled iron *daubières* for this kind of dish, but nowadays one would fill the lid with hot water and put the whole thing into the oven. An important point, particularly with cheaper cuts of beef, is to include some gelatinous pork; this can be a large piece of fresh pork skin, a trotter, or a meaty hock, or several slices of belly of pork, They soothe and enrich the sauce. For this reason I always choose shin of beef, in preference to more expensive stewing cuts, as the nuggets of meat are held together with a jelly-like substance that dissolves into the lean meat, keeping it succulent, preventing stringiness.

> 1–1½ kg (2–3 lb) shin of beef, sliced
> salt, pepper, oregano or marjoram
> 1 large onion, chopped
> 2–3 large cloves garlic, chopped
> 60 g (2 oz) butter
> fresh pork skin, trotter etc.
> 200 ml (7 fl oz) red wine
> ¾ kg (1½ lb) tomatoes, skinned, chopped, or 2 medium (400 g/14 oz) cans
> 1 tablespoon dark brown sugar
> tomato concentrate (see recipe)
> 18 black olives, stoned
> chopped parsley

Season the beef with salt, pepper and oregano or marjoram and set aside for a couple of hours if possible. Brown the onion and garlic lightly in the butter, then put in the beef, raise the heat and seal it. Transfer to a casserole and bury the pork skin, or whatever piece of pork you are using, in the middle. Pour the wine and tomatoes into the frying pan and when they are boiling thoroughly tip into the casserole with the meat. Season with pepper but no extra salt. If necessary, add a little water or stock so that the meat is covered. Put on the lid and simmer gently until the meat is tender; this can take from two to 12 hours according to whether you put the casserole into a moderately slow oven, 150–180°C/300–325°F/Gas Mark 2–3, or whether you wish to make use of an electric slow cooking pot.

When the meat is done, remove it to an earthenware dish. Cut up the pork into pieces, and throw away any bones. Taste the sauce, then boil it down hard to concentrate the flavour and texture. Season with the sugar and concentrate if necessary, add more pepper and salt to taste. Heat the olives in the sauce for five minutes, then pour over the meat, scatter with parsley and serve. Boiled potatoes, noodles, triangles of fried bread all go well with this kind of rich stew.

To me the point of success is the reduction of the sauce. I do this with all the stews I make, serving them eventually as meat with sauce, rather than as an undisguised stew which reminds the family too much of institutional meals and the old-fashioned wash days of the past. The reduction gives one a chance to adjust the seasoning, to boil harder or slower, without overcooking the meat. In this way you get the essence of tomato and wine that sets the beef off to best advantage.

SHAKER TOMATO CUSTARD

A recipe from a cookery book of the American Shakers, that puritanical sect born of the Quaker movement two centuries ago. The collection of favourite dishes is the serene opposite of what might have been expected from these Shaking Quakers, with their ecstatic song and dancing rituals, their gift of tongues, their strong Puritanism.

The first great leader, Mother Ann Lee, had worked as cook in the new Manchester Infirmary before taking a group to America in 1774. I suspect her experiences in the kitchens there must have been happy, because she and her followers developed a tradition of good food lovingly prepared from first class ingredients in clean, well ordered surroundings. Plenty of eggs, cream, wholemeal bread, fruit, vegetables, honey, poultry, maple syrup, puddings flavoured with rose water, all their own produce, gave Shaker mealtimes an air of cheerful comfort – even if men and women sat at separate tables.

'If we find a good thing, we stick to it' was another Shaker principle. So they were still using European 18th-century recipes when they had been forgotten in their original countries. Sometimes a European recipe was adapted to a New World ingredient, as with this tomato custard, a living fossil of the kind one often comes across in American food.

> 1 kg (2 lb) ripe tomatoes
> 4 large eggs, beaten
> 125 ml (4 fl oz) milk
> about 3 tablespoons sugar
> salt, grated nutmeg

Quarter, stew and sieve the tomatoes – add no water. Cool and add the remaining ingredients, adjusting the last three to taste. Pour into six buttered pots. Stand them in a pan of hot water and bake at 180°C/350°F/Gas Mark 4 until the centre is barely firm (it continues to cook as it cools). Serve warm or chilled, with or without cream.

TOMATO JAM

It is not surprising that tomato jam is more of a French taste. Our commercial tomatoes are useless as they dissolve into a poor flavoured slosh. I could not recommend the recipe unless you grow your own firm-fleshed tomatoes out of doors. Then it is delicious, with its light tang of lemon and ginger. Another surprise is that although I have only eaten tomato jam in France, and this recipe comes from a friend there, she swears she had it long ago from an English source.

> *3 kg (6 lb) ripe but really firm tomatoes, skinned*
> *10 tablespoons lemon juice*
> *the rind of 2 large lemons*
> *2½ kg (generous 5 lb) preserving sugar*
> *250 g (8 oz) drained ginger in syrup, chopped*

Slice the tomatoes thickly into a heavy pan. Add the lemon juice. Cut the rind, including the pith, into thin shreds and add that, too. Bring to simmering point and simmer steadily until the tomatoes are pulpy. If there is a lot of liquid either drain it off and keep it for drinking, or raise the heat and boil it away quickly. Meanwhile warm the sugar, then add it to the tomato pulp with the ginger. Lower the heat so that the sugar dissolves slowly (keep stirring), then raise the heat so that the jam boils rapidly to setting-point – 20–30 minutes. Pot and cover in the usual way in sterilized jars.

Delicious with wholemeal bread and butter.

OTHER SUGGESTIONS

Peperonata, Pipérade etc, see pages 382–4.
Ratatouille, see page 52.
Boumiano, see page 53.

TURNIPS and SWEDES

When Arthur Young, the great agriculturist, visited the château of Chambord in 1787, he climbed, like all visitors do, up to the roof. He ignored the spectacular elaboration, the pinnacles, gables, turrets, twirls, balustrades, that make it look more like an ornamental village than the covering for a mansion, and gazed out over the flat landscape of the park. If only, he thought, if only the king of France would take to 'the turnip culture of England' and establish 'one compleat and perfect farm', Chambord would be the ideal place. The château would make a splendid residence for the director and his staff. The barracks would house cattle far more usefully than soldiers.

In fact Chambord was destined for the tourist culture of France – the first *son et lumière* was held here – but the French haven't overlooked turnips, quite the contrary. They have devised several elegant dishes using the young white turnips of early summer, which put our recipes to shame. We stick too much to the agricultural view, regarding the turnip as a coarse, cow-sized vegetable, suitable for the over-wintering of herds, schoolchildren, prisoners and lodgers.

I would not go so far as to say that winter turnips are useless. They have their place in soup and stocks. A nicely flavoured turnip mash can be used as the basis of a delicious soufflé. But unless you are prepared to lavish attention and butter on them, I would suggest waiting until the spring and early summer. Then they can be a delight in a navarin of lamb, or with duck, or caramelized with ham in a port wine sauce, or in a creamy soup. Then their ability to absorb and profit from richness is seen – and tasted – to best advantage.

One of the reasons I love returning to France in the Easter holidays, is to be able to buy the first bunches of young white turnips tied together by their stalks. They have a mat glow to their skins that shows off beautifully against the pale green vigour of the leaves. They smell sweetly peppery, a whole lifetime away from the yellowing turnips of winter with their stringy, waterlogged flesh and a harsh flavour that has to be blanched out. Another point – in spring the young turnip tops make a good green vegetable. Some seedsmen sell special varieties of turnips that provide good green tops in the spring, as well as the usual roots. They are popular in Rome especially, as you will see from the recipe for *broccoletti di rape* in the pages following.

To Rome, ancient Rome, we owe the second part of their name. Like the French *navet*, it derives from the Latin for turnip, *napus*. And in the 16th century our neeps acquired an extra syllable to describe their roundness, turn-neep. A point of confusion – in Scotland when people talk of neeps, bashed neeps with haggis, they mean swedes. The yellowish-orange roots first developed it seems in the 17th century in Bohemia, that came to us via Sweden in 1781. No doubt the new roots' hardy ability to survive in the ground all winter made it a welcome vegetable to northerners. It usurped the place, even the name of the older white-rooted turnip in Scotland, so that a dish called turnip purry is invariably a purée of swede. More confusion – swede is sometimes known, especially in America by the Swedish dialect name of *rotbagga*, ram's root, anglicized to rutabaga.

Although a number of recipes are interchangeable between turnip and swede, remember that the flavours of the two roots are different. The swede tastes far coarser than a young summer turnip and would not partner duck or ham so companionably. As a vehicle for butter, with haggis and whisky, it is exactly right. But after a north country upbringing, I conclude that otherwise swede is a vegetable to be avoided. The watery orange slush of school dinners was unredeemed by drainage or butter.

For agreeable turnip flavour in winter, you would do better to go for kohlrabi, the turnip-rooted cabbage. Usually only the root is sold, though the stalks where the (edible) leaves have been cut away are left to give it a whiskered appearance.

HOW TO CHOOSE AND PREPARE TURNIPS AND SWEDES

At whatever season of the year you are buying turnips and swedes, avoid those that look faded or heavily calloused, or that smell strongly. Young white turnips come in round or long shapes; there is little difference between them in flavour. Swedes come in a variety of colours from purplish brown to yellow and occasionally white; again there is little difference in flavour.

For most recipes, turnips need to be peeled, after they have been topped and tailed. When young, they need only the thinnest layer removing, but once they increase in size you may have to take off quite a thick layer to get rid of hard woody skin. Slice, dice or quarter the vegetables as appropriate for the recipe, or leave them whole, putting the pieces as they are prepared into acidulated water to prevent discoloration.

With turnips a preliminary blanching is usually required. Naturally young turnip dice take only a short time, between five and ten minutes. Tougher older roots can take up to an hour before they come to tenderness, depending on the size.

CREAM OF TURNIP SOUP

> 350 g (12 oz) young turnips, diced
> 250 g (8 oz) potatoes, diced
> 1 leek or 4 spring onions or 1 medium onion, chopped
> 2 tablespoons butter
> 1 tablespoon plain flour
> 1½–2 litres (2½–3½ pt) stock
> salt, black pepper
> 2 large egg yolks
> 4 tablespoons whipping or double cream

Cook the vegetables in the butter in a covered pan over a low heat for ten minutes; shake the pan occasionally, or stir the vegetables about, they must not brown. Add the flour, stir again, then moisten gradually with enough stock to cover the vegetables easily. Season and simmer until the vegetables are tender, from 20–30 minutes.

Blend or sieve the soup through the *mouli-légumes*, adding enough of the remaining stock to make an agreeable consistency. Return the soup to the pan and reheat gently. Mix the yolks and cream, add a ladleful of hot soup, stirring well, then return to the pan and heat through for a few moments without boiling. Keep stirring. Taste again for seasoning. Serve with bread croûtons.

SOYER'S CLEAR TURNIP SOUP

Alexis Soyer, born in France in 1809, came to this country at the age of 21, to become the great star of cookery in the mid-19th century. He began by cooking and writing for 'the higher class of epicures . . . and the easy middle class', in particular at the Reform Club where his splendid kitchens were among the sights of London. He used his fame and inventive skill to help the unfortunate. In the potato famine of 1847 he organized a soup kitchen in Ireland. At his own expense he went out to the Crimea and revolutionized hospital and army catering, while Florence Nightingale was transforming the care of the sick and wounded.

In 1854, he published his *Shilling Cookery for the People* and dedicated it to Lord Shaftesbury, the reformer. Like his grander books, the *Gastronomic Regenerator*, and the huge *Pantropheon* (a history of cooking), it went into many editions. I doubt it did much for the people for whom it was intended. Most of them would have been illiterate. Those who could read it would probably have been put off by the classical allusions, the high style and the anthropological tone: in the dedication Soyer announces his intention of leaving the Reform Club and going off to meet The People, as if he were going to Africa to meet The Pygmies.

What surprised me, when I first read it, was to find the source of the dishes eaten not by the poor, but by the professional middle classes of the thirties. Frugal, plain but delicious – shin of beef stew, sheep's head with brain sauce, wholemeal bread, straightforward roasts, nice puddings and supper dishes.

This recipe for a fine soup is an economical luxury. The ingredients are cheap, the delicious savoury lightness of flavour depending on a simple trick. The method can be used for carrots, Jerusalem artichokes or a mixture of vegetables – Soyer recommends one-third carrot and turnip, two-thirds celery, leek and onion. Be sure to use young, white turnips – 'if turnips are either streaky or spongy, they will not do'. Soyer's recipe for veal stock is given on p. 550; if you do not have the time to make it, use beef cubes well diluted.

> ¼ kg (½ lb) prepared, diced turnips
> 60 g (good 2 oz) butter
> 1 rounded teaspoon sugar
> 1¾ litres (3 pt) veal or light beef stock
> parsley or savory or tarragon, chopped

Stir the turnips and butter together over a low heat until the vegetable is coated, then sprinkle with the sugar and raise the heat. Stir occasionally until the juices are golden-brown and the turnip slightly caramelized but not burnt at the edges (this is the trick). Add stock and simmer for 15–20 minutes until the turnips are tender. Skim and correct the seasoning, and, if you like, remove most of the turnip leaving a few nice pieces. Scatter the soup with a little parsley or savory or tarragon just before serving.

NEEP OR TURNIP PURRY

Scottish mashed swedes that often accompany haggis. 1 kg (2 lb) is ample for six people.

Prepare, cut up and boil the swedes in 2 cm (¾ in) salted water. Drain them well, then put through the *mouli-légumes*, pouring off the extra juices that will flow out of the vegetable. Reheat with a generous amount of butter and correct the seasoning.

In Meg Dods' cookery book of 1826, *The Cook and Housewife's Manual* she recommends an additional seasoning of powdered ginger if the purry is being served with boiled fowl or veal 'or the more insipid meats'. Nutmeg may also be used to add a little spice to the mixture.

To adapt the recipe to young white turnips, peel them thinly and slice them, then blanch them in a good 1 cm (½ in) salted water, with a tightly fitting lid on the pan. Drain them, and complete as above. Chopped parsley and chives make a good flavouring for the purée, rather than the ginger and nutmeg recommended for swedes.

HAGGIS AND BASHED NEEPS

It is quite easy to buy haggis in southern England and it is a good dish. After simmering it steadily in water for an hour and a half, serve it with the neep purry above, potatoes and whisky to drink. If you can manage a fine single malt such as Laphroaig, so much the better.

BUTTERED TURNIPS

Choose young white turnips for this dish. Peel and cut them into slices or 1 cm (½ in) cubes. Blanch them for four or five minutes in boiling salted water until they are half-cooked. Drain them well. Finish them with butter in a frying pan, so that they become juicy and golden yellow, rather than brown. Keep turning them so that they colour evenly. Make a good chopping of green herbs, parsley, chives, tarragon, with a tiny clove of garlic, and stir this into the turnips just before serving them. They make a good dish on their own, embellished perhaps with triangles of fried bread. or they can go with chicken or veal.

To make a change, put a sprig of rosemary into the butter when you are finishing the turnips. Serve scattered with a very little parsley. Remove the sprig of rosemary before you turn the vegetable into its dish.

CREAMED TURNIPS

Follow the recipe above, but when you add the green herbs, mix in a few spoonsful of double cream to make a little sauce. Not too much, just enough to bind the mixture in an appetizing way. A few drops of lemon juice will bring out the flavour.

ROAST TURNIPS OR SWEDES

Quite a good way of serving winter turnips and swedes. Peel them and cut them into fairly thick slices or sticks of about 1½ cm (¾ in). Blanch them for four minutes in plenty of boiling water. Have ready a roasting pan with 1 cm (½ in) depth of dripping or lard or oil, brought to boiling point. Put in the drained pieces and cook at the top of the oven over the roasting joint, just as you do roast potatoes. Turn the pieces occasionally; they need about 40 minutes.

If you like the roast turnips soft and juicy, put the blanched pieces round the joint instead.

You can make a good mixture that goes well with beef, by peeling and blanching parsnips, carrots, turnips or swedes, and potatoes, and by roasting them all together in the pan.

GLAZED TURNIPS

We first came across glazed turnips at our small village hotel in France, the Ariana. *Jambon au porto* had been ordered. This means slices of country ham, heated through in butter, and the cooking pan deglazed with port which is then poured over the ham. It came to table with a dish of glazed turnips, sizzling and caramelized.

Choose young turnips. Peel and cut them into 2 cm (¾ in) dice. Blanch for five minutes in boiling salted water. They should not be quite cooked, so you must watch them: older turnips can do with a little longer. Drain them. Melt a good knob of butter in a heavy frying pan, put in the turnips and sprinkle them with sugar. Fry briskly, so that the turnips caramelize to a delicious golden colour without the sugar turning black. Keep them moving about. If they are not browning in a caramelized way, add a little more sugar to increase the glossy appearance. Serve very hot.

It is difficult to get a proper French country ham here, so you will probably have to use gammon instead. Whatever the Danes may say, I always find it far too salty to be used without any soaking. Allow a thick gammon steak or rasher per person, snip the fat round the edges so that they don't curl up and pour boiling water over them. Leave for half an hour, then cook briefly in butter on both sides and deglaze the pan with 150 ml (¼ pt) stock and about five tablespoons of port. Pour over the gammon and serve surrounded with the glazed turnips. The sauce can be extended with a little cream, but this is not necessary.

GOETHE'S TURNIP WITH CHESTNUTS

'Let us be many-sided! Turnips are good, but they are best mixed with chestnuts. And these two noble products of the earth grow far apart.'

from *Prose Maxims*

The last sentence puts me in mind of the German dish known as *Himmel und Erde*, Heaven and Earth. There the heavenly ingredient, apple, is cooked with the earthbound potato. Here Goethe's chestnuts set off the over-familiar turnip, reminding us of its virtues by the contrast. Such combinations, sometimes with dried fruit, were already an old habit of German and Swiss cookery by Goethe's time. Montaigne comments on it with pleasure in the 16th century. When potatoes became popular two centuries later, they were fitted into the tradition, with apples or pears to make the partnership (see page 408).

Glaze the turnips as above. Add pieces of shelled chestnut to heat through in the pan for the last few moments. A few spoonsful of stock added to the pan juices with the chestnuts help them to heat through more effectively, without risk of burning.

ESCOFFIER'S STUFFED TURNIPS

Pick out even, medium-sized turnips that are still fresh and young. Peel them into a good shape. Then with a sharp knife or round cutter, score a deep circle in the base; this will enable the centres to cook more quickly, and make it easier to hollow the turnips out. Blanch the turnips in boiling salted water until they are almost done. Drain them and rinse under the cold tap. Hollow out the centres and weigh this pulp. Add an equal amount of mashed potato, with seasoning and butter. Fill the turnips with this mixture, piling up the centre in a domed shape. Melt some butter in a deep frying or sauté pan, and set the turnips in it to complete their cooking. Baste frequently with the butter.

Escoffier points out that the emptied turnips could be filled with creamed spinach, chicory or rice. The cooking in butter should be carried out in the same way. These turnips were intended as a 'sightly and excellent garnish' but they can also be served as a course on their own with bread or fried triangular croûtons. If you care to cut out thick bread rings, with a couple of cutters, and fry them, the stuffed turnips can be placed on top, resting in the central hole.

SPRING NAVARIN OF LAMB
NAVARIN PRINTANIER

Navarin is a ragoût of mutton or lamb with turnips and potatoes. Its most delicious variation – *navarin printanier* – is made in the spring, often around Easter, with young vegetables, *les primeurs*, and lamb. It is an old dish of French cookery, being simply a stew or ragoût or haricot or halicot or hotpot of mutton, that was rechristened in the middle of the last century in celebration of the famous victory of Navarino. (The British, French and Russian navies destroyed the Turkish and Egyptian fleets and so ended the war of Greek Independence, on October 20th, 1827.) I imagine the idea came from the similarity in sound between a main ingredient – *navet* being French for turnip – and the place of the battle.

> *1 kg (2 lb) boned shoulder, or 12–16 best end of neck lamb cutlets*
> *salt, pepper, 1 teaspoon sugar*
> *60 g (2 oz) butter*
> *2 tablespoons flour*
> *2 tablespoons tomato concentrate*
> *up to 1 litre (1¾ pt) beef stock*
> *1 large clove garlic, chopped finely*
> *bouquet garni*
> *12–16 new potatoes, scraped*
> *12–16 small new carrots, scraped, left whole*
> *12–16 small onions, skinned*
> *4 young turnips, peeled, diced*
> *about ½ kg (1 lb) shelled peas*
> *chopped parsley*

Cut shoulder into 2 cm (¾ in) pieces. Trim excess fat and skin from the meat. Sprinkle it with the seasonings and leave for several hours. Brown the lamb in butter, remove to a heavy saucepan. Stir the flour into the butter, cook to a pale brown roux, then mix in half the tomato and half the stock. When smooth, add with the garlic and bouquet to the lamb, plus extra stock if needed to cover the meat. Bring to boiling point, put on a lid and leave to simmer for an hour, or until the lamb is almost done. Put in the root vegetables and cook for a further 45 minutes without a lid on the pan, so that the sauce can reduce and thicken. Skim off any surface fat, blotting away any last traces with kitchen paper. Remove the bouquet, correct the seasoning, adding more tomato if you like. Put in the peas and continue to simmer until they are done. Turn the navarin into a hot dish, bringing up the different vegetables to the surface. Scatter with parsley and serve with bread.

Naturally the success of this dish depends on the quality of the lamb and the freshness of the young vegetables. Above all, it is a gardener's dish.

TURNIP TOPS
BROCCOLETTI DI RAPE

Broccoletti di rape, green turnip tops, are a favourite spring vegetable in Rome, ranking with the white and purple flowering broccoli. In this country you have to grow them yourself, as they are not on general sale.

Pick over the greenery, taking special care to remove the tough thick stringy parts of the stalk, and any dead leaves. Wash the rest, then chop the pieces across two or three times. Cook them in fast-boiling salted water until they are just tender.

Now they can be drained and left to cool. Serve them with a lemon and olive oil vinaigrette.

If you want to eat them hot, cook them until they are not quite tender, then drain them well and finish their cooking in olive oil with a chopped clove of garlic. Keep stirring them about and give them time to absorb the flavours. Serve them with fried bread.

If the greens are particularly young and delicate, they can be cooked entirely in the olive oil and garlic. They must be turned and stewed, rather than fried.

OTHER SUGGESTIONS

Turnip soufflé A surprisingly good dish. Turn to the recipe for Jerusalem artichoke soufflé on p. 277. If you are using winter turnips, cinnamon makes an unusual flavouring. In summer, with fresh young turnips, use the parsley only

VEGETABLE MARROW

The Bunter of the kitchen garden has little to be said for it. Some cookery writers define its flavour as 'delicate'. This carries politeness too far. The marrow swells and swells with water, not goodness. Once the dropsy has carried it beyond 1½ kg (3 lb), it is no fun at all, except as a vehicle for other flavours when making jam and chutney. The habit of mushing marrow in with more expensive items such as apricots, to eke them out and disguise the economy, is deplorable. Better by far to make a small quantity of proper apricot jam for special occasions and to make a larger quantity of marrow jam for every day use.

The only well employed giant marrows I have come across were suspended in string bags from the attic beams of a Cotswold manor house. 'Marrow whisky', was the answer. I feel that the resemblance cannot be close, or every home in the kingdom would be festooned with marrows of fermenting liquor.

Marrows have been ruined by Village Show judges who think size is all. They do not have to eat them afterwards. Probably they go home to a nice supper of courgettes from their own garden. When courgettes reach 20 cm (8 in), and begin to swell towards marrowhood, cut them and freeze them before virtue departs. The ones that do get away, should be caught before they are a foot long. You can stuff them as in the first two recipes following. You can slice them and turn to p. 234 to make stuffed marrow rings, or you can peel and seed them, then cube and cook them in a steamer; served with a rich cream sauce, like the one on p. 555, or a cream-enriched béchamel, they will taste their best.

See also recipes for *courgettes*, *pumpkins* and *squashes*.

STUFFED MARROW

A disaster dish of English cookery once the marrow has reached Village Show size. The only thing to do then is to cut it across into rings, and follow the instructions on p. 234.

For a moderate-sized marrow that will fit into a large casserole or pan, try this method: slice off the stalk end diagonally and keep it on one side.

Hollow out the middle with a pointed metal spoon; this is quite hard work and you may well wonder whether it is worth the trouble. Remove long strips of peel to give the outside of the marrow a variegated effect. Now blanch the marrow in boiling, salted water giving it three minutes a side if you have to turn it over in a shallow pan; count the time from when the water returns to the boil. If the whole marrow is submerged, allow four minutes. The point of the blanching is to reduce expensive baking time in the oven. (If you have a solid fuel stove, you can omit this process and leave the marrow to bake gently for two or three hours in a warm oven.)

Now make a well spiced stuffing. You will find that the Greek rice stuffings that absorb water (p. 565–7) taste better than a Western breadcrumb mixture which becomes heavy and sodden. Put it into the well drained marrow, leaving room for it to swell slightly. Replace the lid and push this end of the marrow up against the side of the pot or pan, to keep it more or less in place: the diagonal cut helps too. Dot the top of the marrow with butter and pour some of the basic tomato sauce on p. 511 round the sides. If you add a seeded chilli to the sauce, or some harissa, or cayenne, it will make the dish livelier. An hour in a moderate oven should see it through, though this will depend on size and thickness. It is wise to allow an extra half hour. Baste the marrow from time to time with the cooking juices. Just before the end, say ten minutes before, stir the rest of the tomato sauce into the pan to heat through.

Another way of cooking stuffed marrow is to wrap it in a piece of buttered foil and place it in a roasting pan to prop it up. Start off at a high temperature 220°C/425°F/Gas Mark 7. When you hear the contents of the package bubbling vigorously away, you can lower the heat to 190°C/375°F/Gas Mark 5. 20 minutes before the marrow is cooked, tear back the foil on top so that it can brown slightly. Serve with the tomato sauce, either poured round it or in a jug. The advantage of this method is that the washing up is much reduced. And if you have no convenient pot or pan to hold the marrow, it is the only way you can cook it whole. The disadvantage is that the marrow is not bathed in the rich tomato juices as it cooks, and its wateriness does not evaporate so well.

If you take this amount of trouble, the marrow is bound to taste quite good. But as you hollow out and stuff and baste, you may well wonder why anyone ever lets the delicious courgette fatten and bulge into the watery marrow.

MARROW AND GINGER JAM

A shadow of the pale green melon jam one can sometimes buy in tins from South Africa, this is nonetheless quite a good solution if you are given a monster marrow. Cut off a few rings for the recipe on page 234, peel the rest, remove the seeds and cut it into 2 cm (¾ in) cubes,

> 2 kg (4 lb) prepared marrow cubes
> 1½ kg (3 lb) preserving sugar
> 30 g (1 oz) piece root ginger
> thinly peeled rind of 3 large lemons
> juice of 3 large lemons

Sprinkle the marrow with one-third of the sugar and leave overnight. Bash the ginger with a mallet to bruise the fibrous hardness – this gives the flavour a better chance to emerge. Tie it with the lemon rind in a piece of muslin. Put the marrow and its juices into a preserving pan, add the rest of the sugar, the lemon juice and the muslin bag, tied to the handle but well submerged. Bring slowly to the boil, making sure that the sugar is completely dissolved before boiling point is reached. Boil steadily until setting temperature is reached; the marrow will look transparent. It is a good idea to taste the mixture after 15 minutes: if you think the ginger and lemon flavour is strong enough, remove the little bag, if not, leave it a little longer. If you have any preserved ginger in syrup, drain off a few pieces and cut them into small but recognizable slices, and add them to the jam when it is almost at setting point. Pot and seal as usual.

VARIATION Omit the root ginger and the rind of one lemon. Add the cut up flesh of a medium pineapple or a small tin of crushed pineapple or a small tin of chunks (quarter the chunks, first).

VEGETABLE PEAR, see CHAYOTE

VEGETABLE SPAGHETTI

Like other marrows, the vegetable spaghetti came in from America, where it is known as noodle squash, squash nouvelle, or – a name I find embarrassing to contemplate – squaghetti. It is a newcomer of the last 20 years. When John Organ published his book of *Gourds*, in 1960, it had not yet acquired an English name. By 1964 'vegetable spaghetti' figured in Thompson and Morgan's catalogue as being 'recently introduced'. It was given an entry to itself, though one might have expected to find it under marrow varieties. Quite right from the cook's point of view, as this beautiful golden ochre vegetable contains a mass of thready flesh, like a tangle of pasta. This means it has to be dealt with differently from other marrows.

Publicity these days can mean that a new vegetable will rapidly establish itself in gardens, then creep into commercial favour if it is easy enough to grow, tough enough to survive marketing. Vegetable spaghetti has been lucky. In 1975 it was given just the publicity required, in *The Times* Diary. The editor, Michael Leapman, took on a Brixton allotment. I suspect he started with the intention of making fun of the urban game of 'self-sufficiency', what might more truly be called 'speed the window-box'. As he wrestled with bindweed and fat hen as well as the more desirable plants, satire was tempered with respect, irritation and partial success. He described his struggles in regular Diary bulletins of mock epic style. Vegetable spaghetti, like the Marmande tomato, emerged as a hero. Easy to grow, easy to cook, good to eat. Advice, comment, jeers, recipes were passed on from readers' letters. Other newspapers took pity on non-gardeners and published addresses from which oven-ready samples, rather than seeds, might be obtained.

Whether this will be enough to put vegetable spaghetti on to the regular market lists of *The Fruit Trades Journal*, it is too soon to say. I have my doubts. I remember how enthusiastically, how intelligently, Stephen Switzer promoted celeriac and Florentine fennel at the beginning of the 18th century, yet they are still regarded as 'exotics' by most people. How many of our market gardeners grow them? It would be nice to think that vegetable spaghetti has more of a chance.

HOW TO CHOOSE AND PREPARE VEGETABLE SPAGHETTI

Vegetable spaghetti looks like a short plump marrow. It should be between 20 and 25 centimetres long (8–10 in). The skin is often a rich sunny yellow, but may also be creamy white with a greenish tone. As with all marrows, avoid ones that are bruised or broken.

Break off the stalk and push in a skewer to make a hole. Boil the vegetable spaghetti whole in plenty of salted water. It should be ready in about 30 to 45 minutes. Some catalogues say 20 minutes, but this is rarely long enough. Pierce it with a larding needle to see if the inside is tender. Remove and drain, then cut down in half lengthways from the stalk end to reveal the strings of flesh inside. The seeds can be removed if they seem to be tough. Try one and see.

The vegetable spaghetti can now be finished in various ways. For eating, it should be twirled on to a fork, spaghetti style. It spoils the entertainment to cut up the strings.

According to size, allow one vegetable spaghetti for two or three people. Serve with butter and grated Parmesan, or tomato sauce.

STUFFED VEGETABLE SPAGHETTI

A suggestion from one of Michael Leapman's readers. Boil a vegetable spaghetti until tender, then slice it in half and remove the seeds as usual. Fill the cavity with some delicious mixture such as buttered peas or sweetcorn, or creamed chicken.

It is important that the filling should be good in its own right. Plenty of butter and freshness are necessary in the case of other vegetables. With the chicken, use a good cream sauce pointed up with grated Parmesan. Keep the final result light but richly light, to avoid inspidity.

WATERCRESS

Vivid watercress covering the clear
water of springs, green plump colour,
translucency, caught a poet's eye long
ago in 12th-century Ireland:

> Well of Tráigh Dhá Bahn
> Lovely is your pure-topped cress.

And in 19th-century France, where Victor Hugo came to know a small
town on the Marne, with its mills and *cresson de la fontaine*:

> J'aime Chelles et ses cressonières
> Et le doux tic tac des moulins.

A modern watercress farm is not so romantic as Tráigh Dhá Bahn or
the Chelles cress beds in the 19th century, but the huge oblong basins,
flowing thinly with water or drying out for the next seeding are clean and
cool, the growing watercress as tufted and vigorous as any the poets saw.
Hampshire is the great place for it in England. One firm there grows a
quarter of the whole country's needs. The health of the intensively grown
crop is maintained by treating watercress as an annual rather than the
perennial it is by nature. The process depends on much pure water, no
longer just from natural springs but from deep boreholes, laving the
watercress continually. An eye is kept on its quality and on the quality of
the year round harvest which is maintained by keeping the beds in
rotation. In September you may see people flinging down handfuls of the
cress left over from bunching and packing, on to the drained beds. In a
week the stalks will have rooted and the new crop be on its way. As the
weather grows cold, some firms have taken to protecting the beds with
plastic tunnels. When summer comes round again, the crop is grown from
seed, seed often prepared by the firm itself to encourage particular strains
they have found successful.

Modern methods are really a refinement of the advice given by Philip
Miller in his *Gardener's Dictionary* (1st edition, 1731). He observes that
the watercress in the market is gathered from 'ditches and other standing
Waters' around London, but if you want to grow your own it is quite
simple. Gather a few wild plants, being careful with the roots, plant them
into the mud, then let the water in gradually. They should not be cut the
first season but allowed to run to seed. The seed will fall into the water

and increase the growth. Philip Miller was instructing gentlemen with a taste for salads. Watercress was not grown commercially on any scale for almost another century. The small beginnings seem to have occurred much about the same time in England and Germany, then in France, from about 1800 to 1811. The start of the French trade is worthy of the poets. A Monsieur Cardon who was with Napoleon's headquarters at Erfurt in Upper Thuringia in the winter of 1809–10, was surprised to see the whiteness of a snowy winter's day broken in the country outside the town by patches of brilliant green – a watercress farm, rented out by the municipality. On his return he settled to grow watercress at Saint-Léonard, near Senlis, finding both the flatness and springs necessary in the valley of the Nonette, the small river that a mile or two downstream nourishes the Grand Canal and other waters of the Château of Chantilly, on its way to join the river Oise.

Watercress likes coolness, not cold or heat. It flourishes in the spring and borehole water that comes out of the ground at a temperature of 10°C (51°F). In winter it shrinks into underwater growth from the colder air, so that it has to be pulled. In warmer weather it grows above the water level and can be cut. Then it is bunched, labelled and boxed, the boxes are water-cooled at just above 0°C (34°F) and put into a cold room to await refrigerated transport. Alas at depots and markets this 'cold chain' is broken, which means that some watercress you buy – or don't buy if you are sensible – gets daily more droopy and yellow.

Philip Miller said that people love watercress in the spring of the year in salads, for its 'agreeable warm bitter taste', and because it is an excellent remedy for scurvy and to clear the blood. Modern growers still find, somewhat to their chagrin as watercress is available all the year round, that the big demand is about Eastertime. Odd how the superstitions of earlier times can influence our eating habits now.

BUYING AND PREPARING WATERCRESS

Vacuum packed watercress comes expensive, but there is no waste at all. For salads, sandwiches, for serving with roast and grilled meat, it is ideal. In the salad drawer of the refrigerator it will remain in good condition for a week. When you need it, open the pack and rinse it briefly.

The more usual method of buying watercress is by the bunch. Incidentally all bunches should be labelled, a guarantee of quality. In the old days they were packed into chip baskets; now cardboard boxes do the job more efficiently if less attractively. Do not buy watercress that is wilting and yellowish. There is no need to put up with this and the idea that it will make good soup is an illusion. However you will find that vigorous bunches of watercress are better for soup than the vacuum packed. It is not just a matter of price: in the bunch you get more stalk and this gives body as well as flavour to the soup. Keep some of the leaves to chop in at the end; use the rest for salad and sandwiches.

When you get a bunch of watercress home, slice off the end of the stalks, pick the rest over, rinse it and put it into a plastic bag. Exclude as much air as possible, then store in the salad compartment of the fridge. Or use a plastic box into which the watercress will barely fit; press on the lid and store in the same way. If you have no fridge, treat the bunch like a bunch of flowers – but not for long, as watercress soon flags in a warm room.

Your success with any of these methods will depend entirely on how long the watercress has been out of cold storage. Something you can only judge by eye, as – so far – the labels aren't dated.

In these days of pollution, it is imprudent to eat wild watercress, just as it has become unwise to gather mussels from the sea shore. Fishmonger's mussels have spent time in cleansing tanks. Greengrocer's watercress has been nourished all its life with a pure and controlled water supply. Or it should have been. If you are in any doubt, look for the grower's label.

Watercress can be cooked and used in the same way as spinach. It reduces to a dark green purée, and loses its bulk just as spinach does. On the other hand the strong flavour means that it is best eaten in small quantities, so that it does not work out as expensive as you might expect. It makes a good filling for omelettes and pancakes, and makes a good purée-cum-sauce (see the recipes at the end of the section).

WATERCRESS AS A HERB

In some of the delectable green sauces and butters of French cookery, watercress is used as a herb, along with parsley, tarragon and so on, valued for its pungent flavour and brilliant colour. This is a good point to remember when you want to make a mixture of *fines herbes* to flavour omelettes, stuffings and so on, and have nothing more to hand in the way of garden herbs than parsley and chives.

SAUCE VERTE Take a dozen small spinach leaves, the leaves from a dozen stalks of watercress, eight sprigs of parsley and four of tarragon. Pour boiling water over them and leave for a minute. Drain, refresh under the cold tap and leave to dry. Press out excess moisture. Now reduce the leaves to a purée in the blender. Add a couple of raw egg yolks. When the mixture is smooth, start adding 150 ml ($\frac{1}{4}$ pt) oil, drop by drop at first then more rapidly as the sauce thickens. Finally season with wine vinegar, salt and pepper. Serve with salmon and other cold fish.

If you have no blender, chop and pound the drained herbs to as dry a purée as possible. Press through a sieve and add to the separately made mayonnaise.

SAUCE RAVIGOTE Chop a good handful of parsley, tarragon, watercress, chives, chervil and a thick slice of onion. Mix with a heaped teaspoon of Dijon mustard, a dessertspoon of drained capers and a chopped hard-boiled egg. Gradually mix in three or four tablespoons of olive oil, then wine vinegar to taste. Chopped anchovies and gherkins are often added as well as capers. Serve with cold meat and fish and with hot or cold jellied meat such as calf's head and pig's trotters.

BEURRE MONTPELLIER (Elizabeth David's version) Weigh out 125 g (4 oz) of the following, in approximately equal amounts – watercress, spinach and tarragon leaves, parsley, chervil. If you grow salad burnet, the French *pimprenelle*, add some of that, too. If you have no chervil, double the quantity of parsley.

Pour boiling water over the herbs, leave for a minute, then drain and dry as completely as possible. Pound in a mortar, with six anchovy fillets, two tablespoons capers, four miniature gherkins, the yolks of one raw and three hard-boiled eggs. When all is smooth, mash into 125 g (4 oz) soft butter. Sieve and mix in five or six tablespoons of olive oil gradually. Finally sharpen with a little lemon juice. Store, covered, in the refrigerator where it will keep for several days, but remove in good time or it will be too hard to spread over the cold salmon, or whatever other fish you are eating it with. It can also be served with hot grilled, fried or poached fish.

I find that this process is much speeded by using a blender. Put the herbs and pickles in together, using the olive oil to keep them moving. When smooth add the eggs and mash into the softened butter by hand, or use an electric beater.

WATERCRESS SOUPS

Watercress makes one of the best of all soups, whether you make it with potatoes or as a cream soup. The French have long called it *potage de santé*. The name goes back to the 18th century – no doubt because watercress was regarded, as Philip Miller said, as a spring tonic against scurvy and to clear the blood, but even more I suspect because of the clean true flavour. If you decide to purée the soup with a blender rather than with a *mouli-légumes*, you can reduce the quantities in the first soup by about a third.

WATERCRESS AND POTATO SOUP

> *300 g (10 oz) watercress*
> *350 g (12 oz) potatoes, peeled, cut up*
> *1 medium onion, chopped*
> *60 g (2 oz) butter*
> *¾ litre (1½ pt) water or liquid from green haricot beans*
> *salt, pepper*
> *250 ml (8 fl oz) milk*
> *250 ml (8 fl oz) single or whipping cream*
> *2 egg yolks (optional: see recipe)*
> *chervil if possible, otherwise parsley, chopped*

Set aside and chop some of the best watercress leaves. Cut up the rest roughly and put with potato and onion into a pan with the butter. Cook gently for ten minutes, stirring occasionally so that the vegetables become buttery without browning. Add water or bean liquid, salt and pepper, and simmer for about 20 minutes until the potato is cooked. Put through the *mouli-légumes* or blender, and return to the rinsed out pan. If the purée is too thick, add more water along with the milk and the chopped watercress leaves. Bring to just under boiling point and cook for five minutes*. Beat the cream and yolks together and stir in. Keep over a low heat for a few seconds, so that the flavours blend together well. Check the seasoning. Add chervil or parsley and serve.

Note Yolks can be omitted and cream reduced in quantity; if you intend to do this, use a light chicken or veal stock rather than water, to add richness of flavour.

CREAM OF WATERCRESS

> *300 g (10 oz) watercress*
> *1 medium onion, chopped*
> *60 g (2 oz) butter*
> *1 heaped tablespoon flour*

1 litre (2 pt) water or light veal stock
salt, pepper
250 ml (8 fl oz) single or whipping cream
2 egg yolks (optional: see previous recipe)

Set aside some of the best leaves as in the recipe above. Chop the rest and cook with the onion and butter for about five minutes, without allowing the vegetables to colour. Stir in the flour and cook for a moment or two, then gradually add the water or stock and seasoning. Simmer half an hour, then blend (or put twice through the *mouli-légumes* until very smooth). Return the soup to the pan, with the chopped watercress leaves, and reheat. Judge the consistency and add more water if you like.* Beat the cream and yolks together and pour into the soup. Stir for a couple of minutes over a low heat and correct the seasoning.

Either soup can be served with cheese straws or fingers of grilled cheese on toast, or croûtons of bread fried in butter.

CHILLED WATERCRESS SOUP

Either of these soups can be chilled. Follow the recipes up to the asterisk*, keeping them on the liquid side as they will thicken further as they cool. Season. Pour in the cream – no egg yolks – just before serving, and check the seasoning.

WATERCRESS SANDWICHES

The best of all watercress dishes. Nothing can challenge the perfect combination of good bread, salty butter and peppery crisp watercress. They make a delicious supper when eaten with tomato soup (p. 507).

Good bread is essential, wholemeal, or rye, or white bread, according to taste. And if you can get salty Welsh or West country farm butter, so much the better. Don't go for genteel, triangular mouthfuls – use proper slices and plenty of watercress so that it bursts cheerfully out at the sides.

WATERCRESS SALADS

Watercress moistened with a good vinaigrette makes a fine salad on its own, or mixed with lettuce leaves (preferably a crisp lettuce of the Webb's Wonder type), or with sliced chicory.

MUSHROOM AND WATERCRESS Slice 125 g (4 oz) fresh, tightly closed mushrooms. Sprinkle them with a little lemon juice to stop them turning brown. Mix with a bunch or pack of watercress, and a few tablespoons of olive oil vinaigrette. Do not overdo the quantity of vinaigrette in this, or for that matter in any other green salad: it can make the whole thing too oily. Always start with less than you think you will need.

TURKISH SALAD (Why? The recipe comes from America.) Watercress is topped with strips of skinned red pepper. To skin the pepper, grill it on a turning electric spit (easiest) or under a hot grill. Or spear it on a fork and turn it over a gas flame (most difficult). Remove the blackened skin under the cold tap, drain and cut into pieces: the grilling will have softened the pepper flesh to a delicious texture, that I find far more sympathetic than the harder crunch of raw peppers complete with skin.

WALNUT AND WATERCRESS Add a few walnuts to the mushroom and watercress salad above and dress with a walnut oil vinaigrette, or mix walnuts and watercress only. Or add small cubes of bread that have been fried in walnut oil with garlic, to the walnut and watercress. Use plenty of pepper. If you cannot get walnut oil, use a good olive oil.

MIXED WATERCRESS SALAD Put into a bowl a pack of watercress, or the top part of a bunch of watercress cut off so that the sprigs are about 5 cm (2 in) in length. Add 24 walnut halves, 60 g (2 oz) diced Gruyère cheese, two finely chopped shallots or large spring onion, and an eating apple cored and cut in slices (do not peel). Pour over a little mustardy vinaigrette and turn carefully so that the ingredients are well mixed, and look attractive. Decorate with three hard-boiled eggs – one and a half cut in quarters, the other one and a half crumbled. This French salad should be served as a first course on its own, rather than after the meat course.

ORANGE AND WATERCRESS Slices of peeled and pipped orange, well peppered, and cut in quarters are mixed with sprigs of watercress and some olive oil vinaigrette. Walnuts or a few black olives can be added, too. Delicious and refreshing as a winter salad with duck, ham, veal.

   Instead of mixing the ingredients, arrange them in rings on a plate. Begin with an outer fringe of watercress, then overlapping circles of whole orange slices, and a central eye of black olives. Sprinkle with olive oil vinaigrette and an even sanding of black pepper. This looks spectacular, beautiful, and tastes splendid with either hot or cold duck.

PEAR AND WATERCRESS Sliced, peeled and cored dessert pears arranged on a dish within a border of watercress. Have ready a Roquefort dressing, made by mashing 75 g (scant 3 oz) Roquefort or other creamy blue cheese, with five tablespoons of olive oil vinaigrette. Add a tablespoonful of watercress leaves chopped small, and pour over the pears before they have time to discolour. Pepper the salad well. Serve on its own, or with cold salt pork and ham. Try to use Doyenné du Comice, the finest flavoured of all dessert pears, which was developed at Angers in 1848.

DUMAS'S WATERCRESS SALAD

In his *Grand Dictionnaire de Cuisine*, Alexandre Dumas describes watercress as a winter salad. It should be mixed, he says, with beetroot, and a few olives. Dress it with an olive oil vinaigrette.

LOCKET'S SAVOURY

I make no apology for repeating this fine recipe from *English Food*. It was invented at Locket's Restaurant near the House of Commons.

Toast eight small slices of white bread and cut off the crusts. Arrange in four small ovenproof, buttered dishes, or in one large one. Divide a pack or bunch of watercress between them. Top with four thinly sliced, peeled and cored pears, preferably Doyenné du Comice for their flavour. Cover with slices of Stilton cheese and put into a moderate oven, 180°C/350°F/Gas Mark 4, for five to ten minutes until the cheese begins to melt and the pears release their delicious fragrance. Be careful not to leave the dishes too long, or the watercress will wither. Grind on plenty of black pepper and serve.

VEAL ESCALLOPS WITH WATERCRESS SAUCE
ESCALOPES DE VEAU À LA PURÉE DE CRESSON

Pork slices cut thinly on the bias from the tenderloin, or turkey escallops, can be used instead of veal.

>   *6 veal escallops, beaten, seasoned*
>   *beaten egg*
>   *breadcrumbs*
>   *3 bunches watercress*
>   *100 g (3 oz) butter*
>   *6 tablespoons thick cream*
>   *lemon juice, salt, pepper*

Dip the escallops in beaten egg, then in the breadcrumbs to coat both sides.

Cook the watercress for 15 minutes in boiling salted water, then cool it under the tap, dry it and chop it.

Meanwhile cook the escallops in half the butter. Two minutes a side is usually enough: for thinner pork slices one and a half minutes should do. Remove them from the pan and keep warm. Into the pan juices put the watercress and stir it about to reheat, gradually adding the last of the butter. Finally mix in the cream and lemon and seasoning to taste – not too much lemon, just enough to bring out the flavour. Serve the meat with the watercress sauce poured elegantly over it.

YAMS

Yams in our childhood sounded exotic
and exciting, accompaniments to broiled
missionary or long pig on a South Sea
island, or properties of a story by R. M. Ballantyne. What we never
expected to find was yams on sale as they are to-day, because of the influx
of West Indians. It comes as another surprise that nearly all the yams
originated in Africa, from which they have been carried round the tropics
(the name is Portuguese, probably taken from an African language).

They are not a vegetable to excite chefs and cookery writers, it seems.
Whether African or Caribbean their attitude is lukewarm. 'Treat like
sweet potatoes or potatoes', is all they say.

I have found one idea that works well, from *The Caribbean Cookbook*,
by Rita G. Springer. You peel and boil the yams in salted water. Then you
drain them and put them on to a rack in a grill pan or on a greased gridiron
and brush them all over with melted butter. The yams are then grilled or
barbecued over charcoal and turned until they are brown all over. Serve
them with plenty of butter.

This works well, too, with sweet potatoes, and ordinary potatoes.

APPENDIX

CUTTING UP VEGETABLES

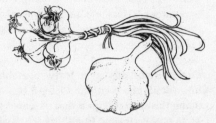

The following terms are much used in cookery books written by chefs, whether French or English. They are a useful shorthand, well worth knowing.

Sometimes the individual vegetables are not specified, because the standard items are well known: for instance a braising recipe may begin by saying, 'Cover the base of the pan with a mirepoix', which means a chop-up of carrots, onions, celery, ham or bacon, with some parsley, thyme and a bay leaf. On the other hand a soup recipe might say, 'Prepare a julienne of carrots, turnips and beetroot', because an extra vegetable beyond the conventional carrot and turnip is included.

MIREPOIX Carrots, celery, onion, bacon or ham, coarsely chopped, with parsley and thyme. There is no need to be too fussy, as a mirepoix is always sieved with the braising liquid. Tuck in a bay leaf when you put the vegetables into the pan.

BRUNOISE A tidier version of the mirepoix, added to sauces that are not going to be sieved and usually consisting of potatoes, carrots, turnips, no bacon. Cut into thin slices, then into tiny cubes.

MACEDOINE Carrots, turnips, haricot beans and peas. Cut the first two into slices ½ cm (¼ in) thick, then cut into cubes. Cut the beans across into cubes or on the slant into diamonds. The peas are left as they are.

DICE Cut slices just over 1 cm (½ in) thick, then cut into cubes.

PAYSANNE is much used for soups. Slice carrots, turnips, potatoes etc, into ½ cm (¼ in) thick slices, then cut the slices into 1 cm (½ in) squares. Or cut them into flat oblong pieces, rather thicker and wider than a julienne.

JULIENNE Carrots, turnips, the white part of leeks etc, sliced thinly then cut into strips the length of a matchstick about 4–5 cm (1½–2 in). With root vegetables, this can be done on a mandolin slicer.

ALLUMETTES, PAILLES, meaning matchsticks or straws, usually refer to potatoes. Trim the potatoes to neat bricks, then cut them into slices the thickness of a matchstick, and cut the slices into even lengths of 4–5 cm (1½–2 in).

CHIFFONADE applies to leafy vegetables, spinach, sorrel or lettuce, cut into shreds the width and length of a matchstick. The simplest way of doing this is to roll up a wad of leaves and cut them across with a sharp knife or scissors into thin slices. They will unroll into shreds. The lengths will be uneven, but this does not matter usually as a chiffonade is such a tumble of greenery.

CHIPS means cutting potatoes, or some other root vegetable, into sticks about 1 cm (½ in) wide and thick, and upwards of 5 cm (2 in) long. This is easiest if you cut the vegetable into oblong brick shapes first. If you make chips often, it is worth buying a special cutter.

LATTICE potatoes have to be cut on a mandolin slicer. Trim the potato to a neat cylindrical shape. Pass it over the ridged grooves of the mandolin, then give it a half turn and repeat. Continue this way and you will rapidly have a pile of nicely latticed potatoes. Check after the first two or three to make sure you have the thickness set right, or you will have shreds instead of lattices. It is a good idea to set the mandolin over a bowl of water, so that the potato slices keep a good colour. Like allumette and chip potatoes, they are always deep-fried.

STEAMING AND BLANCHING VEGETABLES

STEAMING (1) Steaming is not used enough in vegetable cookery, which is a shame as it has several advantages. First the flavour is preserved, as vegetable juices are not lost into the water. Secondly, steaming takes longer than boiling, which in the case of delicate vegetables can be a good thing when you are preparing a meal: it gives you more chance to catch the vegetables when they are perfectly cooked, before they can overcook. And thirdly, if you have a multi-storeyed steamer, two or more vegetables can be cooked on a single burner.

Put plenty of water in the lower pan. Bring it to the boil. Arrange the vegetables in the steamer and season them well. Put the steamer in place – it should be well above boiling water level – and cover.

The time required varies from five minutes for delicate asparagus peas, to 10 or 15 minutes for green beans and cauliflower florets, to 20–25 minutes for young carrots, sliced leeks, seakale, and 30–45 minutes for peeled, cut up root vegetables. Obviously this can only be a rough guide. Everything depends on the size and thickness of the vegetable.

STEAMING (2) A compromise is a method in which green and stalk vegetables are partly steamed, partly boiled. Put a centimetre ($\frac{1}{2}$ in) of water into a pan with salt, bring it to the boil, and put in the vegetables. Cover so tightly that no steam can escape, and lower the heat so that the water boils steadily but not hard. Shake the pan occasionally. Test after five minutes for young or tender vegetables.

BLANC À LÉGUMES AND BLANCHING

Certain pale vegetables are sometimes cooked in a *blanc à légumes* to preserve their light colour and prevent them darkening – celeriac for instance, white asparagus (not the greener English kind), cardoons, seakale, chicory, salsify, hop shoots, the stalks of Swiss chard, the heart or *fond* of globe artichokes, Jerusalem artichokes. When there is a real risk of blackening, say with celeriac, the vegetables are put straight into a bowl of acidulated water, as they are cleaned, to keep the cut surfaces from the damaging air. When the *blanc à légumes* is ready, they are quickly drained and put into it. To acidulate water, add a tablespoon of vinegar to 1 litre (2 pt) water.

A *blanc* is simple to make. Put two level tablespoons of flour into a bowl and gradually mix in 1 litre (2 pt) water until the mixture is smooth. Tip it into a pan with the juice of a lemon, salt, pepper and a heaped tablespoon of butter or chopped suet or two tablespoons of oil (an extra precaution to ensure that vegetables do not come into contact with the air as they rise to the surface). Bring the *blanc* to the boil, leave for two minutes, then add the drained, prepared vegetables.

Blanching means that a vegetable is parboiled in salted water, cooked until it is almost tender. Then it is tipped out into a sieve and 'refreshed' under the cold tap (this helps to preserve colour, in green beans for instance). This practice is essential if you are cooking potatoes in their skins, or salsify or scorzonera, because the skins have to be removed before the finishing process. If the blanching, refreshing and finishing are carried out in rapid succession, you get the most delicious result. Have the liquid at a rolling boil, plunge in the vegetables and time the cooking from the moment the liquid returns to the boil.

LIGHT VEAL STOCK

A simple recipe from Soyer's *Shilling Cookery for the People*, published in 1854. He comments on the dislike of most English people for soup – implying a comparison with the French, his fellow countrymen, who have always depended a great deal on it and still do. He says that this is easy to understand because the recipes for soups in most cookery books are so complicated and expensive, that the poor can afford neither the time, money or attention to prepare them. Then he gives this simple way of preparing a clear stock for soups and sauces. It is ideal for vegetable cookery because the flavour is less obtrusive than beef.

> *1 kg (2 lb) veal knuckle or scrag, cut small*
> *60 g (2 oz) butter or dripping*
> *60 g (2 oz) lean bacon, cut small*
> *3 teaspoons salt*
> *½ teaspoon pepper*
> *175 g (6 oz) onion, sliced*

Put the veal into a large pan with the butter, bacon, salt, pepper and onion, with 150 ml (¼ pt) water. Stand it on the heat and when the liquids boil, stir for about ten minutes until there is a thickish white gravy on the base of the pan. Then add 2½ litres (4½ pt) of water. Bring to the boil and simmer gently for 45 minutes. Skim it well and sieve it.

VARIATIONS

To the ingredients above, add

> *2 cloves*
> *60 g (2 oz) carrot, sliced*
> *60 g (2 oz) turnip, leek and celery, mixed*
> *thyme, or winter savory, or a bay leaf*

TO CLARIFY STOCK

Strain up to 1¾ litres (3 pt) stock into a pan. Add *either* the white of a large egg and its crushed shell, *or* the whites of two eggs. Bring slowly to the boil, whisking with a hand-beater. The surface will soon begin to cloud over with a white foam and the foam will gradually collect into a firm but spongy crust. The whiteness will turn to a murky tone on account of the impurities trapped in it. Allow the stock to boil for three to five minutes, then remove the pan from the heat and allow it to cool for about ten minutes. If you push the egg white gently away from the side of the pan, you will see that the stock below is clear as glass.

Put a damp cloth or double muslin into a sieve and strain the panful through it. The cleared stock is now ready for use in various ways.

TO CLARIFY BUTTER

Small quantities of butter can be clarified and strained into the frying pan for immediate use, but if you cook a lot of vegetable dishes and fish, it is worth making it in quantity. Store it in a covered jar in the refrigerator: it will keep for weeks. The great advantage of clarified butter is that it burns at a higher temperature than unclarified. Any cook will see the advantage of this. It also contributes a particularly pure butter flavour that enhances the quality of simple dishes.

Cut up two or three packets of butter and bring them to the boil in a heavy pan. Boil for one minute, then set aside to cool for ten minutes. Strain through a damp muslin so that the white salty crust is held back, leaving a clear yellow oil. This will solidify when completely cold.

Clarified butter is essential to Indian cookery, as a substitute for *ghee*.

SAVOURY BUTTERS
GARLIC BUTTER

> *250 g (8 oz) unsalted butter*
> *50 g (1½ oz) chopped parsley*
> *30 g (1 oz) chopped shallot, onion, spring onion*
> *3 large cloves garlic, finely chopped*
> *1 teaspoon freshly ground black pepper*
> *1–2 teaspoons fine sea salt*

Soften the butter to a thick cream then mix in the flavouring ingredients.

MAÎTRE D'HÔTEL BUTTER

> *250 g (8 oz) lightly salted butter*
> *60–75 g (2–2½ oz) chopped parsley*
> *lemon juice to taste*

Make as for garlic butter. Mint, savory, tarragon, or chive butter can be made by substituting the herb for all or part of the parsley. In January and February orange butter can be made, whip the grated rind and juice of Seville oranges as flavouring; it goes well with calabrese and cooked anchovies.

BEURRE MONTPELLIER, see page 540.

MUSHROOM BUTTER

> 250 g (8 oz) unsalted butter
> 60 g (2 oz) chopped parsley
> 75 g (2½ oz) finely chopped mushrooms
> slice cooked ham, finely chopped
> 3 cloves garlic, chopped
> pepper, salt

Make as for garlic butter.

BEURRE NOISETTE

Butter cooked to a golden-brown, in other words to a nut or *noisette* colour. Capers, lemon juice and parsley are often added. A last minute sauce, good for pouring over potatoes or parsnips that are being served on their own or with fish.

BEURRE MANIÉ

A useful way of thickening sauces at the last moment. Mash a tablespoon of butter with a tablespoon of flour. Add it to the sauce, which should be just below boiling point, in small knobs. Stop when the sauce seems thick enough. Once you start adding the *beurre manié*, be sure that the sauce doesn't boil again. It takes about five minutes.

BÉCHAMEL SAUCE

This is the most useful sauce of home cookery. If you have a freezer, make it in quantity from time to time and store it in containers of every size from three tablespoons up to half a litre (1 pt). The recipe following gives half a litre more or less, depending on how far you reduce it.

> 600 ml (1 pt) milk
> 1 large shallot, or medium onion, stuck with 2 cloves

1 carrot, quartered
5 cm (2 in) piece celery stalk
bouquet garni
salt, pepper
60 g (2 oz) butter
60 g (2 oz) plain flour

Bring the milk slowly to the boil with the vegetables, bouquet and seasoning. Leave it over a very low heat to infuse for half an hour; it should not boil, just keep very hot.

Melt the butter in a heavy pan, stir in the flour and cook to a roux for two minutes, without browning it. Gradually stir in the flavoured milk, through a strainer. Do this off the heat and use a balloon whisk. Return it to the heat, stir and bring to simmering point, then leave to cook down steadily to the consistency of double cream. Stir occasionally. Making this sauce well can take an hour or even longer. Of course you do not need to stand over it all the time, but you should give it a chance to mature.

QUICK METHOD Make a butter and flour roux as above. Meanwhile bring the milk to the boil, then add it to the roux gradually. Put in vegetables, bouquet and seasoning and simmer steadily, with the occasional stir, for at least 15 minutes, preferably 25. Strain before serving or storing.

CREAM SAUCE This expression is sometimes used as a polite way of saying béchamel, but it should be béchamel sauce with a generous amount of cream added. If this makes the sauce too thin, boil it down. Add a final knob of butter before serving and a squeeze of lemon juice if you like.

MORNAY, MUSTARD AND PARSLEY SAUCES See the velouté variations (page 555), and apply the same seasonings to a béchamel base.

MUSHROOM SAUCE Cook 175 g (6 oz) chopped mushrooms in butter and add them to a béchamel or cream sauce at the end of the cooking. Or stew the mushrooms with a tiny clove of garlic in the roux butter, before adding the flour. A good sauce for potatoes and root vegetables.

SAUCE AURORE It is easy to see how this most beautiful of the béchamel variations was given its name. The flush of colour, from the mixture of ivory béchamel and coral tomato purée, has a Homeric tone of early morning rose. Some recipes are content to achieve this effect with a couple of squirts from a tube of tomato concentrate, but this does not give the right texture at all, or the right delicacy of flavour.

In classic French cookery, sauce aurore partners fine white fish – turbot, sole, monkfish – and chicken and eggs; it goes equally well with vegetables which have a slightly watery character and delicate flavour – for instance cauliflower, seakale, boiled onions, courgettes.

Instead of a béchamel sauce, a mornay or velouté sauce can be used.

> 300 ml (½ pt) béchamel sauce
> up to 300 ml (½ pt) tomato sauce, p. 511 or 512
> 60 g (2 oz) butter

Heat the béchamel sauce to boiling point, then stir in the tomato sauce gradually, stopping when you get to the flavour and colour you prefer. So much depends on the original quality of the tomatoes that one cannot be precise. Bring to the boil again, remove from the heat and whisk in the butter in little bits. Check the seasoning.

It is a good idea to pour this sauce over the cooked, drained vegetable, then to scatter a little grated Parmesan or Gruyère over the top and put it under the grill for a few moments to glaze and colour slightly. If you are using it for seakale, just pour the sauce over the tips, so that the form of the vegetable is not obscured.

VELOUTÉ SAUCE

Velouté sauces are excellent for vegetables, as they can be flavoured with some of the cooking liquor. This adds to the difficulty of timing, as vegetables should not be kept hanging around once they are done or the flavour suffers. The only way round this is to construct a very thick velouté with milk or milk and cream as the only liquid. Then enough of the liquor can be strained into the sauce when the vegetables are ready, to bring it to the consistency and flavour you need. If you have some prepared vegetable stock that harmonizes well, or light veal or chicken stock, you can follow the traditional style:

> 2 tablespoons butter
> 1 small onion, chopped
> 100–125 g (3–4 oz) mushroom trimmings or mushrooms, chopped
> 2 tablespoons flour
> 300 ml (½ pt) vegetable or light meat stock, heated

300 ml (½ pt) milk, or milk and single cream
salt, pepper

Melt the butter and sweat the onion in it until it begins to soften. Add the mushrooms, cover the pan and stew for five minutes. Stir in the flour to take up the juices, then moisten gradually with the stock, then the milk or milk and cream. Cook down to a good consistency – thick for watery vegetables, creamy for the drier kind, thin if it is to be the basis of a vegetable soup. Check the seasoning, and strain the sauce (unless you prefer to keep the bits of mushroom and onion in it).

MUSTARD SAUCE Add Dijon, Meaux, or Urchfont mustard to taste, starting with half a teaspoon. Do not boil again once the mustard is added. For cauliflower with hard-boiled egg, Swiss chard stalks; very thick for vegetable fritters; a good gratin sauce if a little cheese is added, too.

MORNAY SAUCE Add two heaped tablespoons grated Parmesan, two tablespoons grated Gruyère and some grated nutmeg. Dry Cheddar can be substituted for Gruyère, but you will need more of it for a good flavour. Do not allow the sauce to boil once the cheese is added. Useful with many vegetables, most notably with cauliflower, p. 179; an ideal gratin sauce, but keep back a little of the cheese to sprinkle over the top.

PARSLEY SAUCE Make the velouté with stock from cooking the vegetable, milk and cream. Omit the mushrooms. At the end, stir in a good handful of chopped parsley and a knob of butter. This is the traditional English sauce for broad beans, particularly when eaten with ham or other pickled pork. It goes with carrots and green broccoli though with these thinner textured vegetables it should be made from a béchamel base, for extra richness.
See also mustard, mornay and parsley sauces made with béchamel sauce, p. 553.

CREAM SAUCE
SAUCE À LA CRÈME

A lovely sauce of some antiquity, still popular in Normandy, once popular in England. It can be flavoured with nutmeg, or different herbs including tarragon, rosemary, parsley, chives. Serve it with delicate vegetables of the early summer – asparagus, hop shoots, seakale, cucumber that has been blanched, then softened in butter with mint, dwarf *haricots verts*. It transforms young steamed marrow with its ivory richness.

125 g (4 oz) butter
150 g (¼ pt) whipping or double cream
salt, pepper, lemon juice
2–3 tablespoons chopped herbs, or 1 good sprig rosemary, finely chopped
1 heaped teaspoon cornflour (see recipe)

Take a frying pan and melt the butter in it. When it begins to bubble, stir in the cream and keep stirring until the two amalgamate into a thick sauce. This takes a minute or two, no more. Remove from the heat, season with salt and pepper and a squeeze of lemon juice. Add the herbs. Return to the heat for a moment to warm through again.

If you are unlucky and overheat the sauce, it can begin to oil at the edges, or even separate. Try adding a couple of tablespoons of water. Or if this doesn't work, mix the cornflour with three tablespoons water, and amalgamate it with the sauce. This will thicken and bind the mixture, if you give it a little longer on a low to moderate heat for the cornflour to cook. The thing to remember is not to overheat this sauce at any point: the beautiful rich flavours should blend together gently.

SAUCE HOLLANDAISE AND HOT VEGETABLE SALADS

Hollandaise is a hot mayonnaise of egg yolks and melted butter, that goes beautifully with many cooked vegetables or a hot vegetable salad. At a time of the year when many fresh things come into the shops – young peas, broad beans, new carrots, new potatoes, calabrese, spinach – cook a good selection of them and serve them on a large dish with a bowl of hollandaise in the centre. It may sound a lot of trouble, cooking so many different vegetables at once, but with a little ingenuity it is quite easy. Spinach should always be cooked separately, as it needs individual attention. For the rest, put the root vegetables into boiling salted water in a large pan, and tuck in the calabrese wrapped in foil with the flower part upwards: on top of the pan put an expandable steamer with broad beans on one side, peas on the other. Cover with the lid, or a piece of foil. If you have no steamer, wrap the beans and peas in separate pieces of buttered foil, with seasoning, and stand the packages on top of the potatoes and carrots.

3 tablespoons white wine vinegar
3 tablespoons water

12 white peppercorns
3 large egg yolks
250 g (8 oz) unsalted or lightly salted butter
salt, lemon juice

Boil the first three ingredients down to a tablespoon of liquid. Strain into either a pudding basin or the top pan of a double boiler. When cool, beat in the yolks. Stand the basin, or top pan, over a pan of barely simmering water. Keep the heat steady, being careful not to overheat. Now add the butter in chips, beating them in with a wooden spoon gradually. Or, and this works well, melt the butter and beat it in as if you were making mayonnaise; you can do this off the heat if you like, so long as both egg yolks and butter are warm.

When the sauce is very thick, add a final seasoning of salt and lemon juice. If you have to keep it waiting, turn it into a *tepid* bowl or jug or sauceboat and stand it in a pan of *tepid* water. Overheating at this point can curdle the best hollandaise. If at any moment you think the hollandaise is overheating, plunge the base of the bowl or pan into cold water.

Should the sauce curdle, try whisking in a tablespoon of ice cold water. If this doesn't work, break a fresh egg, put the yolk into a clean basin and beat in the curdled sauce gradually. It doesn't take long, so don't give way to despair.

SAUCE MALTAISE Add the grated rind and juice of one or two blood oranges to the sauce, and only the smallest squeeze of lemon. Good with asparagus and seakale, and calabrese, and purple-sprouting broccoli.

SAUCE BÉARNAISE Instead of wine vinegar, use tarragon vinegar; add a heaped tablespoon of chopped shallot and three tablespoons dry white wine to the mixture before reducing it. Finally flavour the sauce with chopped fresh tarragon. For serving with lamb or steak and young fresh vegetables, and with sweetcorn recipe on p. 489.

SAUCE CHORON Flavour a *sauce béarnaise* with a well reduced purée of fresh tomatoes. You will need between four and six tablespoons, according to taste. Start with 250 g (8 oz) skinned, firm tomatoes; chop them coarsely and cook them vigorously for five minutes until they are reduced to an unwatery mass. Quantities are difficult to predict, as different varieties of tomato have different amounts of water to lose, but 250 g is a good starting point.

GREEK EGG AND LEMON SAUCE (1)
AVGOLEMONO SALTSA

A clear flavoured, rich sauce that goes well with broccoli, courgettes, cauliflower and beans. The liquid from cooking the vegetables is often used as the stock.

> *3 large eggs*
> *juice of 2 lemons*
> *¼ litre (scant ½ pt) hot chicken or vegetable stock*

Beat the eggs until they are foamy and thick. Use an electric beater if possible, or a blender; with a hand rotary whisk you will need ten minutes. Add the lemon juice slowly, still beating, then the hot *but not boiling* stock. If it is too hot, the sauce will curdle. Pour the sauce over the vegetables (they should not be boiling hot either) in their serving dish.

GREEK EGG AND LEMON SAUCE (2)

This is an easier version that is less likely to curdle, but it takes longer to make and has a thicker consistency which can seem a little heavy.

> *1 tablespoon butter*
> *1 tablespoon flour*
> *¼ litre (½ pt) hot chicken or vegetable stock*
> *2 large eggs*
> *juice of 2 lemons*
> *2 tablespoons cold water*

Make a velouté sauce in the usual way with the first three ingredients. Beat the eggs until light and fluffy, as in the first recipe. Add the lemon juice gradually, still beating, and the water and finally the hot but not boiling sauce. Return the whole thing to the pan and stir over a moderate heat for a few moments to thicken further.

VINAIGRETTE OR FRENCH DRESSING

Basic cookery books give the ingredients for this sauce as three table-spoons of oil to one tablespoon of wine vinegar, but I find this far too strong. Five to one is a better proportion, at least to start with, although the final quantities will depend on the oil and vinegar used and the opinion of the person who is making it.

The usual seasonings, beyond salt and pepper, are garlic, a hint of sugar, perhaps mustard, and plenty of chopped green herbs such as parsley and chives, with tarragon and basil to add a different note from time to time.

For a plain green salad, or a salad of one vegetable (cooked or raw), olive oil is the best choice. Olive oils vary as much in flavour as wines, but as there is only a limited variety on sale in this country choice is not too bewildering. My own preferences are for the green oils of Tuscany, Umbria and Greece (Minerva brand usually), and for the golden oil of Beaumes-de-Venise in Provence. Walnut oil I use for certain salads, on special occasions. A tasteless oil is good for mixed salads.

Take care with wine vinegar. The best is made in Orléans still. I go for Martin-Pouret brands as I have seen the way they are matured in casks in the old-fashioned manner, which gives their virtues a chance to develop.

Malt vinegar is not suitable for a sauce deriving from wine-growing countries.

For a green salad, mix the dressing in the bowl; cross the salad servers over it to make a platform for the rinsed and dried salad greens. Chill if you like, but do not turn the salad until you are ready to eat it or the softer greenery will collapse unpleasantly.

GREEK DRESSING
LEMONOLATHO

The Greek way of serving cooked vegetables while they are still warm, with an olive oil (*latho*) and lemon dressing, may sound unappetizing. We are brought up to detest food that is neither hot nor cold: it can become too strong a prejudice, in the matter of open tarts both sweet and savoury as in this matter of salads.

The thing to avoid is overcooking the vegetables, so that they flop. Naturally they must be scrupulously drained as well, and of high quality. Cauliflower is a good vegetable to start with, or courgettes. I can recommend broad beans, too, particularly if you are prepared to spend time slitting the white skins of the beans and slipping out the delicious green parts into a bowl. New potatoes, green and purple sprouting broccoli, if cooked when really fresh, give more of their flavour when they are warm rather than chilled.

Greek Minerva oil is worth buying for salads of this kind. To get the most juice out of a lemon, pour boiling water over it, and leave it for five minutes until it softens and becomes more yielding and fragrant. Chopped dill could be used instead of rigani and mint.

> 125 ml (4 fl oz) olive oil
> juice of 1 large lemon
> ½ teaspoon each dried rigani and dried mint
> 1 tablespoon chopped spring onion or chives
> 1 clove garlic, crushed

Beat all the ingredients together and leave for a while before using. Season to taste just before pouring over the salad.

MAYONNAISE

A sauce that rarely fails it you take reasonable precautions beforehand – the ingredients should be at least at room temperature and you can warm the oil, bowl and beater before you start.

> 1 large or 2 standard egg yolks
> wine vinegar or lemon juice
> 1 teaspoon Dijon mustard (optional)
> 150 ml (¼ pt) light olive oil
> salt, pepper, herbs etc.

HAND OR ELECTRIC BEATER METHOD Warm the bowl and spoon or beater with hot water, then dry. Quickly put in the egg yolk with a teaspoon of vinegar or lemon juice, and mustard if appropriate to the dish. Beat thoroughly together, then add the oil, drop by drop at first, then more steadily as the mixture thickens. When the oil is absorbed, add more vinegar or lemon juice to taste, seasonings and herbs.

BLENDER METHOD Use two yolks or one whole large egg. Put the egg into a warmed blender with the vinegar or lemon juice and mustard. Turn to top speed, having covered the blender. After ten seconds, remove the central cover of the lid and gradually pour in the oil. On account of the very high speed, you can put in the oil more rapidly from the start, though it is prudent to go gently at first.

REMEDIES FOR CURDLED MAYONNAISE
1) The moment you suspect curdling, whisk in a tablespoon of boiling water. This is often enough to bring the sauce back.

2) Put another egg yolk into a clean bowl and add the curdled mixture drop by drop at first, then the remainder of the oil, plus extra to bring the quantity up to 250 ml (8 fl oz).

3) Put a tablespoon of Dijon mustard into a clean, warmed bowl, instead of another egg yolk, then continue with method 2. This is handy to know if you are out of eggs, and the dish will stand the extra mustard flavour.

SAUCE TARTARE Is either mayonnaise or vinaigrette seasoned with chopped shallots or spring onions, and *fines herbes*, plus capers and gherkins to taste. The sauce should be thick and speckled with these ingredients.

If you want to serve *sauce tartare* with fritters use mayonnaise as the base. To go with a main course that might include hot potatoes, grilled chicken and gelatinous meats such as calf's head or pig's trotters, a vinaigrette base is better.

SAUCE AILLOLI An olive oil mayonnaise liberally flavoured with garlic, see p. 260.

SKORDALIA

The Greek equivalent to provençal *ailloli*, made without egg yolks as it is a sauce eaten by tradition in Lent, with slices of aubergine and courgette, dusted with flour and fried, and with boiled beetroot and potatoes. On Clean Monday in Greece, the Monday after the last day of the Carnival, when everyone ate themselves to a stupor, it was served with salt cod soaked and deep fried in batter. Nowadays *skordalia* often appears on Greek menus with all kinds of fish fried in batter, but it makes a delicious first course with vegetable fritters. Here is a recipe sent me by a Greek reader:

> *3 or more cloves garlic*
> *5 cm (2 in) slice of stale white bread, from a small loaf*
> *100 g (good 3 oz) blanched, grated almonds*
> *125 ml (4 fl oz) olive oil*
> *wine vinegar, salt*

Crush the garlic well in a mortar. Cut the crusts from the bread, soak it with water and squeeze out any surplus – this makes a thick paste. Add it to the garlic, pounding well, then mix in the almonds gradually, pounding all the time. When you have a homogeneous mixture, start adding the oil drop by drop at first, as for a mayonnaise. Finally sharpen and season to taste with vinegar and salt. The sauce can be made in a blender, or with an electric beater, but the garlic should be crushed by itself before you start, to make sure it is reduced enough to mix completely into the sauce.

BALKAN WALNUT AND GARLIC SAUCE

A sauce akin to *skordalia*, and popular in Romania with cornmeal fritters (p. 491). In Bulgaria it is often served with fried aubergine and courgette slices as it is in Greece, and with the delicious cool cucumber soups and salads. As no bread is used the garlic flavour is more pronounced, making it more of a condiment than a sauce. Unfortunately it also means that the ingredients are inclined to separate, and I confess to using a blender rather than the traditional pestle and mortar. Another snag is that you must blanch the walnuts, or the flavour will be bitter. This is fiddlier than blanching almonds, owing to the twists and convolutions of the nut, but it is well worth the trouble. The mealy flavour of walnuts makes a good earthy sauce. If you are serving it with a yoghurt dish, olive oil must be used: if it's to go with vegetable fritters, you can use oil or yoghurt, as they do in Dobrogea, an eastern district of Romania.

> *6 tablespoons boiling water (blender only)*
> *2–5 large cloves garlic, slightly crushed*
> *125 g (4 oz) walnuts or almonds, blanched*
> *125 ml (4 fl oz) olive oil or yoghurt*
> *wine vinegar, salt*

Blend the water and garlic thoroughly, then add the nuts gradually, alternating them with oil or yoghurt to keep the mixture moving. Flavour to taste with vinegar and salt.

If you use a mortar, you do not need so much liquid, so that the water can be omitted, or reduced. If the nuts start to oil, though, a little very hot water can be added to bring them back. It also helps to grate the nuts first to reduce the pounding labour.

PORK AND HERB STUFFING

A meat stuffing which is light and savoury, ideal for tomatoes, courgettes or cucumbers, or for stuffed cabbage and cabbage rolls. Any pulp scooped out of the vegetable can be added to the frying pan, once the onion is lightly browned, but then the heat must rapidly be raised to evaporate undesirable moisture which would make the stuffing heavy and reduce its flavour. The most important thing for success is to buy sausagemeat or a brand of sausages you know well and can rely on for spiciness and a good percentage of quality meat. A bland, over-cerealized sausage would blunt the dish to a stuffy ordinariness.

The quantity should be enough to fill vegetables for six people, though obviously there is a good margin if you are using courgettes rather than, say, tomatoes or a cabbage.

> 60 g (2 oz) butter
> 175 g (6 oz) chopped onion
> 2 cloves garlic, chopped finely
> 350 g (¾ lb) sausagemeat, or skinned sausages
> 125 g (4 oz) brown or white breadcrumbs, made from slightly
>   stale bread
> 6–8 tablespoons milk
> 1 heaped tablespoon chopped parsley
> ¼ teaspoon dried thyme, or ½ teaspoon fresh
> a pinch of dried marjoram
> salt, pepper

Melt the butter in a frying pan and add the onion and half the garlic. Allow it to fry gently until the onion begins to brown. Add the pulp from the vegetables being stuffed at this stage, and cook the mixture down to a moistly dry purée – avoid wateriness. Break up the sausagemeat or sausages and add to the pan, crushing them into a mixture with a fork. Cook lightly until the sausage loses its raw pink look. In a bowl, mix the breadcrumbs with enough milk to make them a crumbly paste. Stir in the contents of the frying pan, then the herbs, seasoning and remaining garlic. Adjust the herbs to your taste.

HAM AND HERB STUFFING

A stuffing with a slightly sharper, more piquant flavour than the pork and herb recipe. It is a better choice if the vegetables are likely to be eaten cold. And it is more successful, too, with a heavier vegetable such as aubergine.

In the ingredients above, substitute 175–250 g (6–8 oz) chopped ham for the sausagemeat, and dry white wine or dry vermouth of the Chambéry type for the milk.

Choose your ham according to the flavour you like. For spiciness, smoked gammon or lean Danish bacon is the best choice 174 g (6 oz) should be enough. For a milder tone and for the sake of economy, buy 250 g (8 oz) of odds and ends left over from cutting ham on the bone. Trim away any gristle or fatty skin, but leave the nice clear white fat. Be careful not to overcook the ham: it needs to be heated through, that is all.

CHEESE AND HERB STUFFING

A good stuffing for vegetarians who want to avoid meat. It goes especially well with tomatoes or aubergines.

Follow the recipe for pork and herb stuffing, but substitute 250 g (8 oz) cheese for the sausagemeat, and do not cook it; just cut it up into small dice, or grate it, and mix it in with the breadcrumbs. A mixture of diced Gruyère 125 g (4 oz) and grated Parmesan 60 g (2 oz) gives a fine flavour, piquant without heaviness; but Cheddar and Lancashire cheese do well if the vegetables are being served as a course on their own, rather than as an accompaniment to meat or fish.

Use either milk as indicated, or dry white wine, to moisten the breadcrumbs.

Fillets of anchovy, chopped small, make a piquant seasoning for this stuffing. Start with a teaspoonful and add more to taste. Chopped black olives are another good choice – allow 60 g (2 oz) then stone them and cut them into smallish but identifiable pieces.

PROVENÇAL STUFFING

When using this recipe to stuff aubergines, courgettes or tomatoes, scoop out their centre part, chop it and add it to the stuffing ingredients. If you are cooking peppers, add three skinned, chopped tomatoes to the ingredients.

> chopped pulp of vegetables or tomatoes
> 1 large onion, chopped
> 2 large cloves garlic, chopped
> olive oil
> either 125–175 g (4–6 oz) chopped mushrooms
> or 1 tiny packet dried ceps, soaked, drained
> or 1 small sweet pepper, seeded, chopped
> 3 tablespoons chopped parsley
> 125–175 g (4–6 oz) white breadcrumbs
> milk or water

6 fillets of anchovy, chopped (optional)
salt, pepper, sugar

Fry the pulp or tomatoes with the onion and garlic in a little olive oil, until they begin to soften together. Add the fresh mushrooms or chopped pepper, and raise the heat so that the mixture browns slightly but not very much. Taste and season and at this stage add the chopped dried mushrooms if they are being used. Give them a few minutes bubbling to heat through.

Remove the pan from the heat, and add the parsley, the breadcrumbs moistened with milk or water and squeezed out, and the anchovies if they are to be included. Mix everything well together, correct the seasoning and add more parsley if you like. Sometimes a pinch or so of sugar helps to bring out the flavour

Note It is a good idea to pre-cook the vegetables for stuffing (apart from tomatoes). Peppers and courgettes can be blanched for five minutes. Aubergines can be fried or blanched.

GREEK RICE STUFFINGS

1) *250 g (8 oz) chopped onion*
1 large clove garlic, chopped
125 ml (¼ pt) olive oil
250 g (8 oz) long grain rice
1 heaped teaspoon tomato concentrate (optional)
1 heaped tablespoon chopped mint and dill weed
¼ teaspoon ground cinnamon
salt, pepper, sugar

Cook the onion and garlic in the oil until soft and yellow, but not browned. Stir in the rice and the tomato concentrate if used, and about 150 ml (¼ pt) water. Simmer for five minutes, then add remaining ingredients, remove from heat, seasoning to taste.

A good stuffing for young courgettes and courgette flowers. For vegetables that are to be baked rather than simmered on top of the stove add more water and cook a little longer, until the rice begins to be tender.

2) *250 g (8 oz) minced beef or veal or lamb*
125 g (good 4 oz) long grain rice
1 medium onion, chopped finely, or 6 spring onions

>*2 tablespoons tomato concentrate*
>*2 teaspoons salt approx*
>*1 teaspoon freshly ground black pepper*
>*1 heaped tablespoon chopped parsley*
>*1 teaspoon each chopped mint and dill weed*
>*2 tablespoons olive oil or melted butter*
>*salt, pepper*
>*¼ teaspoon ground cinnamon*

Mix the ingredients together with your hands to make a coherent mass. This makes a favourite stuffing for cabbage leaves (p. 135). If the stuffing is for baked vegetables, aubergines for instance, or tomatoes, or courgettes, half-cook it in the manner of the recipe above, or the Byzantine recipe following, using water to moisten the mixture.

An elaborated version of the rice stuffing of Eastern Mediterranean countries. It can be used with aubergines, courgettes, peppers or large tomatoes. Any pulp removed from the vegetables can be chopped and added to the stuffing as it cooks. Aubergines, courgettes, and peppers should be blanched in boiling salted water for five to ten minutes; tomatoes should be hollowed out only, and left upside down to drain. Baked at 180°C/350°F/Gas Mark 4 for 45–60 minutes with a little water in the dish, and the juice of a lemon. Serve hot or cold.

3) *½ kg (1 lb) fairly lean minced beef, or minced shoulder of lamb*
    2 onions, chopped small
    2 large cloves garlic, chopped small
    3 tablespoons olive oil
    ½ litre (1 pt) water
    125 ml (4 fl oz) home-made tomato sauce, or chopped tomato pulp
    250 g (8 oz) long grain rice
    1 dessertspoon each dried mint and fresh parsley, chopped
    ¼ teaspoon ground cinnamon
    125 g (4 oz) currants or raisins
    6 tablespoons port or similar fortified wine
    60 g (good 2 oz) chopped pine kernels or walnuts
    salt and pepper

Fry first three ingredients fast for five minutes in the oil, stirring them about until they are lightly browned. Add the remaining ingredients with salt and pepper to taste. Simmer for ten minutes until the rice is half cooked and the liquids almost absorbed to make a juicy mixture. Remember that the rice will absorb more liquid as it continues to cook in the vegetables: on the other hand you should not end up with a sloppy watery dish. More cinnamon and more herbs can be added: remember that dried mint is much stronger than fresh, and gives a good flavour, and that dried parsley is no use at all.

PANCAKE BATTER

Some cookery books tell you to allow pancake batter to stand for an hour. This can be convenient, but I have never found any other advantage to it. Nor have I met anyone who could tell the difference when the pancakes are cooked. Some Bretons make the batter with warm water and leave the mixture overnight so that it ferments slightly. And you can use ale or beer to lighten the mixture. I find that a tablespoon of rum or brandy, gin or whisky, makes the lightest pancakes of all. Adding melted butter or oil means that you do not need to grease the pan every time some more batter goes into it.

FOR 18 PANCAKES

> 125 g (4 oz) plain flour
> pinch salt
> 2 eggs
> 2 tablespoons melted butter or oil
> 1 tablespoon rum, brandy, gin or whisky
> up to 150 ml ($\frac{1}{4}$ pt) each milk and water

Mix the ingredients together smoothly in the order given, adding the milk and water slowly until the batter has the consistency of thin cream. Rub a 20 cm (8 in) omelette pan with a butter paper, heat up and pour in a little batter – less than two tablespoons. I use a small kitchen ladle holding two good tablespoons, and fill it three-quarters full. This is easier than tipping batter in from a jug, and more accurate.

If the first pancake seems too thick, dilute the batter a little more.

Pancakes can be made in advance. Wrap them in foil and keep them in the fridge overnight. Or layer them with plastic film, and freeze them. The fridge packet can be reheated in a moderate oven, as it is. The freezer packet should be left to thaw partially, then the pancakes can be separated easily and reheated on a baking sheet covered with foil.

GREEK BATTER

A herb-flavoured mixture that gives a light aromatic coating to vegetable fritters.

> *125 g (good 4 oz) flour*
> *1 teaspoon baking powder*
> *1 egg*
> *1 tablespoon olive oil*
> *about 150 ml (¼ pt) water*
> *1 dessertspoon ouzo/pernod (optional)*
> *¼ teaspoon each powdered bay leaf and dried rigani*

Beat the ingredients together in the order given, adding extra water if necessary.

   Dip slices or sprigs of half-cooked vegetables into this batter, and fry them in 1 cm (½ in) depth of olive oil. This sounds expensive, but fritters cooked in olive oil have an extra good flavour and crispness.

PITTA BREAD

On a brief visit to Israel, a friend took me shopping in Tel Aviv market one Friday. An experience of colour and sweetness, light and bustle in the sea air of a brilliant February day. The alleys were lined with stalls, backed by cavernous open-fronted shops. Piles of vegetables and fruit, red, purple, green, orange, white, yellow, buckets of olives, pickled aubergines and peppers, precarious heights of halva blocks wrapped in silver paper and cut across to show the nuts and candied fruit inside. You could run your hands deep into barrels of cinnamon bark, dried beans, chick peas, nuts. There were butchers' shops of a naked and unashamed directness, that I had only seen before in the Caracci painting at Christ Church art gallery in Oxford. By midday I was dazed and loaded with treasures, and found myself propelled to a tiny stall for refreshment. There was a pile of Arab pitta bread, and bowls of *hummus*, *tahina*, tomato and chilli salad, shredded lettuce, deep-fried *felafel*, various pickles. The stallkeeper cut a pitta in half across to make two pouches. Then he filled them cram full with bits from the various bowls, and handed them to us wrapped in a paper napkin. We stood and nibbled steadily, the delicious juices escaping down our chins. The ideal portable meal.

   It would be a shame to keep pitta only for this kind of picnic. You can eat it with many vegetable dishes. Home-made bread is always a good partner for salads and soups, whether it is wholemeal, white or rye, but pitta is best of all. It has a chewy quality. And it comes in soft floury ovals and rounds patched lightly with pale brown. On account of the way it's made, it bubbles up with steam in the oven. Then it collapses, but the bubbles leave behind a central hollow, so that you have this skin of bread like a collapsed football. You can sometimes buy it in specialist shops and bakeries, but it is not always easy to find so you might as well know how to make it at home. It can be kept for several days in a plastic bag in the fridge (revive it under the grill), or for much longer in the deep freeze (heat through in the oven like any other bread). Serve it warm or hot, if possible.

15 g (½ oz) fresh yeast
150 ml (¼ pt) cold water
150 ml (¼ pt) boiling water
pinch sugar
2 tablespoons olive oil
½ kg (1 lb) strong plain flour
½ teaspoon salt

Put the yeast into a bowl. Mix the cold and boiling water with the sugar, and add about a quarter of it to the yeast, creaming the mixture together with a fork. Leave the yeast in a warm place for about ten minutes until it froths up. Meanwhile add the oil to the remaining water, and sift the flour and salt into a warmed bowl. Mix the flour to a dough with the yeast mixture, then the water and oil. Add a fraction more flour if necessary. Turn it on to a board and knead for ten minutes. Wash, dry and oil the mixing bowl lightly. Return the dough to it and put the whole thing into a plastic bag. Fasten tightly. Leave in a warm place for an hour and a half or two hours. Knock down the dough, and divide into lumps each weighing about 80–100 g (around 3 oz). Pat or roll out these lumps into ovals roughly 12 cm (4–5 in) across and 20–22 cm (8–9 in) long, and place them to prove on a lightly floured cloth in a warm place for 20–30 minutes. Turn the oven on to 260°C/500°F/Gas Mark 10, and leave for 15 minutes at least. For the last two minutes, put in two greased baking sheets to heat up. Place as many pitta as you can on these very hot sheets, brush them with water, and put into the oven. After three minutes, lower the heat to 230°C/450°F/Gas Mark 8 and leave for a further three minutes, or a little longer, until they are puffed up and lightly spotted with brown. If there are any pitta left, bake them in the same way.

INDEX

INTRODUCTION TO
THE AMERICAN EDITION

Whoever first observed that the English and the Americans are divided by a common language might have been thinking of cookery.

For me, French and Italian recipes are much easier to work from, more immediately attractive, than American. Even in German, a language I have never learned, it is not difficult to work out a recipe with the aid of a few words and a basic knowledge of European cookery. At a glance I can see how proportions work out, what the result is likely to be, and how to start. This is because we all have the same attitude to constructing dishes. We use avoirdupois weights: the fact that some may be metric, others Imperial, does not much affect the matter – the proportions are expressed in the same way, and have the same relationships.

Now in the United States – and I am sure there is an historical reason for this – you work by volume measurement whether or not your ingredients are liquid. You use cups, and use them in all sorts of ways that we find slightly bizarre. To us, '1 cup butter' seems a tortuous way of measuring. Butter is sold in packages of known weight: it can easily be cut to the right amounts without any kind of measuring at all, if the *weight* is specified. The idea of packing it into a cup seems odd. But of course you don't, I am sure. You, too, use the stick and cut it accordingly, even if, linguistically, you give a picture of cramming recalcitrant bits of butter into a fine porcelain teacup.

To us, volume means liquids. And there again, confusion begins when we look for unanimity. You have kept the old English wine pint of 16 oz, for obvious historical reasons. For less obvious reasons we gave it up in the 1820s and took to Imperial standard, which gave the pint a boost of one quarter. The Imperial pint now equals 1¼ American pints. I wish we had stuck to the old way. It would have made our present change over to metric much easier. Your American pint is a little less than the half litre, just as the pound – which thank heavens we share – is a little less than the half kilo. This seems a lot easier, psychologically, than having a pint which is *more* than half a litre and a pound that is *less* than half a kilo.

All this means that there is a very basic difference between cookbooks that *look* the same merely because they are written in the same language.

Same language? I sometimes wonder about that, too. When Julia Child instructs me to 'mince' a clove of garlic, I automatically start to get out my mincing machine, before I realize that she means 'chop'. To us, mince means 'put through a mincing machine' (which you call a grinder), as we do with meat if we want to make moussaka or shepherd's pie or sausages.

Then, too, the names for things can conceal pitfalls. Your adjectives for sugar prepared in different ways – see glossary – are quite different from ours. And American friends say, though I find this incredible, that you do not have as many kinds of brown sugar as we do. Your cream, too, is different. So is your flour, which is far 'stronger' than ours, because you have the climate for growing durum wheat, hard wheat with a high gluten content; we grow soft wheat so that our plain, i.e. all-purpose, flour, which makes splendid cakes and cookies, does not give us the well-risen bread, éclairs and choux puffs that we like. So for this kind of baking, we have to buy special strong bread flours.

For this reason alone, I am glad that I am not dealing with a book on cakes, pastries, breads. In that area the problem of communication becomes virtually impossible. Nor is it helped by the fact that we like far less sugar in our sweet things than you do. So that we may consider we have a failure on our hands – when it is nothing of the kind – after having followed an American cake recipe.

Vegetables and vegetable cookery are a far more relaxed affair. The number of recipes in the book that depend on exact measurement can be counted on your fingers, I would guess. Well, perhaps fingers and toes. But you can work your way around them – just use your own recipe for shortcrust or bread dough, or for a basic soufflé mixture, taking note of slightly different flavouring ingredients, perhaps, but that is all.

After all, if you go out and buy a pound of any vegetable from the market, who is to judge *exactly* what quantity you will end up with? I am sure everyone takes this in stride – you buy 2 lbs of potatoes to feed the family and do not worry or even notice whether you end up with 1½ lbs cooked weight, or 1¾ lbs.

So relax, and do not allow yourself to be flummoxed by apparent differences. A modest experience of eating and cooking will see you through. I hope so, anyway, because working through the recipes for this book was great fun, often a delight because of the open possibilities.

As you read, you will also see that we share one major problem. The word 'tomatoes' sums it up for both this country and America, as I realized after reading a long article in the *New Yorker* of 24 January

1977. And of course it is beginning to affect the production of certain other vegetables, this supermarket mania for quality-control (which means appearance-control, not flavour-control) and identical size. So often, quality-control is incompatible with fine flavour and aroma, even when the intentions are good – which is by no means always the case, profit being the first intention, sometimes the only intention. Take the persimmon. A lovely fruit if allowed to ripen to an almost bruisy condition, a paradisal fruit – but a nightmare to the greengrocer. The Israelis have cleverly developed a variety that can be eaten skin and all, without a hint of bitterness, even when it feels slightly firm. They call it Sharon fruit, which at least sets it apart from the persimmon so that people are not deceived. Now this new Sharon fruit is good, pleasant to eat, easy for the greengrocer, but it is a long way behind a fine ripe persimmon for subtle fragrance.

In the matter of tomatoes the situation is more serious for you than for us. At least once a year many of us cross the Channel and have the chance to taste a few real tomatoes in Italy, France, Greece and Spain. So we have always at the back of our minds what the real thing should be. And by making a lot of noise, journalists begin to see a modest return; occasionally, a few decent craggy, beefy, firm tomatoes are found on sale in some greengrocery chain, and we pass the word around. In fact we have only to say the words loud enough, and we could be flooded with decent tomatoes within a year. For the very good reason that Spain, Israel, the Canary Islands are growing a *special variety of tasteless tomatoes expressly for the British market*. They would much rather grow the kind of tomato they like themselves, and already supply to France, Germany, and Italy when these countries' tomato seasons are over. I may say that the tasteless variety is appropriately called 'Moneymaker'. And is notorious for its high yield. Quantity, not quality.

We shall watch your battles for good food in America with some concern, knowing that we are next in line for the big battalions. Alas, the concept of tasteless, average, bland, cubic, plastic-wrapped food came to us from America. You let it happen to you. We have let it happen to us in part. And because it has only happened in part, we do not shout loud enough. We would do better to sling our stones at Goliath before he turns his full attention upon us, rather than fight the sort of fight you have on your hands in such matters as tomatoes.

It is for this kind of reason that I hope you will not find *The Vegetable Book* as alien as you might have expected. We share a lot of the same history. We eat a lot of vegetables that originally came from your continent. Many of your typical American dishes, such as pumpkin pie, are adaptations of our older recipes to new but similar ingredients. So, in the matter of vegetables at least, a similar language may not get in the way.

GLOSSARY

A dictionary of British/American terminology for vegetables and other ingredients

NOTE: The vegetables appear in CAPITALS

Anchovy essence: use either a Southeast Asian fish sauce, or anchovy paste

ASPARAGUS CHICORY: not generally grown in America

ASPARAGUS PEA: the winged pea; the goa bean

AUBERGINE: eggplant

Bacon: if no qualifying adjective is given, any bacon can be used

    back: loin bacon with a high proportion of lean to fat

    fat: belly bacon (*see* streaky), or the fat end of bacon hocks, or fetterspeck (*see below*)

    fetterspeck: German and Polish cured bacon fat from the back; does not dissolve in cooking

    geräuchter bauchspeck: smoked belly bacon, imported from Germany, which has been properly cured and smoked to a good old-fashioned flavour, not just injected with smoke-powder or synthetic smoke flavourings

    gammon: bacon side of loin and leg of pork cured in one piece, like ham; can be smoked or unsmoked

    green: unsmoked; substitute salt pork

    hock: forehock or picnic shoulder at the thin end

    streaky: belly bacon; comparable to American commercially packaged bacon

Baps: round Scotch rolls with a soft floury crust

BATAVIAN ENDIVE OR SCAROLE: use escarole or prickly lettuce

Bayonne ham: smoked ham eaten uncooked in wafer-thin slices; substitute Westphalian ham or *prosciutto crudo* (though that is unsmoked) or similarly sliced gammon (*see* bacon)

BEANS

    broad beans: fava beans

    flageolet beans: kidney beans; shell beans

French beans: green beans; snap beans; string beans

haricot beans: dried beans (navy, Boston, etc.)

scarlet runners: Dutch case-knife beans

BEETROOT: beets

Best end of neck: always refers to lamb. Rib chops, from the neck end, such as those used for a crown roast

Boudoir biscuits: long sponge cookies; ladyfingers

CHAYOTE: choyote; christophine; mirliton

CHINESE ARTICHOKE: chorogi, knotroot, Japanese artichoke; crosnes-du-Japon

CHINESE LEAF: Chinese cabbage; celery cabbage; pe-tsai

COURGETTES: zucchini

Cox's Orange Pippin: the favourite English eating apple, with an aromatic flavour and crisp texture. Not dissimilar to the French *reinette* varieties (*see* next page). If you must, substitute a Golden Delicious.

Cream: if no qualifying adjective is given, any cream may be used

single: minimum butterfat 18%; substitute light cream or coffee cream

soured: single cream, which has been ripened at the dairy to give it the right sour taste; sour cream

whipping: minimum butterfat 38%; substitute heavy cream

double: minimum butterfat 48%; no true substitute, use heavy cream

Creamy milk: from Guernsey and Jersey cows, with a good top of cream to the bottle; substitute skimmed milk plus cream to taste

Curd cheese: medium-fat soft cheese that has been well enough drained to produce a curd consistency; not as knobbly as cottage cheese and less cloying than full-fat cream cheese. Substitute ricotta.

CUSTARD MARROW: patty-pan squash; scallop squash

DESIRÉE POTATOES: of Dutch origin, with a waxy, yellowish flesh that does not disintegrate or become floury in cooking. Substitute any other waxy potato variety.

EARTH NUTS OR PIGNUTS: *Conopodium majus;* small edible tubers; the Chufa

Farm chicken: a chicken that has run around on a farm, and been properly fed and fattened, preferably on corn, and hung after killing. No substitute, although an unfrozen, commercial, factory-bred chicken can be used instead at some cost to the flavour of the finished dish.

Fenugreek: the seed of *Trigonella foenum-graecum*, much used – often overused – in commercial curry spice blends. Available at specialty food shops in America.

Groundnut oil: peanut oil. If unavailable, substitute corn or sunflower oil

HAMBURG PARSLEY: turnip-rooted parsley

LAND CRESS: American cress; Belle Isle cress; bitter cress; early winter cress; Upland cress

MANGETOUT: snow peas, sugar peas, edible-podded peas, Chinese pea pods

MARMANDE AND ESHKOL TOMATOES: beefsteak tomatoes. Marmande is a variety developed in France. Eshkol is an Israeli variety from Marmande, especially suited to growing in Israel. Substitute any firm tomato of good flavour that has been grown and ripened out of doors.

MARROW: an overgrown zucchini squash with a hard skin

MARSH SAMPHIRE: samphire; chicken-claws; pigeon-foot

NETTLE: stinging nettle

ORACHE: garden orach

Orléans wine vinegar: the best French wine vinegar is made at Orléans. Substitute any good brand of wine vinegar.

Pig's trotters: pig's feet

Pine kernels: pine nuts

Reinettes: a fine French dessert apple in several closely related varieties. In England, the substitute is a Cox's Orange Pippin (*see* page preceding). In the United States, use any richly aromatic crisp eating apple. Golden Delicious can be used instead, as they remain firm when cooked, but in flavour they are not a good substitute.

Rigani: Dried Greek varieties of marjoram. Substitute oregano.

ROCKET: roquette, rugula, rocket-salad

SALAD BURNET: burnet, garden burnet

Salted pig's hock: the narrow end of the picnic shoulder, a cone-shaped cut that has been salted

Scrag end of neck: always refers to lamb. The round slices cut across the backbone at the neck. Used for soups and stews.

Shin of beef: Leg of beef

Strong flour: bread flour. Substitute all-purpose flour, as in America this has a far higher gluten content than all-purpose plain flour in the United Kingdom.

Sugar: if no qualifying adjective is given, always use white, granulated sugar caster: Superfine granulated sugar

icing: confectioner's or powdered sugar

preserving: large sugar crystals sold in England for jam-making. Substitute lump or granulated.

Demerara: medium-brown grainy sugar. Substitute light brown sugar.

SWEDE: Rutabaga

Tomato concentrate: concentrated tomato paste, sold in tubes and very small cans. If unavailable, substitute boiled-down tomato purée, canned tomatoes or fresh tomatoes reduced to a very strong-tasting paste.

Topside: always refers to beef, unless qualified as 'veal topside'. The top, eye or bottom round of beef.

Tunnyfish: tuna

Veal scrag: neck of veal (see scrag end of neck), used for soups and stock, or for stewing

VEGETABLE MARROW: see MARROW

VEGETABLE SPAGHETTI: spaghetti squash

Worcester: Worcestershire sauce

Young's potted shrimps: a brand of tiny shrimp, boiled, peeled and potted in butter flavoured with mace. The top is sealed with a thin layer of clarified butter. Substitute small, peeled, boiled shrimp, plus extra butter and mace to taste.

TABLE OF EQUIVALENT WEIGHTS AND MEASURES

SPOON MEASUREMENTS

In England, by tradition, all spoon measurements given in recipes were rounded; in other words there was as much above the rim of the spoon as below it. I have followed this:

> 1 tablespoon butter = 2 level tablespoons US
> 1 dessertspoon butter = 4 level teaspoons US
> 1 teaspoon butter = 2 level teaspoons US

By a heaped spoon, I mean as much of the ingredient as you can balance on a spoon-shaped spoon: this can be awkward if you use round, deep plastic spoon sets to cook with – as a rough guide, use 3 level spoons. So we get:

> 1 heaped tablespoon = 3 level tablespoons US
> 1 heaped dessertspoon = 6 level teaspoons US
> 1 heaped teaspoon = 3 level teaspoons US

Naturally, liquid spoonfuls are always level, except that you might wonder for a moment what is meant by:

> 1 tablespoon melted butter = 1 level tablespoon US
> 1 tablespoon butter = 2 level tablespoons US

LIQUID MEASUREMENTS

This is the same for all liquids, when litres, pints, fluid ounces or spoons are specified:

LIQUID EQUIVALENTS

1 litre =	1000 ml	= $1\frac{3}{4}$ Imp. pts	= 2 US pts	= 4 cups
	600 ml	= 1 Imp. pt	= $1\frac{1}{4}$ US pts	= $2\frac{1}{2}$ cups
$\frac{1}{2}$ litre =	500 ml	= $\frac{3}{4}$ Imp. pt	= 1 US pt	= 2 cups
	300 ml	= $\frac{1}{2}$ Imp. pt		= $1\frac{1}{4}$ cups
$\frac{1}{4}$ litre =	250 ml	= 8 fl oz	= $\frac{1}{2}$ US pt	= 1 cup
	150 ml	= 5 fl oz		= $\frac{2}{3}$ cup
	100–125 ml	= $3\frac{1}{2}$–4 fl oz		= $\frac{1}{2}$ cup

With wines and spirits, you sometimes have:

> 1 glass red wine
> 1 glass brandy

In these cases, do not make the mistake of thinking that the glasses are the same size: by European tradition the word 'glass' means the size glass appropriate to the liquid concerned, which in the case of brandy is on the small side. As a general rule, I work like this:

> 1 glass spirits = 2–3 tablespoons
> 1 glass fortified wine (sherry, Madeira, etc.) = ⅓ cup
> 1 glass table wine = ¾ cup

SOME INGREDIENT MEASUREMENT EQUIVALENTS

beans, dried	250 g (8 oz) = 1 cup
	125 g (4 oz) = ½ cup
breadcrumbs, fresh	100 g (3½ oz) = 2 cups
butter	250 g (8 oz) = 1 cup
	100 g (3½ oz) = ⅓ cup
	50–60 g (2 oz) = ¼ cup
cheese, grated	50–60 g (2 oz) = ½ cup
flour, strong, plain *all-purpose*	500 g (1 lb) = 3¾ cups
	150 g (5 oz) = 1 cup
	100 g (3½ oz) = ⅔ cup
	50–60 g (2 oz) = ⅓ cup
fruit, dried, chopped *if large*	125 g (4 oz) = ¾ cup
	50–60 g (2 oz) = ⅓ cup
meat, chopped, minced, *raw, or cooked rare*	500 g (1 lb) = 2 cups, pressed down
peas, shelled	500 g (1 lb) = 2½–2⅔ cups

N B: when buying peas fresh in their pods, it is wise to allow twice the weight of shelled peas given in the recipe.

rice, raw	250 g (8 oz) = 1 cup
rice, raw cooked	250 g (8 oz) = 3 cups
sugar, granulated, *caster, brown*	50–60 g (2 oz) = ⅓ cup
vegetables, chopped, *diced, raw*	125 g (4 oz) = 1 cup

READ MORE IN PENGUIN

In every corner of the world, on every subject under the sun, Penguin represents quality and variety – the very best in publishing today.

For complete information about books available from Penguin – including Puffins, Penguin Classics and Arkana – and how to order them, write to us at the appropriate address below. Please note that for copyright reasons the selection of books varies from country to country.

In the United Kingdom: Please write to *Dept. EP, Penguin Books Ltd, Bath Road, Harmondsworth, West Drayton, Middlesex UB7 ODA*

In the United States: Please write to *Consumer Sales, Penguin Putnam Inc., P.O. Box 12289 Dept. B, Newark, New Jersey 07101-5289.* VISA and MasterCard holders call 1-800-788-6262 to order Penguin titles

In Canada: Please write to *Penguin Books Canada Ltd, 10 Alcorn Avenue, Suite 300, Toronto, Ontario M4V 3B2*

In Australia: Please write to *Penguin Books Australia Ltd, P.O. Box 257, Ringwood, Victoria 3134*

In New Zealand: Please write to *Penguin Books (NZ) Ltd, Private Bag 102902, North Shore Mail Centre, Auckland 10*

In India: Please write to *Penguin Books India Pvt Ltd, 11 Community Centre, Panchsheel Park, New Delhi 110017*

In the Netherlands: Please write to *Penguin Books Netherlands bv, Postbus 3507, NL-1001 AH Amsterdam*

In Germany: Please write to *Penguin Books Deutschland GmbH, Metzlerstrasse 26, 60594 Frankfurt am Main*

In Spain: Please write to *Penguin Books S. A., Bravo Murillo 19, 1° B, 28015 Madrid*

In Italy: Please write to *Penguin Italia s.r.l., Via Benedetto Croce 2, 20094 Corsico, Milano*

In France: Please write to *Penguin France, Le Carré Wilson, 62 rue Benjamin Baillaud, 31500 Toulouse*

In Japan: Please write to *Penguin Books Japan Ltd, Kaneko Building, 2-3-25 Koraku, Bunkyo-Ku, Tokyo 112*

In South Africa: Please write to *Penguin Books South Africa (Pty) Ltd, Private Bag X14, Parkview, 2122 Johannesburg*

READ MORE IN PENGUIN

A SELECTION OF FOOD AND COOKERY BOOKS

The Fratelli Camisa Cookery Book   Elizabeth Camisa

From antipasti to zabaglione, from the origins of gorgonzola to the storage of salami, an indispensable guide to real Italian home cooking from Elizabeth Camisa of the famous Fratelli Camisa delicatessen in Soho's Berwick Street.

Roald Dahl's Cookbook   Felicity and Roald Dahl

Roald Dahl's Cookbook, liberally spiced with lively anecdotes, recreates the many wonderful meals that have been enjoyed by the Dahl family and their friends around the farmhouse table at Gipsy House. 'Full of fun and lovely recipes' – *Sunday Times*

The Best of Floyd   Keith Floyd

Food magician and master chef, Keith Floyd has drawn his favourite recipes from a lifetime devoted to good eating and cooking, inspired by the local ingredients and traditions of the countries he has visited while filming his hugely successful television series.

Classic Cheese Cookery   Peter Graham

From soups and entrées to main courses and desserts, *Classic Cheese Cookery* is a deliciously varied tour of cheese dishes from around the world. 'A satisfying thick bible of cheese . . . Peter Graham clearly knows and loves his subject' – *Sunday Times*

The Dinner Party Book   Patricia Lousada

The Dinner Party Book hands you the magic key to entertaining without days of panic or last-minute butterflies. The magic lies in cooking each course ahead, so that you can enjoy yourself along with your guests.

Easy Cooking in Retirement   Louise Davies

The mouth-watering recipes in this book are delightfully easy to prepare and involve the least possible fuss to cook and serve.

READ MORE IN PENGUIN

A SELECTION OF FOOD AND COOKERY BOOKS

Real Fast Puddings Nigel Slater

'Nigel Slater has produced another winner in *Real Fast Puddings* ...
Slater has great flair for flavour combinations and he talks much sense. The
book is snappy and fun' – *Financial Times*. 'Delectable ... Slater is an
unashamed spoon-licker' – *Daily Telegraph*

Floyd on Italy Keith Floyd

Travelling around Italy in search of authentic local dishes, Keith Floyd has
brought his own inimitable style and expertise to the recipes, which are
interspersed with lively accounts of the places he visited and the people he
encountered.

Simple French Food Richard Olney

'There is no other book about food that is anything like it ... essential and
exciting reading for cooks, of course, but it is also a book for eaters ... its
pages brim over with invention' – *Observer*

Onions Without Tears Lindsey Bareham

Handled with care, alliums – onions, shallots, garlic, leeks and chives –
can bring savour, aroma and harmony to any dish, from the Indian dopiaza
to the French quiche and the South American salsa. 'Calm, measured,
assured, sensible, clever, useful, amusing and jam-packed with spot-on
recipes' – *Guardian*

The Chocolate Book Helge Rubinstein

'Fact-filled celebration of the cocoa bean with toothsome recipes from
turkey in chilli and chocolate sauce to brownies and chocolate grog' – *Mail
on Sunday*

The Rituals of Dinner Margaret Visser

'Margaret Visser's superlative analysis of table manners begins with
the idea that eating together is a terrifying and hazardous ordeal'
– *Independent*

READ MORE IN PENGUIN

FROM THE COOKERY LIBRARY

The Legendary Cuisine of Persia Margaret Shaida
Winner of the 1993 Glenfiddich Food Book of the Year Award

Persian cuisine is one of the oldest in the world and justly famous for its subtlety and fragrance. Central to the meal are the numerous dishes of rice sprinkled with saffron, some containing almonds, pistachios and raisins, others with vegetables and spices or even occasionally meat, creating a delicious selection of sweet and savoury tastes. 'Exquisite ... both a joy and a precious contribution to the world of gastronomy' – Claudia Roden

Summer Cooking Elizabeth David

Summer Cooking contains recipes from all over the world, for table, buffet or picnic. The result is a wonderful selection of summer dishes that are light (not necessarily cold), easy to prepare, and based on the meat, fruit and vegetables in season.

A New Book of Middle Eastern Food Claudia Roden

'This is one of those rare cookery books that is a work of cultural anthropology and Mrs Roden's standards of scholarship are so high as to ensure that it has permanent value' – *Observer*

English Seafood Cookery Richard Stein

'Deserves a place on everyone's kitchen shelf ... There are clear instructions on shopping, varieties, preparation and basic cooking, followed by inspiring recipes that are precise and unfussy' – Sophie Grigson

Pleasures of the Italian Table Burton Anderson

Guided by his appetite, Burton Anderson tracks down truffles in Piedmont, samples chunks of fragrant, freshly baked loaves in the Tuscan hills, sniffs balsamic vinegars in Emilia and enjoys a delicious yet splendidly simple *pizza marinara* in Naples. 'An informative and quietly passionate portrait of the people and their foods' – *Financial Times*

BY THE SAME AUTHOR

Jane Grigson's Fruit Book

A highly acclaimed alphabetical guide to fruit, from apple, apricot and arbutus to sorb apple, strawberry and watermelon. The most colourful, piquant, refreshing and life-enhancing of our foods, fruit is also a versatile and rewarding cooking ingredient. This beautifully written and impeccable research book offers a wealth of information and recipes, ranging through soups, pies, breads, stuffings, compotes, tarts, ice creams, meringue, cakes and jams.

English Food

'Perhaps the most serious and discriminating of her generation of cookery writers, and *English Food* is an anthology all who follow her recipes will want to buy for themselves, as well as for friends who may wish to know about real English food instead of glossified absurdities ... that are so often trotted out in technicolour as traditional "fayre" of olden times ... enticing from page to page' Pamela Vandyke Price, *Spectator*

Good Things

From curried parsnip soup, bouchées à la reine and civet of hare to baked beans Southern-style, gooseberry pie and wine sherbert – these are just a few of the delicious and intriguing dishes in *Good Things*. Jane Grigson emphasizes the delights and solaces of a truly creative activity and shows that cooking is an art to which everyone can aspire.

Mushroom Feast

With simple cooking and serving suggestions to bring out the unique, subtle flavours of many kinds of mushrooms, and more elaborate recipes in which they are used to complement and enhance substantial dishes, this superb book is a celebration of the edible fungus. 'Enjoyable, expert guide through woods, fields and every possible combination of food and fungi' *Observer*